COLLEGE
ALGEBRA

MAX PETERS

Chairman, Mathematics Department
George W. Wingate High School,
New York

BARRON'S EDUCATIONAL SERIES, INC.

New York • London • Toronto • Sydney

© *1962 by Barron's Educational Series, Inc.*
All rights are strictly reserved and no part
of this material may be reproduced in any
form without prior written permission from
the publisher.

All inquiries should be addressed to
Barron's Educational Series, Inc.
250 Wireless Boulevard
Hauppauge, New York 11788

Printed in the United States of America

Library of Congress Catalog Card No. 61–18891

789   510   20 19 18 17 16 15 14 13

Contents

Preface To The Student

Success in the study of College Algebra requires more than regular class attendance and the routine preparation of assignments. Because algebra is a cumulative subject, concepts incompletely understood and skills indifferently mastered will handicap the student throughout. It is the purpose of this book to add a third learning dimension to those offered by the instructor and the textbook.

Although rigorous and correct, the language used in the book is relatively informal and simple. It is expected that the student will encounter no difficulty in following the exposition and in comprehending the solutions of the illustrative examples. The profusion of such illustrative examples and the detailed explanation of each solution will be of inestimable aid to the student. This book covers the subject matter of College Algebra completely. The student need refer to no other book even if he wishes to delve into College Algebra in depth.

On the basis of many years in the classroom, the author suggests a systematic study program. In connection with each topic studied in class, the student should read the corresponding chapter in this book, work through the illustrative examples, and then attempt the practice examples provided. The answers at the end of each chapter will enable the student to check the accuracy of his work. Should the student find that he has not mastered the topic, he should repeat the above process. Students often find that a second or even a third exposure is necessary before comprehensive understanding is achieved. The exercise material is carefully graded. Thus, the average student may find that the last few problems in some exercise sections are too formidable. This need not cause concern.

CHAPTER **1**

Basic Concepts of Algebra

THE SYMBOLS OF ALGEBRA

In algebra, letters of the alphabet, such as n, a, b, x, and y are used to represent numbers.

The following symbols are familiar:

$+$ for addition, $-$ for subtraction, $\times$ for multiplication, $\div$ for division, and $\sqrt{}$ for indicating square root.

The product of the two numbers a and b may be represented as $a \times b$, $a \cdot b$, or ab.

The quotient of two numbers may be represented as $a \div b$, or $\frac{a}{b}$.

The symbols $=$, $\neq$ (unequal), $>$ (is greater than), $<$ (is less than) are used to show relationships between quantities.

Parentheses ( ), and bracket [ ] are symbols of aggregation. They are used to collect quantities which are to be considered as one quantity in performing operations.

The symbol $|\,a\,|$ is read "the absolute value of a."

The absolute value of a number a is the number a itself if a is greater than or equal to zero ($a \geq 0$), and is the number $-a$ if a is less than zero ($a < 0$).

1

Example: If n is a number, represent the following:

Three times the number.	$3n$
The number divided by 5.	$\dfrac{n}{5}$
4 more than the number.	$n + 4$
5 times the square of the number.	$5n^2$
The square root of 7 less than the number.	$\sqrt{n - 7}$
The absolute value of twice the number.	$|2n|$
The quantity 2 more than the number multiplied by 9.	$9(n + 2)$

Example: *Represent the following algebraically:*

The sum of the numbers a and b.	$a + b$
The quotient of the numbers $(x + y)$ and 4.	$\dfrac{x + y}{4}$
The product of the numbers 6 and $(c + d)$.	$6(c + d)$
The sum of three times a and 4 times b.	$3a + 4b$
The difference between 5 times x and y.	$5x - y$
The absolute value of the difference between a and b.	$|a - b|$
The quotient of x and the square root of y.	$\dfrac{x}{\sqrt{y}}$

Example: *Represent the following algebraically:*

The cost (C) of a books at \$$b$ for each book.	$C = \$ab$
The distance (D) covered by a car traveling at the rate of r miles per hour for t hours.	$D = rt$
The sum (S) of 3 consecutive numbers if the first number is n.	$S = 3n + 3$
The perimeter (P) of a rectangle whose length is 5 feet and whose width is a feet.	$P = 10 + 2a$
A two-digit number (N) if its tens digit is t and its units digit is u.	$N = 10t + u$
The number of seats (S) in an auditorium which contains $(n + 3)$ rows and 18 seats in each row.	$S = 18(n + 3)$
The dividend (D) divided by the divisor (d) is equal to the quotient (Q) increased by the remainder (R) divided by the divisor (d).	$\dfrac{D}{d} = Q + \dfrac{R}{d}$

The perimeter (P) of an equilateral triangle
whose sides are each $(2a + b)$. $P = 3(2a + b)$

EXERCISE 1

Represent the following algebraically:

1. The sum of the numbers b and 3.
2. 5 times the number a.
3. The difference between the numbers m and n.
4. The square root of 5 times y.
5. The absolute value of the number x increased by 7.
6. The quotient of the numbers $(a + b)$ and c.
7. The product of the number 9 and the square of the number d.
8. The difference between 12 and the number $(x + y)$.
9. The sum of the square root of a and the square of b.
10. The absolute value of the number d increased by 5 times f.
11. The quotient of 7 decreased by a, and x.
12. The quantity 5 more than the number y, multiplied by the number a.
13. The square root of the quantity $(c + 9)$.
14. The product of the number x and the absolute value of 3 more than y.
15. The square of the quantity $(a + b)$ subtracted from the square of c.
16. The absolute value of 5 less than 3 times q.
17. The perimeter (P) of a square whose side is a.
18. The average (A) of the numbers x, y, and z.
19. The cost (C) of n shirts at $(p + 3)$ dollars each.
20. The area (A) of a rectangle whose length is $(2a - 3)$ and whose width is b.
21. One-quarter of the quantity $(3x + 8)$.
22. The sum (S) of three consecutive even numbers if the smallest number is n.
23. A number (N) if its tens digit is t and its units digit is 3.
24. The number (N) of books in a bookcase if it has p shelves and $(q + 1)$ books on each shelf.
25. The cost (C) of a tie if a dozen ties cost $(2a + 3)$ dollars.
26. The perimeter (P) of a rectangle whose length is $(2x + 7)$ and whose width is 5.

27. The distance (D) covered by a motorist traveling at the average rate of 40 miles per hour for ($t + 4$) hours.

28. The amount (A) left in a bank account of $500 after 3 withdrawals of $b each.

29. The hourly wage (H) if the salary for a 40-hour week is ($3x - 8$) dollars.

30. The amount (A) saved in 4 weeks if the weekly income is I dollars and the weekly expenditure is E dollars.

THE NUMBERS OF ALGEBRA

DEFINITION: *Real numbers* are numbers which can be expressed as decimals. For example, the numbers 7, $2\frac{1}{2}$, and $\sqrt{5}$ are real numbers.

DEFINITION: *Natural numbers* are numbers obtained by counting. For example, the number 3 used in this statement is a natural number: There are 3 men in this room.

DEFINITION: *Rational numbers* are real numbers which can be expressed in the form $\frac{a}{b}$ where a and b are integers ($b \neq 0$).

Examples: $\frac{3}{2}$, $\frac{9}{5}$, $2\frac{1}{3}$, $\frac{6}{1}$ or 6.

DEFINITION: *Irrational numbers* are real numbers which, when expressed as decimals, are neither terminating decimals nor decimals which have a repeating pattern.

Example: $\sqrt{5}$, π

If rational numbers are written as decimals they will take the form of terminating or repeating decimals.

Example: $\frac{1}{5} = .2$ *terminating decimal*
$\frac{5}{6} = .83333 \ldots$ *repeating decimal*
$\frac{3}{8} = .375$ *terminating decimal*
$\frac{5}{11} = .454545 \ldots$ *repeating decimal*

If irrational numbers are written as decimals they will be non-terminating and non-repeating.

Example: $\sqrt{3} = 1.732 \ldots$
$\pi = 3.14159 \ldots$

Rational and irrational numbers may be positive or negative. For example, $+5$, $-\sqrt{3}$, $+\frac{2}{3}$, $-\frac{9}{8}$, -5π, $+\sqrt{17}$.

DEFINITION: *Real Number System:* The system of numbers composed of the positive and negative rational and irrational numbers and zero constitutes the *real number system.*

Real Numbers Represented on a Line

On the horizontal line below select a starting point, 0, called the origin and designate this point by zero. Then select a unit length.

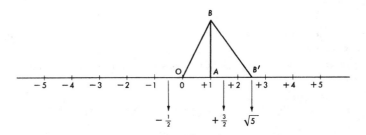

If this unit length is marked off to the right of the zero point, we obtain the points $+1$, $+2$, $+3$, $+4$, etc. If the unit length is marked off to the left of the zero point, we obtain the points -1, -2, -3, -4, etc. Other rational numbers may be located as indicated ($+\frac{3}{2}$, $-\frac{1}{2}$).

Some irrational numbers may also be located. For example, if AB is made 2 units long, then the line segment OB will be equal to $\sqrt{5}$ units. If the length $OB' \stackrel{=}{=} OB$ is marked off to the right of the zero point, then B' will locate the point $+\sqrt{5}$.

Imaginary Numbers

Since a negative real number does not have a real number as its square root, it is necessary to introduce such numbers as $\sqrt{-4}$.

DEFINITION: The square root of a negative number is a *pure imaginary.*

Examples: $\sqrt{-7}$, $\sqrt{-3}$, $\sqrt{-9}$.

In order to work with such numbers we introduce the symbol $i = \sqrt{-1}$ which is called the *imaginary unit.* The number $\sqrt{-4}$ may be written as $\sqrt{4} \cdot \sqrt{-1}$ and consequently as $2\sqrt{-1}$ or $2i$. Thus, pure imaginaries may be represented as the product of a real number and the imaginary unit.

Complex Numbers

A number of the form $a + bi$ where a and b are real numbers is called a complex number.

Examples: $5 + 2i, -4 - \sqrt{3}\,i$.

It follows that complex numbers $(a + bi)$ in which $a = o$, and $b \neq o$, are pure imaginaries.

Imaginary and complex numbers will be treated in greater detail in Chapter 9.

FUNDAMENTAL OPERATIONS AND THEIR LAWS

The fundamental operations are addition $(+)$, subtraction $(-)$, multiplication $(\times)$, and division $(\div)$.

Addition and subtraction are inverse operations since subtraction will undo addition and addition will undo subtraction.

Example: $4 + 3 = 7$ and $7 - 3 = 4$.

Multiplication and division are also inverse operations since division will undo multiplication and multiplication will undo division.

Example: $8 \times 5 = 40$ and $40 \div 5 = 8$.

The addition of real numbers is consistent with the following two postulates.

Commutative Law of Addition

If a and b are real numbers then $a + b = b + a$.   The result of the addition of two real numbers is not affected by the order in which the two numbers are added.

Example: $5 + 7 = 7 + 5$

Associative Law of Addition

If a, b, and c are real numbers then $a + (b + c) = (a + b) + c$. The result of the addition of three numbers is not affected by the way in which the numbers are grouped.

Example: $3 + (5 + 6) = (3 + 5) + 6$
$$3 + 11 = 8 + 6$$
$$14 = 14$$

The multiplication of real numbers is consistent with the following two postulates.

Commutative Law of Multiplication

If a and b are real numbers then $ab = ba$. The result of the multiplication of two real numbers is not affected by the order in which the two numbers are multiplied.

Example: $7 \times 9 = 9 \times 7$

Associative Law of Multiplication

If a, b, and c are real numbers then $a(bc) = ab(c)$. The result of the multiplication of three real numbers is not affected by the order in which the three numbers are multiplied.

Example: $4(3 \cdot 6) = (4 \cdot 3)6$
$$4(18) = (12)6$$
$$72 = 72$$

The addition and multiplication of real numbers are consistent with the following postulate.

Distributive Law

If a, b, and c are real numbers then $a(b + c) = ab + ac$. This law states that multiplication is distributive with respect to addition.

Example: $6(4 + 7) = 6 \cdot 4 + 6 \cdot 7 = 24 + 42 = 66$

In performing operations with real numbers it is to be observed that division by zero is excluded. If an operation like $\frac{5}{0}$ is to have meaning then the quotient $\frac{5}{0} = x$. Clearly, there is no real number x which, when multiplied by 0, will yield the number 5.

OPERATIONS WITH SIGNED NUMBERS

Addition of Signed Numbers

In adding two signed numbers which have the same sign, add their absolute values and place the common sign before the sum.

In adding two signed numbers which have unlike signs, find the difference between their absolute values and before this difference place the sign of the number which has the greater absolute value.

Examples: *Add*

+ 5	− 5	+5	−5
+ 9	− 9	−9	+9
+14	−14	−4	+4

Subtraction of Signed Numbers

In subtracting one signed number from another change the sign of the subtrahend (the number which is being subtracted) and add the result to the other number (the minuend).

Examples: *Subtract*

+18	+15	−12	− 8
+ 7	− 2	+ 4	−15
+11	+17	−16	+ 7

Multiplication of Signed Numbers

In multiplying two signed numbers which have like signs find the product of their absolute values and place a plus sign before the product.

In multiplying two signed numbers which have unlike signs find the product of their absolute values and place a minus sign before the product.

Examples: *Multiply*

$+8$	-9	$+5$	-7
$+4$	-7	-6	$+8$
32	$+63$	-30	-56

Division of Signed Numbers

In dividing one signed number by another signed number which has the same sign, find the quotient of their absolute values and place a plus sign before the quotient.

In dividing one signed number by another signed number which has a different sign, find the quotient of their absolute values and place a minus sign before the quotient.

Examples: *Divide*

$$(+18) \div (+3) = +6 \qquad (-24) \div (-3) = +8$$
$$(+36) \div (-4) = -9 \qquad (-35) \div (+5) = -7$$

EXERCISE 2

1. *Add*

a. $+5$	b. -3	c. -7	d. $+9$	e. -5	f. $+12$
$+7$	$+9$	-8	-7	$+6$	-8

2. *Subtract*

   a. + 15   b. − 12   c. − 9   d. + 2   e. − 3   f. +8
      + 3   + 4   −6   − 10   −9   −6

3. *Multiply*

   a. +3   b. − 8   c. − 9   d. +5   e. − 3   f. − 2
      +7   +6   −2   −6   −9   +7

4. *Divide*

   a. +28 by +7   b. − 16 by +2   c. +32 by −4
   d. −48 by −6   e. +36 by −9   f. − 54 by +6

5. *Perform the indicated operations:*

   a. $(+5)\,(+3)\,(-6)$ b. $(-8) + (+7) + (-15)$
   c. $(-4) + (-3) + (-12)$ d. $(+15) - (+5)$
   e. $(-17) + (-5) - (-1)$ f. $(+6) - (+9) - (+2)$
   g. $(-18) - (+3) + (-7)$ h. $(+10) - (-12) - (+3)$
   i. $(+7)\,(-2)\,(+3)$ j. $(-8)\,(+1)\,(-5)$
   k. $(-3)\,(-6)\,(0)$ l. $(+8)\,(+2)\,(-7)$
   m. $(+12) \cdot (-4) \div (+3)$ n. $(-15)(-8) \div (-6)$
   o. $(+9) \div (-3) \cdot (+4)$ p. $(-20) \div (+5) \cdot (-9)$

EXPONENTS

When a quantity y is multiplied by itself we may write the result as $y \cdot y$ or as y^2. Also, if we wish to multiply y^3 by y^4 we may write the result as $(y \cdot y \cdot y) \times (y \cdot y \cdot y \cdot y)$ or more simply as y^7. The small number which indicates the power to which the base (y in this case) is to be raised is called an *exponent*.

Laws of Exponents

If m and n are positive integers then
$$a^m a^n = a^{m+n}$$

Examples: $b^3 \cdot b^9 = b^{12},$ $3^2 \cdot 3^4 = 3^6 = 729$

$\qquad\qquad (a^m)^n = a^{mn}$

Examples: $(b^3)^4 = b^{12},$ $(2^3)^4 = 2^{12} = 4{,}096$

$\qquad\qquad \dfrac{a^m}{a^n} = a^{m-n}$

Examples: $b^{12} \div b^4 = b^{12-4} = b^8,$ $3^6 \div 3^2 = 3^4 = 81$

$(ab)^n = a^n b^n$

Examples: $(ab)^5 = a^5 b^5,$ $(2 \cdot 5)^3 = 2^3 \cdot 5^3 = (8 \cdot 125) = 1{,}000$

$\left(\dfrac{a}{b}\right)^n = \dfrac{a^n}{b^n} \ (b \neq 0).$

Examples: *Simplify each of the following:*

a. $\left(\dfrac{a}{b}\right)^3 = \dfrac{a^3}{b^3}$

b. $\left(\dfrac{2}{3}\right)^5 = \dfrac{2^5}{3^5} = \dfrac{32}{243}$

c. $\dfrac{a^5 \cdot a^3}{a^2} = \dfrac{a^8}{a^2} = a^6$

d. $\dfrac{(cd)^4}{cd^2} = \dfrac{c^4 d^4}{cd^2} = c^3 d^2$

e. $\dfrac{(b^3)^4}{b^7} = \dfrac{b^{12}}{b^7} = b^5$

f. $\left(\dfrac{a^2}{b}\right)^5 \cdot \dfrac{b}{a^3} = \dfrac{a^{10} \cdot b}{b^5 \cdot a^3} = \dfrac{a^7}{b^4}$

Examples: *Evaluate each of the following:*

a. $3^4 = 3 \cdot 3 \cdot 3 \cdot 3 = 81$ b. $2(5^3) = 2 \cdot 5 \cdot 5 \cdot 5 = 250$

c. $2^2 \cdot 2^5 = 2^7 = 128$ d. $3^2 \cdot 4^3 = 3 \cdot 3 \cdot 4 \cdot 4 \cdot 4 = 576$

e. $\dfrac{7^6}{7^4} = 7^2 = 49$ f. $\dfrac{2^9 \cdot 2^8}{2^{15}} = \dfrac{2^{17}}{2^{15}} = 2^2 = 4$

g. $(4 \cdot 5)^3 = 4^3 \cdot 5^3 = 8{,}000$ h. $\left(\dfrac{1}{2}\right)^5 = \dfrac{1^5}{2^5} = \dfrac{1}{32}$

EXERCISE 3

Simplify each of the following:

1. $a^3 \cdot a^4$

2. $a^{12} \div a^4$

3. $(a^4)^5$

4. $(a^2b)^3$

5. $\left(\dfrac{a^3}{b} \right)^5$

6. $\dfrac{a^7 \cdot a^2}{a^6}$

7. $\dfrac{(pq)^5}{p^2q}$

8. $\dfrac{(c^2)^5}{c}$

9. $\left(\dfrac{c^3}{b^2} \right)^4 \cdot \dfrac{b^2}{c^3}$

10. $\dfrac{(a^2b^3)^5}{ab}$

11. $\dfrac{1}{a^2} (a^2)^5$

12. $\dfrac{c^3}{d^2} \cdot \dfrac{(d^4)^2}{c}$

EXERCISE 4

Evaluate each of the following:

1. 7^3

2. $3^2 \cdot 3^3$

3. $5(2^3)$

4. $5^2 \cdot 4^3$

5. $2^8 \div 2^3$

6. $(3 \cdot 4)^3$

7. $\left(\dfrac{1}{5} \right)^3$

8. $\dfrac{2^6 \cdot 2^5}{2^7}$

9. $\left(\dfrac{2}{3} \right)^4$

10. $\dfrac{3^7 \cdot 3^8}{3^{12}}$

11. $7(5^3)$

12. $\dfrac{4(3)^5}{6^2}$

EVALUATING ALGEBRAIC EXPRESSIONS

It is frequently necessary to obtain the numerical value of an algebraic expression when numbers are substituted for literal expressions. In evaluating algebraic expressions, the order in which operations are performed follows this rule.

RULE: The operations of multiplication and division are performed before the operations of addition and subtraction.

Example: Simplify $19 - 6 \times 2 + 15 \div 3$

▶ First perform multiplication and division obtaining $19 - 12 + 5$

Now perform addition and subtraction obtaining the result 12.

EXERCISE 5

Simplify:

1. $6 + 5 \times 8$
2. $7 - 3 \times 4$
3. $9 + 12 \div 2$
4. $8 \times 6 - 5$

5. $7 + 3 \times 2 - 5$
6. $4 \times 8 - 6 \times 3$
7. $15 - 8 \div 2 + 6$
8. $16 \div 2 - 3 + 5 \times 4$

Example: Find the numerical value of $7x - 2y + 3z$ if $x = 5$, $y = 4$, $z = 9$

▶ $7x - 2y + 3z = 7 \cdot 5 - 2 \cdot 4 + 3 \cdot 9 = 35 - 8 + 27 = 54$

Example: Find the numerical value of $2a^2 - 3ab + 4bc^3$ if $a = 3$, $b = -4$, $c = -5$.

▶ $2a^2 - 3ab + 4bc^3 = 2(3)^2 - 3(3)(-4) + 4(-4)(-5)^3$
$2(9) - 3(-12) + 4(-4)(-125) = 18 + 36 + 2000 = 2,054.$

EXERCISE 6

Find the numerical value of each of the following if
$a = -2, b = 3,$ *and* $c = -4$:

1. $a + b + c$
2. $4a - c$
3. $\dfrac{a + b}{c}$
4. $3a + 2b$
5. $a^2 + b^2$
6. $5a^2b$
7. $20 - ab^2$
8. $\dfrac{c^2 - a^2}{3b}$

9. $2a + 3b^2 - c^2$
10. $c^2 - 2ab$
11. $ab + ac + bc$
12. $ab^2 - bc^2$
13. $\frac{1}{2}ab^2 + \frac{1}{3}bc^2$
14. $\dfrac{a^2 - 4b^2}{2c}$
15. $\dfrac{b^2c^2 - a^2b^2}{a}$
16. $\dfrac{a^2b + b^2c + c^2a}{a + b + c}$

ANSWERS

EXERCISE 1—Page 3

1. $b + 3$
2. $5a$
3. $m - n$
4. $\sqrt{5y}$
5. $|x + 7|$
6. $\dfrac{a + b}{c}$
7. $9d^2$
8. $12 - (x + y)$
9. $\sqrt{a} + b^2$
10. $|d + 5f|$
11. $\dfrac{7 - a}{x}$

12. $a(y + 5)$
13. $\sqrt{c + 9}$
14. $x|y + 3|$
15. $c^2 - (a + b)^2$
16. $|3q - 5|$
17. $P = 4a$
18. $A = \dfrac{x + y + z}{3}$
19. $C = n(p + 3)$
20. $A = b(2a - 3)$
21. $\frac{1}{4}(3x + 8)$
22. $S = n + n +$ $2 + n + 4 = 3n + 6$

23. $N = 10t + 3$
24. $N = p(q + 1)$
25. $C = \dfrac{2a + 3}{12}$
26. $P = 2(2x + 7)$ $+ 2(5) = 4x$ $+ 24$
27. $D = 40(t + 4)$
28. $A = 500 - 3b$
29. $H = \dfrac{3x - 8}{40}$
30. $A = 4I - 4E$

EXERCISE 2—Page 9

1. a. $+12$
   b. $+6$
   c. -15
   d. $+2$
   e. $+1$
   f. $+4$

2. a. $+12$
   b. -16
   c. -3
   d. $+12$
   e. $+6$
   f. $+14$

3. a. $+21$
   b. -48
   c. $+18$
   d. -30
   e. $+27$
   f. -14

4. a. $+4$
   b. -8
   c. -8
   d. $+8$
   e. -4
   f. -9

5. a. -90
   b. -16
   c. -19
   d. $+10$
   e. -21
   f. -5
   g. -28
   h. $+19$
   i. -42
   j. $+40$
   k. 0
   l. -112
   m. -16
   n. -20
   o. -12
   p. $+36$

EXERCISE 3—Page 12

1. a^7
2. a^8
3. a^{20}
4. a^6b^3
5. $\dfrac{a^{15}}{b^5}$
6. a^3
7. p^3q^4
8. c^9
9. $\dfrac{c^9}{b^6}$
10. a^9b^{14}
11. a^8
12. c^2d^6

EXERCISE 4—Page 12

1. 343
2. 243
3. 40
4. 1,600
5. 32
6. 1,728
7. $\dfrac{1}{125}$
8. 16
9. $\dfrac{16}{81}$
10. 27
11. 875
12. 27

EXERCISE 5—Page 13

1. 46
2. -5
3. 15
4. 43
5. 8
6. 14
7. 17
8. 25

EXERCISE 6—Page 13

1. -3
2. -4
3. $-\dfrac{1}{4}$
4. 0
5. 13
6. 60
7. 38
8. $\dfrac{4}{3}$
9. 7
10. 28
11. -10
12. -66
13. 7
14. 4
15. -54
16. $\dfrac{56}{3}$

CHAPTER **2**

Operations With Algebraic Expressions

DEFINITIONS OF IMPORTANT TERMS

ALGEBRAIC EXPRESSION—An algebraic expression is an expression formed from letters and numbers by the use of addition, subtraction, multiplication, division, and the extraction of roots, each operation being used a finite number of times.

Examples:

$$x^2y, \; 2a^2 + 3ab - b^2, \; \frac{a^2 + b^2}{3c}$$

TERM—A term consists of one or more symbols and is formed by the use of any of the operations of algebra except addition and subtraction.

Examples:

$$3a^2b, \; \frac{5x\sqrt{y}}{3}$$

MONOMIAL—A monomial is an algebraic expression which consists of one term.

Examples:

$$-5a^2, \; \frac{7}{2}xy$$

BINOMIAL—A binomial is an algebraic expression which consists of two terms.

Examples: $2a^2 - 3b^2, 4x^3 - 3xy$

TRINOMIAL—A trinomial is an algebraic expression which consists of three terms.

Examples: $a^2 + ab - b^2, 2x^3 - 5xy + y^2$

POLYNOMIAL—A polynomial is an algebraic expression which consists of two or more terms.

Examples: $a^2b - 2b^3, x + y + 5z$

FACTOR—When two or more numbers are multiplied together the result is called a product. The numbers which are multiplied together to form that product are called *factors* of the product.

Example: In the product $5a$, 5 and a are factors of $5a$.

COEFFICIENT—If a term consists of several factors then any one factor or combination of factors is the coefficient of the rest of the term.

Example: For the term $7x^2y$, $7x^2$ is the coefficient of y, $7y$ is the coefficient of x^2, and 7 is the coefficient of x^2y.

NUMERICAL COEFFICIENT—A numerical coefficient is a coefficient consisting of a number as distinct from literal coefficients.

Example: For the term $-6ab^2$, the numerical coefficient is -6.

SIMILAR TERMS—Similar terms are terms which have identical literal factors.

Example: The term $3x^2y$, $-5x^2y$, and $12x^2y$ are similar terms. The terms $3x^2y$ and $5xy^2$ are *not* similar terms.

ADDITION OF ALGEBRAIC EXPRESSIONS

In adding polynomials, similar terms are arranged in vertical columns and each column is added separately.

Example: Add $5a^2b - 6a^2 + 7b^2$, $b^2 + a^2 - 7a^2b$, and $-8a^2 - 6a^2b - 4b^2$.

▶
$$\begin{array}{l} 5a^2b -\ \ 6a^2 + 7b^2 \\ -7a^2b +\ \ \ \ a^2 +\ \ b^2 \\ -6a^2b -\ \ 8a^2 - 4b^2 \\ \hline -8a^2b - 13a^2 + 4b^2 \end{array}$$

Example: Add $3y^3 + 6x^2y - 5xy^2 - 2x^3 + 4xy$, $-3x^2y + 3xy^2 - 5xy + 4x^3$, $xy - 2xy^2 - 7x^3$, $8x^3 - 5y^3 + 8x^2y$.

▶
$$\begin{array}{l} 3y^3 -\ 5xy^2 +\ \ 6x^2y + 4xy - 2x^3 \\ +\ 3xy^2 -\ \ 3x^2y - 5xy + 4x^3 \\ \quad\ -\ 2xy^2 \qquad\quad\ +\ xy - 7x^3 \\ -5y^3 \qquad\quad +\ 8x^2y \qquad\quad +\ 8x^3 \\ \hline -2y^3 - 4xy^2 + 11x^2y + 0 \quad +\ 3x^3 \end{array}$$

EXERCISE 1

Add the following polynomials:

1. $a + 2b - 3c$, $-5a + b + 7c$, $2a - 6b - c$
2. $2y^2 - 3x^2 + xy$, $2x^2 - 5xy - 6y^2$, $7xy - 3x^2 + y^2$
3. $7cd^2 - 5c^3 + 2d^3 - 6c^2d^2$, $8c^3 - cd^2 + 4d^3$, $5c^2d^2 - 7c^3$, $8c^2d^2 + c^3 - cd^2 - 5d^3$
4. $-3a^2b^3 + a^3 - 5a^3b^2$, $6a^3b^2 - 3a^2b^3 + 8b^3$, $5a^3 - a^3b^2 + 2b^3$, $-4b^3 + 2a^3 - 3a^2b^3$
5. $7xy^5 - 3x^4y^2 + 2x^3y^3 - x^2y^4$, $3x^5y - 5xy^5 - 3x^3y^3$, $-8x^4y^2 + 2x^3y^3 - 7x^2y^4$, $3xy^5 - 4x^4y^2 - 2x^2y^4 - x^5y$, $2x^5y - 4x^3y^3 + 9x^2y^4 - 7xy^5$

SUBTRACTION OF ALGEBRAIC EXPRESSIONS

In subtracting polynomials, similar terms are arranged in vertical columns and each column is subtracted separately.

Example: From $5x - 2z + 3y - 8w$ subtract $7y - 2x - 2w + z$.

▶ Arrange similar terms in vertical columns

Change signs of terms in $5x - 2z + 3y - 8w$ (*Minuend*)
subtrahend and add. $2x - z - 7y + 2w$ (*Subtrahend with*
$$\overline{}$$ *Signs Changed*)
        $7x - 3z - 4y - 6w$ (*Difference*)

Example: By how much does $7a - 3b + c - 8d$ exceed $5c - 2d - 9b + 3e$?

▶ $7a - 3b + c - 8d$ (*Minuend*)
             $9b - 5c + 2d - 3e$ (*Subtrahend with*
$$\overline{}$$ *Signs Changed*)
  $7a + 6b - 4c - 6d - 3e$ (*Difference*)

Example: How much must be added to the sum of $2a^2 - 8ab - 9b^2$ and $5a^2 - 3ab + 2b^2$ to yield $a^2 - ab + b^2$?

▶   *Add* $2a^2 - 8ab - 9b^2$
             $5a^2 - 3ab + 2b^2$
$$\overline{}$$
         $7a^2 - 11ab - 7b^2$

Subtract this result from
$$a^2 - ab + b^2$$
$$7a^2 - 11ab - 7b^2$$

The result is $-6a^2 + 10ab + 8b^2$

Example: By how much is the sum of $3x^3 - 2x^2y - y^3$ and $x^2y - 5xy^2 + y^3$ larger than the sum of $4x^3 + 2x^2y - xy^2 - 7y^3$ and $x^3 - 2x^2y + 5xy^2 + 2y^3$?

▶ *Add* $3x^3 - 2x^2y \qquad\quad - \; y^3$

$\qquad\qquad\qquad x^2y - 5xy^2 + \; y^3$

$\qquad\qquad \overline{}$

$\qquad\qquad\; 3x^3 - \; x^2y - 5xy^2$

Add $x^3 - 2x^2y + 5xy \; + 2y^3$

$\qquad\quad 4x^3 + 2x^2y - \; xy^2 - 7y^3$

$\qquad\qquad \overline{}$

$\qquad\quad\; 5x^3 \qquad\qquad + 4xy^2 - 5y^3$

Subtract the second sum from the first sum

$\qquad\qquad\quad 3x^3 - \; x^2y - 5xy^2$

$\qquad\qquad\quad 5x^3 \qquad\quad + 4xy^2 - 5y^3$

$\qquad\qquad \overline{}$

The result is $-2x^3 - x^2y - 9xy^2 + 5y^3$

EXERCISE 2

In examples 1–9, subtract the second polynomial from the first.

1. $3a^2 - 2a - 1, \; -5a^2 - 3a + 4$
2. $6x^2 - 5x + 1, \; 2x + 5x^2 - 8$
3. $b^2 - 9, \; -2b^2 + 3b - 5$
4. $-6d + d^3 - 7d^2, \; 3d^3 - 5 + 6d + d^2$
5. $-x - 4z + 6y + 5, \; 3y + x - 4z$
6. $6bc^2 - 3b^3 - 5b^2c + c^3, \; b^3 - 2c^3 - 5b^2c + 7bc^2$
7. $5x - 7x^3 + 4, \; x^2 - 7 + 3x^3$
8. $8p^2q^2 - 3 + 4p^4, \; 2q^4 - 3p^2q^2 + 1 - 5p^4$
9. $3ab - 4ac + bc - 3cd, \; 5bc - 2bd + 3cd - 4ab$
10. From the sum of $2a + 3b - 5d$, and $3a - 2b + c$, subtract $a - b - c + 3d$.
11. What must be added to the sum of $2b^3 - 3b^2c + d^3$ and $b^3 - 2bc^2 - 5d^3$ to obtain $8b^3 - b^2c + bc^2 - 2d^3$?
12. From the sum of $3ab - 2bc + 3ac$ and $ab - bc + ac$ subtract the sum of $5ab - 7bc + ac$ and $4ab - bc - 6ac$.
13. By how much does the sum of $7xy + 8yz - 6xz$ and $xy - 2xz + 3yz$ exceed $2xy - 3yz + xz$?
14. Subtract $a^2b^2 - 3ab^2 + 2a^2b - b^2$ from the sum of $6a^2b - ab^2 + 7b^2$ and $8a^2b^2 + 5a^2b - 2b^2$.

MULTIPLICATION OF ALGEBRAIC EXPRESSIONS

Multiplication of a Monomial by a Monomial

In multiplying a monomial by a monomial, find the product of the numerical coefficients and multiply the result by the product of the literal factors.

Example: Multiply $-3a^2b$ by $5ab^3$

▶ $(-3a^2b)(5ab^3) = -15a^3b^4$

Example: Find the product of $5x^2y$, $-2xy^2$, and xy

▶ $(5x^2y)(-2xy^2)(xy) = -10x^4y^4$

EXERCISE 3

Find the products:

1. $(3a)(-4ab)$
2. $(-4x)(-7x^2)$
3. $(-5y^2)(3y^3)$
4. $(-5bc)(3c^2)$
5. $(-a^2b)(-2ab^2)$
6. $(7x^2y)(-3xy^3)$
7. $(ab)(-bc)(cd)$
8. $(3xy^2z)(-5xz)$
9. $(-3a^2bc)(7ac^2)(-5b^2c)$
10. $(4x^2yz)(-xy^2z^3)$
11. $(6a^2b)(8b^2c)$
12. $(-c^2d^3)(-8c)(4d^2)$

Multiplication of a Polynomial by Monomial

In multiplying a polynomial by a monomial, multiply each term of the polynomial by the monomial. This is consistent with the Distributive Law.

Example: Multiply $6x^3 - 3x^2 + 2x - 5$ by $-3x$

▶ It is convenient to express this multiplication by the use of parentheses, as follows: $-3x(6x^3 - 3x^2 + 2x - 5)$. The result is $-18x^4 + 9x^3 - 6x^2 + 15x$

Example: Multiply $-6a^2b(3ab - 6a + 2ab^2)$

▶ $-6a^2b(3ab - 6a + 2ab^2)$
The result is $-18a^3b^2 + 36a^3b - 12a^3b^3$

EXERCISE 4

Perform the following multiplications:

1. $-5(6a - 3b)$ 5. $3b(4b^2 - 5b + 2)$
2. $-2a(3a^2 - ab)$ 6. $-4ab(2b^2 - 3a - 6ab)$
3. $-3c^2(8c - 4c^2)$ 7. $-2x^2y(xy^2 - 3x^2y + 5xy)$
4. $xy(2x + 3y - xy)$ 8. $7a^2b(-2ab^3 + a^3b - 5ab^2)$

Multiplying a Polynomial by a Polynomial

In multiplying a polynomial by a polynomial it is convenient to arrange the terms of the multiplicand and the multiplier according to ascending or descending powers of one of the literal factors. Then multiply the multiplicand by each term of the multiplier and add the products.

Example: Multiply $9 - x^2 + 3x$ by $2 - 3x$

▶ Arrange the multiplier and the multiplicand according to descending powers of x.

$$-x^2 + 3x + 9$$
$$-3x + 2$$

Multiply each term of the multiplicand by $-3x$.
Multiply each term by 2.

$$3x^3 - 9x^2 - 27x$$
$$-2x^2 + 6x + 18$$

Combine the results.

$$3x^3 - 11x^2 - 21x + 18$$

Example: Multiply $(a + b - 2c)$ by $(a - b + 2c)$

▶

$$a + b - 2c$$
$$a - b + 2c$$

Multiply each term of the multiplicand by a.

$$a^2 + ab - 2ac$$

Multiply each term of the multiplicand by $-b$.

$$- ab \qquad -b^2 + 2bc$$

Multiply each term of the multiplicand by 2c.

$$2ac \qquad + 2bc - 4c^2$$

Combine results

$$a^2 \qquad\qquad -b^2 + 4bc - 4c^2$$

EXERCISE 5

Perform the following multiplications:

1. $x^2 - 4x + 3$ by $2x - 5$
2. $3y^2 - 2y - 1$ by $y - 2$
3. $3 - x^2 + 2x$ by $6 - 5x$
4. $a^2 - 2ab + 3b^2$ by $2a + 3b$
5. $-4a^2 - 2 + a^4$ by $3a^2 - 1$
6. $6x + 7y$ by $6x - 7y$

7. $2a + 3b$ by $2a + 3b$
8. $c^2 - cd + d^2$ by $c - d$
9. $x^3 - 3x + 2$ by $x - 5$
10. $y^2 - 3y + 4$ by $y^2 - 2$
11. $3x + 2y - z$ by $5x - y + 2z$
12. $a^2 - 2a + 1$ by $2a^2 + a - 3$

Removal of Parentheses

In removing parentheses preceded by a monomial coefficient, multiply each term of the parentheses by the monomial coefficient. If one set of parentheses is included inside another set it is convenient to remove the inside parentheses first and then the outer parentheses.

Example: Remove the parentheses and collect terms:
$5x^2 - 3x(2x - 5)$

▶ $5x^2 - 3x(2x - 5) = 5x^2 - 6x^2 + 15x = -x^2 + 15x$

Example: Remove parentheses and collect terms:
$$2x - 3[x - 5(3x - 7)]$$

▶ $$2x - 3[x - 5(3x - 7)]$$

Removing inner parentheses $2x - 3[x - 15x + 35]$
Removing outer brackets $2x - 3x + 45x - 105$
Collecting similar terms $44x - 105$

EXERCISE 6

Remove parentheses and collect terms:

1. $5 - (x - 3)$
2. $3x - 2(2x + 2)$
3. $2x - 5 + 3(2x - 7)$
4. $9 - 2x - 5(5x + 4)$
5. $8 - x(3x - 2) + 4x$

6. $7x + 3x(2 - 5x) - 3$
7. $5x - 2[3x - 4(x - 3)]$
8. $3x - 5[2x + 3(2x - 1)]$
9. $3[x - 2(1 - 5x)] + 2x$
10. $7[3y - 5(2y - 5)] - 6y$

DIVISION OF ALGEBRAIC EXPRESSIONS

Division of a Monomial by a Monomial

In dividing a monomial by a monomial, find the quotient of the numerical coefficients, find the quotient of the literal factors, and multiply the two quotients.

Example: Divide $24x^4y^3$ by $-6x^2y^2$

▶
$$\frac{24x^4y^3}{-6x^2y^2} = -4x^2y$$

Example: Divide $-48a^7b^5c^4$ by $12a^3b^5c$

▶
$$\frac{-48a^7b^5c^4}{12a^3b^5c} = -4a^4c^3$$

EXERCISE 7

Perform the following divisions:

1. $12x^5 \div 3x^2$
2. $15y^7 \div -5y^3$
3. $-24a^3b^6 \div 6a^2b^4$
4. $36c^5d^7 \div -2c^5d$
5. $-18x^8y^5 \div -2xy^4$

6. $-28a^9b^4c^7 \div -14ab^4c^3$
7. $45c^6d^8 \div -9c^5d^8$
8. $-72x^9y^8z^6 \div 4x^3y^7z^5$
9. $-48ab^5 \div -6a$
10. $-16p^3q^7 \div -16p^3q^6$

Division of a Polynomial by a Monomial

In dividing a polynomial by a monomial divide each term of the polynomial by the monomial.

Example: Divide $16a^4b^2 - 8a^5b^3 - 24ab^2$ by $-4ab$

▶
$$\frac{16a^4b^2}{-4ab}, \frac{-8a^5b^3}{-4ab}, \frac{-24ab^2}{-4ab}$$
$$-4a^3b + 2a^4b^2 + 6b$$

EXERCISE 8

Perform the following divisions:

1. $20x^4 - 8x^3 - 32x^2$ by $-4x^2$

2. $-6y^5 + 9y^3 - 12y$ by $+ 3y$

3. $7a^3 - 9a^2 - 2a$ by $-a$

4. $-8a^2b^2 + 14a^3b^2$ by $2ab$

5. $50x^2y^3 - 45xy^2 + 30x^2y$ by $5xy$

6. $-18c^3d^4 + 24c^4d^3 - 36c^2d^2$ by $-6cd$

7. $81a^3b^3 - 63a^2b^4 + 90a^4b^3$ by $9a^2b^2$

Division of a Polynomial by a Polynomial

In dividing a polynomial by a polynomial proceed as follows:

1. Arrange the terms of both the divisor and the dividend in either descending or ascending powers of the same literal number.

2. Divide the first term in the dividend by the first term in the divisor.   This will yield the first term in the quotient.

3. Multiply the entire divisor by the first term in the quotient and subtract this product from the dividend.

4. Take the difference as the new dividend and repeat steps 2. and 3. until a remainder is obtained which is either one degree lower than the divisor or the remainder is zero.

5. The result may be written by using the relationship

$$\frac{\text{Dividend}}{\text{Divisor}} = \text{Quotient} + \frac{\text{Remainder}}{\text{Divisor}}$$

$$\frac{37\ (\text{Dividend})}{8\ (\text{Divisor})} = 4(\text{Quotient}) + \frac{5\ (\text{Remainder})}{8\ (\text{Divisor})}$$

6. The result may be checked by using the same relationship in the following form

37 (Dividend) = 4 (Quotient) $\times$ 8 (Divisor) + 5 (Remainder)

Example: Divide $2x^3 + 9x^2 - 26x + 12$ by $2x - 3$

▶

$$
\begin{array}{r}
x^2 + 6x - 4 \\
2x - 3 \,\big)\, \overline{2x^3 + 9x^2 - 26x + 12} \\
2x^3 - 3x^2 \\
\hline
12x^2 - 26x \\
12x^2 - 18x \\
\hline
-8x + 12 \\
-8x + 12 \\
\hline
\end{array}
$$

The quotient is $x^2 + 6x - 4$ and the remainder is 0. The division may be checked by multiplying the quotient ($x^2 + 6x - 4$) by the divisor ($2x - 3$) to obtain the dividend ($2x^3 + 9x^2 - 26x + 12$).

Example: Divide $12x + 7x^3 - 3x^2 - 8 + 6x^4$ by $3x + 2x^2 - 2$

▶ Rearrange both dividend and divisor in descending powers of x.

$$
\require{enclose}
\begin{array}{r}
3x^2 - x + 3 \\
2x^2 + 3x - 2 \enclose{longdiv}{6x^4 + 7x^3 - 3x^2 + 12x - 8} \\
\underline{6x^4 + 9x^3 - 6x^2 } \\
-2x^3 + 3x^2 + 12x \\
\underline{-2x^3 - 3x^2 + 2x } \\
6x^2 + 10x - 8 \\
\underline{6x^2 + 9x - 6} \\
x - 2
\end{array}
$$

The quotient is $3x^2 - x + 3$ and the remainder is $x - 2$. The division may be checked by multiplying the quotient ($3x^2 - x + 3$) by the divisor ($2x^2 + 3x - 2$) and adding the remainder ($x - 2$) to the result to obtain the dividend ($6x^4 + 7x^3 - 3x^2 + 12x - 8$).

Example: Divide $31xy^3 - 20y^3 - 10x^2y^2 + 6x^4 - x^3y$ by $3x^2 - 5y^2 + 4xy$.

▶ Rearrange both dividend and divisor in descending powers of x.

$$
\require{enclose}
\begin{array}{r}
2x^2 - 3xy + 4y^2 \\
3x^2 + 4xy - 5y^2 \enclose{longdiv}{6x^4 - x^3y - 10x^2y^2 + 31xy^3 - 20y^4} \\
\underline{6x^4 + 8x^3y - 10x^2y^2 } \\
-9x^3y + 31xy^3 \\
\underline{-9x^3y - 12x^2y^2 + 15xy^3 } \\
12x^2y^2 + 16xy^3 - 20y^4 \\
\underline{12x^2y^2 + 16xy^3 - 20y^4}
\end{array}
$$

The quotient is $2x^2 - 3xy + 4y^2$ and the remainder is 0. The division may be checked by multiplying the quotient ($2x^2 - 3xy + 4y^2$) by the divisor ($3x^2 + 4xy - 5y^2$) to obtain the dividend ($6x^4 - x^3y - 10x^2y^2 + 31xy^3 - 20y^4$).

Example: Divide $x^4 - 16y^4$ by $x - 2y$

▶

$$
\begin{array}{r}
x^3 + 2x^2y + 4xy^2 + 8y^3 \\
x - 2y \enclose{longdiv}{x^4 \qquad\qquad\qquad\quad - 16y^4} \\
\underline{x^4 - 2x^3y} \qquad\qquad\qquad\quad \\
2x^3y \qquad\qquad\qquad \\
\underline{2x^3y - 4x^2y^2} \qquad\qquad \\
4x^2y^2 \qquad\qquad \\
\underline{4x^2y^2 - 8xy^3} \qquad \\
8xy^3 - 16y^4 \\
\underline{8xy^3 - 16y^4}
\end{array}
$$

The quotient is $x^3 + 2x^2y + 4xy^2 + 8y^3$.

Example: Divide $y^6 - 2y^3 + 1$ by $y^2 - 2y + 1$

▶

$$
\begin{array}{r}
y^4 + 2y^3 + 3y^2 + 2y + 1 \\
y^2 - 2y + 1 \enclose{longdiv}{y^6 \qquad\qquad - 2y^3 \qquad\qquad +1} \\
\underline{y^6 - 2y^5 + y^4} \qquad\qquad\qquad\quad \\
2y^5 - y^4 - 2y^3 \qquad\qquad \\
\underline{2y^5 - 4y^4 + 2y^3} \qquad\qquad \\
3y^4 - 4y^3 \qquad\qquad \\
\underline{3y^4 - 6y^3 + 3y^2} \qquad \\
+2y^3 - 3y^2 \qquad \\
\underline{+2y^3 - 4y^2 + 2y} \\
+ y^2 - 2y + 1 \\
\underline{+ y^2 - 2y + 1}
\end{array}
$$

The quotient is $y^4 + 2y^3 + 3y^2 + 2y + 1$, the remainder is 0.

EXERCISE **9**

Perform the following divisions:
1. $3y^2 + y - 14$ by $y - 2$
2. $12x^2 - 11xy - 36y^2$ by $4x - 9y$
3. $14a + 6a^3 - 17a^2 - 3$ by $2a - 3$
4. $-3y^2 + 6y^5 - 11y^3 + 4 + 4y$ by $3y^2 - 4$

5. $12y^4 - y^3 + 18 - 27y^2 - 3y$ by $y - 2 + 4y^2$
6. $1 + 2a^6 - 3a^4$ by $a^2 + 1 + 2a$
7. $x^3 - 27y^3$ by $x - 3y$
8. $x^6 - 4x^4 + 9x^2 - 15$ by $x^2 - 2$

ANSWERS

EXERCISE 1—Page 18

1. $-2a - 3b + 3c$
2. $-3y^2 + 3xy - 4x^2$
3. $-3c^3 + 7c^2d^2 + 5cd^2 + d^3$

4. $8a^3 - 9a^2b^3 + 6b^3$
5. $4x^5y - 15x^4y^2 - 3x^3y^3$
   $-x^2y^4 - 2xy^5$

EXERCISE 2—Page 20

1. $8a^2 + a - 5$
2. $x^2 - 7x + 9$
3. $3b^2 - 3b - 4$
4. $-2d^3 - 8d^2 - 12d + 5$
5. $-2x + 3y + 5$
6. $-4b^3 - bc^2 + 3c^3$
7. $-10x^3 - x^2 + 5x + 11$

8. $9p^4 + 11p^2q^2 - 4 - 2q^4$
9. $7ab - 4ac - 4bc - 6cd$
   $+ 2bd$
10. $4a + 2b + 2c - 8d$
11. $5b^3 + 2b^2c + 3bc^2 + 2d^3$
12. $-5ab + 5bc + 9ac$
13. $6xy + 14yz - 9xz$
14. $7a^2b^2 + 9a^2b + 2ab^2$
    $+ 6b^2$

EXERCISE 3—Page 21

1. $-12a^2b$
2. $28x^3$
3. $-15y^5$
4. $-15bc^3$
5. $2a^3b^3$
6. $-21x^3y^4$

7. $-ab^2c^2d$
8. $-15x^2y^2z^2$
9. $105a^3b^3c^4$
10. $-4x^3y^3z^4$
11. $48a^2b^3c$
12. $32c^3d^5$

EXERCISE 4—Page 22

1. $-30a + 15b$
2. $-6a^3 + 2a^2b$
3. $-24c^3 + 12c^4$
4. $2x^2y + 3xy^2 - x^2y^2$

5. $12b^3 - 15b^2 + 6b$
6. $-8ab^3 + 12a^2b + 24a^2b^2$
7. $-2x^3y^3 + 6x^4y^2 - 10x^3y^2$
8. $-14a^3b^4 + 7a^5b^2 - 35a^3b^3$

EXERCISE 5—Page 23

1. $2x^3 - 13x^2 + 26x - 15$
2. $3y^3 - 8y^2 + 3y + 2$
3. $5x^3 - 16x^2 - 3x + 18$
4. $2a^3 - a^2b + 9b^3$
5. $3a^6 - 13a^4 - 2a^2 + 2$
6. $36x^2 - 49y^2$

7. $4a^2 + 12ab + 9b^2$
8. $c^3 - 2c^2d + 2cd^2 - d^3$
9. $x^4 - 5x^3 - 3x^2 + 17x - 10$
10. $y^4 - 3y^3 + 2y^2 + 6y - 8$
11. $15x^2 + 7xy + xz - 2y^2 + 5yz - 2z^2$
12. $2a^4 - 3a^3 - 3a^2 + 7a - 3$

EXERCISE 6—Page 23

1. $-x + 8$
2. $-x - 4$
3. $8x - 26$
4. $-27x - 11$
5. $-3x^2 + 6x + 8$

6. $-15x^2 + 13x - 3$
7. $7x - 24$
8. $-37x + 15$
9. $35x - 6$
10. $-55y + 175$

EXERCISE 7—Page 24

1. $4x^3$
2. $-3y^4$
3. $-4ab^2$
4. $-18d^6$
5. $9x^7y$

6. $2a^8c^4$
7. $-5c$
8. $-18x^6yz$
9. $8b^5$
10. q

EXERCISE 8—Page 25

1. $-5x^2 + 2x + 8$
2. $-2y^4 + 3y^2 - 4$
3. $-7a^2 + 9a + 2$
4. $-4ab + 7a^2b$

5. $10xy^2 - 9y + 6x$
6. $3c^2d^3 - 4c^3d^2 + 6cd$
7. $9ab - 7b^2 + 10a^2b$

EXERCISE 9—Page 27

1. $3y + 7$
2. $3x + 4y$
3. $3a^2 - 4a + 1$
4. $2y^3 - y - 1$

5. $3y^2 - y - 5 + \dfrac{8}{4y^2 + y - 2}$
6. $2a^4 - 4a^3 + 3a^2 - 2a + 1$
7. $x^2 + 3xy + 9y^2$
8. $x^4 - 2x^2 + 5 - \dfrac{5}{x^2 - 2}$

CHAPTER 3

Special Products
and Factoring

SPECIAL PRODUCTS

It is frequently necessary to find the product of two polynomials. Such work may be facilitated by using the following products as formulas. These products may be verified by performing the multiplications.

The Product of Two Binomials

$$(x + y)^2 = (x + y)(x + y) = x^2 + 2xy + y^2$$

Examples:

$(a + 3)(a + 3) = a^2 + 2(a)(3) + 3^2 = a^2 + 6a + 9$

$(3a + 4)(3a + 4) = (3a)^2 + 2(3a)(4) + 4^2 = 9a^2 + 24a + 16$

$(2b^2 + 5)(2b^2 + 5) = (2b^2)^2 + 2(2b^2)(5) + 5^2 = 4b^4 + 20b^2 + 25$

$(cd + 2)(cd + 2) = (cd)^2 + 2(cd)(2) + 2^2 = c^2d^2 + 4cd + 4$

$$(x - y)^2 = (x - y)(x - y) = x^2 - 2xy + y^2$$

Examples:

$(a - 5)(a - 5) = a^2 - 2(a)(5) + 5^2 = a^2 - 10a + 25$
$(2a - 3)(2a - 3) = (2a)^2 - 2(2a)(3) + 3^2 = 4a^2$
$- 12a + 9$

$(3c^2 - 4)(3c^2 - 4) = (3c^2)^2 - 2(3c^2)(4) + 4^2 = 9c^4$
$- 24c^2 + 16$

$(pq - 7)(pq - 7) = (pq)^2 - 2(pq)(7) + 7^2 = p^2q^2$
$- 14pq + 49$

$$(x + y)(x - y) = x^2 - y^2$$

Examples:

$(a + 7)(a - 7) = a^2 - 49$
$(3a + 4)(3a - 4) = 9a^2 - 16$
$(2b^2 + 5)(2b^2 - 5) = 4b^4 - 25$
$(xy + 9)(xy - 9) = x^2y^2 - 49$

$$(x + b)(x + c) = x^2 + (b + c)x + bc$$

Examples:

$(a + 7)(a + 9) = a^2 + (7 + 9)a + 63 = a^2 + 16a + 63$
$(a + 3)(a - 8) = a^2 + (3 - 8)a - 24 = a^2 - 5a - 24$
$(b^2 - 5)(b^2 + 4) = b^4 + (-5 + 4)b^2 - 20 = b^4 - b^2 - 20$
$(xy + 6)(xy - 10) = x^2y^2 + (+6 - 10)xy - 60 = x^2y^2$
$- 4xy - 60$

$$(ax + b)(cx + d) = acx^2 + (ad + bc)x + bd$$

Examples:

$(3x + 5)(4x + 7) = 12x^2 + (21 + 20)x + 35 = 12x^2$
$+ 41x + 35$

$(2y - 9)(3y + 4) = 6y^2 + (8 - 27)y - 36 = 6y^2$
$- 19y - 36$

$(4a^2 + 1)(2a^2 - 3) = 8a^4 + (-12 + 2)a^2 - 3 = 8a^4$
$- 10a^2 - 3$

$(3ab - 5)(7ab + 6) = 21a^2b^2 + (18 - 35)ab - 30 = 21a^2b^2$
$- 17ab - 30$

Cubes of Binomials

$$(x + y)^3 = x^3 + 3x^2y + 3xy^2 + y^3$$

Examples:

$(a + 2)^3 = a^3 + 3(a)^2(2) + 3(a)(2)^2 + 2^3 = a^3 + 6a^2$
$+ 12a + 8$

$(3a + 5)^3 = (3a)^3 + 3(3a)^2(5) + 3(3a)(5)^2 + 5^3 = 27a^3$
$+ 135a^2 + 225a + 125$

$(ab + 4)^3 = (ab)^3 + 3(ab)^2(4) + 3(ab)(4)^2 + 4^3 = a^3b^3$
$+ 12a^2b^2 + 48ab + 64$

$(a^2 + 7)^3 = (a^2)^3 + 3(a^2)^2(7) + 3(a^2)(7)^2 + 7^3 = a^6$
$+ 21a^4 + 147a^2 + 343$

$(2x + 3y)^3 = (2x)^3 + 3(2x)^2(3y) + 3(2x)(3y)^2 + (3y)^3$
$= 8x^3 + 36x^2y + 54xy^2 + 27y^3$

$$(x - y)^3 = x^3 - 3x^2y + 3xy^2 - y^3$$

Examples:

$(b - 4)^3 = b^3 - 3(b)^2(4) + 3(b)(4)^2 - 4^3 = b^3 - 12b^2$
$+ 48b - 64$

$(2a - 3)^3 = (2a)^3 - 3(2a)^2(3) + 3(2a)(3)^2 - 3^3 = 8a^3$
$- 36a^2 + 54a - 27$

$(xy - 5)^3 = (xy)^3 - 3(xy)^2(5) + 3(xy)(5)^2 - 5^3 = x^3y^3$
$- 15x^2y^2 + 75xy - 125$

$(y^2 - 6)^3 = (y^2)^3 - 3(y^2)^2(6) + 3(y^2)(6)^2 - 6^3$
$= y^6 - 18y^4 + 108y^2 - 216$

$$(3a - 2b)^3 = (3a)^3 - 3(3a)^2(2b) + 3(3a)(2b)^2 - (2b)^3$$
$$= 27a^3 - 54a^2b + 36ab^2 - 8b^3$$

Products of Binomials and Trinomials

$$(x + y)(x^2 - xy + y^2) = x^3 + y^3$$

Examples:

$$(a + 3)(a^2 - 3a + 9) = a^3 + 27$$
$$(2a + 1)(4a^2 - 2a + 1) = 8a^3 + 1$$
$$(3x + 2y)(9x^2 - 6xy + 4y^2) = 27x^3 + 8y^3$$

$$(x - y)(x^2 + xy + y^2) = x^3 - y^3$$

Examples:

$$(b - 2)(b^2 + 2b + 4) = b^3 - 8$$
$$(3a - 1)(9a^2 + 3a + 1) = 27a^3 - 1$$
$$(4x - 3y)(16x^2 + 12xy + 9y^2) = 64x^3 - 27y^3$$

Squares of Trinomials

$$(x + y + z)^2 = x^2 + y^2 + z^2 + 2xy + 2xz + 2yz$$

Examples:

$$(a + 2b + c)^2 = a^2 + (2b)^2 + c^2 + 2(a)(2b) + 2(a)(c)$$
$$+ 2(2b)(c) = a^2 + 4b^2 + c^2 + 4ab + 2ac + 4bc$$

$$(x + 3y - 2)^2 = (x^2) + (3y)^2 + (-2)^2 + 2(x)(3y)$$
$$+ 2(x)(-2) + 2(3y)(-2) = x^2 + 9y^2 + 4 + 6xy$$
$$- 4x - 12y$$
$$(b^2 - 2b - 1)^2 = (b^2)^2 + (-2b)^2 + (-1)^2 + 2(b^2)(-2b)$$
$$+ 2(b^2)(-1) + 2(-2b)(-1) = b^4 + 4b^2 + 1 - 4b^3 - 2b^2$$
$$+ 4b = b^4 - 4b^3 + 2b^2 + 4b + 1$$

Products of Binomials and Polynomials

$$(x + y)(x^{n-1} - x^{n-2}y + x^{n-3}y^2 - x^{n-4}y^3 + \ldots + y^{n-1})$$
$$= x^n + y^n$$

where n is a positive odd integer

Examples:

$(x + y)(x^4 - x^3y + x^2y^2 - xy^3 + y^4) = x^5 + y^5$
$(x + y)(x^6 - x^5y + x^4y^2 - x^3y^3 + x^2y^4 - xy^5 + x^6)$
$= x^7 + y^7$

$$(x - y)(x^{n-1} + x^{n-2}y + x^{n-3}y^2 + x^{n-4}y^3 + \cdots + y^{n-1})$$
$$= x^n - y^n$$

where *n* is a positive odd integer

Examples:

$(x - y)(x^4 + x^3y + x^2y^2 + xy^3 + y^4) = x^5 - y^5$
$(x - y)(x^6 + x^5y + x^4y^2 + x^3y^3 + x^2y^4 + xy^5 + y^6)$
$= x^7 - y^7$

Additional Products

$$(x + y + a)(x + y - a) = [(x + y) + a][(x + y) - a]$$
$$= [(x + y)^2 - a^2] = x^2 + 2xy + y^2 - a^2$$

Examples:

$(b + c + 5)(b + c - 5) = b^2 + 2bc + c^2 - 25$
$(2x + 3y + 4)(2x + 3y - 4) = 4x^2 + 12xy + 9y^2 - 16$

$$[(a + b) + (x + y)][(a + b) - (x + y)] = (a + b)^2$$
$$- (x + y)^2 = a^2 + 2ab + b^2 - x^2 - 2xy - y^2$$

Examples:

$[(x + 4) + (y - 2)][(x + 4) - (y - 2)] = x^2 + 8x + 16$
$- y^2 + 4y - 4 = x^2 + 8x - y^2 + 4y + 12$

$[(2a + 3) + (b - 5)][(2a + 3) - (b - 5)] = 4a^2 + 12a$
$+ 9 - b^2 + 10b - 25 = 4a^2 + 12a - b^2 + 10b - 16$

$$(x + y - a)(x - y + a) = [x + (y - a)][x - (y - a)]$$
$$= x^2 - (y - a)^2 = x^2 - y^2 + 2ay - a^2$$

Examples:

$(c + d - 3)(c - d + 3) = c^2 - d^2 + 6d - 9$
$(3a + 2y - 1)(3a - 2y + 1) = 9a^2 - 4y^2 + 4y - 1$

$$(x + y)^2(x - y)^2 = [(x + y)(x - y)]^2 = (x^2 - y^2)^2$$
$$= x^4 - 2x^2y^2 + y^4$$

Examples:

$(a + 3)^2(a - 3)^2 = a^4 - 18a^2 + 81$
$(2a + 1)^2(2a - 1)^2 = 16a^4 - 8a^2 + 1$

EXERCISE 1

Perform the following multiplications:

1. $(b + y)^2$
2. $(m - n)^2$
3. $(2a + 1)^2$
4. $(3 + c)^2$
5. $(4a - 3)^2$
6. $(ab + 5)^2$
7. $(3x^2 + 7)^2$
8. $(2y^2 - 3)^2$
9. $(2cd - 9)^2$
10. $(ax^2 + 2)^2$
11. $(1 - 3x^2)^2$
12. $(x^3 - 2)^2$

13. $(2x + 3y)(2x - 3y)$
14. $(a + 5b)(a - 5b)$
15. $(7 - xy)(7 + xy)$
16. $(ab + cd)(ab - cd)$
17. $(3c^2 - d)(3c^2 + d)$
18. $(xy^2 - 2)(xy^2 + 2)$
19. $(x + 2)(x + 1)$
20. $(x + 7)(x - 2)$
21. $(y - 5)(y - 7)$
22. $(2x + 1)(3x - 4)$
23. $(2 + 5y)(3 + y)$
24. $(5 - 2a)(6 + 7a)$

25. $(3a + b)(2a - 3b)$
26. $(7x + 2y)(3x + 4y)$
27. $(5c - d)(4c - 3d)$
28. $(a^2 + 2)(a^2 - 5)$
29. $(7 - 2y^2)(8 + y^2)$
30. $(xy + 5)(3xy - 2)$
31. $(ab^2 + 1)(2ab^2 - 5)$
32. $(3 - 4cd^2)(5 + 2cd^2)$
33. $(x + 1)^3$
34. $(y + 4)^3$
35. $(2a + 3)^3$
36. $(4x + 5)^3$
37. $(b^2 + 2)^3$
38. $(xy + 3)^3$

39. $(3a + 2b)^3$
40. $(4x^2 + y)^3$
41. $(y - 1)^3$
42. $(x - 3)^3$
43. $(2x - 5)^3$
44. $(4x - 3)^3$
45. $(a^2 - 4)^3$
46. $(xy - 2)^3$
47. $(3x - 4y)^3$
48. $(2a^2 - b)^3$
49. $(a + 1)(a^2 - a + 1)$
50. $(x + 2)(x^2 - 2x + 4)$
51. $(3y + 1)(9y^2 - 3y + 1)$
52. $(2a + 3b)(4a^2 - 6ab + 9b^2)$

53. $(x - 1)(x^2 + x + 1)$

54. $(y - 3)(y^2 + 3y + 9)$

55. $(4y - 3)(16y^2 + 12y + 9)$

56. $(2x - 3y)(4x^2 + 6xy + 9y^2)$

57. $(a + b + c)^2$

58. $(x + y + 2z)^2$

59. $(a + 3b - 4)^2$

60. $(2x + 3y - 1)^2$

61. $(y^2 - y + 2)^2$

62. $(a^2 - a - 1)^2$

63. $(a + b)(a^4 - a^3b + a^2b^2 - ab^3 + b^4)$

64. $(a + b)(a^6 - a^5b + a^4b^2 - a^3b^3 + a^2b^4 - ab^5 + b^6)$

65. $(a - b)(a^4 + a^3b + a^2b^2 + ab^3 + b^4)$

66. $(a - b)(a^6 + a^5b + a^4b^2 + a^3b^3 + a^2b^4 + ab^5 + b^6)$

67. $(x + y + 5)(x + y - 5)$

68. $(2a + 3b + 7)(2a + 3b - 7)$

69. $(c + 2d - 4)(c + 2d + 4)$

70. $(a + b - 3)(a - b + 3)$

71. $(2x + 3y - 1)(2x - 3y + 1)$

72. $(4a - b + 2)(4a + b - 2)$

73. $[(c + d) + (p + q)]$
$[(c + d) - (p + q)]$

74. $[(y + 4) + (x - 3)]$
$[(y + 4) - (x - 3)]$

75. $[(3a - 1) + (b - 2)]$
$[(3a - 1) - (b - 2)]$

76. $(x + 2)^2(x - 2)^2$

77. $(3a + 1)^2(3a - 1)^2$

78. $(4x + 3)^2(4x - 3)^2$

FACTORING

DEFINITION: Factoring a polynomial is the process of finding two or more polynomials, each of lower degree than the given polynomial, whose product will yield the original polynomial. The polynomials which form the product are called factors.

In this book, we shall restrict factorization to polynomials with integral coefficients.

The factors must also have integral coefficients.

Examples: The algebraic expression $3x^2 + 6xy$ may be written as $3x(x + 2y)$. The two factors, in this case, are the monomial $3x$ and the binomial $(x + 2y)$.

The algebraic expression $x^2 - x - 6$ may be written as $(x - 3)$ $(x + 2)$. The two factors, in this case, are the binomials $(x - 3)$ and $(x + 2)$.

Prime Factors

A prime factor in algebra is one which cannot be divided by any quantity except $\pm$ itself and ± 1.

Examples: $7, x + y, 3a - 4b$

If a polynomial has integral coefficients it will be considered prime if it cannot be expressed as a product of two factors with integral coefficients. For example, the expression $a^2 - 3b^2$ is considered prime although it can be written in the form $(a + \sqrt{3b})(a - \sqrt{3b})$, i.e., with the coefficients of the b terms irrational numbers.

A polynomial is factored completely if it is expressed as a product of prime factors.

It will be noted that the process of factoring is the inverse of that of forming products. Therefore, the student will find it helpful to refer to the special products studied in an earlier section of this chapter.

Common Monomial Factoring

A monomial which can be divided into each term of a polynomial is called a common monomial factor of the polynomial. In the factoring process each term of the polynomial is divided by the common monomial factor.

Example: Factor $6a^2 - 9ab$

▶ The common monomial factor is $3a$.
$$6a^2 - 9ab = 3a(2a - 3b)$$

Example: Factor $12x^2y^3 - 18xy^2 + 24x^2y$

▶ The common monomial factor is $6xy$.
$$12x^2y^3 - 18xy^2 + 24x^2y = 6xy(2xy^2 - 3y + 4x)$$

Example: Find the value of $79 \times 15 - 79 \times 5$

▶ $79 \times 15 - 79 \times 5 = 79(15 - 5) = 79 \times 10 = 790$

The Difference of Two Squares

In working with special products it was observed that $(x + y)$ $(x - y) = x^2 - y^2$. Therefore, the factors of $x^2 - y^2$ are $(x + y)$ and $(x - y)$. In general, to factor the difference of two squares take the square root of each square. Then the factors will be the sum of these two roots and the difference of these two roots.

Example: Factor $a^2 - 25b^2$

▶ $a^2 - 25b^2 = (a + 5b)(a - 5b)$

Example: Factor $100x^2z^4 - 9y^2$

▶ $100x^2z^4 - 9y^2 = (10xz^2 + 3y)(10xz^2 - 3y)$

Example: Factor $16a^2 - 36b^2$

▶   First remove the common monomial factor 4.
   $16a^2 - 36b^2 = 4(4a^2 - 9b^2) = 4(2a + 3b)(2a - 3b)$

In factoring polynomials, the common monomial factor should always be removed before other methods of factoring are applied.

Trinomials That Are Perfect Squares

A trinomial is a perfect square when its first and last terms are perfect squares and positive, and the middle term is twice the product of the square roots of the end terms. The sign of the middle term indicates whether the square root of the trinomial is a sum or a difference. In factoring a trinomial that is a perfect square, take the square roots of the first and last terms, and join these square roots by the sign of the middle term.

Example: Factor $a^2 - 8a + 16$

▶ $a^2 - 8a + 16 = (a - 4)(a - 4) = (a - 4)^2$

Example: Factor $9a^2 + 24ab + 16b^2$

▶ $9a^2 + 24ab + 16b^2 = (3a + 4b)(3a + 4b) = (3a + 4b)^2$

Example: Factor $2b^2 - 20b + 50$

▶   First remove the common monomial factor 2
   $2b^2 - 20b + 50 = 2(b^2 - 10b + 25) = 2(b - 5)$
                                     $= 2(b - 5)^2$

Trinomials That Are Not Perfect Squares

Before devising a method for factoring trinomials of this type let us reexamine the process of finding the product of two binomials:

$$
\begin{array}{r}
3x - 2 \\
2x + 5 \\
\hline
6x^2 - 4x \\
+ 15x - 10 \\
\hline
6x + 11x - 10
\end{array}
$$

The following observations may be made:

1. The first term in the product ($6x^2$) is obtained by multiplying the first terms of the binomials ($3x$)($2x$).

2. The last term in the product (-10) is obtained by multiplying the last terms of the binomials (-2)($+5$).

3. The middle term of the product ($+11x$) is the algebraic sum of the cross-products ($2x$)(-2) + ($3x$)(5).

If the binomials are written horizontally ($3x - 2$)($2x + 5$) the cross-products may be found mentally by obtaining the algebraic sum of the products of the terms joined by the curved lines.

These observations suggest the following procedure for factoring a trinomial. Separate the first term into two such factors, that the algebraic sum of their cross-products is equal to the middle term of the trinomial.

Example: Factor $y^2 + 8y + 15$

▶ $y^2 + 8y + 15 = (y + 5)(y + 3)$

Note that the factors of y^2 are $y \times y$ and that the factors of 15 are 5×3. The algebraic sum of the cross-products is $+8y$.

Example: Factor $y^2 - y - 12$

▶ $y^2 - y - 12 = (y - 4)(y + 3)$

Note that the factors of y^2 are $y \times y$ and that the factors of -12 are $(-4) \times (3)$. The algebraic sum of the cross-products is $-7a$.

Example: Factor $2a^2 - 7a - 30$

▶ $(2a^2 - 7a - 30) = (2a + 5)(a - 6)$

Note that the factors of $2a^2$ are $2a \times a$ and that the factors of -30 are $(+5) \times (-6)$. The algebraic sum of the cross-products is $-7a$.

Example: Factor $12c^2 - 11cd - 15d^2$

▶ $12c^2 - 11cd - 15d^2 = (4c + 3d)(3c - 5d)$

Note that the factors of $12c^2$ are $(4c) \times (3c)$ and that the factors of $-15d^2$ are $(+3d) \times (-5d)$. The algebraic sum of the cross-products is $-11cd$.

Example: Factor $12 - 4x - 5x^2$

▶ $12 - 4x - 5x^2 = (6 - 5x)(2 + x)$

Note that the factors of 12 are $(6) \times (2)$ and that the factors of $-5x^2$ are $(-5x) \times (x)$. The algebraic sum of the cross-products is $-4x$.

Example: Factor $4x^3 - 10x^2 - 6x$

▶   First remove the common monomial factor $2x$, obtaining $2x(2x^2 - 5x - 3)$. Now factor the trinomial. The final result is $2x(2x + 1)(x - 3)$.

EXERCISE 2

1. $16x + 4y$
2. $bx^2 + 9b$
3. $12c^2d^3 - 18cd^4$
4. $45x^2y^2 - 27x^2y$
5. $p + prt$
6. $9a^3b^2c^2 - 15a^2bc^3 + 27a^2bc^4$
7. $54x^3y^3z^3 + 36x^2y^3z^2 - 45x^3y^2z^3$
8. $32a^3b - 24a^2b^2 + 28b^3$
9. $a^2 - 4$
10. $x^2 - 81$
11. $y^2 - 144$
12. $4x^2 - 49y^2$
13. $1 - 16a^2$
14. $100a^2 - 81c^2$
15. $x^2 - .25$
16. $121 - 4y^2$

17. $7x^2 - 7$
18. $x^3 - x$
19. $a - 25a^3$
20. $ab - 36a^3b^3$
21. $25x^2 - 100y^2$
22. $3x^3 - 48x$
23. $4y^2 - 16y^4$
24. $9a^2 - 16x^4$
25. $x^4 - y^4$
26. $16a^4 - b^4$
27. $x^2 + 2x + 1$
28. $y^2 - 4y + 4$
29. $9x^2 - 12x + 4$
30. $4 - 20y + 25y^2$
31. $25x^2 - 30xy + 9y^2$
32. $16a^2 + 24ab + 9b^2$
33. $2y^2 - 12y + 18$
34. $4a^2 - 40a + 100$

35. $a^4 - 6a^2 + 9$

36. $x^4 - 8x^2 + 16$

37. $9a^2b^2 - 42ab + 49$

38. $32x - 48xy + 18xy^2$

39. $7abc^2 - 56abc + 112ab$

40. $ab - 2ab^2 + ab^3$

41. $x^2 + 3x + 2$

42. $y^2 + 5y + 6$

43. $a^2 - 4a + 3$

44. $b^2 - 7b + 12$

45. $x^2 - 4x - 21$

46. $a^2 - 8a + 15$

47. $y^2 - 5y - 14$

48. $x^2 - x - 6$

49. $b^2 - 3b - 4$

50. $a^2 - 19a + 48$

51. $p^2 + 3p - 4$

52. $x^4 - 4x^2 - 21$

53. $2x^2 + 2x - 12$

54. $3x^2 - 6x - 45$

55. $2y^2 - 9y + 4$

56. $3a^2 - a - 2$

57. $5x^2 - 9x - 2$

58. $36 - 16y - y^2$

59. $2a^2 - 5a - 12$

60. $5y^2 - 4y - 12$

61. $30 - x - x^2$

62. $7b^2 - 13b - 2$

63. $4y^2 - 11y - 3$

64. $6x^2 + x - 12$

65. $2x^2 + 3x + 1$

66. $6y^2 - 7y - 5$

67. $2a^2 + 5a - 3$

68. $2c^2 + 5c + 2$

69. $5z^2 + 24z - 5$

70. $20 - 9a - 20a^2$

71. $(x + y)^2 + 5(x + y) - 36$

72. $2y^2 + y - 10$

73. $3c^2 + 7c - 6$

74. $2y^3 - 5y^2 - 3y$

75. $(c + d)^2 + 5x(c + d) + 6x^2$

76. $4 + 13a + 3a^2$

77. $12 - x - x^2$

78. $8x^6 - 38x^3 + 35$

79. $4(a + b)^2 - 4(a + b) - 3$

80. $8a^2b^2 + 22ab^2 - 6b^2$

81. $3y^6 - 83y^3 + 54$

82. $17x^2 - 25x - 18$

83. $a^4 - 3a^3b + 2a^2b^2$

84. $2y^3 + y^4 - 3y^2$

85. $a^5 - 8a^3 + 16a$

86. $9b^2 - 82b + 9$

87. $7y^3 + 77y^2z - 84yz^2$

88. $21x^2 + 26x - 15$

89. $18x^4 - 6x^2 - 4$

90. $12y^2 + 7y - 12$

Sum and Difference of Two Cubes

The following two products appeared earlier in this chapter.

$$(x + y)(x^2 - xy + y^2) = x^3 + y^3$$
$$(x - y)(x^2 + xy + y^2) = x^3 - y^3$$

These products may be used as formulas for factoring the sum of two cubes or the difference of two cubes.

Example: Factor $a^3 + 27b^3$

▶ $a^3 + 27b^3 = (a)^3 + (3b)^3 = (a + 3b)(a^2 - 3ab + 9b^2)$

Example: Factor $8x^3 + 125y^3$

▶ $8x^3 + 125y^3 = (2x)^3 + (5y)^3 = (2x + 5y)$
$$(4x^2 - 10xy + 25y^2)$$

Example: Factor $x^3 - 64y^3$

▶ $x^3 - 64y^3 = (x)^3 - (4y)^3 = (x - 4y)$
$$(x^2 + 4xy + 16y^2)$$

Example: Factor $27c^3 - 8d^3$

▶ $27c^3 - 8d^3 = (3c)^3 - (2d)^3 = (3c - 2d)$
$$(9c^2 + 6cd + 4d^2)$$

Factoring by Grouping

A polynomial may sometimes be factored by grouping its terms in such a manner that the polynomial is divisible by a binomial factor. The following examples will show how this is done.

Example: Factor $ax + bx + ay + by$

▶ The polynomial may be written as $x(a + b) + y(a + b)$. It is now clear that $(a + b)$ is a common factor. Therefore, $ax + bx + ay + by = (a + b)(x + y)$.

Example: Factor $3x^3 - 4x^2y - 4y^3 + 3xy^2$

▶ The polynomial may be written as $x^2(3x - 4y) + y^2(3x - 4y)$. It is now clear that $(3x - 4y)$ is a common factor. Therefore, $3x^3 - 4x^2y - 4y^3 + 3xy^2 = (3x - 4y)(x^2 + y^2)$.

Factoring the Sum or Difference of Two Like Odd Powers

The following two products appeared earlier in this chapter.
$$(x + y)(x^{n-1} - x^{n-2}y + x^{n-3}y - x^{n-4}y^3 + \cdots + y^{n-1}) = x^n + y^n$$
$$(x - y)(x^{n-1} + x^{n-2}y + x^{n-3}y + x^{n-4}y + \cdots + y^{n-1}) = x^n - y^n$$

Where n is a positive odd integer.

These products may be used as formulas for factoring binomials consisting of the sum or difference of two like odd powers.

Example: Factor $a^5 + 32b^5$

▶ $a^5 + 32b^5 = (a)^5 + (2b)^5$
$= (a + 2b)[a^4 - a^3(2b) + a^2(2b)^2 - a(2b)^3 + (2b)^4]$
$= (a + 2b)(a^4 - 2a^3b + 4a^2b^2 - 8ab^3 + 16b^4)$

Example: Factor $128x^7 - y^7$

▶ $128x^7 - y^7 = (2x)^7 - y^7$

$= (2x - y)[(2x)^6 + (2x)^5y + (2x)^4y^2 + (2x)^3y^3$
$+ \ ^-(2x)^2y^2 + (2x)y^5 + y^6]$
$= (2x - y)(64x^6 + 32x^5y + 16x^4y^2 + 8x^3y^3 + 4x^2y^4$
$+ 2xy^5 + y^6)$

Factoring the Sum or Difference of Two Like Even Powers

The sum of two like even powers cannot be factored, in general, by elementary methods unless the expression can be written as the sum of two numbers raised to an odd power.

Example: Factor $x^6 + y^6$

▶ $x^6 + y^6 = (x^2)^3 + (y^2)^3$
$= (x^2 + y^2)(x^4 - x^2y^2 + y^4)$

Expressions such as $x^2 + y^2$, $x^4 + y^4$, and $x^8 + y^8$ cannot be factored. They are prime expressions. However, $x^{10} + y^{10}$ can be factored since it can be written as $(x^2)^5 + (y^2)^5$.

Example: Factor $x^{10} + y^{10}$

▶ $x^{10} + y^{10} = (x^2)^5 + (y^2)^5$
$= (x^2 + y^2)(x^8 - x^6y^2 + x^4y^4 - x^2y^6 + y^8)$

The difference of two like even powers may always be regarded as the difference of two squares.

Example: Factor $x^8 - y^8$

▶ $x^8 - y^8 = (x^4 + y^4)(x^4 - y^4) = (x^4 + y^4)(x^2 + y^2)$
   $(x^2 - y^2)$
   $= (x^4 + y^4)(x^2 + y^2)(x + y)(x - y)$

Additional Methods of Factoring

1. By grouping of terms an algebraic expression may sometimes be arranged in the form of the difference of two perfect squares.

Example: Factor $a^2 - 4ab + 4b^2 - 25x^2$

▶ $a^2 - 4ab + 4b^2 - 25x^2 = (a^2 - 4ab + 4b^2) - 25x^2$
   $= (a - 2b)^2 - 25x^2 = (a - 2b + 5x)(a - 2b - 5x)$

Example: Factor $c^2 - p^2 - 2cd + 2cd + 2py + d^2 - y^2$

▶ Rearranging, we obtain $c^2 - 2cd + d^2 - y^2 + 2py - p^2$
   $= (c^2 - 2cd + d^2) - (y^2 - 2py + p^2) = (c - d)^2$
   $- (y - p)^2 = (c - d + y - p)(c - d - y + p)$

2. By adding or subtracting a perfect square, an algebraic expression may sometimes assume the form of the difference of two squares.

Example: Factor $x^4 + x^2y^2 + y^4$

▶ To the expression $x^4 + x^2y^2 + y^4$ add and subtract x^2y^2. We now have $x^4 + 2x^2y^2 + y^4 - x^2y^2$. This may be written $(x^2 + y^2)^2 - x^2y^2 = (x^2 + y^2 + xy)(x^2 + y^2 - xy)$.

Example: Factor $a^4 - 14a^2b^2 + b^4$

▶ To the expression $a^4 - 14a^2b^2 + b^4$ add and subtract $16a^2b^2$. We now have $a^4 + 2a^2b^2 + b^4 - 16a^2b^2$. This may be written $(a^2 + b^2)^2 - 16a^2b^2 = (a^2 + b^2 + 4ab)(a^2 + b^2 - 4ab)$

3. The product $(x + y + z)^2 = x^2 + y^2 + z^2 + 2xy + 2yz + 2xz$ suggests another method of factoring.

Example: Factor $a^2 + b^2 + c^2 + 2ab + 2bc + 2ac$

▶ We observe that this expression is the square of the trinomial $(a + b + c)$. Therefore, $a^2 + b^2 + c^2 + 2ab + 2bc + 2ac = (a + b + c)^2$.

Example: Factor $x^2 + 4y^2 + z^2 + 2xy + xz + 2yz$

▶ We observe that this expression is the square of the trinomial $x + 2y + z$.

Therefore, $x^2 + 4y^2 + z^2 + 2xy + xz + 2yz = (x + 2y + z)^2$

REMEMBER—In factoring an algebraic expression it is important to remove a common monomial factor, if it is present, before applying other methods of factoring.

EXERCISE 3

Factor the following:

1. $x^3 - 1$
2. $y^3 + 64$
3. $a^3 - 8b^3$
4. $27x^3 - y^3$
5. $a^3b^3 + 64$
6. $(x + y)^3 + 27$
7. $125 - y^3$
8. $(a - b)^3 - 64c^3$
9. $27y^4 + b^3y$
10. $8x^3 - 125y^3$
11. $2a^4 - 128a$
12. $81x^4 - 24x$
13. $ab + bc + ad + cd$
14. $y^2 + cy + by + bc$

15. $2pq - 3tq + 2ps - 3ts$
16. $2x^4 - x^3 + 4x - 2$
17. $a^3b^3 - a^2b^2 - ab + 1$
18. $6x^2 + 3xy - 2bx - by$
19. $1 - y - y^2 + y^3$
20. $c^3 - c - c^2xy + xy$
21. $cx + dy - cy - dx$
22. $3a^3 - 7a^2 + 3a - 7$
23. $z^3 - 21 + 3z - 7z^2$
24. $px + qx + py + qy + tx + ty$
25. $a^3 - 5a^2 + 2a - 10$
26. $x^2y^2 + z^2y^2 - x^2a^2 - z^2a^2$
27. $x^5 - y^5$

28. $a^7 + 128b^7$
29. $c^5 + 243d^5$
30. $p^7 - q^7$
31. $x^6 - y^6$
32. $a^4 - b^4$
33. $p^6 + y^6$
34. $c^8 - d^8$
35. $64x^6 - b^6$
36. $a^6 + 729b^6$
37. $x^2 + 2xy + y^2 - a^2$
38. $a^2 + b^2 + 2ab - 9x^2$
39. $16c^2 - a^2 + 2ab - b^2$
40. $a^2 + 2ab + b^2 - x^2 - 2xy - y^2$

41. $y^4 + y^2 + 1$
42. $4a^4 - 13a^2 + 1$
43. $16b^4 - 9b^2 + 1$
44. $x^4 + 4y^4$
45. $1 + 64c^4$
46. $y^4 + b^2y^2 + b^4$
47. $p^2 + q^2 + r^2 + 2pq + 2pr + 2rq$
48. $x^2 + y^2 + 4z^2 + 2xy + 4xz + 4yz$
49. $a^2 + b^2 + c^2 - 2ab - 2ac + 2bc$
50. $x^2 + 4y^2 + z^2 - 4xy + 2xz - 4yz$

Highest Common Factor

DEFINITION: The highest common factor (H.C.F.) of two or more algebraic expressions is the algebraic expression of highest degree which is a factor of each of the original expressions.

Examples: The H.C.F. of $12a^2b$ and $18ab^2$ is $6ab$

The H.C.F. of $(a^2 - b^2)$ and $3(a + b)$ is $(a + b)$

In finding the H.C.F. the following steps are suggested.

a. Write each expression in factored form.

b. The H.C.F. is the product obtained by taking each common factor to the lowest degree.

Example: Find the H.C.F. of $12a^3b^4c^5$, $18a^4b^5c^3$, $24a^3b^6c^2$

▶   Write the monomials as $2^2 \cdot 3 \cdot a^3 \cdot b^4 \cdot c^5$, $2 \cdot 3^2 \cdot a^4 \cdot b^5 \cdot c^3$, $2^3 \cdot 3 \cdot a^3 \cdot b^6 \cdot c^2$.

The lowest common powers of the factors are $2 \cdot 3 \cdot a^3 \cdot b^4 \cdot c^2$. Therefore, the H.C.F. is $6a^3b^4c^2$.

Example: Find the H.C.F. of $a^2 - 4ab + 4b^2$, $a^2 - 3ab + 2b^2$, and $a^2 - 7ab + 10b^2$

▶   Writing each polynomial in factored form

$$a^2 - 4ab + 4b^2 = (a - 2b)^2$$
$$a^2 - 3ab + 2b^2 = (a - 2b)(a - b)$$
$$a^2 - 7ab + 10b^2 = (a - 2b)(a - 5b)$$

Therefore, the H.C.F. is $(a - 2b)$

Example: Find the H.C.F. of $2x^2 + 9x + 4$, $2x^2 + 11x + 5$, and $2x^2 - 3x - 2$.

▶   Writing each polynomial in factored form

$$2x^2 + 9x + 4 = (2x + 1)(x + 4)$$
$$2x^2 + 11x + 5 = (2x + 1)(x + 5)$$
$$2x^2 - 3x - 2 = (2x + 1)(x - 2)$$

Therefore, the H.C.F. is $(2x + 1)$

EXERCISE 4

In each case, find the H.C.F.
1. $24x^2yz^3$, $36x^3y^2z^5$, $48x^2y^5z^4$
2. $50a^3b^2c$, $75ab^2c^2$, $20a^4b^3c^3$
3. $6a^3b^4$, $12a^2b^3 - 18a^3b^3$
4. $48p^2q^3$, $12p^3q^2 + 24p^2q^3$
5. $a^2 - 4b^2$, $a^2 + 2ab$
6. $4a^3 + 12a^2b + 9ab^2$, $16ab + 24b^2$
7. $y^2 - 7y + 12$, $y^2 - 8y + 15$
8. $x^2 + 4xy + 3y^2$, $x^2 + 2xy - 3y^2$, $x^2 + 9xy + 18y^2$

Lowest Common Multiple

DEFINITION: The lowest common multiple (L.C.M.) of two or more algebraic expressions is the expression of lowest degree into which each of the original expressions can be divided without a remainder.

Examples: The L.C.M. of $6x^2y^3$ and $9xy^4$ is $18x^2y^4$
The L.C.M. of $(x - y)^2$, $5(x + y)$, and $10(x + y)$ $(x - y)$ is $10(x + y)(x - y)^2$

In finding the L.C.M. the following steps are suggested.

a. Write each expression in factored form.

b. The L.C.M. is the product obtained by taking each factor to the highest degree.

Example: Find the L.C.M. of $8x^2yz$, $12x^5y^3$, and $14z^3$

▶ Write the monomials as $2^3 \cdot x^2 \cdot y \cdot z$, $2^2 \cdot 3 \cdot x^5 \cdot y^3$, $2 \cdot 7 \cdot z^3$. Taking each factor to the highest degree we have $2^3 \cdot 3 \cdot 7 \cdot x^5 \cdot y^3 \cdot z^3$. Therefore, the L.C.M. is $168x^5y^3z^3$.

Example: Find the L.C.M. of $9a$, $15b$, $a^2b - ab^2$

▶ Write the expressions as $3^2 \cdot a$, $3 \cdot 5 \cdot b$, $ab(a - b)$. The L.C.M. is $3^2 \cdot 5 \cdot a \cdot b(a - b) = 45ab(a - b)$.

Example: Find the L.C.M. of $y^2 + 5y + 6$, $y^2 + 4y + 3$, $y^3 + y^2$

▶ Writing each expression in factored form
$$y^2 + 5y + 6 = (y + 3)(y + 2)$$
$$y^2 + 4y + 3 = (y + 3)(y + 1)$$
$$y^3 + y^2 \qquad = y^2(y + 1)$$
The L.C.M. is $y^2(y + 1)(y + 2)(y + 3)$.

EXERCISE 5

In each case, find the L.C.M.
1. $6a^3b^2c^1, 8a^2b^3c^2, 10a^1bc^2$
2. $16x^2yz^2, 21x^4y^3z, 18x^3y^3z$
3. $12x, 8y, 3x^2 - 3xy^2$
4. $5a, 7b^2, 10ab - 20b^2$
5. $9a + 9b, 3(a - b), 6ab + 6b^2$
6. $3x^2 - 12y^2, 4x^2 - 8xy, 6x + 12y$
7. $4x^2 + 2x - 2, 4x^2 - 1, 4x^2 + 4x - 3$
8. $2x^2 - 3x - 2, 2x^2 - 15x - 8, x^2 - 10x + 16$

ANSWERS

EXERCISE 1—Page 35

1. $b^2 + 2by + y^2$
2. $m^2 - 2mn + n^2$
3. $4a^2 + 4a + 1$
4. $9 + 6c + c^2$
5. $16a^2 - 24a + 9$
6. $a^2b^2 + 10ab + 25$
7. $9x^4 + 42x^2 + 49$
8. $4y^4 - 12y^2 + 9$
9. $4c^2d^2 - 36cd + 81$
10. $a^2x^4 + 4ax^2 + 4$
11. $1 - 6x^2 + 9x^4$
12. $x^6 - 4x^3 + 4$
13. $4x^2 - 9y^2$
14. $a^2 - 25b^2$
15. $49 - x^2y^2$
16. $a^2b^2 - c^2d^2$
17. $9c^4 - d^2$
18. $x^2y^4 - 4$
19. $x^2 + 3x + 2$
20. $x^2 + 5x - 14$
21. $y^2 - 12y + 35$
22. $6x^2 - 5x - 4$
23. $6 + 17y + 5y^2$
24. $30 + 23a - 14a^2$
25. $6a^2 - 7ab - 3b^2$
26. $21x^2 + 34xy + 8y^2$
27. $20c^2 - 19cd + 3d^2$
28. $a^4 - 3a^2 - 10$
29. $56 - 9y^2 - 2y^4$
30. $3x^2y^2 + 13xy - 10$

31. $2a^2b^4 - 3ab^2 - 5$

32. $15 - 14cd^2 - 8c^2d^4$

33. $x^3 + 3x^2 + 3x + 1$

34. $y^3 + 12y^2 + 48y + 64$

35. $8a^3 + 36a^2 + 54a + 27$

36. $64x^3 + 240x^2 + 300x$
    $+ 125$

37. $b^6 + 6b^4 + 12b^2 + 8$

38. $x^3y^3 + 9x^2y^2 + 27xy + 27$

39. $27a^3 + 54a^2b + 36ab^2$
    $+ 8b^3$

40. $64x^6 + 48x^4y + 12x^2y^2$
    $+ y^3$

41. $y^3 - 3y^2 + 3y - 1$

42. $x^3 - 9x^2 + 27x - 27$

43. $8x^3 - 60x^2 + 150x - 125$

44. $64x^3 - 144x^2 + 108x$
    $- 27$

45. $a^6 - 12a^4 + 48a^2 - 64$

46. $x^3y^3 - 6x^2y^2 + 12xy - 8$

47. $27x^3 - 108x^2y + 144xy^2$
    $- 64y^3$

48. $8a^6 - 12a^4b + 6a^2b^2$
    $- b^3$

49. $a^3 + 1$

50. $x^3 + 8$

51. $27y^3 + 1$

52. $8a^3 + 27b^3$

53. $x^3 - 1$

54. $y^3 - 27$

55. $64y^3 - 27$

56. $8x^3 - 27y^3$

57. $a^2 + b^2 + c^2 + 2ab$
    $+ 2ac + 2bc$

58. $x^2 + y^2 + 4z^2 + 2xy$
    $+ 4xz + 4yz$

59. $a^2 + 9b^2 + 16 + 6ab$
    $- 8a - 24b$

60. $4x^2 + 9y^2 + 1 + 12xy - 4x$
    $- 6y$

61. $y^4 - 2y^3 + 5y^2 - 4y + 4$

62. $a^4 - 2a^3 - a^2 + 2a + 1$

63. $a^5 + b^5$

64. $a^7 + b^7$

65. $a^5 - b^5$

66. $a^7 - b^7$

67. $x^2 + 2xy + y^2 - 25$

68. $4a^2 + 12ab + 9b^2 - 49$

69. $c^2 + 4cd + 4d^2 - 16$

70. $a^2 - b^2 + 6b - 9$

71. $4x^2 - 9y^2 + 6y - 1$

72. $16a^2 - b^2 + 4b - 4$

73. $c^2 + 2cd + d^2 - p^2$
    $- 2pq - q^2$

74. $y^2 + 8y + 16 - x^2 + 6x$
    $- 9$

75. $9a^2 - 6a + 1 - b^2 + 4b$
    $- 4$

76. $x^4 - 8x^2 + 16$

77. $81a^4 - 18a^2 + 1$

78. $256x^4 - 288x^2 + 81$

EXERCISE 2—Page 40

1. $4(4x + y)$

2. $b(x^2 + 9)$

3. $6cd^3(2c - 3d)$

4. $9x^2y(5y - 3)$

5. $p(1 + rt)$

6. $3a^2bc^2(3ab - 5c + 9c^2)$

7. $9x^2y^2z^2(6xyz + 4y - 5xz)$

8. $4b(8a^3 - 6a^2b + 7b^2)$

9. $(a + 2)(a - 2)$

10. $(x + 9)(x - 9)$

11. $(y + 12)(y - 12)$

12. $(2x + 7y)(2x - 7y)$

13. $(1 + 4a)(1 - 4a)$

14. $(10a + 9c)(10a - 9c)$

15. $(x + .5)(x - .5)$

16. $(11 + 2y)(11 - 2y)$

17. $7(x + 1)(x - 1)$

18. $x(x + 1)(x - 1)$

19. $a(1 + 5a)(1 - 5a)$

20. $ab(1 + 6ab)(1 - 6ab)$

21. $25(x + 2y)(x - 2y)$

22. $3x(x + 4)(x - 4)$

23. $4y^2(1 + 2y)(1 - 2y)$

24. $(3a + 4x^2)(3a - 4x^2)$

25. $(x^2 + y^2)(x + y)(x - y)$

26. $(4a^2 + b^2)(2a + b)(2a - b)$

27. $(x + 1)^2$

28. $(y - 2)^2$

29. $(3x - 2)^2$

30. $(2 - 5y)^2$

31. $(5x - 3y)^2$

32. $(4a + 3b)^2$

33. $2(y - 3)^2$

34. $4(a - 5)^2$

35. $(a^2 - 3)^2$

36. $(x + 2)^2(x - 2)^2$

37. $(3ab - 7)^2$

38. $2x(4 - 3y)^2$

39. $7ab(c - 4)^2$

40. $ab(1 - b)^2$

41. $(x + 2)(x + 1)$

42. $(y + 3)(y + 2)$

43. $(a - 3)(a - 1)$

44. $(b - 4)(b - 3)$

45. $(x - 7)(x + 3)$

46. $(a - 5)(a - 3)$

47. $(y - 7)(y + 2)$

48. $(x - 3)(x + 2)$

49. $(b - 4)(b + 1)$

50. $(a - 16)(a - 3)$

51. $(p + 4)(p - 1)$

52. $(x^2 - 7)(x^2 + 3)$

53. $2(x + 3)(x - 2)$

54. $3(x - 5)(x + 3)$

55. $(2y - 1)(y - 4)$

56. $(3a + 2)(a - 1)$

57. $(5x + 1)(x - 2)$

58. $(18 + y)(2 - y)$

59. $(2a + 3)(a - 4)$

60. $(5y + 6)(y - 2)$

61. $(5 - x)(6 + x)$

62. $(7b + 1)(b - 2)$

63. $(4y + 1)(y - 3)$

64. $(3x - 4)(2x + 3)$

65. $(2x + 1)(x + 1)$

66. $(3y - 5)(2y + 1)$

67. $(2a - 1)(a + 3)$

68. $(2c + 1)(c + 2)$

69. $(5z - 1)(z + 5)$
70. $(4 - 5a)(5 + 4a)$
71. $(x + y + 9)(x + y - 4)$
72. $(2y + 5)(y - 2)$
73. $(3c - 2)(c + 3)$
74. $y(2y + 1)(y - 3)$
75. $(c + d + 3x)(c + d + 2x)$
76. $(4 + a)(1 + 3a)$
77. $(4 + x)(3 - x)$
78. $(4x^3 - 5)(2x^3 - 7)$
79. $[2(a + b) - 3][2(a + b) + 1]$

80. $2b^2(4a - 1)(a + 3)$
81. $(3y^3 - 2)(y^3 - 27)$
82. $(17x + 9)(x - 2)$
83. $a^2(a - 2b)(a - b)$
84. $y^2(y + 3)(y - 1)$
85. $a(a + 2)^2(a - 2)^2$
86. $(9b - 1)(b - 9)$
87. $7y(y + 12z)(y - z)$
88. $(7x - 3)(3x + 5)$
89. $2(3x^2 + 1)(3x^2 - 2)$
90. $(4y - 3)(3y + 4)$

EXERCISE 3—Page 45

1. $(x - 1)(x^2 + x + 1)$
2. $(y + 4)(y^2 - 4y + 16)$
3. $(a - 2b)(a^2 + 2ab + 4b^2)$
4. $(3x - y)(9x^2 + 3xy + y^2)$
5. $(ab + 4)(a^2b^2 - 4ab + 16)$
6. $(x + y + 3)(x^2 + 2xy + y^2 - 3x - 3y + 9)$
7. $(5 - y)(25 + 5y + y^2)$
8. $(a - b - 4c)(a^2 - 2ab + b^2 + 4ac - 4bc + 16c^2)$
9. $y(3y + b)(9y^2 - 3by + b^2)$
10. $(2x - 5y)(4x^2 + 10xy + 25y^2)$
11. $2a(a - 4)(a^2 + 4a + 16)$
12. $3x(3x - 2)(9x^2 + 6x + 4)$
13. $(a + c)(b + d)$

14. $(y + b)(y + c)$
15. $(q + s)(2p - 3t)$
16. $(x^3 + 2)(2x - 1)$
17. $(ab - 1)^2(ab + 1)$
18. $(3x - b)(2x + y)$
19. $(1 - y)^2(1 + y)$
20. $(c - xy)(c + 1)(c - 1)$
21. $(c - d)(x - y)$
22. $(a^2 + 1)(3a - 7)$
23. $(z - 7)(z^2 + 3)$
24. $(x + y)(p + q + t)$
25. $(a^2 + 2)(a - 5)$
26. $(x^2 + z^2)(y + a)(y - a)$
27. $(x - y)(x^4 + x^3y + x^2y^2 + xy^3 + y^4)$
28. $(a + 2b)(a^6 - 2a^5b + 4a^4b^2 - 8a^3b^3 + 16a^2b^4 - 32ab^5 + 64b^6)$
29. $(c + 3d)(c^4 - 3c^3d + 9c^2d^2 - 27cd^3 + 81d^4)$
30. $(p - q)(p^6 + p^5q + p^4q^2 + p^3q^3 + p^2q^4 + pq^5 + q^6)$

31. $(x + y)(x - y)(x^2 + xy + y^2)(x^2 - xy + y^2)$

32. $(a^2 + b^2)(a + b)(a - b)$

33. $(p^2 + q^2)(p^4 - p^2q^2 + q^4)$

34. $(c^4 + d^4)(c^2 + d^2)(c + d)(c - d)$

35. $(2x + b)(2x - b)(4x^2 + 2bx + b^2)(4x^2 - 2bx + b^2)$

36. $(a^2 + 9b^2)(a^4 - 9a^2b^2 + 81b^4)$

37. $(x + y + a)(x + y - a)$

38. $(a + b + 3x)(a + b - 3x)$

39. $(4c + a - b)(4c - a + b)$

40. $(a + b + x + y)(a + b - x - y)$

41. $(y^2 + y + 1)(y^2 - y + 1)$

42. $(2a^2 + 3a - 1)(2a^2 - 3a - 1)$

43. $(4b^2 + b - 1)(4b^2 - b - 1)$

44. $(x^2 + 2xy + 2y^2)(x^2 - 2xy + 2y^2)$

45. $(8c^2 + 4c + 1)(8c^2 - 4c + 1)$

46. $(y^2 + by + b^2)(y^2 - by + b^2)$

47. $(p + q + r)^2$

48. $(x + y + 2z)^2$

49. $(a - b - c)^2$

50. $(x - 2y + z)^2$

EXERCISE 4—Page 47

1. $12x^2 y z^3$

2. $5ab^2c$

3. $6a^2b^3$

4. $12p^2q^2$

5. $(a + 2b)$

6. $(2a + 3b)$

7. $(y - 3)$

8. $(x + 3y)$

EXERCISE 5—Page 48

1. $120a^4b^3c^4$

2. $1,008x^4y^3z^2$

3. $24xy (x - y^2)$

4. $70ab^2 (a - 2b)$

5. $18b (a + b)(a - b)$

6. $12x (x + 2y)(x - 2y)$

7. $2(2x + 1)(2x - 1)(x + 1)(2x + 3)$

8. $(2x + 1)(x - 2)(x - 8)$

CHAPTER 4

Fractions

DEFINITION: A fraction is an indicated quotient e.g. $\dfrac{x}{y}$ $(y \neq 0)$
The dividend x is called the *numerator* and the divisor y is called the *denominator*. The denominator may not be zero since division by zero is undefined.

DEFINITION: A rational algebraic fraction is the quotient of two polynomials. For example, $\dfrac{3x + 4}{x^2 + x + 5}$ is a rational algebraic fraction.

In general, manipulation of algebraic fractions follows the principles of manipulation with arithmetic fractions.

1. If the numerator and denominator of a fraction are divided by the same quantity (zero excluded) the value of the fraction is unchanged.

Examples: $\dfrac{12}{15} = \dfrac{4}{5}$. The numerator and denominator of the frac-

tion $\dfrac{12}{15}$ were each divided by 3.

$\dfrac{6y}{8y^2} = \dfrac{3}{4y}$. The numerator and denominator of the

fraction $\dfrac{6y}{8y^2}$ were each divided by $2y$.

2. If the numerator and denominator of a fraction are each multiplied by the same quantity (zero excluded) the value of the fraction
is unchanged.

Examples: $\dfrac{2}{3} = \dfrac{14}{21}$. The numerator and denominator of the frac-

tion $\dfrac{2}{3}$ were each multiplied by 7.

$\dfrac{4}{x} = \dfrac{20x}{5x^2}$. The numerator and denominator of the

fraction $\dfrac{4}{x}$ were each multiplied by $5x$.

3. Each fraction has 3 signs associated with it. There is a sign in
front of the fraction, there is a sign attached to the numerator, and
there is a sign attached to the denominator. If any two of these signs
are changed, then the value of the fraction is unchanged.

Examples: $+\dfrac{+x}{+y} = +\dfrac{-x}{-y} = -\dfrac{+x}{-y} = -\dfrac{-x}{+y}$

$-\dfrac{+x}{+y} = +\dfrac{-x}{+y} = +\dfrac{+x}{-y}$

REDUCTION OF FRACTIONS

In reducing fractions, write the numerator and denominator in
factored form. Then divide the numerator and denominator of the
fraction by all common factors.

Example: Reduce the fraction $\dfrac{360}{756}$

▶ $$\frac{360}{756} = \frac{2^3 \cdot 3^2 \cdot 5}{2^2 \cdot 3^3 \cdot 7}$$

The numerator and denominator of this fraction have the common factor $2^2 \cdot 3^2$. Dividing the numerator and denominator of the fraction by $2^2 \cdot 3^2$ we have $\dfrac{2 \cdot 5}{3 \cdot 7}$ or $\dfrac{10}{21}$.

Example: Reduce the fraction $\dfrac{36x^3y^2}{48x^2y^4}$

▶ $$\frac{36x^3y^2}{48x^2y^4} = \frac{2^2 \cdot 3^2 \cdot x^3 \cdot y^2}{2^4 \cdot 3 \cdot x^2 \cdot y^4}$$

Dividing the numerator and denominator of the fraction by the common factor $2^2 \cdot 3 \cdot x^2 \cdot y^2$ we obtain the result $\dfrac{3x}{4y^2}$.

Example: Reduce the fraction $\dfrac{120ab^3c^2}{-72a^3b^2c^4}$.

▶ $$\frac{120ab^3c^2}{-72a^3b^2c^4} = \frac{2^3 \cdot 5 \cdot 3 \cdot ab^3c^2}{-2^3 \cdot 3^2 \cdot a^3b^2c^4}$$

Dividing the numerator and denominator of the fraction by the common factor $2^3 \cdot 3ab^2c^2$ we obtain the result $\dfrac{5b}{-3a^2c^2}$ or $-\dfrac{5b}{3a^2c^2}$

Example: Reduce the fraction $\dfrac{9x - 18}{4x - 8}$

▶ Writing the fraction with numerator and denominator in factored form $\dfrac{9x - 18}{4x - 8} = \dfrac{9(x - 2)}{4(x - 2)}$

Dividing the numerator and denominator by the common factor $(x - 2)$ we obtain the result $\dfrac{9}{4}$.

Example: Reduce the fraction $\dfrac{2a^2 - 8b^2}{4a^2 - 2ab - 12b^2}$

▶ Writing the fraction with numerator and denominator in factored form $\dfrac{2a^2 - 8b^2}{4a^2 - 2ab - 12b^2} = \dfrac{2(a + 2b)(a - 2b)}{2(2a + 3b)(a - 2b)}$

Dividing the numerator and denominator by the common factor $2(a - b)$ we obtain the result $\dfrac{a + 2b}{2a + 3b}$.

Example: Reduce the fraction $\dfrac{3 - 6y}{2y^2 + 5y - 3}$

▶ $\dfrac{3 - 6y}{2y^2 + 5y - 3} = -\dfrac{6y - 3}{2y^2 + 5y - 3}$

(Multiplying the numerator of the fraction by -1 and changing the sign in front of the fraction does not change the value of the fraction.)

$$-\dfrac{6y - 3}{2y^2 + 5y - 3} = -\dfrac{3(2y - 1)}{(2y - 1)(y + 3)} = -\dfrac{3}{y + 3}$$

Example: Reduce the fraction $\dfrac{4x^3 + 8x^2 + 16x}{2x^3 - 16}$

▶ $\dfrac{4x^3 + 8x^2 + 16x}{2x^3 - 16} = \dfrac{4x(\cancel{x^2 + 2x + 4})}{2(x - 2)(\cancel{x^2 + 2x + 4})} = \dfrac{2x}{x - 2}$

Example: Reduce the fraction $\dfrac{3ab^3 - 81a}{12a^2b^2 - 6a^2b - 90a^2}$

▶ $\dfrac{3ab^3 - 81a}{12a^2b^2 - 6a^2b - 90a^2} = \dfrac{3a(b^3 - 27)}{6a^2(2b^2 - b - 15)}$

$$= \dfrac{3a(b - 3)(b^2 + 3b + 9)}{6a^2(2b + 5)(b - 3)}$$

$$= \dfrac{b^2 + 3b + 9}{2a(2b + 5)}$$

Example: Reduce the fraction $\dfrac{cd - 2c + 3d - 6}{cd - 2c - 3d + 6}$

▶ $\dfrac{cd - 2c + 3d - 6}{cd - 2c - 3d + 6} = \dfrac{c(d - 2) + 3(d - 2)}{c(d - 2) - 3(d - 2)}$

$$= \frac{(c + 3)(d - 2)}{(c - 3)(d - 2)} = \frac{c + 3}{c - 3}$$

EXERCISE 1

Reduce the following fractions:

1. $\dfrac{8x^3y^4}{12x^2y^5}$

2. $\dfrac{72a^2b^3c^4}{84a^3b^2c^5}$

3. $\dfrac{120p^8q^6t^4}{-45p^4q^6t^8}$

4. $\dfrac{-132d^5e^3g^7}{22d^2g^8}$

5. $\dfrac{3x - 9y}{6x - 18y}$

6. $\dfrac{2x}{4x^2 - 2x}$

7. $\dfrac{3b^2x}{6b^2 - 9b^3x}$

8. $\dfrac{2x - 8}{4 - x}$

9. $\dfrac{y^2 - 4}{5y - 10}$

10. $\dfrac{8a - 10}{16a^2 - 25}$

11. $\dfrac{2y^2 - 18}{18 - 6y}$

12. $\dfrac{16a^2 - 36b^2}{16a + 24b}$

13. $\dfrac{x^2 - 3x}{x^2 - 4x + 3}$

14. $\dfrac{a^2 - 25}{a^2 - 2a - 15}$

15. $\dfrac{4x - 2}{2x^2 - 5x + 2}$

16. $\dfrac{4 - 4b}{8b^2 - 16b + 8}$

17. $\dfrac{a^2 - 2a + 1}{a^2 + 2a - 3}$

18. $\dfrac{x^2 - 7x + 12}{x^2 - 2x - 3}$

19. $\dfrac{2y^2 + 9y + 4}{2y^2 + 11y + 5}$

20. $\dfrac{2a^2 - ab - 6b^2}{3a^2 - 7ab + 2b^2}$

21. $\dfrac{15b^2 + 2bc - c^2}{12b^2 + bc - c^2}$

22. $\dfrac{3y^4 - 21y^3 + 36y^2}{y^3 - 8y^2 + 15y}$

23. $\dfrac{x^4 - y^4}{x^2 - y^2}$

24. $\dfrac{a^3 - b^3}{7a - 7b}$

25. $\dfrac{4y^3 - 32}{8y^2 + 16y + 32}$

26. $\dfrac{16x^3y - 54y^4}{4x^2y + 6xy^2 + 9y^3}$

27. $\dfrac{ay + a - y - 1}{a^3 - 1}$

28. $\dfrac{ax + ay - cx - cy}{ax + ay + cx + cy}$

29. $\dfrac{2ac - 2ad + bc - bd}{c^2 - d^2}$

30. $\dfrac{9a - 3y}{3ax + 3ab - xy - by}$

MULTIPLICATION OF FRACTIONS

If two fractions are multiplied the result is the product of their numerators divided by the product of their denominators.   If the numerators and denominators are written in factored form the result may be simplified by dividing the numerators and denominators by common factors.

Example: Multiply $\dfrac{8}{21} \times \dfrac{49}{112}$

▶ $\dfrac{8}{21} \times \dfrac{49}{112} = \dfrac{2^3}{7 \cdot 3} \times \dfrac{7^2}{7 \cdot 2^4}$

The numerators and denominators can be divided by the common factor $2^3 \cdot 7^2$.

The result is $\dfrac{1}{3 \cdot 2} = \dfrac{1}{6}$

Example: Multiply $\dfrac{20x^3}{18y^2} \cdot \dfrac{6y}{5x}$

▶ $\dfrac{20x^3}{18y^2} \cdot \dfrac{6y}{5x} = \dfrac{2^2 \cdot 5 \cdot x^3}{2 \cdot 3^2 \cdot y^2} \cdot \dfrac{3 \cdot 2 \cdot y}{5x} = \dfrac{4x^2}{3y}$

Example: Multiply $\dfrac{8y}{y^2 - 25} \cdot \dfrac{5y + 25}{24y^2}$

▶ $\dfrac{8y}{y^2 - 25} \cdot \dfrac{5y + 25}{24y^2} = \dfrac{8y}{(y + 5)(y - 5)} \cdot \dfrac{5(y + 5)}{24y^2}$

$= \dfrac{5}{3y(y - 5)}$

Example: Multiply $\dfrac{y^2 - 7y + 10}{6 - 3y} \cdot \dfrac{9y + 45}{y^2 - 25}$

▶ $\dfrac{y^2 - 7y + 10}{6 - 3y} \cdot \dfrac{9y + 45}{y^2 - 25} = \dfrac{(y - 5)(y - 2)}{3(2 - y)}$

$\cdot \dfrac{9(y + 5)}{(y + 5)(y - 5)} = \dfrac{3(y - 2)}{(2 - y)} = \dfrac{-3(y - 2)}{(y - 2)} = -3$

Example: Multiply $\dfrac{2ax - 2ay - bx + by}{x^2 - y^2} \cdot \dfrac{4x + 4y}{16a - 8b}$

▶ $\dfrac{2ax - 2ay - bx + by}{x^2 - y^2} \cdot \dfrac{4x + 4y}{16a - 8b} = \dfrac{(2a - b)(x - y)}{(x + y)(x - y)}$

$\dfrac{4(x + y)}{8(2a - b)} = \dfrac{1}{2}$

EXERCISE 2

Multiply the following fractions:

1. $\dfrac{14}{135} \times \dfrac{9}{98}$

2. $48 \times \dfrac{y}{6}$

3. $6a^2 \times \dfrac{8}{3a}$

4. $\dfrac{48x^3y^2}{7z^2} \cdot \dfrac{21z^3}{12xy}$

5. $\dfrac{5a^2b^3}{24x^3y^2} \cdot \dfrac{36x^2y^2}{15ab^2}$

6. $\dfrac{9}{4x^2 - 16} \cdot \dfrac{2x - 4}{45}$

7. $\dfrac{8a - 24}{5b} \cdot \dfrac{20b^3}{6a^2 - 54}$

8. $\dfrac{6y + 30}{6y + 18} \cdot \dfrac{5y + 15}{5y^2 - 125}$

9. $\dfrac{y + 4}{3y^2 - 27} \cdot \dfrac{12y - 36}{6y}$

10. $\dfrac{x^2 + x - 2}{x^2 - 6x} \cdot \dfrac{x^2 - 7x + 6}{x + 2}$

11. $\dfrac{4y^2 - 9}{y^2 + y} \cdot \dfrac{y^2 - 1}{2y - 3}$

12. $\dfrac{2a^2 - a - 1}{2a^2 + a - 1} \cdot \dfrac{4a^2 - 1}{a^2 - 1}$

13. $\dfrac{x^3 - 1}{5x - 5} \cdot \dfrac{10}{3x^2 + 3x + 3}$

14. $\dfrac{xy + ax - 2y - 2a}{4x - 8} \cdot \dfrac{y^2 - ay}{y^2 - a^2}$

15. $\dfrac{2ax + 4ay + bx + 2by}{x^2 - xy - 6y^2} \cdot \dfrac{x^2 - 9y^2}{10a + 5b}$

DIVISION OF FRACTIONS

In dividing one fraction by another, first invert the divisor and then multiply the dividend by the inverted divisor. If the numerators and denominators are written in factored form the result may be simplified by dividing the numerators and denominators by common factors.

Example: Divide $\dfrac{18}{25}$ by $\dfrac{63}{50}$

▶ $\dfrac{18}{25} \div \dfrac{63}{50} = \dfrac{18}{25} \times \dfrac{50}{63} = \dfrac{2 \cdot 3^2}{5^2} \cdot \dfrac{2 \cdot 5^2}{7 \cdot 3^2}$

The numerators and denominators can be divided by the common factor $3^2 \cdot 5^2$. The result is $\dfrac{2 \cdot 2}{7} = \dfrac{4}{7}$.

Example: Divide $\dfrac{8a^3}{3b^4}$ by $\dfrac{2a}{9b^2}$

▶ $\dfrac{8a^3}{3b^4} \div \dfrac{2a}{9b^2} = \dfrac{8a^3}{3b^4} \times \dfrac{9b^2}{2a}$

$= \dfrac{2^3 \cdot a^3}{3b^4} \cdot \dfrac{3^2 b^2}{2a}$

The numerators and denominators can be divided by the common factor $3 \cdot 2 \cdot b^2 \cdot a$.

The result is $\dfrac{12a^2}{b^2}$.

Example: Divide $\dfrac{2x^2 - 72}{7b^3} \div \dfrac{4x - 24}{14b^4}$

▶ $\dfrac{2x^2 - 72}{7b^3} \div \dfrac{4x - 24}{14b^4} = \dfrac{2x^2 - 72}{7b^3} \cdot \dfrac{14b^4}{4x - 24}$

$= \dfrac{2(x + 6)(x - 6)}{7b^3} \cdot \dfrac{14b^4}{4(x - 6)} = b(x + 6)$

Example: Divide $\dfrac{x^2 - 16}{9x^2 - 25} \div \dfrac{4 - x}{6x + 10}$

$$\blacktriangleright \quad \frac{x^2 - 16}{9x^2 - 25} \div \frac{4 - x}{6x + 10} = \frac{x^2 - 16}{9x^2 - 25} \times \frac{6x + 10}{4 - x}$$

$$= \frac{(x + 4)(x - 4)}{(3x + 5)(3x - 5)} \cdot \frac{2(3x + 5)}{-(x - 4)} = \frac{-2(x + 4)}{(3x - 5)}$$

Example: Divide $\dfrac{2ac - ad + 2bc - bd}{10c - 5d} \div \dfrac{3a^2 - 3b^2}{10a - 10b}$

$$\blacktriangleright \quad \frac{2ac - ad + 2bc - bd}{10c - 5d} \div \frac{3a^2 - 3b^2}{10a - 10b}$$

$$= \frac{2ac - ad + 2bc - bd}{10c - 5d} \cdot \frac{10a - 10b}{3a^2 - 3b^2}$$

$$= \frac{(2c - d)(a + b)}{5(2c - d)} \cdot \frac{10(a - b)}{3(a + b)(a - b)}$$

The result is $\dfrac{2}{3}$.

EXERCISE 3

Divide the following fractions:

1. $\dfrac{21}{81} \div \dfrac{28}{135}$

2. $\dfrac{56}{45} \div \dfrac{35}{27}$

3. $\dfrac{3a^2}{5b^3} \div \dfrac{21a}{40b}$

4. $\dfrac{7xy^2}{20p^2q^3} \div \dfrac{28y^3}{15pq^4}$

5. $\dfrac{2y^2 - 50}{9} \div \dfrac{4y^2 - 20y}{27}$

6. $\dfrac{x}{y^2 - 49} \div \dfrac{3x^3}{4y + 28}$

7. $\dfrac{x + 5y}{x^2 + 6xy} \div \dfrac{xy + 5y^2}{x^3 + 6x^2y}$

8. $\dfrac{b^3y - bx^2y}{b^3x^2 + b^2x^2y} \div \dfrac{b^2y - 2bxy + x^2y}{b^2 + by}$

9. $\dfrac{y^2 + 14y - 15}{y^2 + 4y - 5} \div \dfrac{y^2 + 12y - 45}{y^2 + 6y - 27}$

10. $\dfrac{2ax - bx + 4ay - 2by}{4x^2 - 16y^2} \div \dfrac{5}{2x - 4y}$

ADDITION AND SUBTRACTION OF FRACTIONS

Fractions with Identical Denominators

In combining fractions which have identical denominators, add the numerators algebraically and write the result over the common denominator. Reduce the resulting fraction to lowest terms.

Example: Combine $\dfrac{5}{12} + \dfrac{11}{12}$

▶ $\dfrac{5}{12} + \dfrac{11}{12} = \dfrac{16}{12}. \ \dfrac{16}{12} = \dfrac{4}{3}$

Example: Combine $\dfrac{4}{9x} + \dfrac{7}{9x} - \dfrac{5}{9x}$

▶ $\dfrac{4}{9x} + \dfrac{7}{9x} - \dfrac{5}{9x} = \dfrac{4 + 7 - 5}{9x} = \dfrac{6}{9x}, \ \dfrac{6}{9x} = \dfrac{2}{3x}$

Example: Combine $\dfrac{8}{y + 3} - \dfrac{2}{y + 3}$

▶ $\dfrac{8}{y + 3} - \dfrac{2}{y + 3} = \dfrac{8 - 2}{y + 3} = \dfrac{6}{y + 3}$

Example: Combine $\dfrac{3x - 4}{8} - \dfrac{2x - 5}{8}$

▶ $\dfrac{3x - 4}{8} - \dfrac{2x - 5}{8} = \dfrac{(3x - 4) - (2x - 5)}{8}$

$= \dfrac{3x - 4 - 2x + 5}{8} = \dfrac{x + 1}{8}$

EXERCISE 4

Combine the following fractions:

1. $\dfrac{1}{8} + \dfrac{3}{8}$

2. $\dfrac{11}{15} - \dfrac{2}{15}$

3. $\dfrac{2}{3y} + \dfrac{7}{3y}$

4. $\dfrac{6a}{5} + \dfrac{3a}{5} - \dfrac{a}{5}$

5. $\dfrac{9}{4x} - \dfrac{5}{4x} + \dfrac{3}{4x}$

6. $\dfrac{7}{x - 2} - \dfrac{3}{x - 2}$

7. $\dfrac{4}{3x - 1} - \dfrac{7}{3x - 1}$

8. $\dfrac{2x + 3}{5} + \dfrac{4x - 1}{5}$

9. $\dfrac{3x - 2}{6} - \dfrac{x - 5}{6}$

10. $\dfrac{4a + 3}{9} - \dfrac{3a - 1}{9}$

11. $\dfrac{2x - 5}{x} - \dfrac{5x - 2}{x}$

12. $\dfrac{3x + 5}{2x - 1} - \dfrac{x - 7}{2x - 1}$

Fractions with Unlike Denominators

Fractions which have unlike denominators cannot be combined directly. Before they may be combined each fraction must be so transformed that all will have the same denominator. This identical denominator is called the Least Common Denominator (L.C.D.). The L.C.D. is the Least Common Multiple of the denominators of the fractions to be added. In the process of transforming a fraction so that its denominator will be changed we make use of the following principle. The numerator and denominator of a fraction may be multiplied by the same number (except zero) without changing the value of the fraction.

Example: Combine $\dfrac{1}{3} + \dfrac{2}{5}$

▶ $\dfrac{1}{3} + \dfrac{2}{5}$

The L.C.D. is 15. Therefore, we must transform the two fractions so that they will both have the same denominator, 15. In order to accomplish this, we multiply the numerator and denominator of the fraction $\dfrac{1}{3}$ by 5 and the numerator and denominator of the fraction $\dfrac{2}{5}$ by 3. Thus, $\dfrac{1}{3} + \dfrac{2}{5} = \dfrac{5}{15} + \dfrac{6}{15} = \dfrac{11}{15}$.

Example: Combine $\dfrac{5}{6a^2} + \dfrac{1}{3a} - \dfrac{3}{4a}$

▶ $\dfrac{5}{6a^2} + \dfrac{1}{3a} - \dfrac{3}{4a}$

The L.C.D. is $12a^2$. Therefore, we must transform the three fractions so that each will have the denominator $12a^2$. In order to accomplish this, we multiply the numerator and the denominator of the fraction $\dfrac{5}{6a^2}$ by 2, the numerator and denominator of the fraction $\dfrac{1}{3a}$ by $4a$, and the numerator and denominator of the fraction $\dfrac{3}{4a}$ by $3a$. Thus, $\dfrac{5}{6a^2} + \dfrac{1}{3a} - \dfrac{3}{4a} = \dfrac{10}{12a^2} + \dfrac{4a}{12a^2} - \dfrac{9a}{12a^2}$

$= \dfrac{10 + 4a - 9a}{12a^2} = \dfrac{10 - 5a}{12a^2}$.

Example: Combine $\dfrac{2x + 7}{3} - \dfrac{4x - 1}{5}$

▶ $\dfrac{2x + 7}{3} - \dfrac{3x - 1}{5}$

The L.C.D. is 15. Therefore, we must transform the fractions so that each will have the denominator 15. In order to accomplish this, we multiply the numerator and denominator of $\dfrac{2x + 7}{3}$ by 5 and the numerator and denominator of $\dfrac{4x - 1}{5}$ by 3. Thus,

$\dfrac{2x + 7}{3} - \dfrac{4x - 1}{5} = \dfrac{5(2x + 7)}{15} - \dfrac{3(4x - 1)}{15}$

$= \dfrac{5(2x + 7) - 3(4x - 1)}{15} = \dfrac{10x + 35 - 12x + 3}{15}$

$= \dfrac{-2x + 38}{15}$.

Example: Combine $\dfrac{2a - b}{2a} - \dfrac{a - 2b}{7b}$

▶ $\dfrac{2a - b}{2a} - \dfrac{a - 2b}{7b}$

The L.C.D. is $14ab$. Therefore, multiply the numerator and denominator of $\dfrac{2a - b}{2a}$ by $7b$ and the numerator and denominator of

$\dfrac{a - 2b}{7b}$ by $2a$. Thus, $\dfrac{2a - b}{2a} - \dfrac{a - 2b}{7b}$

$= \dfrac{14ab - 7b^2}{14ab} - \dfrac{2a^2 - 4ab}{14ab}$

$= \dfrac{14ab - 7b^2 - 2a^2 + 4ab}{14ab}$

$= \dfrac{18ab - 7b^2 - 2a^2}{14ab}$

Example: Combine $\dfrac{5}{x - 2} - \dfrac{3}{2x - 5}$

▶ The L.C.D. is $(x - 2)(2x - 5)$. Therefore, multiply the numerator and denominator of the fraction $\dfrac{5}{x - 2}$ by $2x - 5$ and the numerator and denominator of the fraction $\dfrac{3}{2x - 5}$ by $x - 2$.

Thus, $\dfrac{5}{x - 2} - \dfrac{3}{2x - 5} = \dfrac{5(2x - 5)}{(x - 2)(x - 5)} - \dfrac{3(x - 3)}{(x - 2)(x - 5)}$

$= \dfrac{5(2x - 5) - 3(x - 2)}{(x - 2)(2x - 5)} = \dfrac{10x - 25 - 3x + 6}{(x - 2)(2x - 5)}$

$= \dfrac{7x - 19}{(x - 2)(2x - 5)}$

Example: Combine $\dfrac{2y + 5}{3y - 2} - \dfrac{y - 7}{2y - 1}$

▶ $\dfrac{2y + 5}{3y - 2} - \dfrac{y - 7}{2y - 1}$

The L.C.D. is $(3y - 2)(2y - 1)$. Therefore, multiply the numerator and denominator of the fraction $\dfrac{2y + 5}{3y - 2}$ by $2y - 1$ and the numerator and denominator of the fraction $\dfrac{y - 7}{2y - 1}$ by $3y - 2$.

Thus, $\dfrac{(2y + 5)(2y - 1)}{(3y - 2)(2y - 1)} - \dfrac{(y - 7)(3y - 2)}{(3y - 2)(2y - 1)}$

$$= \frac{(2y + 5)(2y - 1) - (y - 7)(3y - 2)}{(3y - 2)(2y - 1)}$$

$$= \frac{(4y^2 + 8y - 5) - (3y^2 - 23y + 14)}{(3y - 2)(2y - 1)}$$

$$= \frac{4y^2 + 8y - 5 - 3y^2 + 23y - 14}{(3y - 2)(2y - 1)}$$

$$= \frac{y^2 + 31y - 19}{(3y - 2)(2y - 1)}$$

Example: Combine $\dfrac{11x}{6x - 12} + \dfrac{7x}{5x - 10}$

▶ $\dfrac{11x}{6x - 12} + \dfrac{7x}{5x - 10}$

Before obtaining the L.C.D. it is necessary to factor the denominators of the two fractions

$$\frac{11x}{6x - 12} + \frac{7x}{5x - 10} = \frac{11x}{6(x - 2)} + \frac{7x}{5(x - 2)}$$

The L.C.D. is $30(x - 2)$. Therefore, multiply the numerator and denominator of the fraction $\dfrac{11x}{6(x - 2)}$ by 5 and the numerator and denominator of the fraction $\dfrac{7x}{5(x - 2)}$ by 6. Thus,

$$\frac{11x}{6(x - 2)} + \frac{7x}{5(x - 2)} = \frac{(5)(11x)}{30(x - 2)} + \frac{(6)(7x)}{30(x - 2)}$$

$$= \frac{55x}{30(x - 2)} + \frac{42x}{30(x - 2)} = \frac{97x}{30(x - 2)}$$

Example: Combine $\dfrac{5x - 1}{12x - 8} - \dfrac{9}{3x^2 + x - 2}$

▶ $\dfrac{5x - 1}{12x - 8} - \dfrac{9}{3x^2 + x - 2}$

Factoring the denominators we obtain

$$\frac{5x - 1}{4(3x - 2)} - \frac{9}{(3x - 2)(x + 1)} \, .$$

The L.C.D. is $4(3x - 2)(x + 1)$. Therefore, multiply the numerator and denominator of the fraction $\dfrac{5x - 1}{4(3x - 2)}$ by $(x + 1)$ and the numerator and denominator of the fraction $\dfrac{9}{(3x - 2)(x + 1)}$ by 4. Thus,

$$\frac{(5x - 1)(x + 1)}{4(3x - 2)(x + 1)} - \frac{(9)(4)}{4(3x - 2)(x + 1)}$$

$$= \frac{(5x - 1)(x + 1) - 36}{4(3x - 2)(x + 1)} = \frac{5x^2 + 4x - 1 - 36}{4(3x - 2)(x + 1)}$$

$$= \frac{5x^2 + 4x - 37}{4(3x - 2)(x + 1)}$$

Example: Combine $\dfrac{7}{9a^2 - 1} + \dfrac{2a - 5}{2 - 6a}$

▶ $\dfrac{7}{9a^2 - 1} + \dfrac{2a - 5}{2 - 6a}$

Factoring the denominators we obtain $\dfrac{7}{(3a + 1)(3a - 1)}$ $+ \dfrac{2a - 5}{2(1 - 3a)}$. We note that, if the denominator $2(1 - 3a)$ is multiplied by -1, we obtain $2(3a - 1)$. This may be done if we change another sign of the fraction $\dfrac{2a - 5}{2(1 - 3a)}$. It is convenient to change the sign in front of the fraction and to balance this change by multiplying the denominator by -1. When this is done the result is $\dfrac{7}{(3a + 1)(3a - 1)} - \dfrac{2a - 5}{2(3a - 1)}$.

The L.C.D. is $2(3a + 1)(3a - 1)$.

$$\frac{(7)(2)}{2(3a + 1)(3a - 1)} - \frac{(3a + 1)(2a - 5)}{2(3a + 1)(3a - 1)}$$

$$= \frac{14 - (6a^2 - 13a - 5)}{2(3a + 1)(3a - 1)} = \frac{14 - 6a^2 + 13a + 5}{2(3a + 1)(3a - 1)}$$

$$= \frac{-6a^2 + 13a + 19}{2(3a + 1)(3a - 1)}$$

Example: Combine $\dfrac{x-5}{3x^2+12x} - \dfrac{2x-7}{4x^2-64}$

▶ $\dfrac{x-5}{3x^2+12x} - \dfrac{2x-7}{4x^2-64}$

Factoring the denominators, we obtain

$$\frac{x-5}{3x(x+4)} - \frac{2x-7}{4(x+4)(x-4)}.$$

The L.C.D. is $12x(x+4)(x-4)$.

Multiply the numerator and denominator of the fraction $\dfrac{x-5}{3x(x+4)}$ by $4(x-4)$ and the numerator and denominator of the fraction $\dfrac{2x-7}{4(x+4)(x-4)}$ by $3x$.

$$\frac{4(x-5)(x-4)}{12x(x+4)(x-4)} - \frac{3x(2x-7)}{12x(x+4)(x-4)}$$

$$= \frac{4x^2 - 36x + 80 - 6x^2 + 21x}{12x(x+4)(x-4)} = \frac{-2x^2 - 15x + 80}{12x(x+4)(x-4)}$$

Example: Combine $\dfrac{3a}{a^2-4a+3} - \dfrac{a}{a^2-a-6}$

▶ $\dfrac{3a}{a^2-4a+3} - \dfrac{a}{a^2-a-6}$

Factoring the denominators we have

$$\frac{3a}{(a-1)(a-3)} - \frac{a}{(a-3)(a+2)}.$$

The L.C.D. is $(a-1)(a-3)(a+2)$.

Therefore, we have $\dfrac{3a(a+2)}{(a-1)(a-3)(a+2)}$

$$- \frac{a(a-1)}{(a-1)(a-3)(a+2)} = \frac{3a(a+2) - a(a-1)}{(a-1)(a-3)(a+2)}$$

$$= \frac{3a^2 + 6a - a^2 + a}{(a-1)(a-3)(a+2)} = \frac{2a^2 + 7a}{(a-1)(a-3)(a+2)}$$

Example: Combine $\dfrac{3y - 1}{5y^2 + 15y + 10} - \dfrac{7y - 4}{2y^2 - 8}$

▶ $\dfrac{3y - 1}{5y^2 + 15y + 10} - \dfrac{7y - 4}{2y^2 - 8}$

Factoring the denominators we have

$$\frac{3y - 1}{5(y + 2)(y + 1)} - \frac{7y - 4}{2(y + 2)(y - 2)}.$$

The L.C.D. is $10(y + 2)(y + 1)(y - 2)$.

Therefore, we have

$$\frac{(3y - 1)(2)(y - 2)}{10(y + 2)(y + 1)(y - 2)} - \frac{(7y - 4)(5)(y + 1)}{10(y + 2)(y + 1)(y - 2)}$$

$$= \frac{6y^2 - 14y + 4}{10(y + 2)(y + 1)(y - 2)} - \frac{35y^2 + 15y - 20}{10(y + 2)(y + 1)(y - 2)}$$

$$= \frac{6y^2 - 14y + 4 - 35y^2 - 15y + 20}{10(y + 2)(y + 1)(y - 2)} = \frac{-29y^2 - 29y + 24}{10(y + 2)(y + 1)(y - 2)}$$

EXERCISE 5

Combine the following fractions:

1. $\dfrac{3}{4} + \dfrac{1}{3}$

2. $\dfrac{5}{6} + \dfrac{1}{2} - \dfrac{1}{4}$

3. $\dfrac{3}{2x} + \dfrac{1}{5x^2} - \dfrac{7}{10x}$

4. $\dfrac{5a}{8b} - \dfrac{7a}{2b^2} + \dfrac{3a}{4b}$

5. $\dfrac{7x + 3}{5} - \dfrac{2x - 1}{4}$

6. $\dfrac{2y - 5}{7} + \dfrac{3y - 1}{6}$

7. $\dfrac{3c - 8}{2} - \dfrac{c - 4}{9}$

8. $\dfrac{x - y}{3x} - \dfrac{2x - 3y}{6y}$

9. $\dfrac{3a - 2b}{5a} - \dfrac{a - 5b}{2b}$

10. $\dfrac{4x - 5y}{3x} - \dfrac{2x + y}{7y}$

11. $\dfrac{6}{y - 3} - \dfrac{5}{3y - 2}$

12. $\dfrac{8}{3a - 4} + \dfrac{2}{4a - 7}$

13. $\dfrac{y + 2}{2y - 5} + \dfrac{y - 1}{y + 3}$

14. $\dfrac{3y - 4}{4y + 1} - \dfrac{2y - 5}{y + 5}$

15. $\dfrac{2y + 9}{3y - 5} - \dfrac{2y + 1}{2y - 7}$

16. $\dfrac{3a}{2a - 10} + \dfrac{4a}{7a - 35}$

17. $\dfrac{5y}{9y - 15} - \dfrac{7y}{12y - 20}$

18. $\dfrac{2b - 5}{4b - 3} - \dfrac{5b}{8b - 6}$

19. $\dfrac{3x - 2}{10x - 5} + \dfrac{x - 3}{4x - 2}$

20. $\dfrac{5 - 3a}{6a - 4} - \dfrac{2}{9a - 6}$

21. $\dfrac{x - 3}{2x + 1} - \dfrac{5x^2}{2x^2 - x - 1}$

22. $\dfrac{3x^2 - 5}{3x^2 + 2x - 1} - \dfrac{x - 6}{4x + 4}$

23. $\dfrac{x^2 + x - 3}{2x^2 + 7x - 4} - \dfrac{2x - 5}{6x - 3}$

24. $\dfrac{5}{a - b} + \dfrac{2}{b - a}$

25. $\dfrac{7}{3a - b} + \dfrac{5}{2b - 6a}$

26. $\dfrac{8}{4y^2 - 1} + \dfrac{4y - 1}{3 - 6y}$

27. $\dfrac{a + 7}{4 - 6a} + \dfrac{a^2 + 2}{9a^2 - 4}$

28. $\dfrac{2y + 7}{12y - 9} - \dfrac{3 + 2y^2}{9 - 16y^2}$

29. $\dfrac{5}{4x^2 - 8x} + \dfrac{2x - 9}{3x^2 - 12}$

30. $\dfrac{3a - 2}{5a^2 - 45} - \dfrac{2a + 3}{2a^2 - 6a}$

31. $\dfrac{2x}{x^2 - 3x + 2} - \dfrac{4x}{x^2 - x - 2}$

32. $\dfrac{6a}{2a^2 + 5a - 3} - \dfrac{a}{3a^2 + 8a - 3}$

33. $\dfrac{3x - 5}{x^2 - 16} - \dfrac{x - 8}{2x^2 - 7x - 4}$

34. $\dfrac{7a - 1}{12a^2 - 27} + \dfrac{a - 6}{2a^2 + 7a - 15}$

REDUCING MIXED EXPRESSIONS TO IMPROPER FRACTIONS

In dealing with fractions it is often necessary to combine an integral expression with a fraction. An integral expression may be regarded as a fraction with denominator 1. Therefore, combining an integral expression with a fraction is similar to the addition of fractions.

Example: Combine $x - \dfrac{y}{z}$

▶ $x - \dfrac{y}{z} = \dfrac{x}{1} - \dfrac{y}{z}$

The L.C.D. is z. Therefore, $\dfrac{x}{1} - \dfrac{y}{z} = \dfrac{xz - y}{z}$.

Example: Combine $2 - \dfrac{a}{a - 3}$

▶ $2 - \dfrac{a}{a - 3} = \dfrac{2}{1} - \dfrac{a}{a - 3}$

The L.C.D. is $a - 3$.

Therefore $\dfrac{2}{1} - \dfrac{a}{a - 3} = \dfrac{2(a - 3) - a}{a - 3} = \dfrac{2a - 6 - a}{a - 3} = \dfrac{a - 6}{a - 3}$.

Example: Combine $x - y - \dfrac{x^2 + y^2}{x - y}$

▶ $x - y - \dfrac{x^2 + y^2}{x - y} = \dfrac{x - y}{1} - \dfrac{x^2 + y^2}{x - y}$

The L.C.D. is $x - y$.

Therefore, $\dfrac{x - y}{1} - \dfrac{x^2 + y^2}{x - y} = \dfrac{(x - y)^2 - (x^2 + y^2)}{x - y}$

$= \dfrac{x^2 - 2xy + y^2 - x^2 - y^2}{x - y} = \dfrac{-2xy}{x - y}$

EXERCISE 6

Combine the following:

1. $a - \dfrac{b}{c}$

2. $2x + \dfrac{y}{3}$

3. $3y - 5 - \dfrac{2}{y}$

4. $5 - \dfrac{x}{x - 2}$

5. $y + \dfrac{3}{2y - 1}$

6. $7 + x - \dfrac{5x}{9}$

7. $a - b + \dfrac{2ab}{a - b}$

8. $2y - 3 - \dfrac{3y}{y + 3}$

9. $x + y - \dfrac{x^2 + y^2}{x + y}$

COMPLEX FRACTIONS

DEFINITION: A complex fraction is a fraction whose numerator, or denominator, or both, contain fractions.

Examples: $\dfrac{3}{5 + \dfrac{1}{6}}$, $\dfrac{a - \dfrac{2}{3}}{a + \dfrac{2}{3}}$, $\dfrac{\dfrac{x-1}{5}}{\dfrac{x+2}{10}}$

In simplifying a complex fraction, combine the terms in the numerator into a single fraction, combine the terms in the denominator into a single fraction, and then divide the resulting numerator by the resulting denominator.

An alternative method, which is sometimes simpler, is to multiply the numerator and denominator of the complex fraction by the L.C.D. of the denominators of the complex fraction.

Example: Simplify $\dfrac{6}{2 + \dfrac{1}{4}}$

▶ $6 \div \left(2 + \dfrac{1}{4}\right) = 6 \div \left(\dfrac{9}{4}\right) = 6 \times \dfrac{4}{9} = \dfrac{8}{3}$

Example: Simplify $\dfrac{4 - \dfrac{3}{7}}{5}$

▶ $\left(4 - \dfrac{3}{7}\right) \div 5 = \dfrac{28 - 3}{7} \div 5 = \dfrac{25}{7} \times \dfrac{1}{5} = \dfrac{5}{7}$

Example: Simplify $\dfrac{2 + \dfrac{2}{3}}{2 - \dfrac{2}{3}}$

▶ $\left(2 + \dfrac{2}{3}\right) \div \left(2 - \dfrac{2}{3}\right) = \left(\dfrac{8}{3}\right) \div \left(\dfrac{4}{3}\right) = \dfrac{8}{3} \times \dfrac{3}{4} = 2$

Example: Simplify $\left(1 + \dfrac{2}{x}\right) \div \left(x - \dfrac{4}{x}\right)$

▶ $\left(1 + \dfrac{2}{x}\right) \div \left(x - \dfrac{4}{x}\right) = \dfrac{x + 2}{x} \div \dfrac{x^2 - 4}{x}$

$= \dfrac{(\cancel{x + 2})}{\cancel{x}} \cdot \dfrac{\cancel{x}}{(\cancel{x + 2})(x - 2)} = \dfrac{1}{x - 2}$

Example: Simplify $\dfrac{\dfrac{1}{y} - \dfrac{1}{x}}{1 - \dfrac{x}{y}}$

▶ $\left(\dfrac{1}{y} - \dfrac{1}{x}\right) \div \left(1 - \dfrac{x}{y}\right) = \dfrac{x - y}{xy} \div \dfrac{y - x}{y}$

$= \dfrac{\overset{-1}{\cancel{(x - y)}}}{\cancel{xy}} \cdot \dfrac{\cancel{y}}{\cancel{(y - x)}} = -\dfrac{1}{x}$

Example: Simplify $\dfrac{\dfrac{a}{b} - \dfrac{b}{a}}{\dfrac{a}{b} + \dfrac{b}{a} - 2}$

▶ $\dfrac{\dfrac{a}{b} - \dfrac{b}{a}}{\dfrac{a}{b} + \dfrac{b}{a} - 2} = \dfrac{\dfrac{a^2 - b^2}{ab}}{\dfrac{a^2 + b^2 - 2ab}{ab}} = \dfrac{a^2 - b^2}{ab} \div \dfrac{a^2 - 2ab + b^2}{ab}$

$= \dfrac{(a + b)(\cancel{a - b})}{ab} \cdot \dfrac{ab}{(a - b)\cancel{}} = \dfrac{a + b}{a - b}$

Example: Simplify $\dfrac{\dfrac{1}{y - 2} - \dfrac{1}{y - 3}}{1 + \dfrac{1}{y^2 - 5y + 6}}$

▶ $\dfrac{\dfrac{1}{y - 2} - \dfrac{1}{y - 3}}{1 + \dfrac{1}{y^2 - 5y + 6}} = \dfrac{\dfrac{(y - 3) - (y - 2)}{(y - 2)(y - 3)}}{\dfrac{y^2 - 5y + 6 + 1}{y^2 - 5y + 6}}$

$$= \frac{y - 3 - y + 2}{(y - 2)(y - 3)} \div \frac{y^2 - 5y + 7}{y^2 - 5y + 6}$$

$$= \frac{-1}{(y - 2)(y - 3)} \cdot \frac{(y - 2)(y - 3)}{y^2 - 5y + 7} = -\frac{1}{y^2 - 5y + 7}$$

Example: Simplify $\dfrac{x + \dfrac{1}{x + 1} - 1}{x + \dfrac{1}{x - 1} + 1}$

▶ $\dfrac{x + \dfrac{1}{x + 1} - 1}{x + \dfrac{1}{x - 1} + 1} = \dfrac{\dfrac{x^2 + x + 1 - x - 1}{x + 1}}{\dfrac{x^2 - x + 1 + x - 1}{x - 1}} = \dfrac{\dfrac{x^2}{x + 1}}{\dfrac{x^2}{x - 1}}$

$$= \frac{x^2}{x + 1} \cdot \frac{x - 1}{x^2} = \frac{x - 1}{x + 1}$$

Example: Simplify $\left(\dfrac{x^2}{y^2} - 4 \right) \left(\dfrac{y}{x + 2y} \right) \left(2 - \dfrac{x}{y} \right)$

▶ $\left(\dfrac{x^2}{y^2} - 4 \right) \left(\dfrac{y}{x + 2y} \right) \left(2 - \dfrac{x}{y} \right) =$

$\left(\dfrac{x^2 - 4y^2}{y^2} \right) \left(\dfrac{y}{x + 2y} \right) \left(\dfrac{2y - x}{y} \right)$

$\dfrac{x}{(x + 2y)} \cdot \dfrac{2y - x}{x} = \dfrac{(x + 2y)(x - 2y)}{y^2} \cdot$

Note that $2y - x = -(x - 2y)$.

Therefore, the result is $\dfrac{-(x - 2y)^2}{y^2}$.

Example: Simplify $3 - \dfrac{1}{1 - \dfrac{2x}{3x - \dfrac{3x}{x + 1}}}$

▶ $3 - \cfrac{1}{1 - \cfrac{2x}{3x - \cfrac{3x}{x+1}}} = 3 - \cfrac{1}{1 - \cfrac{2x}{\cfrac{3x^2 + \cancel{3x} - \cancel{3x}}{x+1}}}$

$= 3 - \cfrac{1}{1 - \cfrac{2x}{\cfrac{3x^2}{x+1}}}$

$= 3 - \cfrac{1}{1 - \cfrac{2x(x+1)}{3x^2}} = 3 - \cfrac{3x^2}{x^2 - 2x}$

$= \cfrac{3x^2 - 6x - 3x^2}{x^2 - 2x} = \cfrac{-6x}{x(x-2)} = -\cfrac{6}{x-2}$

Example: $\cfrac{1 - \cfrac{y^3}{8}}{1 + \cfrac{y}{2} + \cfrac{y^2}{4}} \times \cfrac{1}{y-2}$

▶ $\cfrac{1 - \cfrac{y^3}{8}}{1 + \cfrac{y}{2} + \cfrac{y^2}{4}} \times \cfrac{1}{y-2} = \cfrac{\cfrac{8 - y^3}{8}}{\cfrac{4 + 2y + y^2}{4}} \times \cfrac{1}{y-2}$

$= \cfrac{(8 - y^3)}{8} \times \cfrac{4}{(4 + 2y + y^2)} \times \cfrac{1}{y-2}$

$= \cfrac{\overset{-1}{\cancel{(2 - y)}}(\cancel{4 + 2y + y^2})}{8} \times \cfrac{4}{\cancel{(4 + 2y + y^2)}} \times \cfrac{1}{\cancel{y - 2}}$

Note that $2 - y = -(y - 2)$.

Therefore, the result is $-\cfrac{1}{2}$.

EXERCISE 7

Simplify the following:

1. $\dfrac{5}{2 - \dfrac{1}{3}}$

2. $\dfrac{9}{3 + \dfrac{6}{7}}$

3. $\dfrac{7 - \dfrac{2}{5}}{6}$

4. $\dfrac{6 + \dfrac{7}{8}}{11}$

5. $\dfrac{5 + \dfrac{5}{6}}{5 - \dfrac{5}{6}}$

6. $\dfrac{8 - \dfrac{1}{5}}{8 + \dfrac{2}{5}}$

7. $\dfrac{y + \dfrac{1}{4}}{y - \dfrac{1}{4}}$

8. $\left(1 + \dfrac{x}{y}\right) \div \left(\dfrac{y^2 - x^2}{y^2}\right)$

9. $\left(\dfrac{a - b}{2ab}\right) \div \left(\dfrac{1}{b^2} - \dfrac{1}{a^2}\right)$

10. $\left(\dfrac{x}{x + y}\right) \div \left(1 - \dfrac{x}{x + y}\right)$

11. $\left(\dfrac{a^2}{b^2} - 9\right) \div \left(\dfrac{a}{b} - 3\right)$

12. $\left(\dfrac{1}{16} - y^2\right) \div \left(y - \dfrac{1}{4}\right)$

13. $\dfrac{1 - \dfrac{1}{x + 1}}{1 + \dfrac{1}{x - 1}}$

14. $\dfrac{\dfrac{a}{b} - \dfrac{b}{a}}{\dfrac{a}{b} + \dfrac{b}{a} + 2}$

15. $\left(2 - \dfrac{a}{a + b}\right)\left(1 + \dfrac{2b}{a - b}\right)$

16. $\left(\dfrac{ab}{a^2 - 4}\right) \div \left(\dfrac{4ab + 14b}{2a^2 + 3a - 14}\right)$

17. $\left(x - \dfrac{1}{x}\right)\left(2 + \dfrac{2}{x - 1}\right)$

18. $\dfrac{1}{1 - \dfrac{1}{1 + y}} + \dfrac{1}{1 - \dfrac{1}{1 - y}}$

19. $\dfrac{\dfrac{x}{x - 1} + \dfrac{x}{x + 1}}{\dfrac{x}{x - 1} - \dfrac{x}{x + 1}}$

20. $\dfrac{1}{a - 1 + \dfrac{1}{1 + \dfrac{a}{4 - a}}}$

21. $\dfrac{\dfrac{x}{y^2} + \dfrac{y}{x^2}}{\dfrac{1}{x^2} - \dfrac{1}{xy} + \dfrac{1}{y^2}}$

22. $\dfrac{\dfrac{3}{a + 2b} - \dfrac{2b}{a^2 + 2ab}}{\dfrac{3b}{a^2 + 2ab} + \dfrac{5}{a}}$

23. $\dfrac{\dfrac{1}{y + 1} + \dfrac{1}{y - 1}}{\dfrac{1}{y - 1} - \dfrac{1}{y + 1}}$

24. $\dfrac{\dfrac{x - 3y}{x + y} + \dfrac{x}{y}}{\dfrac{x + 3y}{y} \cdot \dfrac{x}{x + y}}$

25. $\dfrac{b^2 - \dfrac{a^3}{b}}{a + b + \dfrac{a^2}{b}}$

26. $\dfrac{1}{x^2 - \dfrac{x + 1}{1 + \dfrac{1}{x - 1}}}$

27. $\dfrac{\dfrac{x + 1}{x} - 1}{\dfrac{x - 1}{x} + 1} \div \dfrac{1}{4x^2 - 1}$

ANSWERS

EXERCISE 1—Page 57

1. $\dfrac{2x}{3y}$

2. $\dfrac{6b}{7ac}$

3. $\dfrac{-8p^4}{3t^4}$

4. $\dfrac{-6d^3 e^3}{g}$

5. $\dfrac{1}{2}$

6. $\dfrac{1}{2x - 1}$

7. $\dfrac{x}{2 - 3bx}$

8. -2

9. $\dfrac{y + 2}{5}$

10. $\dfrac{2}{4a + 5}$

11. $-\dfrac{y + 3}{3}$

12. $\dfrac{2a - 3b}{2}$

13. $\dfrac{x}{x - 1}$

14. $\dfrac{a + 5}{a + 3}$

15. $\dfrac{2}{x - 2}$

16. $-\dfrac{1}{2(b - 1)}$

17. $\dfrac{a - 1}{a + 3}$

18. $\dfrac{x - 4}{x + 1}$

19. $\dfrac{y + 4}{y + 5}$

20. $\dfrac{2a + 3b}{3a - b}$

21. $\dfrac{5b - c}{4b - c}$

22. $\dfrac{3y(y - 4)}{(y - 5)}$

23. $x^2 + y^2$

24. $\dfrac{a^2 + ab + b^2}{7}$

25. $\dfrac{y - 2}{2}$

26. $2(2x - 3y)$

27. $\dfrac{y + 1}{a^2 + a + 1}$

28. $\dfrac{a - c}{a + c}$

29. $\dfrac{2a + b}{c + d}$

30. $\dfrac{3}{x + b}$

EXERCISE 2—Page 59

1. $\dfrac{1}{105}$

2. $8y$

3. $16a$

4. $12x^2 yz$

5. $\dfrac{ab}{2x}$

6. $\dfrac{1}{10(x + 2)}$

7. $\dfrac{16b^2}{3(a + 3)}$

8. $\dfrac{1}{y - 5}$

9. $\dfrac{2(y + 4)}{3y(y + 3)}$

10. $\dfrac{(x - 1)^2}{x}$

11. $\dfrac{(2y + 3)(y - 1)}{y}$

12. $\dfrac{(2a + 1)^2}{(a + 1)^2}$

13. $\dfrac{2}{3}$

14. $\dfrac{y}{4}$

15. $\dfrac{x + 3y}{5}$

EXERCISE 3—Page 61

1. $\dfrac{5}{4}$

2. $\dfrac{24}{25}$

3. $\dfrac{8a}{7b^2}$

4. $\dfrac{3xq}{16yp}$

5. $\dfrac{3(y+5)}{2y}$

6. $\dfrac{4}{3x^2(y-7)}$

7. $\dfrac{x}{y}$

8. $\dfrac{b+x}{x^2(b-x)}$

9. $\dfrac{y+9}{y+5}$

10. $\dfrac{2a-b}{10}$

EXERCISE 4—Page 62

1. $\dfrac{1}{2}$

2. $\dfrac{3}{5}$

3. $\dfrac{3}{y}$

4. $\dfrac{8a}{5}$

5. $\dfrac{7}{4x}$

6. $\dfrac{4}{x-2}$

7. $\dfrac{-3}{3x-1}$

8. $\dfrac{6x+2}{5}$

9. $\dfrac{2x+3}{6}$

10. $\dfrac{a+4}{9}$

11. $\dfrac{-3x-3}{x}$

12. $\dfrac{2x+12}{2x-1}$

EXERCISE 5—Page 69

1. $\dfrac{13}{12}$

2. $\dfrac{13}{12}$

3. $\dfrac{4x+1}{5x^2}$

4. $\dfrac{11ab-28a}{8b^2}$

5. $\dfrac{18x+17}{20}$

6. $\dfrac{33y-37}{42}$

7. $\dfrac{25c-64}{18}$

8. $\dfrac{5xy-2y^2-2x^2}{6xy}$

9. $\dfrac{31ab-4b^2-5a^2}{10ab}$

10. $\dfrac{25xy-35y^2-6x^2}{21xy}$

11. $\dfrac{13y + 3}{(y - 3)(3y - 2)}$

12. $\dfrac{38a - 64}{(3a - 4)(4a - 7)}$

13. $\dfrac{3y^2 - 2y + 11}{(2y - 5)(y + 3)}$

14. $\dfrac{-5y^2 + 29y - 15}{(4y + 1)(y + 5)}$

15. $\dfrac{-2y^2 + 11y - 58}{(3y - 5)(2y - 7)}$

16. $\dfrac{29a}{14(a - 5)}$

17. $\dfrac{-y}{12(3y - 5)}$

18. $\dfrac{-b - 10}{2(4b - 3)}$

19. $\dfrac{11x - 19}{10(2x - 1)}$

20. $\dfrac{11 - 9a}{6(3a - 2)}$

21. $\dfrac{-4x^2 - 4x + 3}{(2x + 1)(x - 1)}$

22. $\dfrac{9x^2 + 19x - 26}{4(3x - 1)(x + 1)}$

23. $\dfrac{x^2 + 11}{3(2x - 1)(x + 4)}$

24. $\dfrac{3}{a - b}$

25. $\dfrac{9}{2(3a - b)}$

26. $\dfrac{-8y^2 - 2y + 25}{3(2y + 1)(2y - 1)}$

27. $\dfrac{-2a^2 - 23a - 10}{2(3a + 2)(3a - 2)}$

28. $\dfrac{14y^2 + 34y + 30}{3(4y + 3)(4y - 3)}$

29. $\dfrac{8x^2 - 21x + 30}{12x(x + 2)(x - 2)}$

30. $\dfrac{-4a^2 - 49a - 45}{10a(a + 3)(a - 3)}$

31. $\dfrac{-2x^2 + 6x}{(x + 1)(x - 1)(x - 2)}$

32. $\dfrac{16a^2 - 5a}{(2a - 1)(a + 3)(3a - 1)}$

33. $\dfrac{5x^2 - 3x + 27}{(2x + 1)(x + 4)(x - 4)}$

34. $\dfrac{13a^2 + 7a - 59}{3(2a + 3)(2a - 3)(a + 5)}$

EXERCISE **6**—Page 71

1. $\dfrac{ac - b}{c}$

2. $\dfrac{6x + y}{3}$

3. $\dfrac{3y^2 - 5y - 2}{y}$

4. $\dfrac{4x - 10}{x - 2}$

5. $\dfrac{2y^2 - y + 3}{(2y - 1)}$

6. $\dfrac{63 + 4x}{9}$

7. $\dfrac{a^2 + b^2}{a - b}$

8. $\dfrac{2y^2 - 9}{y + 3}$

9. $\dfrac{2xy}{x + y}$

EXERCISE 7—Page 76

1. 3

2. $\dfrac{7}{3}$

3. $\dfrac{11}{10}$

4. $\dfrac{5}{8}$

5. $\dfrac{7}{5}$

6. $\dfrac{13}{14}$

7. $\dfrac{4y + 1}{4y - 1}$

8. $\dfrac{y}{y - x}$

9. $\dfrac{ab}{2(a + b)}$

10. $\dfrac{x}{y}$

11. $\dfrac{a + 3b}{b}$

12. $\dfrac{-(1 + 4y)}{4}$

13. $\dfrac{x - 1}{x + 1}$

14. $\dfrac{a - b}{a + b}$

15. $\dfrac{a + 2b}{a - b}$

16. $\dfrac{a}{2(a + 2)}$

17. $2(x + 1)$

18. 2

19. x

20. $\dfrac{4}{3a}$

21. $x + y$

22. $\dfrac{3a - 2b}{5a + 13b}$

23. y

24. $\dfrac{x - y}{x}$

25. $b - a$

26. $\dfrac{x}{x^3 - x^2 + 1}$

27. $(2x + 1)$

CHAPTER 5

Exponents and Radicals

POSITIVE INTEGRAL EXPONENTS

It will be recalled that if n is a positive integer then x^n represents the product $x \cdot x \cdot x \cdots x$ (x taken n times). In this notation, x is called the *base* and n is called the exponent of the power or index of the power.

Examples:

$$x^3 = x \cdot x \cdot x$$
$$3^6 = 3 \cdot 3 \cdot 3 \cdot 3 \cdot 3 \cdot 3 = 729$$
$$(-2)^5 = (-2)(-2)(-2)(-2)(-2) = -32$$
$$\left(\frac{4}{7}\right)^3 = \left(\frac{4}{7}\right)\left(\frac{4}{7}\right)\left(\frac{4}{7}\right) = \frac{64}{343}$$

LAWS OF EXPONENTS

The following are the five basic laws of exponents. If m and n are positive integers, then

1. $x^m \cdot x^n = x^{m+n}$

Example: $2^3 \cdot 2^4 = 2^7 = 128$. If we write 2^3 as $2 \cdot 2 \cdot 2$ and 2^4 as $2 \cdot 2 \cdot 2 \cdot 2$ then $2^3 \cdot 2^4 = 2 \cdot 2 \cdot 2 \cdot 2 \cdot 2 \cdot 2 \cdot 2 = 128$

2. $(x^m)^n = x^{mn}$

Example: $(3^4)^2 = 3^8 = 6{,}561$. If we write $(3^4)^2$ as $(3^4)(3^4)$ then $(3^4)(3^4) = 3^8 = 6{,}561$

3. $(xy)^n = x^n y^n$

Example: $(5 \cdot 7)^3 = 5^3 \cdot 7^3 = 42{,}875$. If we write $(5 \cdot 7)^3$ as $(35)^3$ then $(35)^3 = 35 \cdot 35 \cdot 35 = 42{,}875$

4. $\dfrac{x^m}{x^n} = x^{m-n}$ (If $m > n$ and $x \neq 0$)

$\dfrac{x^m}{x^n} = \dfrac{1}{x^{n-m}}$ (If $n > m$ and $x \neq 0$)

Examples: $\dfrac{5^7}{5^4} = 5^{7-4} = 5^3 = 125$. If we write 5^7 as $5 \cdot 5 \cdot 5 \cdot$
$5 \cdot 5 \cdot 5 \cdot 5$ and 5^4 as $5 \cdot 5 \cdot 5 \cdot 5$ then $\dfrac{5^7}{5^4} = \dfrac{\cancel{5} \cdot \cancel{5} \cdot \cancel{5} \cdot \cancel{5} \cdot 5 \cdot 5 \cdot 5}{\cancel{5} \cdot \cancel{5} \cdot \cancel{5} \cdot \cancel{5}}$
$= 5^3 = 125$

$\dfrac{3^2}{3^5} = \dfrac{1}{3^{5-2}} = \dfrac{1}{3^3} = \dfrac{1}{3 \cdot 3 \cdot 3} = \dfrac{1}{27}$. If we write 3^2 as $3 \cdot 3$ and 3^5
as $3 \cdot 3 \cdot 3 \cdot 3 \cdot 3$ then $\dfrac{3^2}{3^5} = \dfrac{\cancel{3} \cdot \cancel{3}}{\cancel{3} \cdot \cancel{3} \cdot 3 \cdot 3 \cdot 3} = \dfrac{1}{3^3} = \dfrac{1}{27}$

5. $\left(\dfrac{x}{y}\right)^n = \dfrac{x^n}{y^n}$ (if $y \neq 0$)

Example: $\left(\dfrac{2}{5}\right)^4 = \dfrac{2^4}{5^4} = \dfrac{2 \cdot 2 \cdot 2 \cdot 2}{5 \cdot 5 \cdot 5 \cdot 5} = \dfrac{16}{625}$.
If we write $\left(\dfrac{2}{5}\right)^4$ as $\dfrac{2}{5} \cdot \dfrac{2}{5} \cdot \dfrac{2}{5} \cdot \dfrac{2}{5}$ then $\left(\dfrac{2}{5}\right)^4 = \dfrac{16}{625}$

Examples: Simplify the following:
1. $b^{2x} \cdot b^{x+1} = b^{2x+x+1} = b^{3x+1}$
2. $(y^2)^{a+1} = y^{2(a+1)} = y^{2a+2}$
3. $(c^{2p} y^q)^3 = c^{(2p)(3)} y^{(q)(3)} = c^{6p} y^{3q}$
4. $\dfrac{a^{3n+1}}{a^{n-1}} = a^{(3n+1)-(n-1)} = a^{3n+1-n+1} = a^{2n+2}$

84 *Exponents and Radicals*

5. $\left(\dfrac{b^2 c^3}{d^4}\right)^a = \dfrac{b^{(2)(a)} c^{(3)(a)}}{d^{(4)(a)}} = \dfrac{b^{2a} c^{3a}}{d^{4a}}$

6. Express 64 as a power of 2. $64 = 2 \cdot 2 \cdot 2 \cdot 2 \cdot 2 \cdot 2 = 2^6$.

7. Express $9^x \cdot 27^x$ as a power of 3. $9^x = (3^2)^x = 3^{2x}$.
$(27)^x = (3^3)^x = 3^{3x}$. $9^x \cdot 27^x = 3^{2x} \cdot 3^{3x} = 3^{5x}$.

8. Express $\dfrac{49^y}{2{,}401^y}$ as a power of 7. $49^y = (7^2)^y = 7^{2y}$.

$2{,}401^y = (7^4)^y = 7^{4y}$. $\dfrac{49^y}{2{,}401^y} = \dfrac{7^{2y}}{7^{4y}} = \dfrac{1}{7^{2y}}$

EXERCISE 1

Simplify the following:

1. $p^5 \cdot p^6$

2. $x^5 \cdot x$

3. $y^a \cdot y^{2a}$

4. $z^5 \cdot z^{b+1}$

5. $2^2 \cdot 2^3$

6. $a^{2n} \cdot a^3$

7. $5^4 \cdot 5$

8. $(-7)^3$

9. $(b^2)^5$

10. $(-5b)^3$

11. $(2x^3)^4$

12. $(-3a^3)^2$

13. $(-2a^2 b^3)^3$

14. $\dfrac{y^8}{y^2}$

15. $\dfrac{x^{2n+3}}{x^2}$

16. $9^8 \div 9^4$

17. $x^{3p+1} \div x^{p-2}$

18. $(\tfrac{1}{2}y)^5$

19. $(-\tfrac{3}{4}b^2)^3$

20. $(-\tfrac{1}{2}x^2 y^3)^3$

21. $(a^{2x+3})^4$

22. $(x^2 y^{a+2})^3$

23. $\left(\dfrac{3x}{4y}\right)^2$

24. $\left(\dfrac{a}{2b}\right)^x$

25. $\left(\dfrac{-3a}{b^3}\right)^4$

26. $\left(\dfrac{2c^3}{-a}\right)^5$

27. Express as a power of 2 (*a*) 16, (*b*) 8^a, (*c*) $4^b \cdot 16^{2b}$, (*d*) $\dfrac{2^a}{32^a}$.

28. Express as a power of 3 (*a*) 27, (*b*) 9^b, (*c*) $27^{2a} \cdot 9^a$, (*d*) $\dfrac{81^b}{729^b}$.

29. Express as a power of 5 (*a*) 625, (*b*) 25^b, (*c*) $125^a \cdot 25^{3a}$, (*d*) $\dfrac{625^a}{3{,}125^a}$.

30. Express as a power of 7 (*a*) 343^{2a}, (*b*) $49^{6a} \div 7^a$.

NEGATIVE AND ZERO EXPONENTS

According to the Fourth Law of Exponents, $\dfrac{x^m}{x^n} = \dfrac{1}{x^{n-m}}$, if $n > m$ and $x \neq 0$. If, in this Law $m = n$ then $\dfrac{x^m}{x^n} = x^0$. Also, if $m = n$, then $\dfrac{x^m}{x^n} = 1$. Therefore, it is logical to define $x^0 = 1$ (provided $x \neq 0$).

Examples: $(-29)^0 = 1, 7y^0 = 7(1) = 7$.

Again, according to the Fourth Law of Exponents, $\dfrac{x^m}{x^n} = \dfrac{1}{x^{n-m}}$, if $n > m$ and $x \neq 0$. The expression $\dfrac{x^3}{x^5} = \dfrac{1}{x^2}$. However, if we simply retain the base x and subtract exponents we obtain $\dfrac{x^3}{x^5} = x^{-2}$. In order that these results be consistent we define x^{-2} as $\dfrac{1}{x^2}$. In general, we define $x^{-n} = \dfrac{1}{x^n}$ $(x \neq 0)$.

Examples: $a^2 = \dfrac{1}{a^2}, 2^{-3} = \dfrac{1}{2^3} = \dfrac{1}{8}$.

Examples: $\dfrac{5x^2}{x^{-4}} = 5x^2 \cdot x^4 = 5x^6$

$$\frac{cd^{-4}}{c^{-2}d^3} = \frac{c \cdot c^2}{d^3 \cdot d^4} = \frac{c^3}{d^7}$$

Note that the Laws of Exponents apply to both negative and zero exponents.

Example: Evaluate $\dfrac{2^{-3}}{3^{-2}}$

▶ $\dfrac{2^{-3}}{3^{-2}} = \dfrac{3^2}{2^3} = \dfrac{9}{8}$

Example: Evaluate $\dfrac{4^{-2} + 5^{-3}}{4^{-2} \cdot 5^{-3}}$

▶ $\dfrac{4^{-2} + 5^{-3}}{4^{-2} \cdot 5^{-3}} = \dfrac{\dfrac{1}{4^2} + \dfrac{1}{5^3}}{\dfrac{1}{4^2} \cdot \dfrac{1}{5^3}} = \dfrac{\dfrac{1}{16} + \dfrac{1}{125}}{\dfrac{1}{16} \cdot \dfrac{1}{125}}$

$\qquad = \dfrac{\dfrac{125}{2,000} + \dfrac{16}{2,000}}{\dfrac{1}{2,000}} = \dfrac{\dfrac{141}{2,000}}{\dfrac{1}{2,000}} = 141$

Example: Simplify: $\dfrac{x^0 - x^{-2}}{x^0 + x^{-1}}$

▶ $\dfrac{x^0 - x^{-2}}{x^0 + x^{-1}} = \dfrac{1 - \dfrac{1}{x^2}}{1 + \dfrac{1}{x}} = \dfrac{\dfrac{x^2 - 1}{x^2}}{\dfrac{x + 1}{x}} = \dfrac{x^2 - 1}{x^2} \cdot \dfrac{x}{x + 1}$

$= \dfrac{(x + 1)(x - 1)}{\cancel{x^2}} \cdot \dfrac{\cancel{x}}{(x + 1)} = \dfrac{x - 1}{x}$

Example: Evaluate $3(4)^{-2} + \left(\dfrac{2}{3}\right)^{-3} + 7^0 \cdot 8^{-1}$

▶ $3(4)^{-2} + \left(\dfrac{2}{3}\right)^{-3} + 7^0 \cdot 8^{-1} = \dfrac{3}{4^2} + \left(\dfrac{3}{2}\right)^3 + 1 \cdot \dfrac{1}{8}$

$= \dfrac{3}{16} + \dfrac{27}{8} + \dfrac{1}{8} = \dfrac{59}{16}$

Example: Simplify $\left(\dfrac{x^{2a+1}}{x} - \dfrac{y^{2b+1}}{y}\right) \cdot (x^{-a} + y^{-b})^{-1}$

▶ $\left(\dfrac{x^{2a+1}}{x} - \dfrac{y^{2b+1}}{y}\right) \cdot (x^{-a} + y^{-b})^{-1}$

$= (x^{2a} - y^{2b}) \cdot \left(\dfrac{1}{x^a} + \dfrac{1}{y^b}\right)^{-1} = (x^{2a} - y^{2b})\left(\dfrac{y^b + x^a}{x^a y^b}\right)^{-1}$

$= (x^a + y^b)(x^a - y^b) \cdot \dfrac{x^a y^b}{(y^b + x^a)} = x^a y^b (x^a - y^b)$

Example: Simplify $(a^0 + b^{-1})^{-2}(c^0 - b^{-2})$

$$(a^0 + b^{-1})^{-2}(c^0 - b^{-2})$$

$$= \left(1 + \frac{1}{b}\right)^{-2} \cdot \left(1 - \frac{1}{b^2}\right) = \left(\frac{b+1}{b}\right)^{-2}\left(\frac{b^2-1}{b^2}\right)$$

$$= \left(\frac{b}{b+1}\right)^{2}\left(\frac{b^2-1}{b^2}\right) = \frac{\cancel{b^2}}{(b+1)^{\cancel{2}}} \cdot \frac{(b+1)(b-1)}{\cancel{b^2}}$$

$$= \frac{b-1}{b+1}$$

EXERCISE 2

Evaluate or simplify:

1. 9^0

2. $9x^0$

3. $(9x)^0$

4. a^0b

5. 6^{-2}

6. $(2a)^{-1}$

7. $-8(2)^{-2}$

8. $\dfrac{1}{a^{-3}}$

9. $a^0 + b^{-1}$

10. $\dfrac{x^0 + 4^{-2}}{y^0 + 2^{-3}}$

11. $\dfrac{(a+b)^0}{9^{-1}}$

12. $c^{-2}(a^0 + c^2)$

13. $(-3b)^{-2}(b)^0$

14. $\dfrac{3^{-2} + 5^{-3}}{3^{-2} \cdot 5^{-3}}$

15. $2(3)^{-2} + \left(\dfrac{3}{2}\right)^{-3} + a^0 \cdot 9^{-1}$

16. $3(5)^{-2} + \left(\dfrac{5}{3}\right)^{-3} + (a+5)^0$

17. $\dfrac{a^{-2} - a^0}{a^{-1} - a^0}$

18. $\dfrac{4(a)^0 - x^{-2}}{2(b)^0 + x^{-1}}$

19. $(x^{-2} - y^{-2})(x^{-1} - y^{-1})^{-1}$

20. $(a^{-1} + 3^{-1})(a^2 - 9)^{-1}$

21. $(ab^{-1} - c^0)(ab^{-1} + c^0)(a+b)^{-1}$

22. $\left(\dfrac{s^{-3}t}{s^2t^{-3}}\right) \cdot \left(\dfrac{s^{-2}t^{-1}}{s^3t^3}\right)^2$

23. $(x+y)^{-2} \cdot (a^0 + yx^{-1})$

24. $(ab^{-1} + ba^{-1})(a^2 + b^2)^{-1}$

SCIENTIFIC NOTATION

Scientists have frequent occasion to deal with numbers that are very large or very small. The use of positive exponents for very large numbers and negative exponents for very small numbers simplifies the writing of such numbers and computation with such numbers. It is found convenient to write very large or very small numbers as the product of a number between 1 and 10 and a power of 10. This is known as *scientific notation.*

Example: The distance between two planets is 450,000,000,000 miles. Express this distance in scientific notation.

▶   The starting number is between 1 and 10. In this case it is 4.5. To go from 4.5 to 450,000,000,000 we must move the decimal point 11 positions to the right. Each movement to the right represents multiplication by 10. Therefore, a movement of 11 positions to the right represents a multiplication of 10^{11}.

Therefore, $450,000,000,000 = 4.5 \times 10^{11}$.

Example: The length of a wave of violet light is .000,016 inches. Write this number in scientific notation.

▶   The starting number is 1.6. To go from 1.6 to .000,016 we must move the decimal point 5 positions to the left. Each movement to the left represents division by 10, or multiplication by 10^{-1}. Therefore, a movement of 5 positions to the left represents a multiplication of 10^{-5}.

Therefore, $.000,016 = 1.6 \times 10^{-5}$.

Example: Write the following in ordinary decimal notation: (a) 3.42×10^{7}, (b) 5×10^{-6}

▶   (a) 3.42×10^{7} Multiplication by 10^{7} means that we are to move the decimal point 7 positions to the right. Therefore, $3.42 \times 10^{7} = 34200000 = 34,200,000$.

(b) 5×10^{-6} Multiplication by 10^{-6} means that we are to move the decimal point 6 positions to the left. Therefore, $5 \times 10^{-6} = .000005$.

EXERCISE 3

Write the following numbers in scientific notation.

1. .00003   **4.** .000000079

2. .00000000018   **5.** 97,200,000,000,000,000

3. 6,530,000,000   **6.** .0000000000045

Write the following numbers in ordinary decimal notation.

7. 3.5×10^{-4} **10.** 7.2×10^{-6}
8. 6.84×10^{7} **11.** 5.46×10^{5}
9. 8.6×10^{9} **12.** 9.8×10^{-8}

THE PRINCIPAL ROOTS OF A NUMBER

If n is a positive integer then the square root of n is a number which will yield n when multiplied by itself. For example, the number 25 has two square roots $+5$ and -5. In general, every positive number has two real square roots. The notation used is $\sqrt{25} = +5$ and -5.

In a similar manner, we can see that every positive number has two real fourth roots. A fourth root of n is a number which when used 4 times as a factor will yield a product equal to the original number. For example, the number 16 has two real fourth roots, $+2$ and -2, as well as two imaginary roots. The notation used is $\sqrt[4]{16} = +2$ and -2. In general, every positive number has two real fourth roots. Also, every positive number has two real sixth roots, etc.

A negative number has no real roots if n is even. For example, the fourth roots of -16, $\sqrt[4]{-16}$ are imaginary.[1]

If n is odd then both positive and negative numbers have one real root. For example, $\sqrt[5]{32} = 2$, $\sqrt[5]{-32} = -2$.

By the *principal nth root* of the positive number a is meant the *positive real number* represented by $\sqrt[n]{a}$. By the *principal nth root* of the negative number b is meant the *negative real number* represented by $\sqrt[n]{b}$ when n is odd.

Examples:

NUMBER	ROOT	PRINCIPAL ROOT
(a) -64	$\sqrt[3]{-64}$	-4 (principal cube root)
(b) $+81$	$\sqrt[4]{81}$	$+3$ (principal fourth root)
(c) $+32$	$\sqrt[5]{32}$	$+2$ (principal fifth root)

[1] Note: For detailed treatment of imaginary numbers see Chapter 9. This chapter may be fully comprehended without a knowledge of imaginary numbers.

FRACTIONAL EXPONENTS

By $\sqrt[3]{a}$ we indicate a number which when used 3 times as a factor will yield as a product the number a, i.e., $? \times ? \times ? = a$. If the basic laws of exponents are to hold in this case then each $?$ must be equal to $a^{1/3}$ *because* $a^{1/3} \cdot a^{1/3} \cdot a^{1/3} = a^1$. By a similar line of reasoning $\sqrt[5]{y} = y^{1/5}$, $\sqrt[6]{b} = b^{1/6}$, $\sqrt[4]{c} = c^{1/4}$, provided that b, and c, are positive or zero.

By $\sqrt[3]{a^2}$ we indicate a number which when used 3 times as a factor will yield as a product the number a^2, i.e. $? \times ? \times ? = a^2$. If the basic laws of exponents are to hold in this case then each $?$ must be equal to $a^{2/3}$ because $a^{2/3} \cdot a^{2/3} \cdot a^{2/3} = a^{6/3} = a^2$. By a similar line of reasoning $\sqrt[5]{x^3} = x^{3/5}$, $\sqrt[3]{y^4} = y^{4/3}$, $\sqrt[7]{a^2} = a^{2/7}$.

In general, we define $a^{p/q}$ as $\sqrt[q]{a^p}$ $(q \neq 0)$. It is useful to observe that q represents a *root* whereas p represents a power. For example, $5^{2/3} = \sqrt[3]{5^2} = \sqrt[3]{25}$. Also, just as we defined x^{-n} as $\dfrac{1}{x^n}$ we define $a^{-p/q}$ as $\dfrac{1}{a^{p/q}}$. In performing calculations with fractional exponents it is usually best to *take the indicated root first* and *then* raise the result to the indicated power.

Example: Find the value of $16^{3/4} + 27^{-2/3}$

▶ $16^{3/4}$ To find the value of $16^{3/4}$ first take the 4th root of 16 and then raise the result to the 3rd power. The 4th root of 16 is 2, and 2 raised to the 3rd power is 8.

Therefore, $16^{3/4} = 8$.

$$27^{-2/3} = \frac{1}{27^{2/3}}$$

To find the value of $27^{2/3}$ take the cube root of 27 and raise the result to the 2nd power.

Therefore, $27^{-2/3} = \dfrac{1}{27^{2/3}} = \dfrac{1}{9}$.

$$16^{3/4} + 27^{-2/3} = 8 + \frac{1}{9} = 8\frac{1}{9}$$

Example: Find the value of $8^{2/3} - 25^{-1/2} + \left(\dfrac{1}{16}\right)^{-3/4}$

▶ $8^{2/3}$ = 3rd root of 8 raised to the 2nd power = 4

$25^{-1/2} = \dfrac{1}{25^{1/2}}$ = square root of $\dfrac{1}{25} = \dfrac{1}{5}$

$\left(\dfrac{1}{16}\right)^{-3/4} = (16)^{3/4}$ = 4th root of 16 raised to 3rd power = 8

Therefore, $8^{2/3} - 25^{-1/2} + \left(\dfrac{1}{16}\right)^{-3/4} = 4 - \dfrac{1}{5} + 8 = 11\dfrac{4}{5}$

Example: Find the value of $(-27)^{-2/3} + \left(\dfrac{1}{9}\right)^{-1/2} - 4^{5/2}$

▶ $(-27)^{-2/3} = \dfrac{1}{(-27)^{2/3}}$

$= \dfrac{1}{\text{3rd root of } -27 \text{ raised to 2nd power}} = \dfrac{1}{9}$

$\left(\dfrac{1}{9}\right)^{-1/2} = (9)^{1/2} = 3$

$-(4)^{5/2} = -(2)^5 = -32$

$(-27)^{-2/3} + \left(\dfrac{1}{9}\right)^{-1/2} - 4^{5/2} = \dfrac{1}{9} + 3 - 32 = -28\dfrac{8}{9}$

EXERCISE 4

Find the value of the following:

1. $9^{3/2} + 4^{-3/2}$
2. $(-8)^{2/3} - (-27)^{2/3}$
3. $16^{3/4} + 2(25)^{-1/2}$
4. $(32)^{3/5} - 5(64)^{2/3}$
5. $8^{2/3} - 8^0 + 8(2)^{-2}$
6. $4(-8)^{-1/3} + (-32)^{2/5}$
7. $(.04)^{1/2} + (.027)^{2/3}$
8. $16^{3/4} + (-64)^{1/3}$
9. $(81)^{-3/4} - (9)^{-3/2}$
10. $(.81)^{-1/2} + (-4)^{-1}$
11. $125^{2/3} - (-8)^{2/3}$
12. $3\left(\dfrac{1}{9}\right)^{3/2} - \left(\dfrac{1}{4}\right)^{-2}$

GENERAL THEORY OF EXPONENTS

The following examples illustrate the general theory of exponents:

Example: Simplify $6^{1/2} \cdot 6^{1/3} \cdot 6^{1/4} \cdot 6^{1/6}$

▶ $6^{1/2} \cdot 6^{1/3} \cdot 6^{1/4} \cdot 6^{1/6} = 6^{1/2+1/3+1/4+1/6} = 6^{15/12} = 6^{5/4}$

Example: Simplify $a^{-1/2} \cdot a^{-3/2} \cdot a^2$

▶ $a^{-1/2} \cdot a^{-3/2} \cdot a^2 = a^{-1/2-3/2+2} = a^0 = 1$

Example: Simplify $x^{n+3} \cdot x^{n-2} \cdot x^{5-n}$

▶ $x^{n+3} \cdot x^{n-2} \cdot x^{5-n} = x^{n+3+n-2+5-n} = x^{n+6}$

Example: Simplify $4a^{-2/3} \cdot 3a^{-1/3}$

▶ $4a^{-2/3} \cdot 3a^{-1/3} = (4)(3)(a^{-2/3-1/3}) = 12a^{-1}$

Example: Simplify $\left(\sqrt[3]{(a+b)} \right) \left(\sqrt[6]{(a+b)^5} \right)$

▶ $\left(\sqrt[3]{a+b} \right) \left(\sqrt[6]{(a+b)^5} \right) = (a+b)^{1/3} \cdot (a+b)^{5/6}$

$= (a+b)^{1/3+5/6} = (a+b)^{7/6}$

Example: Simplify $7\sqrt[4]{8} \cdot 6\sqrt[6]{4}$

▶ $7\sqrt[4]{8} \cdot 6\sqrt[6]{4} = 7(2^3)^{1/4} \cdot 6(2^2)^{1/6} = 7(2^{3/4}) \cdot 6(2^{2/6})$

$= 7(2^{3/4}) \cdot 6(2^{1/3}) = 42(2)^{3/4+1/3} = 42(2)^{13/12} = 42 \cdot 2 \cdot 2^{1/12}$

$= 84 \cdot 2^{1/12}$

Example: Simplify $(a^2 + b^2)^{-2/3} \cdot \left(\sqrt[4]{(a^2+b^2)^3} \right)$

$\left(\sqrt{a^2 + b^2} \right)$

▶ $(a^2 + b^2)^{-2/3} \left(\sqrt[4]{(a^2 + b^2)^3} \right) \left(\sqrt{a^2 + b^2} \right)$

$= (a^2 + b^2)^{-2/3}$

$(a^2 + b^2)^{3/4}(a^2 + b^2)^{1/2} = (a^2 + b^2)^{-2/3+3/4+1/2} = (a^2 + b^2)^{7/12}$

Example: Simplify $\sqrt[3]{xy^2} \cdot \sqrt[3]{x^2y} \cdot \sqrt[6]{x^2y^2}$

▶ $\sqrt[3]{xy^2} \cdot \sqrt[3]{x^2y} \cdot \sqrt[6]{x^2y^2} = x^{1/3}y^{2/3} \cdot x^{2/3} \cdot y^{1/3} \cdot x^{2/6} \cdot y^{2/6}$

$= x^{1/3+2/3+2/6}y^{2/3+1/3+2/6} = x^{4/3}y^{4/3}$

Example: Simplify $3^{5/2} \cdot 3^{3/4} \div 3^{3/8}$

▶ $3^{5/2} \cdot 3^{3/4} \div 3^{3/8} = 3^{5/2+3/4-3/8} = 3^{23/8}$

Example: Simplify $\dfrac{24 \cdot 10^4}{6 \cdot 10^{-3}}$

▶ $$\frac{24 \cdot 10^4}{6 \cdot 10^{-3}} = \frac{24}{6} \cdot 10^{4+3} = 4 \cdot 10^7$$

Example: Simplify $\dfrac{x^2 \cdot y^{-2/3}}{x^{-3} \cdot y^{1/3}}$

▶ $$\frac{x^2 \cdot y^{-2/3}}{x^{-3} \cdot y^{1/3}} = x^{2+3} \cdot y^{-2/3-1/3} = x^5 \cdot y^{-1}$$

Example: Simplify $\dfrac{(2a+1)^{3x+5}}{(2a+1)^{x-3}}$

▶ $$\frac{(2a+1)^{3x+5}}{(2a+1)^{x-3}} = (2a+1)^{3x+5-x+3} = (2a+1)^{2x-8}$$

Example: Simplify $\dfrac{15a^3 b^{-1/2} c^{1/3}}{5a^{1/2} b^{-3/2} c^{4/3}}$

▶ $$\frac{15a^3 b^{-1/2} c^{1/3}}{5a^{1/2} b^{-3/2} c^{4/3}} = 3a^{3-1/2} b^{-1/2+3/2} c^{1/3-4/3} = 3a^{5/2} bc^{-1}$$

Example: Simplify $\dfrac{\sqrt{a^5} \cdot \sqrt[3]{b^2} \cdot \sqrt[4]{c^2}}{\sqrt{a^3} \cdot \sqrt[3]{b^5} \cdot \sqrt[4]{\dfrac{1}{c^2}}}$

▶ $$\frac{\sqrt{a^5} \cdot \sqrt[3]{b^2} \cdot \sqrt[4]{c^2}}{\sqrt{a^3} \cdot \sqrt[3]{b^5} \cdot \sqrt[4]{\dfrac{1}{c^2}}} = \frac{a^{5/2} \cdot b^{2/3} \cdot c^{2/4}}{a^{3/2} \cdot b^{5/3} \cdot c^{-2/4}} = a^{5/2-3/2} \cdot b^{2/3-5/3}$$

$\cdot c^{2/4+2/4} = ab^{-1} c$

Example: Simplify $\dfrac{a^{2x+1} \cdot \sqrt[3]{a^2}}{a^{x-1} \cdot \sqrt{a^3}}$

▶ $$\frac{a^{2x+1} \cdot \sqrt[3]{a^2}}{a^{x-1} \cdot \sqrt{a^3}} = \frac{a^{2x+1} \cdot a^{2/3}}{a^{x-1} \cdot a^{3/2}} = a^{2x+1+2/3-(x-1)-3/2} = a^{2x+1+2/3-x+1-3/2}$$

$= a^{x+\frac{1}{6}}$

Example: Simplify the following.
a. $(a^3)^5 = a^{15}$
b. $(x^2)^{n+1} = x^{2n+2}$
c. $(b^{2p-3})^4 = b^{8p-12}$
d. $(3^{x+1})^{x-1} = 3^{(x+1)(x-1)} = 3^{x^2-1}$

e. $[(x + y)^2]^5 = (x + y)^{10}$

f. $(a^{2x-10})^{-1/2} = a^{-x+5}$

g. $\left(\dfrac{8}{27}\right)^{4/3} = \left(\dfrac{2^3}{3^3}\right)^{4/3} = \dfrac{2^{3(4/3)}}{3^{3(4/3)}} = \dfrac{2^4}{3^4} = \dfrac{16}{81}$

h. $(\sqrt[4]{y + z})^6 = [(y + z)^{1/4}]^6 = (y + z)^{6/4} = (y + z)^{3/2}$

i. $\sqrt[4]{\sqrt[3]{a^4}} = \sqrt[4]{a^{4/3}} = (a^{4/3})^{1/4} = a^{1/3}$

j. $\sqrt[6]{\dfrac{1}{b^3}} = \sqrt[6]{\sqrt{b^{-3}}} = \sqrt[6]{b^{-3/2}} = (b^{-3/2})^{1/6} = b^{-1/4}$

k. $(x^2y)^3 = x^6y^3$

l. $(3a)^4 = 3^4 \cdot a^4 = 81a^4$

m. $(16a^4b^6)^{1/2} = (16^{1/2})(a^4)^{1/2}(b^6)^{1/2} = 4a^2b^3$

n. $(-27x^{3/2}y^{-6/5})^{1/3} = (-27)^{1/3}(x^{3/2})^{1/3}(y^{-6/5})^{1/3} = -3x^{1/2}y^{-2/5}$

o. $(a^{2n+4}b^{4n+6}c^8)^{1/2} = a^{1/2(2n+4)} \cdot b^{1/2(4n+6)} \cdot c^{1/2(8)} = a^{n+2}b^{2n+3}c^4$

p. $\sqrt{\dfrac{x^5y^2}{z^6}} = \sqrt{x^5y^2z^{-6}} = (x^5y^2z^{-6})^{1/2} = x^{5/2}yz^{-3}$

q. $(8 \cdot 10^6)^{1/3} = 8^{1/3} \cdot (10^6)^{1/3} = 8^{1/3} \cdot 10^2 = 2 \cdot 100 = 200$

r. $\left(\dfrac{x^2y}{z}\right)^a = \dfrac{x^{2a}y^a}{z^a}$

s. $\left(\dfrac{3a}{2b}\right)^3 = \dfrac{3^3 \cdot a^3}{2^3 \cdot b^3} = \dfrac{27a^3}{8b^3}$

t. $\left(\dfrac{x^{2n+1}}{y^n}\right)^2 = \dfrac{x^{2(2n+1)}}{y^{2n}} = \dfrac{x^{4n+2}}{y^{2n}}$

u. $\left(\dfrac{4a^2}{9b^2}\right)^{3/2} = \dfrac{(4^{3/2})(a^2)^{3/2}}{9^{3/2} \cdot (b^2)^{3/2}} = \dfrac{8a^3}{27b^3}$

v. $\left(\dfrac{27}{125}\right)^{-2/3} = \left(\dfrac{125}{27}\right)^{2/3} = \dfrac{(125)^{2/3}}{(27)^{2/3}} = \dfrac{25}{9}$

w. $\sqrt{\dfrac{49x^4}{16y^6}} = \left(\dfrac{49x^4}{16y^6}\right)^{1/2} = \dfrac{49^{1/2} \cdot x^{1/2(4)}}{16^{1/2} \cdot y^{1/2(6)}} = \dfrac{7x^2}{4y^3}$

x. $\sqrt{\dfrac{216x^9}{343y^3}} = \dfrac{(216x^9)^{1/3}}{(343y^3)^{1/3}} = \dfrac{216^{1/3} \cdot x^{9/3}}{343^{1/3} \cdot y^{3/3}} = \dfrac{6x^3}{7y}$

Example: Simplify the following.

a. $2 \cdot 10^3 + 2 \cdot 10^2 - 5 \cdot 10 + 6 \cdot 10^0 - 5 \cdot 10^{-2}$
$= 2(1,000) + 2(100) - 50 + 6(1) - 5(.01)$
$= 2,000 + 200 - 50 + 6 - .05$
$= 2,155.95$

b. $(2.5 \times 10^8)(.8 \times 10^{-6}) = (2.5)(.8)(10^{8-6}) = 2(10^2)$
$= 200$

c. $\dfrac{19.2 \times 10^3}{3.2 \times 10^4} = \dfrac{19.2}{3.2} \cdot 10^{3-4} = 6.10^{-1} = \dfrac{6}{10} = \dfrac{3}{5}$

d. Write the numerical value of $\dfrac{(5a)^0}{a^{-2/3}}$ when $a = 8$.

$\dfrac{(5a)^0}{a^{-2/3}} = \dfrac{(5 \cdot 8)^0}{8^{-2/3}} = 1 \cdot 8^{2/3} = 1 \cdot 4 = 4$

e. Write the numerical value of $3x^{1/2} - (2x)^0 + 16x^{-2}$ when $x = 4$.
$3x^{1/2} - (2x)^0 + 16x^{-2} = 3(4)^{1/2} - (2 \cdot 4)^0 + 16(4)^{-2}$

$= 3(2) - 1 + 16\left(\dfrac{1}{16}\right) = 6 - 1 + 1 = 6.$

f. $\dfrac{8^{-1/3} + 2^{-2}}{16^{-3/4} + 2^{-1}} = \dfrac{\dfrac{1}{8^{4/3}} + \dfrac{1}{2^2}}{\dfrac{1}{16^{3/4}} + \dfrac{1}{2}} = \dfrac{\dfrac{1}{16} + \dfrac{1}{4}}{\dfrac{1}{8} + \dfrac{1}{2}} = \dfrac{\dfrac{5}{16}}{\dfrac{5}{8}} = \dfrac{1}{2}$

g. $\dfrac{x^{a-b}}{\sqrt[b]{x^{b^2-ab}}} \cdot \sqrt[a]{x^{2ab-a^2}} = \dfrac{\left(x^{a-b}\right)\left(x^{2ab-a^2}\right)^{1/a}}{\left(x^{b^2-ab}\right)^{1/b}}$

$= \dfrac{x^{a-b} \cdot x^{2b-a}}{x^{b-a}} = x^{a-b+2b-a-b+a} = x^a$

h. $\dfrac{8^n(2^{n-1})^n}{4^{n+1} \cdot 2^3} = \dfrac{(2^3)^n(2^{n-1})^n}{(2^2)^{n+1} \cdot 2^3} = \dfrac{2^{3n} \cdot 2^{n^2-n}}{2^{2n+2} \cdot 2^3}$

$= \dfrac{2^{n^2+2n}}{2^{2n+5}} = 2^{n^2+2n-2n-5} = 2^{n^2-5}$

i. $\left(\sqrt{b^y} \cdot \sqrt[4]{b^{3y}}\right)^{-2/3} = \left(b^{y/2} \cdot b^{3y/4}\right)^{-2/3} = \left(b^{5y/4}\right)^{-2/3} = b^{-5y/6}$

j. $\sqrt[p]{a^{-2}b^{-3}c^{-1}} \div \sqrt[2p]{a^{-4}b^8} = \dfrac{\left(a^{-2/p}\,b^{-3/p}\,c^{-1/p}\right)}{\left(a^{-2/p}\,b^{4/p}\right)} = b^{-7/p}c^{-1/p}$

EXERCISE 5

Simplify the following:

1. $3^{1/2} \cdot 3^{1/3} \cdot 3^{1/4} \cdot 3^{1/12}$
2. $2^{1/4} \cdot 2^{1/8} \cdot 2^{5/6} \cdot 2^{5/12}$
3. $b^{-1/3} \cdot b^{-1/2} \cdot b^{1/4} \cdot b^{1/6}$
4. $x^{p+1} \cdot x^{p-2} \cdot x^{7-p}$
5. $8x^{-1/2} \cdot 9x^{5/2}$
6. $4a^{2/3} \cdot 3a^{-1/6}$
7. $\sqrt[4]{(x+y)^3} \cdot \sqrt{(x+y)}$
8. $\sqrt[5]{(2a-b)^2} \cdot \sqrt[3]{(2a-b)^4}$
9. $4\sqrt[3]{9} \cdot 5\sqrt[4]{27}$
10. $3\sqrt[4]{25} \cdot 7\sqrt{125}$
11. $(x+y)^{-1/3} \cdot \sqrt[3]{(x+y)^2} \cdot \sqrt{(x+y)}$
12. $(a+b)^{-1} \cdot \sqrt[4]{(a+b)^3} \cdot \sqrt{(a+b)^3}$
13. $\sqrt{ab^3} \cdot \sqrt[3]{a^3b} \cdot \sqrt[4]{a^2b^2}$
14. $\sqrt[3]{x^4y^2} \cdot \sqrt{x^3y^3} \cdot \sqrt[4]{xy^2}$
15. $5^{2/3} \cdot 5^{5/6} \div 5^{1/2}$
16. $7^{3/8} \cdot 7^{1/2} \div 7^{3/4}$
17. $\dfrac{18 \cdot 10^5}{3 \cdot 10^{-2}}$
18. $\dfrac{21 \cdot 10^{-1}}{7 \cdot 10^{-3}}$
19. $\dfrac{a^2 \cdot b^{-3/4}}{a^{-1/2} \cdot b^2}$
20. $\dfrac{x^{-5} \cdot y^{-1/2}}{x^{-6} \cdot y^{3/2}}$
21. $\dfrac{(a+b)^{2x-1}}{(a+b)^{x+5}}$
22. $\dfrac{(t+v)^{3a+b}}{(t+v)^{a-b}}$
23. $\dfrac{24x^2y^{-1/3}z^{1/2}}{4x^{1/2}y^{-2/3}z^{-5/2}}$
24. $\dfrac{36a^3b^{-1/2}c^{2/3}}{18a^{-1}b^{3/2}c^{5/6}}$
25. $\dfrac{\sqrt{x^3} \cdot \sqrt[3]{x} \cdot \sqrt[4]{x^3}}{\sqrt{x} \cdot \sqrt[3]{x^4} \cdot \sqrt[4]{x}}$

26. $\dfrac{\sqrt[6]{b^5} \cdot \sqrt{b^3} \cdot \sqrt[3]{b}}{\sqrt[6]{b} \cdot \sqrt{b^7} \cdot \sqrt[3]{b^4}}$
27. $\dfrac{b^{3y+1}\sqrt{b}}{b^{2y-1}\sqrt{b^3}}$
28. $\dfrac{x^{2a+2}\sqrt[3]{y^2}}{x^{2a-1}\sqrt[6]{y}}$
29. $(x^2)^4$
30. $(a^3)^{2n-1}$
31. $(y^{2p-3})^4$
32. $(z^{2a+1})^{2a-1}$
33. $[(m+n)^3]^4$
34. $\left(\sqrt[4]{(a+b)^3}\right)^6$
35. $(b^{4x-6})^{-1/2}$
36. $\left(\dfrac{25}{16}\right)^{3/2}$
37. $\left(\dfrac{81}{16}\right)^{3/4}$
38. $\left(\sqrt[5]{(x+y)^4}\right)^{10}$
39. $\sqrt[6]{\sqrt[3]{x^4}}$
40. $\sqrt[8]{\sqrt{y^4}}$
41. $\sqrt[10]{\dfrac{1}{x^5}}$
42. $\sqrt[12]{\dfrac{1}{y^8}}$
43. $(a^2b^3)^4$
44. $(x^{1/2}y^{2/3})^{12}$
45. $(2x)^7$
46. $(4y^8)^{5/2}$
47. $(64x^9y^{15})^{4/3}$
48. $(-125a^{6/5}b^{9/2})^{1/3}$
49. $(x^{3n+6}y^{6n+3}z^{9n})^{1/3}$
50. $(a^{8n+4}b^{-12n}c^{-4n})^{-1/4}$

51. $\sqrt{\dfrac{a^7b^4}{z^2}}$

52. $\sqrt[3]{\dfrac{x^9y^{10}}{z^6}}$

53. $(27 \cdot 10^6)^{2/3}$

54. $(4 \cdot 10^2)^{3/2}$

55. $\left(\dfrac{a^3b^2}{c}\right)^x$

56. $\left(\dfrac{x^ay^c}{z^{b+1}}\right)^p$

57. $\left(\dfrac{5x}{3y}\right)^3$

58. $\left(\dfrac{2a}{3b}\right)^5$

59. $\left(\dfrac{a^{2n+1}}{b^n}\right)^3$

60. $\left(\dfrac{25a^4}{49b^2}\right)^{3/2}$

61. $\left(\dfrac{8a^6}{27b^9}\right)^{2/3}$

62. $\left(\dfrac{125}{8}\right)^{-1/3}$

63. $\left(\dfrac{81}{16}\right)^{-3/4}$

64. $\sqrt{\dfrac{64x^6}{9y^4}}$

65. $\sqrt[3]{\dfrac{27a^9}{64b^6}}$

66. $\sqrt[4]{\dfrac{16a^8}{81b^{12}}}$

67. $4 \cdot 10^3 - 2 \cdot 10^2 + 3 \cdot 10 - 7 \cdot 10^0 + 3 \cdot 10^{-2}$

68. $3 \cdot 10^3 + 7 \cdot 10^2 - 9 \cdot 10 - 4 \cdot 10^0 - 8 \cdot 10^{-1}$

69. $(3.5 \times 10^6)(.2 \times 10^{-2})$

70. $(2.4 \times 10^7)(1.5 \times 10^{-6})$

71. $\dfrac{11.2 \times 10^5}{2.8 \times 10^7}$

72. $\dfrac{28.8 \times 10^9}{3.6 \times 10^8}$

73. Write the numerical value of $\dfrac{2x^{-1/3}}{3x^{5/3}}$ when $x = 4$.

74. Write the numerical value of $y^{2/3} - (4y)^0 + 15y^{-2}$ when $y = 8$.

75. $\dfrac{9^{-3/2} + 3^{-2}}{27^{-1/3} + 3^{-1}}$

76. $\dfrac{25^{-1/2} + 125^{-2/3}}{5^{-2} + 25^{-3/2}}$

77. $\dfrac{9^n(3^{n-1})^n}{27^{n+1} \cdot 3^2}$

78. $\left(\sqrt[3]{a^x} \cdot \sqrt{a^{3x}}\right)^{-1/2}$

79. $\sqrt[3]{x^{-3}y^{-4}c^{-2}} \div \sqrt[3a]{x^{-3}y^{-1}c^6}$

80. $\left(\dfrac{27x^{-3}}{64y^{-3}}\right)^{-1/3}$

81. $\left(\dfrac{81a^{-8}b^6}{25a^4b^{-10}}\right)^{-1/2}$

82. $\left(\dfrac{a^{-3}b}{a^2b^{-3}}\right)^{-3} \cdot \left(\dfrac{a^{-2}b^{-1}}{a^35^3}\right)^5$

83. $\dfrac{5^{n+1}}{7^{n+1}} \div \dfrac{(25)^{n-1}}{(49)^{n-2}}$

84. $36^{1/2} + \left(\frac{1}{16}\right)^{-1/2} + 9^{-1/2} + (.25)^{-1} - (5x^0)^{-1}$

85. $\dfrac{8^{-4/3} \cdot 27^{2/3} \cdot 81^{-1/2}}{3^{-3} \cdot 4^{-2} \cdot 6^2 \cdot 12^{-1}}$

94. $\left(\dfrac{x^{-2}b^2}{a^3c^{-4}}\right)^{-2} \div \left(\dfrac{x^2b^{-1}}{a^{-2}c^3}\right)^2$

86. $\dfrac{8x^2y^{-2/3}}{6(xy)^{1/2}}$

95. $\left(\dfrac{27x^3y^{-3/2}}{8y^3x^{-1}}\right)^{-2/3}$

87. $\left(\dfrac{ab^{-2}}{a^{-1/3}}\right)^{3/4}$

96. $\left(\dfrac{y^{n+1}}{y}\right)^{n-1} \div \left(\dfrac{y}{y^{n-1}}\right)$

88. $\left(\dfrac{x^{-2}y^2}{x^5y^{-3}}\right)^{1/5} \left(\dfrac{x^{-2}y^2}{x^2 \cdot y^{-2}}\right)^{-2}$

89. $\dfrac{x^{-a}}{y^{-b}} \div \dfrac{y^b}{x^{a-3}}$

97. $\left(\dfrac{a^{-6}}{b^{-2}c}\right)^{-3/4} \times \left(\dfrac{a^{-1}bc^{-3/2}}{ab^{-2}}\right)^{1/2}$

90. $x^{1/3}(x^{-1/3} + x^{1/3})$

98. $\dfrac{x^n(x^{n+1})^n}{x^{n+1} \cdot x^{n-2}}$

91. $a^{-3/2} \cdot y^{-3/4} \div a^{1/2}y^{1/4}$

92. $(9b^{-2}x^{1/2}y^{-6})^{-3/2}$

99. $\dfrac{a}{b}\left(\dfrac{a^{-2}b^0}{ab^{-6}}\right)^{1/3}$

93. $x^{n(n-2)} \div x^{n^2}$

100. $\left(\dfrac{a + b^{-1}}{a - b^{-1}}\right)^{n+2} \cdot \left(\dfrac{ab + 1}{ab - 1}\right)^{-2}$

RADICALS

Quantities involving fractional exponents may be written by using *radicals*. For example, the quantity $a^{1/n}$ may be written as $\sqrt[n]{a}$. This notation is to be interpreted as the nth principal root of a. n is called the *index* of the radical.

DEFINITION: A radical is a root of a quantity indicated by the use of the radical sign ($\sqrt{}$).

DEFINITION: The *radicand* is the quantity under the radical sign.

DEFINITION: A *surd* is an indicated root which cannot be exactly extracted, e.g., $\sqrt{2}$, $\sqrt[5]{7}$, etc.

DEFINITION: The *order* or *degree* of a surd is indicated by the index of the root.

$\sqrt{a}$ is of the second order or degree
$\sqrt[3]{a}$ is of the third order or degree

DEFINITION: *Similar radicals* are radicals which have the same quantity under the radical sign and the same index. For example, $2\sqrt[3]{5}$ and $-7\sqrt[3]{5}$ are similar radicals.

Since a quantity under a radical sign may be written with a fractional exponent, operations with radicals obey the laws of exponents.

1. $x^m \cdot x^n = x^{m+n}$

$\sqrt[m]{a} \cdot \sqrt[n]{a} = a^{1/m} \cdot a^{1/n} = a^{1/m + 1/n}$

Example: $\sqrt{2} \cdot \sqrt[3]{2} = 2^{1/2} \cdot 2^{1/3} = 2^{1/2 + 1/3} = 2^{5/6} = \sqrt[6]{2^5}$
$= \sqrt[6]{32}$

2. $(x^m)^n = x^{mn}$

$(\sqrt[n]{a})^m = (a^{1/n})^m = a^{m/n} = \sqrt[n]{a^m}$

Example: $(\sqrt[5]{3})^2 = (3^{1/5})^2 = 3^{2/5} = \sqrt[5]{3^2} = \sqrt[5]{9}$

$\sqrt[4]{\sqrt[3]{7}} = \sqrt[4]{7^{1/3}} = (7^{1/3})^{1/4} = 7^{1/12} = \sqrt[12]{7}$

3. $(xy)^n = x^n y^n$

$\sqrt[n]{ab} = (ab)^{1/n} = a^{1/n} \cdot b^{1/n} = \sqrt[n]{a} \cdot \sqrt[n]{b}$ (provided that $\sqrt[n]{a}$

and $\sqrt[n]{b}$ are real numbers)

Example: $\sqrt[3]{24} = (24)^{1/3} = 8^{1/3} \cdot 3^{1/3} = \sqrt[3]{8} \cdot \sqrt[3]{3} = 2\sqrt[3]{3}$

4. $\dfrac{x^m}{x^n} = x^{m-n}$

$\dfrac{\sqrt[n]{a}}{\sqrt[m]{a}} = \dfrac{a^{1/n}}{a^{1/m}} = a^{1/n - 1/m} = a^{\frac{m-n}{mn}}$

Example: $\dfrac{\sqrt{32}}{\sqrt[4]{32}} = \dfrac{(32)^{1/2}}{(32)^{1/4}} = 32^{1/2 - 1/4} = (32)^{1/4} = (2^5)^{1/4}$

$= 2^{5/4} = \sqrt[4]{2^5} = \sqrt[4]{2^4} \cdot \sqrt{2} = 2\sqrt[4]{2}$

5. $\left(\dfrac{x}{y}\right)^n = \dfrac{x^n}{y^n}$

$\sqrt[n]{\dfrac{a}{b}} = \left(\dfrac{a}{b}\right)^{1/n} = \dfrac{a^{1/n}}{b^{1/n}} = \dfrac{\sqrt[n]{a}}{\sqrt[n]{b}}$

Example: $\sqrt[3]{\dfrac{16}{5}} = \dfrac{16^{1/3}}{5^{1/3}} = \dfrac{(2^4)^{1/3}}{5^{1/3}} = \dfrac{\sqrt[3]{2^4}}{\sqrt[3]{5}} = \dfrac{2\sqrt[3]{2}}{\sqrt[3]{5}}$

Transformation of Radicals

The laws of exponents extended to include radicals provide the basic theory for the transformation of radicals. By a transformation is meant a change in the form of a radical without changing its value. Transformations are performed in order to write a radical in simplest form. A radical is in simplest form when the expression under the radical sign is integral, and contains no factor whose power is equal to or greater than the index.

Examples: $\sqrt{175} = \sqrt{25} \cdot \sqrt{7} = 5\sqrt{7}$

$\sqrt[3]{16} = \sqrt[3]{8} \cdot \sqrt[3]{2} = 2\sqrt[3]{2}$

Removal of a Perfect Nth Power Factor

1. Separate the quantity under the radical sign into two factors, one of which is the greatest perfect power of the same degree as the radical.

2. Extract the required root of this factor, and multiply the coefficient of the radical by the extracted root. The remaining factor is left under the radical sign.

Example: Simplify $4\sqrt{75}$

▶ $4\sqrt{75} = 4\sqrt{25 \cdot 3} = 4 \cdot 5\sqrt{3} = 20\sqrt{3}$

Example: Simplify $3\sqrt[3]{56}$

▶ $3\sqrt[3]{56} = 3\sqrt[3]{8 \cdot 7} = 3 \cdot 2\sqrt[3]{7} = 6\sqrt[3]{7}$

Example: Simplify $\dfrac{1}{3}\sqrt[3]{1,080}$

▶ $\dfrac{1}{3}\sqrt[3]{1,080} = \dfrac{1}{3}\sqrt[3]{8 \cdot 27 \cdot 5} = \dfrac{1}{\cancel{3}} \cdot 2 \cdot \cancel{3}\sqrt[3]{5} = 2\sqrt[3]{5}$

Example: Simplify $\sqrt[3]{250a^3b^6}$

▶ $\sqrt[3]{250a^3b^6} = \sqrt[3]{125 \cdot 2 \cdot a^3b^6} = 5ab^2\sqrt[3]{2}$

Example: Simplify $\sqrt[3]{-81a^4b^3}$

▶ $\sqrt[3]{-81a^4b^3} = \sqrt[3]{-27 \cdot 3 \cdot a^3 \cdot a \cdot b^3} = -3ab\sqrt[3]{3a}$

Example: Simplify $\sqrt[n]{x^{4n}y^{3n+2}}$

▶ $\sqrt[n]{x^{4n}y^{3n+2}} = (x^{4n}y^{3n} \cdot y^2)^{1/n} = x^4y^3\sqrt[n]{y^2}$

Example: Simplify $\sqrt[3]{375a^{-6}b^7}$

▶ $\sqrt[3]{375a^{-6}b^7} = \sqrt[3]{3 \cdot 125 \cdot a^{-6} \cdot b^6 \cdot b} = 5a^{-2}b^2\sqrt[3]{3b}$

$= \dfrac{5b^2}{a^2}\sqrt[3]{3b}$

Example: Simplify $\sqrt{3a^2 - 6ab + 3b^2}$

▶ $\sqrt{3a^2 - 6ab + 3b^2} = \sqrt{3(a^2 - 2ab + b^2)} = \sqrt{3(a - b)^2}$

$= (a - b)\sqrt{3}$

EXERCISE 6

Simplify the following:

1. $\sqrt{8}$
2. $\sqrt{48}$
3. $2\sqrt{98}$
4. $\dfrac{1}{10}\sqrt{125}$
5. $2\sqrt{18a^2b}$
6. $5\sqrt{27x^5y^3}$
7. $\sqrt[3]{16}$
8. $7\sqrt[3]{24}$
9. $3\sqrt[3]{250}$
10. $\dfrac{7}{8}\sqrt[3]{128}$
11. $y\sqrt[3]{243}$
12. $\sqrt[3]{32x^3y^9}$

13. $\sqrt[3]{216a^4b^5}$
14. $\sqrt[3]{-54c^2d^5}$
15. $\sqrt[3]{-320x^{3a}}$
16. $\sqrt[3]{-1,024y^{6n}}$
17. $\sqrt[3]{875a^{-9}b^8}$
18. $\dfrac{2x}{3}\sqrt[3]{189x^{-6}y^{10}}$
19. $\sqrt{5x^2 - 10xy + 5y^2}$
20. $\sqrt{2a^2 - 8ab + 8b^2}$
21. $\sqrt[a]{x^{6a}y^{2a+1}}$
22. $\sqrt[n]{a^{2n+1}b^{5n}}$
23. $\sqrt[m]{x^{3m+1}y^{2m+1}}$
24. $\sqrt[p]{5a^{2p+3}b^{4p}}$

Simplifying a Radical Whose Radicand Is a Fraction

When the quantity under the radical sign is a fraction the radicand may be simplified by multiplying both numerator and denominator of the fraction by a quantity which will make the denominator a perfect power of the same degree as the radical.

Example: Simplify $\sqrt{\dfrac{7}{6}}$

▶ $\qquad \sqrt{\dfrac{7}{6}} = \sqrt{\dfrac{7}{6} \cdot \dfrac{6}{6}} = \sqrt{\dfrac{42}{6^2}} = \dfrac{1}{6}\sqrt{42}$

Example: Simplify $10\sqrt{\dfrac{3}{5}}$

▶ $\quad 10\sqrt{\dfrac{3}{5}} = 10\sqrt{\dfrac{3}{5} \cdot \dfrac{5}{5}} = 10\sqrt{\dfrac{15}{5^2}} = \dfrac{10}{5}\sqrt{15}$

$= 2\sqrt{15}$

Example: Simplify $8\sqrt[3]{\dfrac{3}{4}}$

▶ $\quad 8\sqrt[3]{\dfrac{3}{4}} = 8\sqrt[3]{\dfrac{3}{4} \cdot \dfrac{2}{2}} = 8\sqrt[3]{\dfrac{6}{8}} = \dfrac{8}{2}\sqrt[3]{6} = 4\sqrt[3]{6}$

Example: Simplify $\dfrac{2x}{y}\sqrt{\dfrac{5y}{27x}}$

▶ $\quad \dfrac{2x}{y}\sqrt{\dfrac{5y}{27x} \cdot \dfrac{3x}{3x}} = \dfrac{2x}{y}\sqrt{\dfrac{15xy}{81x^2}} = \dfrac{2x}{y} \cdot \dfrac{1}{9x}\sqrt{15xy}$

$= \dfrac{2}{9y}\sqrt{15xy}$

Example: Simplify $\sqrt{\dfrac{18(a-b)}{5(a+b)}}$

▶ $\qquad \sqrt{\dfrac{18(a-b)}{5(a+b)}} = \sqrt{\dfrac{9 \cdot 2(a-b)}{5(a+b)} \cdot \dfrac{5(a+b)}{5(a+b)}}$

$= \sqrt{\dfrac{3^2 \cdot 2 \cdot 5(a-b)(a+b)}{5^2(a+b)^2}} = \dfrac{3}{5(a+b)}\sqrt{10(a-b)(a+b)}$

EXERCISE 7

Express in simplest form:

1. $\sqrt{\dfrac{3}{5}}$ $\qquad\qquad\qquad$ 3. $\sqrt{\dfrac{1}{2}}$

2. $\sqrt{\dfrac{7}{4}}$ $\qquad\qquad\qquad$ 4. $\sqrt{\dfrac{2}{3}}$

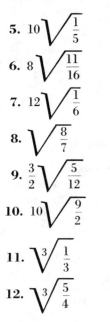

5. $10\sqrt{\dfrac{1}{5}}$

6. $8\sqrt{\dfrac{11}{16}}$

7. $12\sqrt{\dfrac{1}{6}}$

8. $\sqrt{\dfrac{8}{7}}$

9. $\dfrac{3}{2}\sqrt{\dfrac{5}{12}}$

10. $10\sqrt{\dfrac{9}{2}}$

11. $\sqrt[3]{\dfrac{1}{3}}$

12. $\sqrt[3]{\dfrac{5}{4}}$

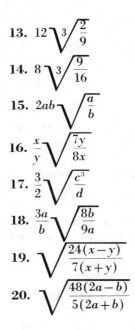

13. $12\sqrt[3]{\dfrac{2}{9}}$

14. $8\sqrt[3]{\dfrac{9}{16}}$

15. $2ab\sqrt{\dfrac{a}{b}}$

16. $\dfrac{x}{y}\sqrt{\dfrac{7y}{8x}}$

17. $\dfrac{3}{2}\sqrt{\dfrac{c^3}{d}}$

18. $\dfrac{3a}{b}\sqrt{\dfrac{8b}{9a}}$

19. $\sqrt{\dfrac{24(x-y)}{7(x+y)}}$

20. $\sqrt{\dfrac{48(2a-b)}{5(2a+b)}}$

Reduction of Indices

In cases where the exponent of the quantity under the radical sign and the index of the radical sign have a common factor, this factor may be removed and the radical simplified.

Example: Simplify $\sqrt[6]{x^3}$
$$\sqrt[6]{x^3} = (x^3)^{1/6} = x^{3/6} = x^{1/2} = \sqrt{x}$$

Example: Simplify $\sqrt[4]{49x^2y^4}$
$$\sqrt[4]{49x^2y^4} = (49x^2y^4)^{1/4} = (7^2x^2y^4)^{1/4}$$
$$= 7^{2/4}x^{2/4}y^{4/4} = 7^{1/2}x^{1/2}y = y\sqrt{7x}$$

Example: Simplify $\sqrt[9]{27a^3b^6}$
$$\sqrt[9]{27a^3b^6} = (27a^3b^6)^{1/9} = (3^3a^3b^6)^{1/9}$$
$$= 3^{3/9}a^{3/9}b^{6/9} = 3^{1/3}a^{1/3}b^{2/3} = \sqrt[3]{3ab^2}$$

Example: Simplify $\sqrt[6]{x^2-2xy+y^2}$
$$\sqrt[6]{x^2-2xy+y^2} = \sqrt[6]{(x-y)^2} = (x-y)^{2/6} = (x-y)^{1/3}$$
$$= \sqrt[3]{(x-y)}$$

EXERCISE 8

Simplify the following:

1. $\sqrt[4]{4}$

2. $\sqrt[6]{8}$

3. $\sqrt[4]{a^2 b^2}$

4. $\sqrt[8]{y^4}$

5. $\sqrt[6]{9c^2 d^2}$

6. $\sqrt[9]{8x^3 y^6}$

7. $\sqrt[10]{32b^5}$

8. $\sqrt[4]{81c^2 d^6}$

9. $\sqrt[12]{16a^4 b^8}$

10. $\sqrt[20]{243x^5}$

11. $\sqrt[6]{125a^3 b^3}$

12. $\sqrt[6]{343x^3 y^3}$

Transforming Radicals to Radicals of the Same Index

If it is desired to compare the magnitudes of two radicals it is usually necessary to transform the radicals to radicals which have the same index. For example, it is not feasible to compare $\sqrt[3]{15}$ with $\sqrt{6}$ as to magnitude. If these radicals are transformed to radicals of index 6 the comparison may be made directly.

Example: Express $\sqrt[3]{15}$ and $\sqrt{6}$ as radicals of index 6 and compare their magnitudes.

▶ $$\sqrt[3]{15} = (15)^{1/3} = (15)^{2/6} = \sqrt[6]{15^2} = \sqrt[6]{225}$$
$$\sqrt{6} = (6)^{1/2} = (6)^{3/6} = \sqrt[6]{6^3} = \sqrt[6]{216}$$
Therefore, $\sqrt[3]{15}$ is greater than $\sqrt{6}$.

Example: Express $6\sqrt[4]{3}$ and $3\sqrt{7}$ as radicals of the same index and compare their magnitudes.

▶ Since $6 = \sqrt[4]{(6)^4}$, $6\sqrt[4]{3} = \sqrt[4]{3(6)^4}$
$= \sqrt[4]{3(1296)} = \sqrt[4]{3,888}$
Since $3 = \sqrt{9}$, $3\sqrt{7} = \sqrt{7 \cdot 9} = \sqrt{63} = (63)^{1/2} = (63)^{2/4}$
$= \sqrt[4]{(63)^2} = \sqrt[4]{3,969}$
Therefore, $3\sqrt{7}$ is greater than $6\sqrt[4]{3}$.

Example: Arrange the following radicals in order of magnitude $\sqrt{5}$, $\sqrt[3]{10}$, $2\sqrt[6]{2}$

▶ In order to compare the magnitudes of these radicals they will be expressed as radicals of index 6.

$$\sqrt{5} = 5^{1/2} = 5^{3/6} = \sqrt[6]{5^3} = \sqrt[6]{125}$$
$$\sqrt[3]{10} = 10^{1/3} = 10^{2/6} = \sqrt[6]{10^2} = \sqrt[6]{100}$$
$$2\sqrt[6]{2} = \sqrt[6]{2(2)^6} = \sqrt[6]{2 \cdot 64} = \sqrt[6]{128}$$
Therefore, $2\sqrt[6]{2} > \sqrt{5} > \sqrt[3]{10}$

EXERCISE 9

1. Express as a radical of index 8, $\sqrt{2}$.
2. Express as a radical of index 6, $\sqrt[3]{7}$.
3. Express as a radical of index 9, $\sqrt[3]{4}$.
4. Express as a radical of index 12, $\sqrt[6]{5}$.
5. Express as a radical of index 4, $5\sqrt{3}$.
6. Express as a radical of index 6, $2\sqrt{3}$.
7. Compare the magnitudes of the following sets of radicals.
   (a) $\sqrt{2}$, $\sqrt[3]{3}$ (b) $\sqrt[3]{5}$, $\sqrt[4]{7}$ (c) $2\sqrt[3]{3}$, $\sqrt{8}$ (d) $2\sqrt{5}$, $2\sqrt[3]{11}$
   (e) $\sqrt{11}$, $3\sqrt[6]{2}$, $2\sqrt[3]{3}$ (f) $\sqrt{3}$, $2\sqrt[3]{2}$, $\sqrt[4]{10}$

OPERATIONS WITH RADICALS

Addition and Subtraction of Radicals

The addition and subtraction of similar radicals may be performed like the addition and subtraction of similar algebraic terms i.e. by taking the algebraic sum of the coefficients of the terms.

Example: $3\sqrt{2} + 5\sqrt{2} - 6\sqrt{2} = 2\sqrt{2}$
If radicals are not similar it may be feasible to add or subtract them by transforming them into similar radicals.

Example: $\sqrt{12} - 2\sqrt{75} = \sqrt{4\cdot 3} - 2\sqrt{25\cdot 3} = 2\sqrt{3}$
$- 2(5)\sqrt{3} = 2\sqrt{3} - 10\sqrt{3} = -8\sqrt{3}$
If radicals cannot be made similar their addition or subtraction may only be indicated.

Example: Simplify $3\sqrt{8} - \sqrt{32} + 2\sqrt{72}$

$$3\sqrt{8} = 3\sqrt{4\cdot 2} = 3\cdot 2\sqrt{2} = 6\sqrt{2}$$
$$-\sqrt{32} = -\sqrt{16\cdot 2} = -4\sqrt{2}$$
$$2\sqrt{72} = 2\sqrt{36\cdot 2} = 2\cdot 6\sqrt{2} = 12\sqrt{2}$$
Therefore, $3\sqrt{8} - \sqrt{32} + 2\sqrt{72} = 6\sqrt{2} - 4\sqrt{2} + 12\sqrt{2}$
$= 14\sqrt{2}$.

Example: Simplify $3\sqrt{150} - \frac{7}{3}\sqrt{54} + 5\sqrt{24}$

▶ $3\sqrt{150} = 3\sqrt{25 \cdot 6} = 3 \cdot 5\sqrt{6} = 15\sqrt{6}$

$-\dfrac{7}{3}\sqrt{54} = -\dfrac{7}{3}\sqrt{9 \cdot 6} = -\dfrac{7}{3} \cdot 3\sqrt{6} = -7\sqrt{6}$

$5\sqrt{24} = 5\sqrt{4 \cdot 6} = 5 \cdot 2\sqrt{6} = 10\sqrt{6}$

The result is $18\sqrt{6}.$

Example: Simplify $2\sqrt[3]{16} - \sqrt[3]{54} + \sqrt[3]{250}$

▶ $2\sqrt[3]{16} = 2\sqrt[3]{8 \cdot 2} = 2 \cdot 2\sqrt[3]{2} = 4\sqrt[3]{2}$

$-\sqrt[3]{54} = -\sqrt[3]{27 \cdot 2} = -3\sqrt[3]{2}$

$\sqrt[3]{250} = \sqrt[3]{125 \cdot 2} = 5\sqrt[3]{2}$

The result is $6\sqrt[3]{2}.$

Example: Simplify $5\sqrt[3]{24} - \dfrac{7}{2}\sqrt[3]{192} + 2\sqrt[3]{375}$

▶ $5\sqrt[3]{24} = 5\sqrt[3]{8 \cdot 3} = 5 \cdot 2\sqrt[3]{3} = 10\sqrt[3]{3}$

$-\dfrac{7}{2}\sqrt[3]{192} = -\dfrac{7}{2}\sqrt[3]{64 \cdot 3} = -\dfrac{7}{2} \cdot 4\sqrt[3]{3} = -14\sqrt[3]{3}$

$2\sqrt[3]{375} = 2\sqrt[3]{125 \cdot 3} = 2 \cdot 5\sqrt[3]{3} = 10\sqrt[3]{3}$

The result is $6\sqrt[3]{3}.$

Example: Simplify $2\sqrt{28} + \sqrt[3]{40} - 3\sqrt[3]{320} + 4\sqrt{63}$

▶ $2\sqrt{28} = 2\sqrt{4 \cdot 7} = 2 \cdot 2\sqrt{7} = 4\sqrt{7}$

$\sqrt[3]{40} = \sqrt[3]{8 \cdot 5} = 2\sqrt[3]{5}$

$-3\sqrt[3]{320} = -3\sqrt[3]{64 \cdot 5} = -3 \cdot 4\sqrt[3]{5} = -12\sqrt[3]{5}$

$4\sqrt{63} = 4\sqrt{9 \cdot 7} = 4 \cdot 3\sqrt{7} = 12\sqrt{7}$

The result is the sum of $4\sqrt{7} + 2\sqrt[3]{5} - 12\sqrt[3]{5} + 12\sqrt{7}.$
Collecting similar terms, $16\sqrt{7} - 10\sqrt[3]{5}.$

Example: Simplify

$$16\sqrt{\dfrac{5}{8}} + \dfrac{2}{5}\sqrt{40} + 6\sqrt{\dfrac{5}{2}} - 4\sqrt{\dfrac{125}{8}}$$

▶ $16\sqrt{\dfrac{5}{8}} = 16\sqrt{\dfrac{5}{8} \cdot \dfrac{2}{2}} = 16\sqrt{\dfrac{10}{4^2}} = \dfrac{16}{4}\sqrt{10}$

$= 4\sqrt{10}$

$\dfrac{2}{5}\sqrt{40} = \dfrac{2}{5}\sqrt{4 \cdot 10} = \dfrac{2}{5} \cdot 2\sqrt{10} = \dfrac{4}{5}\sqrt{10}$

$$6\sqrt{\frac{5}{2}} = 6\sqrt{\frac{5}{2} \cdot \frac{2}{2}} = 6\sqrt{\frac{10}{2^2}} = \frac{6}{2}\sqrt{10} = 3\sqrt{10}$$

$$-4\sqrt{\frac{125}{8}} = -4\sqrt{\frac{25 \cdot 5}{4 \cdot 2}} = -4 \cdot \frac{5}{2}\sqrt{\frac{5}{2}} = -10\sqrt{\frac{5}{2}}$$

$$= -10\sqrt{\frac{5}{2} \cdot \frac{2}{2}} = -5\sqrt{10}$$

The result is $\frac{14}{5}\sqrt{10}$

Example: Simplify $7y\sqrt{\dfrac{x^3}{y}} + 2x^2\sqrt{\dfrac{y}{x}} - \dfrac{5x}{y}\sqrt{xy^3}$

▶ $7y\sqrt{\dfrac{x^3}{y}} = 7y\sqrt{\dfrac{x^2 x \cdot y}{y\ \ y}} = 7y\sqrt{\dfrac{x^2 xy}{y^2}} = 7x\sqrt{xy}$

$2x^2\sqrt{\dfrac{y}{x}} = 2x^2\sqrt{\dfrac{y}{x} \cdot \dfrac{x}{x}} = 2x^2\sqrt{\dfrac{xy}{x^2}} = 2x\sqrt{xy}$

$-\dfrac{5x}{y}\sqrt{xy^3} = -\dfrac{5x}{y}\sqrt{x \cdot y^2 \cdot y} = -\dfrac{5x}{y} \cdot y\sqrt{xy} = -5x\sqrt{xy}$

The result is $4x\sqrt{xy}$.

EXERCISE 10

Simplify the following:

1. $2\sqrt{32} - 5\sqrt{8} + 2\sqrt{50}$
2. $3\sqrt{54} - 2\sqrt{24} - \sqrt{96}$
3. $2\sqrt{28} - 3\sqrt{175} + 4\sqrt{63}$
4. $\dfrac{3}{2}\sqrt{20} - \dfrac{7}{2}\sqrt{80} + \dfrac{3}{4}\sqrt{320}$
5. $2\sqrt{147} - \dfrac{1}{2}\sqrt{48} - \dfrac{4}{5}\sqrt{75}$
6. $2\sqrt[3]{81} - 5\sqrt[3]{24} - \sqrt[3]{375}$
7. $4\sqrt[3]{40} - 5\sqrt[3]{320} + 2\sqrt[3]{625}$
8. $2\sqrt[3]{56} + 3\sqrt{44} - 2\sqrt{99} + 3\sqrt[3]{189}$
9. $\dfrac{1}{2}\sqrt{112} - \dfrac{3}{4}\sqrt[3]{192} + \sqrt{343} - \dfrac{2}{3}\sqrt[3]{648}$
10. $12\sqrt{\dfrac{1}{3}} - 4\sqrt{\dfrac{1}{12}} + 18\sqrt{\dfrac{1}{27}}$

11. $9\sqrt[3]{\dfrac{1}{9}} - \sqrt[3]{375} + 8\sqrt[3]{\dfrac{1}{4}}$

12. $12\sqrt{\dfrac{2}{3}} - 14\sqrt{\dfrac{3}{2}} + 24\sqrt{\dfrac{1}{6}}$

13. $10\sqrt{\dfrac{4}{5}} - 15\sqrt{\dfrac{16}{5}} + 3\sqrt{45}$

14. $2a\sqrt{\dfrac{b^3}{a}} - 2b^2\sqrt{\dfrac{a}{b}} - \dfrac{3b}{a}\sqrt{a^3b}$

15. $6\sqrt{\dfrac{xy}{3}} - 9\sqrt{\dfrac{xy}{27}} + 54\sqrt{\dfrac{xy}{243}}$

Multiplication of Radicals

Radicals of the same order are multiplied by multiplying the product of the coefficients by the product of the irrational factors provided that the irrational factors are real numbers. In order to avoid large numbers it is desirable to simplify each radical before performing the multiplication.

Radicals of different orders are first transformed into radicals of the same order and then multiplied.

Example: Multiply $4\sqrt{3} \times 2\sqrt{6}$

▶ $(4\sqrt{3})(2\sqrt{6}) = (4 \times 2\sqrt{3} \cdot \sqrt{6}) = 8\sqrt{18} = 8\sqrt{9 \cdot 2}$
$= 24\sqrt{2}$

Example: Multiply $\frac{2}{3}\sqrt{35}$ by $\frac{3}{4}\sqrt{28}$

▶ $(\frac{2}{3}\sqrt{35})(\frac{3}{4}\sqrt{28}) = (\frac{2}{3} \times \frac{3}{4})\sqrt{35 \times 28}$
$= \frac{1}{2}\sqrt{7 \times 5 \times 7 \times 4} = \frac{1}{2}\sqrt{7^2 \cdot 2^2 \cdot 5} = \dfrac{7 \times 2}{2}\sqrt{5} = 7\sqrt{5}$

Example: Multiply $\dfrac{2}{3}\sqrt{\dfrac{5}{6}}$ by $\dfrac{1}{5}\sqrt{\dfrac{9}{10}}$

▶ $\left(\dfrac{2}{3}\sqrt{\dfrac{5}{6}}\right)\left(\dfrac{1}{5}\sqrt{\dfrac{9}{10}}\right) = \left(\dfrac{2}{3} \cdot \dfrac{1}{5}\right)\sqrt{\dfrac{5}{6} \cdot \dfrac{9}{10}}$
$= \dfrac{2}{15}\sqrt{\dfrac{3}{4}} = \dfrac{2}{15} \cdot \dfrac{1}{2}\sqrt{3} = \dfrac{1}{15}\sqrt{3}$

Example: Multiply $2\sqrt[3]{4}$ by $5\sqrt[3]{6}$

▶ $(2\sqrt[3]{4})(5\sqrt[3]{6}) = (2 \times 5\sqrt[3]{4 \times 6}) = 10\sqrt[3]{2^3 \cdot 3}$
$= 20\sqrt[3]{3}$

Example: Multiply $\sqrt[3]{3a^2} \times \sqrt[3]{6a^2} \times \sqrt[3]{15a^2}$

▶ $\sqrt[3]{3a^2} \times \sqrt[3]{6a^2} \times \sqrt[3]{15a^2} = \sqrt[3]{3 \cdot 6 \cdot 15a^6}$
$= 3a^2\sqrt[3]{10}$

Example: Multiply $8\sqrt[3]{\dfrac{2}{3}}$ by $\dfrac{7}{2}\sqrt[3]{\dfrac{4}{3}}$

▶ $\left(8\sqrt[3]{\dfrac{2}{3}}\right)\left(\dfrac{7}{2}\sqrt[3]{\dfrac{4}{3}}\right) = \left(8 \times \dfrac{7}{2}\right)\sqrt[3]{\dfrac{2}{3} \cdot \dfrac{4}{3}}$

$= 28\sqrt[3]{\dfrac{2^3}{3^2} \cdot \dfrac{3}{3}} = \dfrac{56}{3}\sqrt[3]{3}$

Example: Multiply $2\sqrt{6}$ by $\sqrt[3]{4}$

▶ $(2\sqrt{6})(\sqrt[3]{4}) = (\sqrt{24} \times \sqrt[3]{4}) = (24^{1/2})(4^{1/3})$
$= 24^{3/6} \cdot 4^{2/6} = \sqrt[6]{(24)^3(4)^2} = \sqrt[6]{(2^3 \cdot 3)^3(2^2)^2} = \sqrt[6]{2^9 \cdot 3^3 \cdot 2^4}$
$= 4\sqrt[6]{54}$

Example: Multiply $\sqrt[3]{16}$ by $\sqrt[4]{12}$

▶ $\sqrt[3]{16} \times \sqrt[4]{12} = (16)^{1/3}(12)^{1/4} = (16)^{4/12} \cdot (12)^{3/12}$
$= \sqrt[12]{(16)^4(12)^3} = \sqrt[12]{(2^4)^4(2^2 \cdot 3)^3} = \sqrt[12]{2^{16} \cdot 2^6 \cdot 3^3} = \sqrt[12]{2^{22} \cdot 3^3}$
$= 2\sqrt[12]{2^{10} \cdot 3^3} = 2\sqrt[12]{27,648}$

Example: Multiply $(2\sqrt{3} + 3\sqrt{5})^2$

▶
$$\begin{array}{l} 2\sqrt{3} + \ \ 3\sqrt{5} \\ 2\sqrt{3} + \ \ 3\sqrt{5} \\ \hline 4 \cdot 3 + \ \ 6\sqrt{3 \cdot 5} \\ + \ \ 6\sqrt{3 \cdot 5} + \ \ 9 \cdot 5 \\ \hline 12 \ \ \ \ + \ 12\sqrt{15} \ \ + \ 45 \ \ \ = 57 + 12\sqrt{15} \end{array}$$

Example: Multiply $(5\sqrt{2} - 2\sqrt{3})$ by $(4\sqrt{2} + 3\sqrt{3})$

▶
$$\begin{array}{r} 5\sqrt{2} - 2\sqrt{3} \\ 4\sqrt{2} + 3\sqrt{3} \\ \hline 20\cdot 2 - 8\sqrt{3\cdot 2} \\ + 15\sqrt{3\cdot 2} - 6\cdot 3 \\ \hline 40 \quad + 7\sqrt{6} \quad - 18 \quad = 22 + 7\sqrt{6} \end{array}$$

Example: Multiply $(2\sqrt{7} + 3\sqrt{5})$ by $(2\sqrt{7} - 3\sqrt{5})$

▶
$$\begin{array}{r} 2\sqrt{7} + 3\sqrt{5} \\ 2\sqrt{7} - 3\sqrt{5} \\ \hline 4\cdot 7 + 6\sqrt{5\cdot 7} \\ - 6\sqrt{5\cdot 7} - 9\cdot 5 \\ \hline 28 \quad\quad - 45 \quad = -17 \end{array}$$

Example: Multiply $(\sqrt{a + b} + \sqrt{c})$ by $(\sqrt{a + b} - \sqrt{c})$

▶
$$\begin{array}{r} \sqrt{a + b} \quad + \sqrt{c} \\ \sqrt{a + b} \quad - \sqrt{c} \\ \hline \sqrt{(a + b)^2} + \sqrt{c(a + b)} \\ - \sqrt{c(a + b)} - \sqrt{c^2} \\ \hline a + b \quad\quad - c \quad = a + b - c \end{array}$$

Example: Multiply $(b + \sqrt{1 - b^2})^2$

▶
$$\begin{array}{r} b + \sqrt{1 - b^2} \\ b + \sqrt{1 - b^2} \\ \hline b^2 + b\sqrt{1 - b^2} \\ + b\sqrt{1 - b^2} + \sqrt{(1 - b^2)^2} \\ \hline b^2 + 2b\sqrt{1 - b^2} + 1 - b^2 = 2b\sqrt{1 - b^2} + 1 \end{array}$$

Example: Multiply $(\sqrt{x + y} + \sqrt{x - y})^2$

▶
$$\begin{array}{r} \sqrt{x + y} \quad + \sqrt{x - y} \\ \sqrt{x + y} \quad + \sqrt{x - y} \\ \hline \sqrt{(x + y)^2} + \sqrt{(x + y)(x - y)} \\ + \sqrt{(x + y)(x - y)} + \sqrt{(x - y)^2} \\ \hline x + y \quad + 2\sqrt{x^2 - y^2} + x - y = 2x + 2\sqrt{x^2 - y^2} \end{array}$$

Example: Multiply $(\sqrt{3} + \sqrt{5} + \sqrt{7})^2$

▶

$$
\begin{array}{l}
\sqrt{3} + \ \ \sqrt{5} + \ \ \sqrt{7} \\
\underline{\sqrt{3} + \ \ \sqrt{5} + \ \ \sqrt{7}} \\
3 + \ \ \sqrt{15} + \ \ \sqrt{21} \\
\ \ \ \ + \ \ \sqrt{15} \ \ \ \ \ \ \ \ \ \ \ \ + 5 + \ \ \sqrt{35} \\
\ \ \ \ \ \ \ \ \ \ \ \ \ \ + \ \ \sqrt{21} \ \ \ \ \ + \ \ \sqrt{35} + 7 \\
\hline
15 + 2\sqrt{15} + 2\sqrt{21} \ \ \ \ \ + 2\sqrt{35}
\end{array}
$$

Example: Multiply $(\sqrt{2} - \sqrt{3} + \sqrt{5})(\sqrt{2} + \sqrt{3} - \sqrt{5})$

▶

$$
\begin{array}{l}
\sqrt{2} - \sqrt{3} + \sqrt{5} \\
\underline{\sqrt{2} + \sqrt{3} - \sqrt{5}} \\
2 - \sqrt{6} + \sqrt{10} \\
\ \ \ \ + \sqrt{6} \ \ \ \ \ \ \ \ - 3 + \ \ \sqrt{15} \\
\ \ \ \ \ \ \ \ - \sqrt{10} \ \ \ \ + \ \ \sqrt{15} - 5 \\
\hline
-6 \ \ \ \ \ \ \ \ \ \ \ \ \ \ \ \ + 2\sqrt{15}
\end{array}
$$

EXERCISE 11

Perform the following multiplications:

1. $4\sqrt{10}$ by $5\sqrt{6}$

2. $\dfrac{3}{2}\sqrt{12}$ by $8\sqrt{2}$

3. $12\sqrt{15}$ by $\dfrac{2}{3}\sqrt{6}$

4. $4\sqrt{27}$ by $\dfrac{1}{12}\sqrt{96}$

5. $\dfrac{2}{3}\sqrt{\dfrac{1}{2}}$ by $7\sqrt{54}$

6. $\dfrac{4}{5}\sqrt{10}$ by $\dfrac{3}{2}\sqrt{24}$

7. $\dfrac{5}{4}\sqrt{\dfrac{2}{5}}$ by $\dfrac{10}{3}\sqrt{\dfrac{1}{2}}$

8. $\dfrac{6}{5}\sqrt{\dfrac{3}{8}}$ by $\dfrac{2}{3}\sqrt{\dfrac{7}{9}}$

9. $\sqrt[3]{9}$ by $\sqrt[3]{6}$

10. $3\sqrt[3]{2}$ by $4\sqrt[3]{54}$

11. $\dfrac{1}{3}\sqrt[3]{\dfrac{3}{4}}$ by $\dfrac{1}{2}\sqrt[3]{\dfrac{1}{12}}$

12. $3\sqrt{2}$ by $\sqrt[3]{6}$

13. $2\sqrt{10}$ by $\sqrt[3]{2}$

14. $3\sqrt{6}$ by $\sqrt[3]{3}$

15. $2\sqrt{6}$ by $\sqrt[4]{3}$

16. $(3\sqrt{2} + 2\sqrt{3})^2$

17. $(2\sqrt{6} - 4\sqrt{10})^2$

18. $(3\sqrt{2} - 5\sqrt{3})$ by $(6\sqrt{2} + 7\sqrt{3})$

19. $(2\sqrt{5} - 3\sqrt{6})(4\sqrt{5} + 2\sqrt{6})$

20. $(\sqrt{2x + y} + \sqrt{z})(\sqrt{2x + y} - \sqrt{z})$

21. $(\sqrt{a + b} + \sqrt{a})(\sqrt{a + b} - \sqrt{a})$

22. $(c + \sqrt{4 - c^2})^2$

23. $(\sqrt{2a + b} + \sqrt{2a - b})^2$

24. $(\sqrt{4 - x^2} - \sqrt{4 + x^2})^2$

25. $\left(\sqrt{y} - \dfrac{1}{\sqrt{y}} \right)^2$

26. $(\sqrt{x} + \sqrt{y} + \sqrt{z})^2$

27. $(\sqrt{2} + \sqrt{5} + \sqrt{7})^2$

28. $(\sqrt{10} + \sqrt{3} - \sqrt{5})(\sqrt{10} - \sqrt{3} + \sqrt{5})$

29. $(\sqrt{3} - \sqrt{5} - \sqrt{6})^2$

30. $(\sqrt{2} - \sqrt{7} - \sqrt{11})(\sqrt{2} + \sqrt{7} + \sqrt{11})$

Division of Radicals and Rationalization of Denominators

Monomial radicals of the same order may be divided by multiplying the quotient of the coefficients by the quotient of the radicands provided that the radicands are real numbers.

Example: Divide $\sqrt{5}$ by $\sqrt{2}$

▶ $$\frac{\sqrt{5}}{\sqrt{2}} = \sqrt{\frac{5}{2}} = \sqrt{\frac{5}{2} \cdot \frac{2}{2}} = \frac{1}{2}\sqrt{10}$$

Example: Divide $2\sqrt{24}$ by $4\sqrt{3}$

▶ $$\frac{2\sqrt{24}}{4\sqrt{3}} = \frac{1}{2}\sqrt{8} = \frac{1}{2}\sqrt{4 \cdot 2} = \frac{2}{2}\sqrt{2} = \sqrt{2}$$

Example: Divide $8\sqrt[3]{15}$ by $2\sqrt[3]{5}$

▶ $$\frac{8\sqrt[3]{15}}{2\sqrt[3]{5}} = 4\sqrt[3]{3}$$

Example: Divide $(12\sqrt{15} + 20\sqrt{27})$ by $4\sqrt{3}$

▶ $$\frac{12\sqrt{15} + 20\sqrt{27}}{4\sqrt{3}} = 3\sqrt{5} + 5\sqrt{9} = 3\sqrt{5} + 15$$

Example: Divide $\sqrt{a^2 - b^2}$ by $\sqrt{a - b}$

▶ $$\frac{\sqrt{a^2 - b^2}}{\sqrt{a - b}} = \frac{\sqrt{(a + b)(a - b)}}{\sqrt{a - b}} = \sqrt{a + b}$$

If the denominator of a fraction is a radical, the fraction may be simplified by transforming the fraction so that the denominator becomes rational. For example, the value of the fraction $\frac{1}{\sqrt{2}}$ is equal to $\frac{1}{1.4142\ldots}$, an awkward calculation to make. However, if the numerator and denominator of the fraction $\frac{1}{\sqrt{2}}$ are both multiplied by $\sqrt{2}$ we have $\frac{1}{\sqrt{2}} \cdot \frac{\sqrt{2}}{\sqrt{2}} = \frac{\sqrt{2}}{\sqrt{4}} = \frac{\sqrt{2}}{2} = \frac{1.4142\ldots}{2} = .7071\ldots$ This process is called *rationalizing the denominator.*

Example: Rationalize the denominator of the fraction $\frac{8}{\sqrt{6}}$.

▶ $$\frac{8}{\sqrt{6}} = \frac{8}{\sqrt{6}} \cdot \frac{\sqrt{6}}{\sqrt{6}} = \frac{8\sqrt{6}}{6} = \frac{4}{3}\sqrt{6}$$

Example: Rationalize the denominator of the fraction $\frac{4\sqrt{5} - 2\sqrt{3} + 7\sqrt{2}}{\sqrt{3}}$

▶ $$\frac{4\sqrt{5} - 2\sqrt{3} + 7\sqrt{6}}{\sqrt{3}} = \frac{4\sqrt{5} - 2\sqrt{3} + 7\sqrt{6}}{\sqrt{3}} \cdot \frac{\sqrt{3}}{\sqrt{3}}$$

$$= \frac{4\sqrt{15} - 2(3) + 7\sqrt{18}}{3} = \frac{4\sqrt{15} - 6 + 21\sqrt{2}}{3}$$

Example: Rationalize the denominator of the fraction $\frac{5\sqrt{2}}{\sqrt[3]{3}}$.

▶ $$\frac{5\sqrt{2}}{\sqrt[3]{3}} = \frac{5\sqrt{2}}{\sqrt[3]{3}} \cdot \frac{\sqrt[3]{9}}{\sqrt[3]{9}} = \frac{5(2^{1/2})(9^{1/3})}{(3^{1/3})(9^{1/3})} = \frac{5(2^{3/6})(9^{2/6})}{(27)^{1/3}}$$

$$= \frac{5\sqrt[6]{2^3 \cdot 9^2}}{3} = \frac{5\sqrt[6]{8 \cdot 81}}{3} = \frac{5\sqrt[6]{648}}{3}$$

Two binomial quadratic radicals are called *conjugate* if they differ only in the sign which connects their terms. For example, $3 + \sqrt{5}$ and $3 - \sqrt{5}$ are conjugates, as are $3\sqrt{7} - \sqrt{2}$ and $3\sqrt{7} + \sqrt{2}$.

To rationalize the denominator of a fraction whose denominator is a binomial quadratic radical, multiply the numerator and denominator of the fraction by the conjugate of the denominator.

Example: Rationalize the denominator of $\dfrac{5}{2 + 3\sqrt{6}}$.

▶ $\dfrac{5}{2 + 3\sqrt{6}} = \dfrac{5}{2 + 3\sqrt{6}} \cdot \dfrac{2 - 3\sqrt{6}}{2 - 3\sqrt{6}} = \dfrac{5(2 - 3\sqrt{6})}{4 - 9(6)}$

$= \dfrac{5(2 - 3\sqrt{6})}{4 - 54} = \dfrac{5(2 - 3\sqrt{6})}{-50} = \dfrac{2 - 3\sqrt{6}}{-10}$

If we multiply the numerator and denominator of this result by -1 we obtain $\dfrac{3\sqrt{6} - 2}{10}$.

Example: Rationalize the denominator of $\dfrac{3\sqrt{15}}{10 - 3\sqrt{5}}$.

▶ $\dfrac{3\sqrt{15}}{10 - 3\sqrt{5}} = \dfrac{3\sqrt{15}}{10 - 3\sqrt{5}} \cdot \dfrac{10 + 3\sqrt{5}}{10 + 3\sqrt{5}} = \dfrac{30\sqrt{15} + 9\sqrt{75}}{100 - 9(5)}$

$= \dfrac{30\sqrt{15} + 45\sqrt{3}}{55} = \dfrac{5(6\sqrt{15} + 9\sqrt{3})}{55} = \dfrac{6\sqrt{15} + 9\sqrt{3}}{11}$

Example: Rationalize the denominator of the fraction $\dfrac{\sqrt{2xy}}{\sqrt{2xy} + y\sqrt{x}}$.

▶ $\dfrac{\sqrt{2xy}}{\sqrt{2xy} + y\sqrt{x}} = \dfrac{\sqrt{2xy}}{\sqrt{2xy} + y\sqrt{x}} \cdot \dfrac{\sqrt{2xy} - y\sqrt{x}}{\sqrt{2xy} - y\sqrt{x}}$

$= \dfrac{\sqrt{2xy}(\sqrt{2xy} - y\sqrt{x})}{2xy - y^2 x} = \dfrac{2xy - y\sqrt{2x^2 y}}{2xy - y^2 x}$

$= \dfrac{2xy - xy\sqrt{2y}}{2xy - y^2 x} = \dfrac{xy(2 - \sqrt{2y})}{xy(2 - y)} = \dfrac{2 - \sqrt{2y}}{2 - y}$

Example: Rationalize the denominator of the fraction

$$\frac{\sqrt{y + a} - \sqrt{y}}{\sqrt{y + a} + \sqrt{y}}$$

▶ $$\frac{\sqrt{y + a} - \sqrt{y}}{\sqrt{y + a} + \sqrt{y}} = \frac{\sqrt{y + a} - \sqrt{y}}{\sqrt{y + a} + \sqrt{y}} \cdot \frac{\sqrt{y + a} - \sqrt{y}}{\sqrt{y + a} - \sqrt{y}}$$

$$= \frac{y + a - 2\sqrt{y(y + a)} + y}{y + a - y} = \frac{2y + a - 2\sqrt{y^2 + ay}}{a}$$

Example: Rationalize the denominator of the fraction

$$\frac{2}{1 + \sqrt{3} - \sqrt{2}}.$$

▶ $$\frac{2}{1 + \sqrt{3} - \sqrt{2}} = \frac{2}{(1 + \sqrt{3}) - \sqrt{2}} \cdot \frac{(1 + \sqrt{3}) + \sqrt{2}}{(1 + \sqrt{3}) + \sqrt{2}}$$

$$= \frac{2(1 + \sqrt{3} + \sqrt{2})}{(1 + \sqrt{3})^2 - 2} = \frac{2(1 + \sqrt{3} + \sqrt{2})}{1 + 2\sqrt{3} + 3 - 2}$$

$$= \frac{2(1 + \sqrt{3} + \sqrt{2})}{2 + 2\sqrt{3}} = \frac{\cancel{2}(1 + \sqrt{3} + \sqrt{2})}{\cancel{2}(1 + \sqrt{3})} = \frac{1 + \sqrt{3} + \sqrt{2}}{1 + \sqrt{3}}$$

$$\frac{1 + \sqrt{3} + \sqrt{2}}{1 + \sqrt{3}} = \frac{1 + \sqrt{3} + \sqrt{2}}{1 + \sqrt{3}} \cdot \frac{1 - \sqrt{3}}{1 - \sqrt{3}}$$

$$= \frac{-2 + \sqrt{2} - \sqrt{6}}{-2} \text{ or } \frac{2 - \sqrt{2} + \sqrt{6}}{2}$$

Example: Rationalize the denominator of the fraction

$$\frac{\sqrt{2} + \sqrt{3}}{\sqrt{2} - \sqrt{3} - \sqrt{6}}.$$

▶ $$\frac{\sqrt{2} + \sqrt{3}}{\sqrt{2} - \sqrt{3} - \sqrt{6}} = \frac{\sqrt{2} + \sqrt{3}}{(\sqrt{2} - \sqrt{3}) - \sqrt{6}} \cdot$$

$$\frac{(\sqrt{2} - \sqrt{3}) + \sqrt{6}}{(\sqrt{2} - \sqrt{3}) + \sqrt{6}} = \frac{-1 + \sqrt{12} + \sqrt{18}}{-1 - 2\sqrt{6}}$$

$$= \frac{-1 + 2\sqrt{3} + 3\sqrt{2}}{-1 - 2\sqrt{6}}$$

$$\frac{-1 + 2\sqrt{3} + 3\sqrt{2}}{-1 - 2\sqrt{6}} = \frac{-1 + 2\sqrt{3} + 3\sqrt{2}}{-1 - 2\sqrt{6}} \cdot \frac{-1 + 2\sqrt{6}}{-1 + 2\sqrt{6}}$$

$$= \frac{1 + 10\sqrt{3} + 9\sqrt{2} - 2\sqrt{6}}{1 - 24} = \frac{1 + 10\sqrt{3} + 9\sqrt{2} - 2\sqrt{6}}{-23}$$

or $\dfrac{-1 - 10\sqrt{3} - 9\sqrt{2} + 2\sqrt{6}}{23}$

EXERCISE 12

Perform the following divisions:

1. $\sqrt{6} \div \sqrt{2}$
2. $\sqrt{10} \div \sqrt{3}$
3. $\sqrt{7} \div \sqrt{5}$
4. $12\sqrt{18} \div 4\sqrt{3}$
5. $48\sqrt{32} \div 6\sqrt{20}$
6. $54\sqrt{40} \div 9\sqrt{24}$
7. $12\sqrt[3]{48} \div 4\sqrt[3]{16}$
8. $36\sqrt[3]{50} \div 4\sqrt[3]{25}$
9. $(21\sqrt{32} + 42\sqrt{28}) \div 3\sqrt{2}$
10. $(16\sqrt{45} - 28\sqrt{51}) \div 4\sqrt{3}$
11. $\sqrt{x^2 - 4y^2} \div \sqrt{x + 2y}$
12. $\sqrt{9b^2 - 1} \div \sqrt{3b + 1}$

Rationalize the denominators of the following fractions

13. $\dfrac{9}{\sqrt{6}}$

14. $\dfrac{4}{\sqrt{2}}$

15. $\dfrac{5}{\sqrt{10}}$

16. $\dfrac{3}{\sqrt{32}}$

17. $\dfrac{3\sqrt{2} - 4\sqrt{5} + 8\sqrt{6}}{\sqrt{2}}$

18. $\dfrac{8\sqrt{10} - 3\sqrt{3} + 2\sqrt{6}}{\sqrt{5}}$

19. $\dfrac{8}{3 + \sqrt{2}}$

20. $\dfrac{12}{3 - \sqrt{5}}$

21. $\dfrac{1}{2 - \sqrt{2}}$

22. $\dfrac{4 - 5\sqrt{2}}{8 + 3\sqrt{2}}$

23. $\dfrac{2\sqrt{5} - 1}{2\sqrt{5} + 1}$

24. $\dfrac{2\sqrt{7} + 3}{3\sqrt{7} - 5}$

25. $\dfrac{\sqrt{6} + \sqrt{3}}{\sqrt{6} - \sqrt{3}}$

26. $\dfrac{3\sqrt{5} + 3\sqrt{2}}{2\sqrt{5} + \sqrt{2}}$

27. $\dfrac{\sqrt{7} - 3\sqrt{6}}{\sqrt{7} - 2\sqrt{3}}$

28. $\dfrac{\sqrt{6}}{2\sqrt{10} - \sqrt{5}}$

29. $\dfrac{x + \sqrt{x^2 + a^2}}{x - \sqrt{x^2 + a^2}}$

30. $\dfrac{\sqrt{a + b} + \sqrt{a - b}}{\sqrt{a + b} - \sqrt{a - b}}$

31. $\dfrac{\sqrt{x + y} + z}{\sqrt{x + y} - z}$

32. $\dfrac{b - \sqrt{a + b}}{b + \sqrt{a + b}}$

33. $\dfrac{5}{1 - \sqrt{2} + \sqrt{3}}$

34. $\dfrac{1 - \sqrt{2}}{\sqrt{6} + \sqrt{5} - \sqrt{3}}$

ANSWERS

EXERCISE 1

1. p^{11}

2. x^6

3. y^{3a}

4. z^{b+6}

5. $2^5 = 32$

6. a^{2n+3}

7. $5^5 = 3{,}125$

8. -343

9. b^{10}

10. $-125b^3$

11. $16x^{12}$

12. $9a^6$

13. $-8a^6b^9$

14. y^6

15. x^{2n+1}

16. $9^4 = 6{,}561$

17. x^{2p+3}

18. $\dfrac{y^5}{32}$

19. $\dfrac{-27b^6}{64}$

20. $-\dfrac{1}{8}x^6y^9$

21. a^{8x+12}

22. x^6y^{3a+6}

23. $\dfrac{9x^2}{16y^2}$

24. $\dfrac{a^x}{(2b)^x} = \dfrac{a^x}{2^x b^x}$

25. $\dfrac{81a^4}{b^{12}}$

26. $\dfrac{32c^{15}}{-a^5}$

27a. 2^4

  b. 2^{3a}

  c. 2^{10b}

  d. 2^{-4a}

28a. 3^3 **29a.** 5^4 **30a.** 7^{6a}
  b. 3^{2b} **b.** 5^{2b} **b.** 7^{11a}
  c. 3^{8a} **c.** 5^{9a}
  d. 3^{-2b} **d.** 5^{-a}

EXERCISE 2

1. 1

2. 9

3. 1

4. b

5. $\dfrac{1}{36}$

6. $\dfrac{1}{2a}$

7. -2

8. a^3

9. $1 + \dfrac{1}{b} = \dfrac{b+1}{b}$

10. $\dfrac{17}{18}$

11. 9

12. $\dfrac{1}{c^2} + 1 = \dfrac{1+c^2}{c^2}$

13. $\dfrac{1}{9b^2}$

14. 134

15. $\dfrac{17}{27}$

16. $\dfrac{167}{125}$

17. $\dfrac{1+a}{a}$

18. $\dfrac{2x-1}{x}$

19. $\dfrac{x+y}{xy}$

20. $\dfrac{1}{3a(a-3)}$

21. $\dfrac{a-b}{b^2}$

22. $\dfrac{1}{s^{15}t^4}$

23. $\dfrac{1}{x(x+y)}$

24. $\dfrac{1}{ab}$

EXERCISE 3

1. 3×10^{-5}

2. 1.8×10^{-10}

3. 6.5×10^9

4. 7.9×10^{-8}

5. 9.72×10^{16}

6. 4.5×10^{-12}

7. $.00035$

8. $68,400,000$

9. $8,600,000,000$

10. $.0000072$

11. $546,000$

12. $.000000098$

EXERCISE 4

1. $27\frac{1}{8}$
2. -5
3. $8\frac{2}{5}$
4. -72

5. 5
6. 2
7. $.29$
8. 4

9. 0
10. $\frac{31}{36}$
11. 21
12. $-15\frac{8}{9}$

EXERCISE 5

1. $3^{7/6}$
2. $2^{39/24}$
3. $2^{-5/12}$
4. x^{p+6}
5. $72x^2$
6. $12a^{1/2}$
7. $(x+y)^{5/4}$
8. $(2a-b)^{26/15}$
9. $20 \cdot 3^{17/12}$
10. 525
11. $(x+y)^{5/6}$
12. $(a+b)^{5/4}$
13. $a^2b^{7/3}$
14. $x^{37/12}y^{8/3}$
15. 5

16. $7^{1/8}$
17. $6 \cdot 10^7$
18. $3 \cdot 10^2$
19. $a^{5/2} \cdot b^{-11/4}$
20. $x \cdot y^{-2}$
21. $(a+b)^{x-6}$
22. $(t+v)^{2a+2b}$
23. $6x^{3/2}y^{1/3}z^3$
24. $2a^4b^{-2}c^{-1/6}$
25. $x^{1/2}$
26. $b^{-7/3}$
27. b^{y+1}
28. $x^3y^{1/2}$
29. x^8
30. a^{6n-3}

31. y^{8p-12}
32. z^{4a^2-1}
33. $(m+n)^{12}$
34. $(a+b)^{9/2}$
35. b^{-2x+3}
36. $\frac{125}{64}$
37. $\frac{27}{8}$
38. $(x+y)^8$
39. $x^{2/9}$
40. $y^{1/4}$
41. $x^{-1/2}$
42. $y^{-2/3}$
43. a^8b^{12}

44. x^6y^8
45. $128x^7$
46. $32y^{20}$
47. $256x^{12}y^{20}$
48. $-5a^{2/5}b^{3/2}$
49. $x^{n+2}y^{2n+1}z^{3n}$
50. $a^{-2n-1}b^{3n}c^n$
51. $\frac{a^3b^2}{z}\sqrt{a}$
52. $\frac{x^3y^3}{z^2}\sqrt[3]{y}$
53. $9 \cdot 10^4$
54. $8 \cdot 10^3$

55. $\frac{a^{3x}b^{2x}}{c^x}$
56. $\frac{x^{ap}y^{cp}}{z^{bp+p}}$
57. $\frac{125x^3}{27y^3}$
58. $\frac{32a^5}{243b^5}$
59. $\frac{a^{6n+3}}{b^{3n}}$
60. $\frac{125a^6}{343b^3}$
61. $\frac{4a^4}{9b^6}$

62. $\frac{2}{5}$
63. $\frac{8}{27}$
64. $\frac{8x^3}{3y^2}$
65. $\frac{3a^3}{4b^2}$
66. $\frac{2a^2}{3b^3}$
67. $3,823.03$
68. $3,605.2$

69. 7,000

70. 36

71. .04

72. 80

73. $\dfrac{1}{24}$

74. $3\frac{15}{64}$

75. $\dfrac{2}{9}$

76. 5

77. 3^{n^2-2n-5}

78. $a^{-11x/12}$

79. $x^{-2/a}y^{-11/3a}c^{-4/a}$

80. $\dfrac{4x}{3y}$

81. $\dfrac{5a^6}{9b^8}$

82. $a^{-10}b^{-32}$

83. $5^{-n-3} \cdot 7^{n-5}$

84. $14\frac{2}{15}$

85. 9

86. $\dfrac{4x^{3/2}}{3y^{7/6}}$

87. $a \cdot b^{-3/2}$

88. $\dfrac{x^{33/5}}{y^7}$

89. x^{-3}

90. $1 + x^{2/3}$

91. $\dfrac{1}{a^2 y}$

92. $\dfrac{b^3 y^9}{27 x^{3/4}}$

93. x^{-2n}

94. $\dfrac{a^2}{b^2 c^2}$

95. $\dfrac{4y^3}{9x^{8/3}}$

96. y^{n^2-2}

97. $a^{7/2}$

98. x^{n^2+1}

99. b

100. $\left(\dfrac{ab + 1}{ab - 1} \right)^n$

EXERCISE 6

1. $2\sqrt{2}$

2. $4\sqrt{3}$

3. $14\sqrt{2}$

4. $\frac{1}{2}\sqrt{5}$

5. $6a\sqrt{2b}$

6. $15x^2y\sqrt{3xy}$

7. $2\sqrt[3]{2}$

8. $14\sqrt[3]{3}$

9. $15\sqrt[3]{2}$

10. $\frac{7}{2}\sqrt[3]{2}$

11. $3y\sqrt[3]{9}$

12. $2xy^3\sqrt[3]{4}$

13. $6ab\sqrt[3]{ab^2}$

14. $-3d\sqrt[3]{2c^2d^2}$

15. $-4x^a\sqrt[3]{5}$

16. $-8y^{2n}\sqrt[3]{2}$

17. $5a^{-3}b^2\sqrt[3]{7b^2}$

18. $\dfrac{2y^3}{x}\sqrt[3]{7y}$

19. $(x - y)\sqrt{5}$

20. $(a - 2b)\sqrt{2}$

21. $x^6y^2\sqrt[4]{y}$

22. $a^2b^5\sqrt[n]{a}$

23. $x^3y^2\sqrt[m]{xy}$

24. $a^2b^4\sqrt[p]{5a^3}$

EXERCISE 7

1. $\frac{1}{5}\sqrt{15}$

2. $\frac{1}{2}\sqrt{7}$

3. $\frac{1}{2}\sqrt{2}$

4. $\frac{1}{3}\sqrt{6}$

5. $2\sqrt{5}$

6. $2\sqrt{11}$

7. $2\sqrt{6}$

8. $\frac{2}{7}\sqrt{14}$

9. $\frac{1}{4}\sqrt{15}$

10. $15\sqrt{2}$

11. $\frac{1}{3}\sqrt[3]{9}$

12. $\frac{1}{2}\sqrt[3]{10}$

13. $4\sqrt[3]{6}$

14. $2\sqrt[3]{36}$

15. $2a\sqrt{ab}$

16. $\dfrac{1}{4y}\sqrt{14xy}$

17. $\dfrac{3c}{2d}\sqrt{cd}$

18. $\dfrac{2}{b}\sqrt{2ab}$

19. $\dfrac{2}{7(x + y)}\sqrt{42(x + y)(x - y)}$

20. $\dfrac{4}{5(2a + b)}\sqrt{15(2a - b)(2a + b)}$

EXERCISE 8

1. $\sqrt{2}$　　　　　**5.** $\sqrt{3cd}$　　　　　**9.** $\sqrt[3]{2ab^2}$

2. $\sqrt{2}$　　　　　**6.** $\sqrt[3]{2xy^2}$　　　**10.** $\sqrt[3]{3x}$

3. $\sqrt{ab}$　　　　**7.** $\sqrt{2b}$　　　　　**11.** $\sqrt{5ab}$

4. $\sqrt{y}$　　　　　**8.** $3d\sqrt{cd}$　　　　**12.** $\sqrt{7xy}$

EXERCISE 9

1. $\sqrt[8]{16}$

2. $\sqrt[6]{49}$

3. $\sqrt[9]{64}$

4. $\sqrt[12]{25}$

5. $5\sqrt[4]{9}$ or $\sqrt[4]{5,625}$

6. $2\sqrt[6]{27}$ or $\sqrt[6]{1,728}$

7a. $\sqrt{2} = \sqrt[6]{8}$, $\sqrt[3]{3} = \sqrt[6]{9}$, $\sqrt[3]{3} > \sqrt{2}$

b. $\sqrt[4]{5} = \sqrt[12]{625}$, $\sqrt[6]{7} = \sqrt[12]{343}$, $\sqrt[4]{5} > \sqrt[6]{7}$

c. $2\sqrt[3]{3} = \sqrt[6]{576}$, $\sqrt{8} = \sqrt[6]{512}$, $2\sqrt[3]{3} > \sqrt{8}$

d. $2\sqrt{5} = \sqrt[6]{8000}$, $2\sqrt[3]{11} = \sqrt[6]{7744}$, $2\sqrt{5} > 2\sqrt[3]{11}$

e. $\sqrt{11} = \sqrt[6]{1,331}$, $3\sqrt[6]{2} = \sqrt[6]{1,458}$, $2\sqrt[3]{3} = \sqrt[6]{576}$
　　$3\sqrt[6]{2} > \sqrt{11} > 2\sqrt[3]{3}$

f. $\sqrt{3} = \sqrt[12]{729}$, $2\sqrt[3]{2} = \sqrt[12]{65,536}$, $\sqrt[4]{10} = \sqrt[12]{1,000}$
　　$2\sqrt[3]{2} > \sqrt[4]{10} > \sqrt{3}$

EXERCISE 10

1. $8\sqrt{2}$
2. $\sqrt{6}$
3. $\sqrt{7}$
4. $-5\sqrt{5}$
5. $8\sqrt{3}$

6. $-9\sqrt[3]{3}$
7. $-2\sqrt[3]{5}$
8. $13\sqrt[3]{7}$
9. $9\sqrt{7} - 7\sqrt[3]{3}$
10. $5\frac{1}{3}\sqrt{3}$

11. $-2\sqrt[3]{3} + 2\sqrt[3]{2}$
12. $\sqrt{6}$
13. $\sqrt{5}$
14. $-3b\sqrt{ab}$
15. $\sqrt[3]{3xy}$

EXERCISE 11

1. $40\sqrt{15}$
2. $24\sqrt{6}$
3. $24\sqrt{10}$
4. $12\sqrt{2}$
5. $14\sqrt{3}$
6. $\frac{24}{5}\sqrt{15}$
7. $\frac{5}{6}\sqrt{5}$

8. $\frac{1}{15}\sqrt{42}$
9. $3\sqrt[3]{2}$
10. $36\sqrt[3]{4\cdot}$
11. $\frac{1}{24}\sqrt[3]{4}$
12. $3\sqrt[6]{288}$
13. $2\sqrt[6]{4000}$
14. $3\sqrt[6]{1944}$
15. $2\sqrt[4]{108}$

16. $30 + 12\sqrt{6}$
17. $184 - 32\sqrt{15}$
18. $-69 - 9\sqrt{6}$
19. $4 - 8\sqrt{30}$
20. $2x + y - z$
21. b
22. $4 + 2c\sqrt{4 - c^2}$

23. $4a + 2\sqrt{(2a + b)(2a - b)}$
24. $8 - 2\sqrt{(4 + x^2)(4 - x^2)}$
25. $y - 2 + \dfrac{1}{y}$
26. $x + y + z + 2\sqrt{xy} + 2\sqrt{yz} + 2\sqrt{xz}$
27. $14 + 2\sqrt{10} + 2\sqrt{14} + 2\sqrt{35}$
28. $2 + 2\sqrt{15}$
29. $14 - 2\sqrt{15} - 6\sqrt{2} + 2\sqrt{30}$
30. $-16 - 2\sqrt{77}$

EXERCISE 12

1. $\sqrt{3}$
2. $\frac{1}{3}\sqrt{30}$
3. $\frac{1}{5}\sqrt{35}$
4. $3\sqrt{6}$
5. $\frac{16}{5}\sqrt{10}$

6. $2\sqrt{15}$
7. $3\sqrt[3]{3}$
8. $9\sqrt[3]{2}$
9. $28 + 14\sqrt{14}$
10. $4\sqrt{15} - 7\sqrt{17}$

11. $\sqrt{x - 2y}$
12. $\sqrt{3b - 1}$
13. $\frac{3}{2}\sqrt{6}$
14. $2\sqrt{2}$
15. $\frac{1}{2}\sqrt{10}$

16. $\frac{3}{8}\sqrt{2}$

17. $3 - 2\sqrt{10} + 8\sqrt{3}$

18. $8\sqrt{2} - \frac{3}{5}\sqrt{15} + \frac{2}{5}\sqrt{30}$

19. $\dfrac{8(3 - \sqrt{2})}{7}$

20. $3(3 + \sqrt{5})$

21. $\dfrac{2 + \sqrt{2}}{2}$

22. $\dfrac{31 - 26\sqrt{2}}{23}$

23. $\dfrac{21 - 4\sqrt{5}}{19}$

24. $\dfrac{3 + \sqrt{7}}{2}$

25. $3 + 2\sqrt{2}$

26. $\dfrac{8 + \sqrt{10}}{6}$

27. $\dfrac{7 - 3\sqrt{42} + 2\sqrt{21} - 18\sqrt{2}}{-5}$

28. $\dfrac{4\sqrt{15} + \sqrt{30}}{35}$

29. $\dfrac{2x^2 + a^2 + 2x\sqrt{x^2 + a^2}}{-a^2}$

30. $\dfrac{a + \sqrt{a^2 - b^2}}{b}$

31. $\dfrac{x + y + z^2 + 2z\sqrt{x + y}}{x + y - z^2}$

32. $\dfrac{b^2 + a + b - 2b\sqrt{a + b}}{b^2 - a - b}$

33. $\dfrac{5\sqrt{2} - 10 - 5\sqrt{6}}{-4}$

34. $\dfrac{4\sqrt{5} + 6\sqrt{3} - \sqrt{10} - 5\sqrt{6}}{-28}$

CHAPTER 6

Functions and Graphs

FUNCTIONS

DEFINITION: A *constant* is a symbol which represents one particular number during a discussion. For example, the numbers 5, −3, π, and $2\frac{6}{7}$ are constants.

Constants are usually designated by the first few letters of the alphabet, such as a, b, c, d.

DEFINITION: A *variable* is a symbol to which may be assigned any of several numbers during the course of a discussion.

Variables are usually designated by the last few letters of the alphabet, such as v, w, x, y, z.

DEFINITION: If two variables, x and y, are so related that to each permissible value of x, there corresponds one or more values of a second variable y, then y is said to be a *function of x*.

In this formulation, x is called the *independent* variable. Permissible values may be assigned to the independent variable at will. For each such value assigned to the independent variable there is determined one or more values of the *dependent* variable y.

If, with each value of the independent variable x there is associated a single value of y, then y is said to be a *single-valued function* of x. If several values of y correspond to each value of x, then y is called a *multiple-valued* function of x.

124

Example: $y = x^2 - 3x + 7$ is a *single-valued* function of x, since each value of x determines a unique value of y.

Example: $y^2 = x^2 - 9x + 12$ is a double-valued function of x, since each value of x determines two values of y (except when $x^2 - 9x + 12 = 0$).

Functional relationships may be expressed in a number of ways.

1. By a statement or rule

The distance (d) covered by a motorist traveling at the uniform rate of 40 miles per hour is equal to the product of 40 and the time (t) expressed in hours.

In this statement, the number 40 is a constant, and the distance (d) and the time (t) are variables. The time (t) is the independent variable and the distance (d) is the dependent variable.

2. By a formula

$C = 20 + 5 (n - 3)$

This is a formula for the cost (C) of borrowing a book from a circulating library which charges $.20 for the first 3 days and $.05 per day thereafter.

In this formula, n is the independent variable and C is the dependent variable. Note that in this case, the independent variable (n) must be a positive integer whose value is 3 or more.

3. By an algebraic expression. In this book, most of the functions discussed will be defined by algebraic expressions.

$y = 2x^2 - 7x + 3$

In this relationship, x is the independent variable and y is the dependent variable.

4. By other mathematical expressions

$y = \cos x$

In this relationship, x is the independent variable and y is the dependent variable.

5. By a table

Day of week	Mon.	Tues.	Wed.	Thurs.	Fri.	Sat.
Number of sales	12	16	15	18	21	38

This table shows the number of car sales made by the Star Auto Sales Corp. during a given week. In this table, the independent variable is the set of days (Monday through Saturday), and the dependent variable is the set of numbers indicating sales (12, 16, 15, 18, 21, 38).

6. By a graph (This method will be discussed later in this chapter.).

EXERCISE 1

For the functions defined below identify the independent variable and the dependent variable.

1. The cost (C) of a number (N) of pairs of shoes is obtained by multiplying the number of pairs bought by $15.

2. The formula $A = 10 + 5W$ indicates the amount (A) in a Christmas fund at the end of number (W) of weeks.

3. $Y = 5x^2 + 2x - 8$

4.

Month of year	Jan.	Feb.	Mar.	Apr.	May	June	July	Aug.
Attendance	27	23	32	28	26	24	19	17

The above table gives the attendance at the Barton Theatre for the months indicated. Attendance figures are given to the nearest thousand.

Functional Notation

In order to express the fact that y is a function of x the symbol used is $y = f(x)$. This is read "y is equal to a function of x." In this notation, x is the independent variable and y is the dependent variable.

Let us consider the area (y) of a square in terms of its side (x). This relationship may be written as $y = x^2$. In this case, $f(x) = x^2$.

Let us consider the perimeter (P) of a square in terms of its side (x). This relationship may be written as $P = 4x$. In this case, $g(x) = 4x$. Note that letters other than f may be used to state the functional relationship. In the latter case, the letter g was used.

If $f(x)$ or $g(x)$ is defined by an algebraic expression involving x then $f(2)$ or $g(2)$ represents the value of the dependent variable (y) when 2 is substituted for x.

Example: If $f(x) = x^2 + 3x - 7$, find $f(2)$

$f(2)$ indicates that 2 is to be substituted for x.

Example: If $f(x) = 2x^3 - 3x - 7$ find (a) $f(-2)$; (b) $f(-\frac{1}{2})$; (c) $f(0)$; (d) $f(3)$; (e) $f\left(\dfrac{1}{a}\right)$.

▶ $f(x) = 2x^3 - 3x - 7$

(a) $f(-2)$ indicates that -2 is to be substituted for x in $f(x)$.

$f(-2) = 2(-2)^3 - 3(-2) - 7 = -16 + 6 - 7$
$= -17$

(b) $f(-\frac{1}{2})$ indicates that $-\frac{1}{2}$ is to be substituted for x in $f(x)$.

$f(-\tfrac{1}{2}) = 2(-\tfrac{1}{2})^3 - 3(-\tfrac{1}{2}) - 7 = -\dfrac{23}{4}$

(c) $f(0) = 2(0)^3 - 3(0) - 7 = 0 - 0 - 7 = -7$

(d) $f(3) = 2(3)^3 - 3(3) - 7 = 54 - 9 - 7 = 38$

(e) $f\left(\dfrac{1}{a}\right) = 2\left(\dfrac{1}{a}\right)^3 - 3\left(\dfrac{1}{a}\right) - 7$

$= \dfrac{2}{a^3} - \dfrac{3}{a} - 7 = \dfrac{2 - 3a^2 - 7a^3}{a^3}$

Example: If $f(x) = x^2 - 5x + 2$ find
(a) $f(y + 2)$;
(b) $f(a + h)$.

▶ $f(x) = x^2 - 5x + 2$

(a) $f(y + 2)$ indicates that $y + 2$ is to be substituted for x in $f(x)$

$f(y + 2) = (y + 2)^2 - 5(y + 2) + 2 = y^2$
$+ 4y + 4 - 5y - 10 + 2 = y^2 - y - 4$

(b) $f(a + h) = (a + h)^2 - 5(a + h) + 2$
$= a^2 + 2ah + h^2 - 5a - 5h + 2$

Example: If $f(y) = \dfrac{2-y}{2+y}$ find $\dfrac{f(y+k) - f(y)}{k}$

▶ $f(y) = \dfrac{2-y}{2+y}$

$f(y+k) = \dfrac{2 - (y+k)}{2 + (y+k)}$

$f(y+k) - f(y) = \dfrac{2 - y - k}{2 + y + k} - \dfrac{2-y}{2+y}$

$= \dfrac{(4 - 2k - y^2 - ky) - (4 + 2k - y^2 - ky)}{(2 + y + k)(2 + y)}$

$= \dfrac{4 - 2k - y^2 - ky - 4 - 2k + y^2 + ky}{(2 + y + k)(2 + y)}$

$= \dfrac{-4k}{(2 + y + k)(2 + y)}$

$\dfrac{f(y+k) - f(y)}{k}$

$= \dfrac{-4k}{(2 + y + k)(2 + y)} \cdot \dfrac{1}{k}$

$= \dfrac{-4}{(2 + y + k)(2 + y)}$

Example: If $h(x) = \dfrac{3x}{2x - 1}$ find

(a) $h\left(\dfrac{y-3}{y+3}\right)$

(b) $\dfrac{h(x+a) - h(x)}{a}$

▶ (a) $h\left(\dfrac{y-3}{y+3}\right) = \dfrac{3\left(\dfrac{y-3}{y+3}\right)}{2\left(\dfrac{y-3}{y+3}\right) - 1} = \dfrac{\dfrac{3(y-3)}{y+3}}{\dfrac{2y - 6 - y - 3}{y+3}}$

$= \dfrac{\dfrac{3(y-3)}{y+3}}{\dfrac{y-9}{y+3}} = \dfrac{3(y-3)}{\cancel{y+3}} \cdot \dfrac{\cancel{y+3}}{y-9} = \dfrac{3(y-3)}{y-9}$

(b) $h(x + a) = \dfrac{3(x + a)}{2(x + a) - 1} = \dfrac{3(x + a)}{2x + 2a - 1}$

$h(x + a) - h(x) = \dfrac{3(x + a)}{2x + 2a - 1} - \dfrac{3x}{2x - 1}$

$= \dfrac{3(x + a)(2x - 1) - 3x(2x + 2a - 1)}{(2x + 2a - 1)(2x - 1)}$

$= \dfrac{-3a}{(2x + 2a - 1)(2x - 1)}$

$\dfrac{h(x + a) - h(x)}{a} = \dfrac{-3a}{(2x + 2a - 1)(2x - 1)} \cdot \dfrac{1}{a}$

$= \dfrac{-3}{(2x + 2a - 1)(2x - 1)}$

Example: If $g(x) = \dfrac{x^2 + 3x}{x - 1}$ find the value of $g(3) - g(2)$

$$g(3) = \frac{3^2 + 3 \cdot 3}{3 - 1} = \frac{9 + 9}{2} = 9$$

$$g(2) = \frac{2^2 + 3 \cdot 2}{2 - 1} = \frac{4 + 6}{1} = 10$$

$$g(3) - g(2) = 9 - 10 = -1$$

Functions of Several Variables

If one variable (z) is so related to several other variables $(x, y, w,$ etc.) that to each set of permissible values of x, y, w, etc., there corresponds one or more values of z, then z is said to be a function of the variables $(x, y, w,$ etc.) This may be written as $z = f(x, y, w, \ldots)$.

Examples: The area (A) of a triangle is equal to one-half the product of its base (b) and its altitude (h). Therefore, $A = \frac{1}{2} bh$ or $A = f(b, h)$.

The volume (V) of a rectangular solid is equal to the product of its length (l), its width (w), and its height (h). Therefore, $V = lwh$, or $V = f(l, w, h)$.

$z = 2x^2 - 3xy + y^2$ may be written $f(x,y) = 2x^2 - 3xy + y^2$.
The symbol $f(1, -2)$ means that 1 is to be substituted for x and that -2 is to be substituted for y.

Example: If $f(x,y) = 3x^2 + y^2 - 2xy - 3$, find the values of
(a) $f(2,1)$ (b) $f(0, -2)$.

▶   (a) $f(2,1)$ indicates that 2 is to be substituted for x and 1 is to be substituted for y in $f(x,y)$.

$$f(2,1) = 3(2)^2 + (1)^2 - 2(2)(1) - 3 = 12 + 1 - 4 - 3 = 6$$

Example: If $f(x,y) = \dfrac{x + y}{x - y}$ find the value of $f(c,d) + f(d,c)$

▶ $f(c,d) = \dfrac{c + d}{c - d}$

$f(d,c) = \dfrac{d + c}{d - c}$

$f(c,d) + f(d,c) = \dfrac{c + d}{c - d} + \dfrac{d + c}{d - c} = \dfrac{c + d}{c - d} - \dfrac{d + c}{c - d} = 0$

EXERCISE 2

1. If $f(x)$ $= 4x^3 - 5x - 3$ find   a. $f(-3)$ b. $f(-\tfrac{1}{2})$ c. $f(0)$
   d. $f(1)$ e. $f(2)$

2. If $h(y)$ $= 3y^2 + 2y - 7$ find   a. $h(4)$ b. $h(2)$ c. $h(0)$
   d. $h(-\tfrac{1}{3})$ e. $h(-1)$

3. If $f(x)$ $= x^2 - 3x + 4$ find   a. $f(y - 1)$ b. $f(a + b)$

4. If $h(y)$ $= 3y^2 + 4y - 2$ find   a. $h(x + a)$ b. $h(y - .3)$

5. If $g(x)$ $= \dfrac{2x}{3x - 2}$ find   a. $g(a - 2)$ b. $g\left(\dfrac{a}{b}\right)$

6. If $f(x)$ $= \dfrac{2x - 1}{5x}$ find   a. $f(y - 3)$ b. $f\left(\dfrac{a - b}{b}\right)$

7. If $g(y)$ $= \dfrac{2y - 1}{2y + 1}$ find $\dfrac{g(y + k) - g(y)}{k}$

8. If $f(x)$ $= \dfrac{3x - 2}{3x + 2}$ find $\dfrac{f(x + a) - f(x)}{a}$

9. If $f(y)$ $= \dfrac{5y}{3y - 4}$ find   a. $f\left(\dfrac{x - 1}{x + 1}\right)$ b. $\dfrac{f(x + a) - f(x)}{a}$

10. If $g(x)$ $= \dfrac{3x^2 - 5x}{2x - 1}$ find the value of $g(4) - g(1)$.

11. If $f(x,y) = 2x^2 + 3y^2 - 5xy - 8$ find the values of a. $f(3,-1)$
  b. $(-2,0)$

12. If $f(x,y) = \dfrac{2x + y}{2x - y}$ find the value of $f(a,b) + f(b,a)$

Setting Up Functions

As noted above, functional relationships may be expressed in a variety of ways. However, it is frequently necessary to express a functional relationship by a formula or by an equation. The following illustrative examples will indicate how this is done.

Example: Express the product of 3 consecutive integers as a function of n in which n is the smallest integer.

▶ If n is the smallest integer, the next two consecutive integers are $(n + 1)$ and $(n + 2)$. The product of the three integers is $n(n + 1)(n + 2)$.
Thus, $P(\text{product}) = f(n) = n(n + 1)(n + 2)$.

Example: The hypotenuse of a right triangle is 8. Express one side a as a function of the other side b.

▶ In a right triangle, the square of the hypotenuse is equal to the sum of the squares of the two legs.
Therefore, $a^2 + b^2 = 64$
$a^2 = 64 - b^2$
$a = \sqrt{64 - b^2}$
Thus, $a = f(b) = \sqrt{64 - b^2}$

Example: A rectangle is inscribed in a circle of diameter 12 inches. Express the area (A) as a function of the base (b) of the rectangle.

▶ In the diagram at the right,
$b^2 + h^2 = 144$ or $h = \sqrt{144 - b^2}$.
The area (A) of the rectangle $= bh$.
Therefore, $A = f(b) = b\sqrt{144 - b^2}$.

Example: A right circular cylinder is inscribed in a sphere of radius 10. Express the volume (V) of the cylinder in terms of the altitude (h) of the cylinder.

▶ In the diagram, the altitude of the cylinder is denoted by h and the radius of the base of the cylinder by r.

Using the Pythagorean relationship, $r^2 + \dfrac{h^2}{4}$

$= 100, 4r^2 + h^2 = 400, r^2 = \dfrac{400 - h^2}{4}$.

The volume (V) of the cylinder $= \pi r^2 h$.

Thus, $V = f(h) = \pi \left(\dfrac{400 - h^2}{4} \right) h = \dfrac{\pi h (400 - h^2)}{4}$.

Example: If $V = e^3$ and $d = e\sqrt{3}$, express V as a function of d.

▶ $\qquad V = e^3, e = \sqrt[3]{V}$

$\qquad\qquad d = e\sqrt{3}, e = \dfrac{d}{\sqrt{3}}$

Therefore, $\sqrt[3]{V} = \dfrac{d}{\sqrt{3}}, V = \dfrac{d^3}{(\sqrt{3})^3}, V = \dfrac{d^3}{3\sqrt{3}}$

Rationalizing the denominator, $V = \dfrac{d^3\sqrt{3}}{9}$.

Example: A rectangular cardboard box, 10 inches in height, has a square base and no top. The total amount of cardboard used is S square inches. Express the total surface (S) as a function of the volume (V).

▶ Let each edge of the box be x.

Then the area of each side of the box $= 10x$.

The area of the base $= x^2$.

Therefore, $S = 40x + x^2$.

The volume of the box, $V = 10x^2$, and

$x = \sqrt{\dfrac{V}{10}}$.

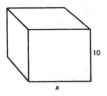

Therefore, $S = 40\sqrt{\dfrac{V}{10}} + \dfrac{V}{10}$

Example: A rectangle whose dimensions are x and y is inscribed in a triangle whose base is 25 and whose altitude is 15. (a) Express y as a function of x. (b) Express the area (A) of the rectangle as a function of x.

▶ (a) In the figure, $\triangle ADE$ is similar to $\triangle ABC$. In these similar triangles, the bases have the same ratio as corresponding altitudes. Therefore,

$$\frac{DE(x)}{BC(25)} = \frac{AF(15-y)}{A'C(15)}$$
$$25(15-y) = 15x$$
$$375 - 25y = 15x$$
$$y = \frac{375 - 15x}{25} = \frac{75 - 3x}{5}$$

(b) Area $= \tfrac{1}{2}bh = \tfrac{1}{2}(25)\left(\frac{75-3x}{5}\right) = \tfrac{5}{2}(75 - 3x)$

EXERCISE 3

1. Express the sum of 3 consecutive odd integers as a function of n in which n is the smallest integer.

2. Express the area (A) of a square as a function of its perimeter (P).

3. The equal sides of an isosceles triangle are each 12. Express the area (A) of the triangle as a function of its altitude (h).

4. One side of a rectangle is 12. Express the other side b as a function of the diagonal (d).

5. A square of side a is inscribed in a circle. Express the area of the circle (A) as a function of the side of the square.

6. Express the diagonal (d) of a cube as a function of the edge (e) of the cube.

7. If $S = \tfrac{1}{2}gt^2$ and $V = gt$ express S as a function of V. g is a constant.

8. A rectangle has a base b and an altitude 10. The rectangle is surmounted by a semicircle. Express the area (A) of the figure as a function of the base (b).

9. A right circular cylinder is inscribed in a sphere of radius 8. Express the volume (V) of the cylinder in terms of the altitude (h) of the cylinder.

10. A piece of wire of length a is bent in the form of a rectangle of width y.

   a. Express the length (1) of the rectangle as a function of y.

   b. Express the area (A) of the rectangle as a function of y.

11. Express the total surface (S) of a cube as a function of its volume (V).

12. A rectangular cardboard box has a square base 5 inches in length, and a closed top. The total amount of cardboard used in the box is A square inches. Express the volume (V) of the box as a function of its total surface, (A).

13. The radius of the base of a closed right circular cylinder is 6. Express the total surface (S) of the cylinder as a function of the altitude (h) of the cylinder.

14. A rectangle whose length is x and width is y is inscribed in a triangle whose base is 24 and whose altitude is 18. Express y as a function of x.

15. The radius of a sphere is 12. A right circular cone of base radius (a) is inscribed in the sphere. Express the volume (V) of the cone as a function of h and a.

GRAPHS

Rectangular Coordinates

Let xx' and yy' be two straight lines which intersect at right angles at point O. The line xx' is usually taken in a horizontal position and the line yy' is usually taken in a vertical position. Starting with point O, we take a convenient unit of length and lay off a scale, writing positive numbers to the right of O and negative numbers to the left of O. In the same manner, we take a convenient

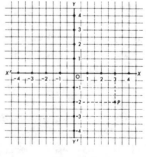

Graph 1

unit of length and lay off a scale, writing positive numbers above O and negative numbers below O. The unit of length used vertically need not necessarily be identical with the unit of length used horizontally.

If we take the point P in the plane we note that it is 3 units to the right of the yy' line and two units below the xx' line. Thus, the point P is located by writing the numbers $(3, -2)$ in parentheses. It can be seen that each point in the plane may be located by using a unique pair of numbers. The following important terms are used in connection with graphs.

X-axis = The line xx'

Y-axis = The line yy'

O—The origin

ABSCISSA. or X-coordinate—The number indicating how many units a point is to the right or to the left of the Y-axis. In the case of point P, the abscissa is 3.

ORDINATE. or Y-coordinate—The number indicating how many units a point is above or below the X-axis. In the case of point P, the ordinate is -2.

COORDINATES—The pair of numbers used to locate a point in the plane. In the case of point P, the coordinates are $(3, -2)$. Coordinates are written in parentheses.

QUADRANT—It will be noticed that the X-axis and the Y-axis divide the plane into 4 parts. Each of these parts is called a quadrant.

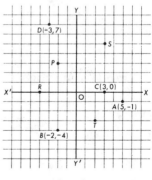

Graph 2

Example: Locate the following points on a graph. A $(5, -1)$, $B(-2, -4)$, $C(3, 0)$, $D(-3, 7)$.

▶ On graph

Example: Read the coordinates of points $P, S, T,$ and R. (See graph 2).

▶ $P(-2, 3)$, $S(3, 5)$, $T(2, -3)$, $R(-4, 0)$

EXERCISE 4

1. Locate the following points on a graph:

$A(-5,-2)$, $B(3,-1)$, $C(0,4)$, $D(-4,6)$, $E(-1,0)$

2. What are the coordinates of a point 5 units to the left of the Y-axis and 2 units below the X-axis.

3. The straight line joining the points $(-2,5)$ and $(2,3)$ crosses the Y-axis. What are the coordinates of the point where this line cuts the Y-axis?

4. What are the coordinates of the point where the line joining the points $(3,5)$ and $(-1,-3)$ cuts the Y-axis?

5. The line joining the points $(-1,2)$ and $(2,5)$ is extended in either direction. Find the coordinates of 3 additional points on the line.

6. The points $A(-1,2)$ and $B(5,2)$ are the end-points of the base of an isosceles triangle. Locate a point C which will make triangle ABC isosceles. (There are an infinite number of such points.)

Graph of a Function:

DEFINITION: The graph of a function $y = f(x)$ consists of all the points and only those points whose coordinates satisfy the relationship $y = f(x)$.

Although a functional relationship may be expressed in several ways we shall construct only graphs of functions given by algebraic expressions. The process consists of obtaining a series of points whose coordinates satisfy the relationship $y = f(x)$. If a sufficient number of such points are plotted and joined by a smooth line the graph, or geometric figure will appear.

Zeros of a Function

A zero of a function $y = f(x)$ is a value of the independent variable, x, which will yield a corresponding value of 0 for the dependent variable y.

Example: If $y = f(x) = x^2 - 2x - 15$ the values $x = 5$ and $x = -3$ will yield corresponding values of 0 for the dependent variable y. Therefore, 5 and -3 are zeros of $f(x)$.

Example: Plot the graph of $3x + 2y = 8$

▶ It is convenient to express y as a function of x

$$3x + 2y = 8, y = f(x) = \frac{8 - 3x}{2}$$

In order to plot the graph we must obtain number pairs which satisfy the above relationship. We may arbitrarily assume permissible values for x and find the corresponding values for y. For example, if $x = 0$, then $y = \frac{8 - 0}{2} = 4$.

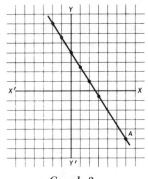

Graph 3

The table below gives a set of such values.

x	-2	-1	1	2	3
y	7	$5\frac{1}{2}$	$2\frac{1}{2}$	1	$-\frac{1}{2}$

When these points are plotted, we have the graph shown.

Note that, in this case, the geometric figure formed by joining the points whose coordinates satisfy the original relationship is a straight line. The function $y = \frac{8 - 3x}{2}$ is called a *linear function.*

Obviously, the graph of a linear function may be obtained by plotting only 2 points.

The line representing the graph of the linear function contains all the points and only the points whose coordinates satisfy the original relationship. For example, if we take any point on the graph such as point A, we will find that its coordinates $(6, -5)$ satisfy the equation $3x + 2y = 8$.

Example: Plot the graph of $y = x^2 - 2x - 3$

▶ In order to plot the graph we must obtain number of pairs which satisfy the given relationship. We will arbitrarily assume permissible values for x and find the corresponding values for y.

The table below gives a set of such values.

x	4	3	2	1	0	-1	-2
y or $f(x)$	5	0	-3	-4	-3	0	5

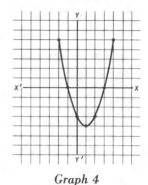

Graph 4

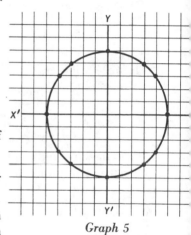

Example: Plot the graph of $x^2 + y^2 = 25$

▶ We may express this relationship as $y = \pm\sqrt{25 - x^2}$.

In preparing the table it can be seen that x cannot have values greater than

Graph 5

$+5$ or less than -5. For values of x greater than $+5$ or less than -5, y has imaginary values. The table below gives a set of values.

x	5	4	3	2	1	0	-1	-2	-3	-4	-5
y or $f(x)$	0	±3	±4	$\pm\sqrt{21}$	$\pm\sqrt{24}$	±5	$\pm\sqrt{24}$	$\pm\sqrt{21}$	±4	±3	0

In this case, the geometric figure formed by joining the points whose coordinates satisfy the original relationship is a circle. Both the table and the graph indicate that the zeros of this function are $+5$ and -5.

Example: Plot the graph of $4x^2 + y^2 = 16$

▶ We may express this relationship as

$$y = \pm \sqrt{16 - 4x^2}$$

In preparing the table, it can be seen that x cannot have values greater than $+2$ or less than -2. For values of x greater than $+2$ or less than -2, y has imaginary values. The table below gives a set of values.

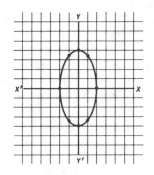

Graph 6

x	2	1	0	-1	-2
y or $f(x)$	0	$\pm\sqrt{12}$	± 4	$\pm\sqrt{12}$	0

In this case, the geometric figure formed by joining the points whose coordinates satisfy the original relationship is an ellipse. If additional points are required to obtain a smooth curve values for x such as $1\frac{1}{2}$, $\frac{3}{4}$, $-\frac{1}{2}$, etc., may be taken.

The zeroes of the function are $+2$ and -2.

Example: Plot the graph of $xy - 2y = 6 - x$

▶ We may express this relationship as

$$y = \frac{6 - x}{x - 2}$$

In preparing the table, it can be seen that x cannot have

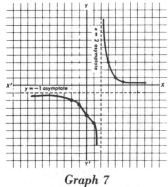

Graph 7

the value 2. When $x = 2$ the value of $y = \frac{4}{0}$ which has no meaning.

x	5	4	3	2	1	0	-1	-2	-3	-4
y or $f(x)$	$\frac{1}{3}$	1	3	No Value	-5	-3	$-2\frac{1}{2}$	-2	$-1\frac{4}{5}$	$-1\frac{2}{3}$

In this case, the geometric figure formed by joining the points whose coordinates satisfy the original relationship is a hyperbola. If additional points are taken between the values $x = 2$ and $x = 3$ the graph will be extended as shown by the broken line in the first quadrant. If additional points are taken between the values $x = 1$ and $x = 2$ the graph will be extended as shown by the broken line in the fourth quadrant. Thus, the graph will approach the line $x = 2$ but the graph will not intersect this line. The line $x = 2$ is known as an *asymptote*. The line $y = -1$ is also an asymptote.

The zero of this function is $+6$.

Example: Plot the graph of $y = x^3 + 2x^2 - 5x - 1$

▶   The following table contains the coordinates of points on the graph.

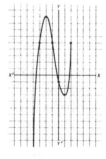

Graph 8

x	$+2$	$+1$	0	-1	-2	-3	-4
y or $f(x)$	$+5$	-3	-1	$+5$	$+9$	$+5$	-13

The zeros of this function may be approximated from the graph by reading the values of x where the graph cuts the X-axis at points A, B and C.

The zeros of the function are approximately $+1.6$, $-.2$, and -3.4.

Using Graphs to Solve Problems

Graphs are often used to solve problems as illustrated by the following.

Example: A farmer has 90 rods of fencing. What are the dimensions of the largest rectangular field that he can enclose with this fencing?

▶   Let $l =$ the number of rods in the length

And $w =$ the number of rods in the width

Then $2l + 2w = 90$

$A = lw$

It is desired to find values for l and w which will give A a maximum value.

Expressing A as a function of l

$2l + 2w = 90$

$l + w = 45, w = 45 - l$

$A = l(45 - l) = 45l - l^2$

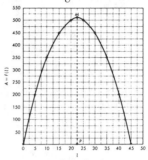

Graph 9

To draw a graph of the function $A = 45l - l^2$ make a table of values at intervals of 5 for l.

l	0	5	10	15	20	25	30	35	40	45
A or $f(l)$	0	200	350	450	500	500	450	350	200	0

Plot l on the vertical axis and let each interval represent $2\frac{1}{2}$ units. Plot $A = f(l)$ on the horizontal axis and let each interval represent 25 units.

On the graph, the point M represents the maximum ordinate, or the maximum area. The abscissa (P) corresponding to the maximum ordinate is $22\frac{1}{2}$. Therefore, the area of the field will be a maximum when the length is $22\frac{1}{2}$ rods. Since the perimeter is 90 rods, the width will also be $22\frac{1}{2}$ rods and the field will be a square.

Example: A rectangular box is to be made with a square base and open top. The base costs \$2 a square foot and the sides \$3 a square foot. What should be the length of the base if the volume is to be 27 cubic feet and the cost is to be a minimum?

▶ Let each side of the base $= x$

And let the height $= y$

Volume $(V) = x^2y$

Total Area $= x^2 + 4xy$

Cost $= 2x^2 + 3(4xy) = 2x^2 + 12xy$

$V = x^2y = 27$ or $y = \dfrac{27}{x^2}$

$C = 2x^2 + 12x\left(\dfrac{27}{x^2}\right) = 2x^2 + \dfrac{324}{x}$

In order to find the value of x which will yield minimum cost we plot the graph of the function

$$C = 2x^2 + \frac{324}{x}.$$

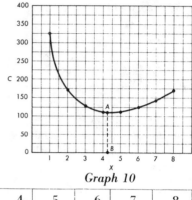

Graph 10

x	1	2	3	4	5	6	7	8
$C = f(x)$	326	170	126	113	$114\frac{4}{5}$	126	$144\frac{2}{7}$	$168\frac{1}{2}$

Plot x on the vertical axis and C on the horizontal axis. From the graph we see that point A represents the minimum cost. We wish to find the length of the base (x) when the cost is a minimum. The point B, corresponding to point A, represents the desired value. The value of x which will yield a minimum cost is approximately 4.3.

Example: A business man estimates that, if he charges 8 cents a pound for a certain product, he will sell 600 pounds per week and that for each decrease of $\frac{1}{2}$ cent per pound on the selling price, he will increase his sales by 150 pounds per week. Find the selling price that will yield greatest gross returns.

▶ Let P = Total income from sales

And x = Number of $\frac{1}{2}$ cent decreases made

Then P = Selling Price $\times$ Number Sold

$$P = \left(8 - \frac{x}{2}\right)$$
$$(600 + 150x)$$
$$P = 4{,}800 + 900x - 75x^2$$

Graph 11

x	1	2	3	4	5	6	7	8
$P = f(x)$	5,625	6,300	6,825	7,200	7,425	7,500	7,425	7,200

On the graph, point A represents the maximum value for P. Corresponding to point A we have point B on the X-axis.

Therefore, the maximum selling price is $75.00, and this is obtained after 6 reductions of $\frac{1}{2}$ cent each. Therefore, the selling price is $8 - \frac{6}{2}$ or $8 - 3 = 5$ cents per pound.

Statistical Graphs

A function is frequently expressed by a table of collected statistical data. A graph of such a function is useful in tracing overall trends. In plotting such a function the values of the independent variable are usually taken as abscissas and those of the dependent variable as ordinates.

Example: The monthly deposits in savings accounts in the National Bank are given in the following table.

Month	Jan.	Feb.	Mar.	Apr.	May	June	July	Aug.	Sept.	Oct.	Nov.	Dec.
Deposits (000 Omitted)	$769	$643	$802	$915	$712	$1,054	$894	$873	$853	$1,117	$946	$869

In this functional relationship, the independent variable is represented by the set of months and will be plotted as abscissas. The dependent variable is represented by the set of deposits and will be plotted as ordinates.

In this type of graph, values between those given in the table have no meaning. Therefore, interpolation on the graph will be meaningless.

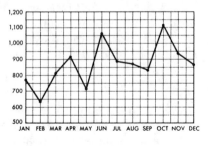

Graph 12

Graphs Plotted from Formulas

A functional relationship may be stated by a formula. In plotting, a table of values is calculated from the formula and this table is used as the basis of the graph. The graph of a formula is useful in making a series of readings which yield the value of one variable when the value of the other variable is given.

Example: The velocity (V) of a point in feet per second, on the rim of a flywheel t seconds after starting is given by the formula $V = 18t^2 - t^3$ where t is time in seconds and

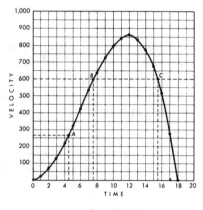

a. Draw the graph of the formula
b. Find the velocity at $4\frac{1}{2}$ seconds.
c. After how many seconds will the velocity be 600 feet per second?

Graph 13

b. The velocity at $4\frac{1}{2}$ seconds is indicated by the point A and is about 270 feet per second.

c. Velocity of 600 feet per second will be reached at approximately 7.6 seconds and again at approximately 15.5 seconds. These values may be read from the graph at points on the times axis corresponding to points B and C.

EXERCISE 5

1. A man has 60 yards of fencing. What are the dimensions of the largest rectangular garden that he can enclose with this fencing?

2. A rectangular box is to be made with a square base and open top. The base costs $3 a square foot and the sides $4 a square foot. What should be the length of the base if the volume is to be 24 cubic feet and the cost is to be a minimum?

3. A farmer has 90 bushels of tomatoes worth 75 cents a bushel. The tomatoes are spoiling at the rate of 4 bushels a day but the price is increasing at the rate of 5 cents per bushel per day. In how many days should the farmer sell?

4. How large a rectangular play area may be enclosed by 50 feet of fencing if part of the side of a house is used for one side?

5. Find the dimensions of the rectangle of largest area that can be inscribed in a circle of diameter 8.

6. A sheet of cardboard for a poster contains 12 square feet. The margins at top and bottom are each $\frac{1}{3}$ of a foot and the margins on each side are $\frac{1}{4}$ of a foot. What is the height of the printed area if this printed area is to be a maximum?

7. Illustrate the following table of data graphically. The daily sales of the Empire Department Store are:

Day	Mon.	Tues.	Wed.	Thurs.	Fri.	Sat.
Sales (000 Omitted)	48.3	61.2	68.7	59.5	74.9	87.2

8. Illustrate the following table of data graphically. The average daily circulation of the *Evening Record* given by months is:

Month	Jan.	Feb.	Mar.	Apr.	May	June	July	Aug.	Sept.	Oct.	Nov.	Dec.
Circulation (000 Omitted)	372.4	345.3	312.0	402.7	408.6	425.3	382.6	325.6	365.4	391.9	405.3	412.8

9. The formula used to convert from Centigrade to Fahrenheit temperatures is $C = \frac{5}{9}(F - 32)$.

    a. Plot the graph of this formula

    b. From the graph, find the value of C when $F = 68°$

    c. From the graph, find the value of F when $C = 35°$

10. The time (t) in seconds, of vibration of a pendulum consisting of a slender uniform rod suspended at one end is given by the formula $t = 2\pi \sqrt{\dfrac{2l}{3g}}$ where l is the length in feet, g is the constant 32.16 and π is the constant 3.14.

    a Plot the graph of this formula

    b. From the graph, find the value of t when $l = 200$ feet

    c. From the graph, find the value of l when $t = 8$ seconds

ANSWERS

EXERCISE 1—Page 126

1. N is independent variable, C is dependent variable
2. W is independent variable, A is dependent variable
3. X is independent variable, Y is dependent variable
4. Month of year is independent variable, attendance is dependent variable

EXERCISE 2—Page 130

1. a. -96
   b. -1
   c. -3
   d. -4
   e. 19
2. a. 49
   b. 9
   c. -7
   d. $-7\frac{1}{3}$
   e. -6
3. a. $y^2 - 5y + 8$
   b. $a^2 + 2ab + b^2 - 3a - 3b + 4$
4. a. $3x^2 + 6ax + 3a^2 + 4x + 4a - 2$
   b. $3y^2 - 14y + 13$
5. a. $\dfrac{2a - 4}{3a - 8}$
   b. $\dfrac{2a}{3a - 2b}$

6. a. $\dfrac{2y - 7}{5y - 15}$
   b. $\dfrac{2a - 3b}{5a - 5b}$
7. $\dfrac{4}{(2y + 2k + 1)(2y + 1)}$
8. $\dfrac{12}{(3x + 3a + 2)(3x + 2)}$
9. a. $\dfrac{5(x - 1)}{-x - 7}$
   b. $\dfrac{-20}{(3x + 3a - 4)(3x - 4)}$
10. 6
11. a. 28
    b. 0
12. $\dfrac{6ab}{(2a - b)(2b - a)}$

EXERCISE 3—Page 133

1. $S = 3n + 6$
2. $A = \dfrac{p^2}{16}$
3. $A = h\sqrt{144 - h^2}$
4. $b = \sqrt{d^2 - 144}$
5. $A = \dfrac{\pi a^2}{2}$
6. $d = e\sqrt{3}$
7. $S = \dfrac{V^2}{2g}$

8. $A = 10b + \dfrac{\pi b^2}{8}$
9. $V = \pi h\left(64 - \dfrac{h^2}{4}\right)$
10. a. $l = \dfrac{a}{2} - y$
    b. $A = y\left(\dfrac{a}{2} - y\right)$

11. $S = 6V^{2/3}$

12. $V = \dfrac{5(A - 50)}{4}$

13. $S = 72\pi + 12\pi h$

14. $y = \dfrac{72 - 4x}{3}$

15. $V = \frac{1}{3}\pi a^2 (\sqrt{144 - a^2} + 12)$

EXERCISE 4—Page 136

2. $(-5, -2)$

3. $(0, 4)$

4. $(0, -1)$

5. Any set of values satisfying the equation $y = x + 3$. For example, $(1,4)$, $(3,6)$, $(-4, -1)$, etc.

6. Any point whose abscissa is 2

EXERCISE 5—Page 144

1. 15 yards by 15 yards

2. 4 ft. by 4 ft. by $1\frac{1}{2}$ ft.

3. $3\frac{3}{4}$ days

4. 25 ft. by $12\frac{1}{2}$ ft.

5. $4\sqrt{2}$ by $4\sqrt{2}$

6. 4 ft.

9. a. $20°$

   b. $95°$

10. b. 12.6 approximately

   c. 79 ft. approximately

CHAPTER **7**

Linear Equations in One Unknown

Introduction

DEFINITION: An equation is a statement of the equality of two mathematical expressions. The two expressions are called members of the equation and are often identified as the *left member* and the *right member*.

Example: $3x - 7 = x + 1$

In this equation, $3x + 7$ is the left member and $x - 1$ is the right member. The equation states the fact that the two members are equal.

A *solution* of an equation is a value or set of values which, when substituted for the unknown or unknowns, will make the left and right members equal. For example, the number 4 is a solution of the equation $3x - 7 = x + 1$. A number which is a solution of an equation is called a root of the equation. A root of an equation is said to satisfy the equation.

There are two types of equations:

a. Identities: An *identity* is an equation which is satisfied by all permissible values of the variable or variables.

For example, the equation $3x + 2x = 5x$ is an identity since it satisfied for all permissible values of x such as $x = 2, 3, -\frac{1}{2}, \sqrt{5}$, etc.

b. Conditional equation: A conditional equation is an equation which is satisfied by only certain values of the variable or variables.

The term equation refers to both identities and conditional equations. However, conditional equations are usually called equations.

For example, the equation $3x + 2x = 10$ is a conditional equation since it is satisfied by only one value of the unknown, $x = 2$.

EQUIVALENT EQUATIONS—Two equations are equivalent if they have the same solutions. For example, the equations, $3x = 6$, $5x = 10$, and $3x + 1 = 7$, are equivalent equations, since the 3 equations have the solution $x = 2$.

OPERATIONS ON EQUATIONS—In the solution of equations, operations may be performed which will yield equivalent equations.

1. The same quantity may be added to both members of an equation.

Example: $\begin{aligned} x - 5 &= 7 \\ + 5 &= + 5 \\ \hline x &= 7 + 5 = 12 \end{aligned}$

This operation is equivalent to *transposing* the negative number, -5.

2. The same quantity may be subtracted from both members of an equation.

Example: $\begin{aligned} y + 3 &= 11 \\ - 3 &= -3 \\ \hline y &= 8 \end{aligned}$

This operation is equivalent to transposing the positive number, 3.

3. Both members of an equation may be multiplied by the same quantity (except zero), provided that the multiplier does not contain the unknown.

Example: $\dfrac{x}{2} = 4$

Multiplying both members of the equation by 2, yields

$$2\left(\frac{x}{2}\right) = 4(2) \text{ or } x = 8$$

4. Both members of an equation may be divided by the same quantity (except zero), provided that the divisor does not contain the unknown.

Example: $3y = 21$
Dividing both members of the equation by 3, yields
$$\frac{3y}{3} = \frac{21}{3} = 7$$

Redundant Equations

DEFINITION: If a derived equation contains all the roots of the original equation and some others, it is called a redundant equation.

If each member of an equation is multiplied by an expression which contains the variable, the resulting equation may have more solutions than the original equation. A value of the variable which satisfies the resulting equation but not the original equation is called an *extraneous root*.

Example: The equation $y - 2 = 0$ has one solution, $y = 2$. If each member of the equation is multiplied by $(y - 3)$ the resulting equation is $(y - 2)(y - 3) = 0$. This equation has two roots, $y = 2$ and $y = 3$. The root $y = 3$ was introduced by multiplying both sides of the equation by $(y - 3)$. It is an extraneous root.

Example: Consider the equation $3x - 5 = 2x$. This equation has one solution, $x = 5$. If we square both sides of the equation and transpose we obtain the equation $5x^2 - 30x + 25 = 0$. This equation has two roots, $x = 5$ and $x = 1$. The root $x = 1$ was introduced by squaring both sides of the equation. It is an extraneous root.

Defective Equations

DEFINITION: If a derived equation lacks some roots of the original equation, it is called a defective equation.

If each member of an equation is divided by an expression which contains the variable, the resulting equation may have fewer roots than the original equation.

Example: The equation $x^2 - 3x = 0$ has two roots, 3 and 0. If both members of the equation are divided by x we have $x - 3 = 0$. The resulting equation has only one root, $x = 3$.

Solution of Linear Equations

Linear equations may be solved by applying the 4 fundamental laws which yield equivalent equations.

Example: Solve and check the equation $7x - 1 = 5x + 9$.

▶ $7x - 1 = 5x + 9$

Subtract 5x from both members	$2x - 1 = 9$
Add 1 to both members	$2x = 10$
Divide both members by 2	$x = 5$

Check:

$$7(5) - 1 = 5(5) + 9$$
$$35 - 1 = 25 + 9$$
$$34 = 34$$

Example: Solve and check the equation $\dfrac{3y}{2} - \dfrac{4y}{5} = 7$

▶ $\dfrac{3y}{2} - \dfrac{4y}{5} = 7$

Multiply both members by 10	$15y - 8y = 70$
	$7y = 10$
Divide both members by 7	$y = 10$

Check:

$$\frac{3(10)}{2} - \frac{4(10)}{5} = 7$$
$$\frac{30}{2} - \frac{40}{5} = 7$$
$$15 - 8 = 7$$

Example: Solve and check the equation $3(x - 2) - 5(x - 3) = 17$.

▶

$$3(x - 2) - 5(x - 3) = 17$$
$$3x - 6 - 5x + 15 = 17$$
$$-2x + 9 = 17$$

Subtract 9 from both members $\quad -2x = 8$
Divide both members by -2 $\qquad x = -4$

   Check:

$$3(-4 - 2) - 5(-4 - 3) = 17$$
$$3(-6) - 5(-7) = 17$$
$$-18 + 35 = 17$$
$$17 = 17$$

Example: Solve and check the equation $(x + 1)(x - 3)$ $= x^2 + 4x + 9$

▶

$$(x + 1)(x - 3) = x^2 + 4x + 9$$
$$x^2 - 2x - 3 = x^2 + 4x + 9$$

Subtract $x^2 + 4x$ from both members $\quad -6x - 3 = 9$
Add 3 to both members $\qquad\qquad\qquad -6x = 12$
Divide both members by -6 $\qquad\qquad x = -2$

   Check:

$$(x + 1)(x - 3) = x^2 + 4x + 9$$
$$(-2 + 1)(-2 - 3) = (-2)^2 + 4(-2) + 9$$
$$(-1)(-5) = 4 - 8 + 9$$
$$5 = 5$$

Example: Solve and check the equation $\sqrt{2x + 3} = 5$

▶

$$\sqrt{2x + 3} = 5$$

Square both members $\qquad 2x + 3 = 25$
Subtract 3 from both members $\qquad 2x = 22$
Divide both members by 2 $\qquad\quad x = 11$

   Check:

$$\sqrt{2x + 3} = 5$$
$$\sqrt{2(11) + 3} = 5$$
$$\sqrt{22 + 3} = 5$$
$$\sqrt{25} = 5$$
$$5 = 5$$

Example: Solve and check the equation $\sqrt{x + 3} + 5 = 0$

▶ $$\sqrt{x + 3} + 5 = 0$$

Subtract 5 from both members $\sqrt{x + 3} = -5$

Square both members $x + 3 = 25$

Subtract 3 from both members $x = 22$

Check:

$$\sqrt{x + 3} + 5 = 0$$
$$\sqrt{22 + 3} + 5 = 0$$
$$\sqrt{25} + 5 = 0$$
$$5 + 5 = 0$$

Therefore, 5 is not a root of the equation. It is an *extraneous root*. The equation has no solution.

Example: Solve and check the equation $\dfrac{2y}{y - 5} = -2 + \dfrac{10}{y - 5}$

▶ $$\frac{2y}{y - 5} = -2 + \frac{10}{y - 5}$$

Multiplying both members by $y - 5$ $2y = -2(y - 5) + 10$
$$2y = -2y + 10 + 10$$

Adding 2y to both members $4y = 20$

Dividing both members by 4 $y = 5$

Check:

$$\frac{2y}{y - 5} = -2 + \frac{10}{y - 5}$$
$$\frac{2(5)}{5 - 5} = -2 + \frac{10}{5 - 5}$$

The denominators in the check become zero. Since division by zero is not a permissible operation, 5 is not a root. The original equation has no root.

Example: Solve and check the equation $\dfrac{y + 2}{3y} - \dfrac{2}{y - 1} = \dfrac{1}{3}$

▶ $$\frac{y + 2}{3y} - \frac{2}{y - 1} = \frac{1}{3}$$

Multiply both members of the equation by the L.C.D., $3y(y - 1)$
$$(y + 2)(y - 1) - (6y) = y(y - 1)$$
$$y^2 + y - 2 - 6y = y^2 - y; \quad -4y = 2$$
$$y = -\frac{1}{2}$$

Check:

$$\frac{-\dfrac{1}{2} + 2}{3\left(-\dfrac{1}{2}\right)} - \frac{2}{-\dfrac{1}{2} - 1} = \frac{1}{3}$$

$$\frac{\dfrac{3}{2}}{\dfrac{-3}{2}} - \frac{2}{\dfrac{-3}{2}} = \frac{1}{3}; \quad \frac{1}{3} = \frac{1}{3}$$

Example: Solve and check the equation

$$\frac{8}{x + 2} - \frac{3}{x - 2} = \frac{13}{x^2 - 4}$$

▶ *Multiply both members of the equation by the L.C.D.* $(x + 2)(x - 2)$

$$8(x - 2) - 3(x + 2) = 13$$
$$8x - 16 - 3x - 6 = 13$$
$$5x = 35$$
$$x = 7$$

Check:

$$\frac{8}{7 + 2} - \frac{3}{7 - 2} = \frac{13}{49 - 4}$$

$$\frac{8}{9} - \frac{3}{5} = \frac{13}{45}$$

$$\frac{40}{45} - \frac{27}{45} = \frac{13}{45}, \quad \frac{13}{45} = \frac{13}{45}$$

Example: Solve and check the equation

$$\frac{10}{x^2 - 9} + \frac{3x}{x + 3} - \frac{7}{x - 3} = 3$$

▶ *Multiply both members of the equation by the L.C.D.* $(x + 3)(x - 3)$

$$10 + 3x(x - 3) - 7(x + 3) = 3(x^2 - 9)$$
$$10 + 3x^2 - 9x - 7x - 21 = 3x^2 - 27$$
$$-16x = -16$$
$$x = 1$$

Check:

$$\frac{10}{1-9} + \frac{3}{1+3} - \frac{7}{1-3} = 3$$

$$\frac{10}{-8} + \frac{3}{4} - \frac{7}{-2} = 3$$

$$-\frac{5}{4} + \frac{3}{4} + \frac{14}{4} = 3$$

$$3 = 3$$

EXERCISE 1

1. Determine whether the following equations are identities or conditional equations.

a. $3x + 5(x - 2)(x + 1)$
   $= 4x^2 + x(x - 2) - 10$

b. $\dfrac{2x + 3}{x - 1} = \dfrac{2x - 5}{x + 3}$

c. $x(x - 1)(x - 2)$
   $= x^2(x - 3) + 1$

d. $(x - 5)(x + 3)(x - 1)$
   $= x^3 - 3x(x + 4)$
   $- x + 15$

e. $4(x - 3)^2 = (2x + 1)^2$

2. Are the following pairs of equations equivalent? If not, is the second one redundant or defective with respect to the first?

a. $2x = x + 3$
   $4x^2 = x^2 + 6x + 9$

b. $2x^3 = 18x$
   $x^2 = 9$

c. $3x = x + 2$
   $3x(x - 5) = (x + 2)(x - 5)$

d. $\sqrt{1 - y} = y - 1$
   $1 - y = y^2 - 2y + 1$

3. Solve and check the following equations:

a. $3y - 7 = 14 - 4y$

b. $8x + 2 = 5x + 17$

c. $9x - 1 = 3x + 23$

d. $7y - (2y - 1) = 41$
e. $2x - 4 = 5x - 25$
f. $y^2 - y(y + 5) = y + 12$
g. $3 - 2(3x + 2) = 11$
h. $(x - 8)(x + 12)$
   $= (x + 1)(x - 6)$
i. $4(x - 3) - 5(6 - x)$
   $= x - 2$
j. $\dfrac{5x}{3} - \dfrac{x}{4} = 5$
k. $\dfrac{2x - 3}{2} - \dfrac{5x - 2}{3} = 7$
l. $\sqrt{3y - 1} = 4$
m. $\sqrt{x - 2} + 3 = 0$
n. $2 - \dfrac{3}{x - 1} = 8$
o. $\dfrac{3x - 1}{x + 2} = \dfrac{6x - 5}{2x + 7}$
p. $\dfrac{x + 5}{5x} - \dfrac{3}{2x - 7} = \dfrac{1}{5}$
q. $\dfrac{5}{3y + 1} - \dfrac{y + 2}{4y} = -\dfrac{1}{4}$
r. $\dfrac{7}{2x - 1} - \dfrac{5}{2x + 1} = \dfrac{8}{4x^2 - 1}$
s. $\dfrac{6}{2 - 3y} - \dfrac{5}{2 + 3y} = \dfrac{13}{4 - 9y^2}$
t. $\dfrac{24}{4y^2 - 9} + \dfrac{4y}{2y + 3} - \dfrac{9}{2y - 3} = 2$

Literal Equations and Formulas

DEFINITION: A *literal equation* is an equation in which some or all of the known quantities are denoted by letters.

Examples: $ax + bx = c$
                  $c^2x - 2cdx = 5 - 3c^2x$

Example: Solve the equation $ax + c = cx + 2c$

▶
$$ax + c = cx + 2c$$
$$ax - cx = 2c - c$$
$$x(a - c) = 2c - c$$
$$x = \frac{c}{a - c}$$

Example: Solve the equation $\dfrac{a - bx}{c} + b = \dfrac{bc - x}{c}$

▶
$$\frac{a - bx}{c} + b = \frac{bc - x}{c}$$
$$a - bx + bc = bc - x$$
$$x - bx = -a$$
$$x(1 - b) = -a$$
$$x = \frac{-a}{1 - b} = \frac{a}{b - 1}$$

DEFINITION: A *formula* is a literal equation which states a general rule or principle.

Examples: $A = lw$, $E = IR$ (Ohm's Law)

There are two important operations with formulas, (1) finding the value of one letter of the formula when the others are known, (2) expressing one letter of the formula in terms of the others.

Example: In the formula, $\dfrac{D}{d} = q + \dfrac{R}{d}$. find the value of q when $D = 89$, $d = 7$, and $R = 5$.

▶
$$\frac{D}{d} = q + \frac{R}{d}$$
$$\frac{89}{7} = q + \frac{5}{7}$$
$$\frac{89}{7} - \frac{5}{7} = q$$
$$q = 12$$

Example: In the formula $\dfrac{1}{f} = \dfrac{1}{p} + \dfrac{1}{p'}$, find the value of p when $f = 9$ and $p' = 15$.

▶
$$\frac{1}{f} = \frac{1}{p} + \frac{1}{p'}$$
$$\frac{1}{9} = \frac{1}{p} + \frac{1}{15}$$
$$\frac{1}{p} = \frac{1}{9} - \frac{1}{15} = \frac{2}{45}$$
$$2p = 45$$
$$p = 22\tfrac{1}{2}$$

Example: In the formula $K = \sqrt{s(s-a)(s-b)(s-c)}$, find the value of a when $K = 54, s = 18, b = 12$ and $c = 15$.

▶
$$K = \sqrt{s(s-a)(s-b)(s-c)}$$
$$54 = \sqrt{18(18-a)(18-12)(18-15)}$$
$$54 = \sqrt{18(18-a)(6)(3)}$$
$$54 = \sqrt{(18)(18)(18-a)}$$
$$54 = 18\sqrt{18-a}$$
$$3 = \sqrt{18-a}, 9 = 18 - a, a = 9$$

Example: In the formula $A = \dfrac{rs}{r+s}$, solve for s

▶
$$A = \frac{rs}{r+s}$$
$$A(r+s) = rs; \quad Ar + As = rs$$
$$As - rs = -Ar; \quad s(A-r) = -Ar$$
$$s = -\frac{Ar}{A-r} = \frac{Ar}{r-A}$$

Example: In the formula $\dfrac{1}{R} = \dfrac{1}{R_1} + \dfrac{1}{R_2}$ solve for R_2

▶
$$\frac{1}{R} = \frac{1}{R_1} + \frac{1}{R_2}$$
$$R_1R_2 = RR_2 + RR_1$$
$$R_1R_2 - RR_2 = RR_1$$
$$R_2(R_1 - R) = RR_1$$
$$R_2 = \frac{RR_1}{R_1 - R}$$

Example: In the formula $E = \dfrac{1}{t} \left(\dfrac{T^2}{R-1} \right)$, solve for T

▶
$$E = \frac{1}{t} \left(\frac{T^2}{R} - 1 \right)$$

$$Et = \frac{T^2}{R} - 1$$

$$EtR = T^2 - R$$
$$T = \pm\sqrt{EtR + R}$$

EXERCISE 2

1. Solve the following literal equations.

a. $a(c - x) = c$

b. $ay = p - by$

c. $(c - d)x = 2c - (c + d)x$

d. $ax - a^2 = 7a^2 - 2ax$

e. $\dfrac{ax}{b} - \dfrac{bx}{a} = \dfrac{a}{b} - \dfrac{b}{a}$

f. $\dfrac{x - a}{x + b} = \dfrac{x + b}{x - a}$

g. $\dfrac{c}{d - y} = \dfrac{d}{c - y}$

h. $\dfrac{y + m}{y - m} + \dfrac{y + n}{y - n} = 2$

i. $\dfrac{2}{3} \left(\dfrac{y}{a} - 1 \right) = \dfrac{4}{5} \left(\dfrac{y}{a} + 1 \right)$

j. $\dfrac{3y - a}{y + a} - 2 = \dfrac{y + a}{y - a}$

2. If $C = \frac{5}{9}$ (F $- 32$) find C when $F = 68$.

3. If $V = \pi r^2 h$ find r when $V = 29\frac{1}{3}$, $h = 2\frac{1}{3}$, and $\pi = \frac{22}{7}$.

4. If $S = 2\pi r(r + h)$ find the value of h when $S = 1760$, $r = 14$, and $\pi = \frac{22}{7}$.

5. In the formula $F = \dfrac{mv^2}{gr}$, find the positive value of v if $F = 160$, $m = 1200$, $g = 32$, and $r = 15$.

6. In the formula $V = \dfrac{H}{3}(B + b + \sqrt{Bb})$, find the value of H if $V = 266$, $B = 18$, and $b = 8$.

7. In the formula $A = P(1 + RT)$ find the value of R when $A = 855$, $P = 750$, and $T = 3\frac{1}{2}$.

8. In each of the following formulas, solve for the letter indicated.

a. $V = lwh$ for w

b. $S = \dfrac{a}{1 - r}$ for r

c. $S = \dfrac{n}{2}(a + l)$ for n

d. $A = \dfrac{h}{2}(b + b')$ for b

e. $S = \dfrac{rl - a}{r - 1}$ for r

f. $C = \dfrac{nE}{R + nr}$ for n

g. $t = \pi\sqrt{\dfrac{l}{g}}$ for l

h. $\dfrac{1}{f} = \dfrac{1}{p} + \dfrac{1}{q}$ for q

i. $\dfrac{D}{d} = q + \dfrac{r}{d}$ for r

j. $E = \dfrac{Fs}{(P + R)d}$ for R

k. $S = \frac{1}{2}gt^2$ for t

l. $E = \dfrac{WV^2}{2a}$ for V

m. $A = 2wh + 2hl + 2lw$ for l

n. $u = kd\sqrt{\dfrac{q}{R}}$ for q

SOLUTION OF VERBAL PROBLEMS

In solving verbal problems it is advisable to follow the steps outlined below.

1. Translate the verbal statements in the problem into symbolic language, representing the unknown quantities by letters which represent numbers.

Example: The sum of two numbers is 48. If x represents one of the numbers, then $48 - x$ represents the other number.

2. Use the relationships stated in the problem to set one quantity equal to another, i.e. to set up an equation.

Example: Separate 48 into 2 parts such that the larger one exceeds 3 times the smaller one by 4.

$$\text{Larger one} = 3 \text{ times the smaller one} + 4$$
$$48 - x = 3x + 4$$

3. Solve the equation.

Example: $48 - x = 3x + 4, x = 11$

4. Check the answer in the original verbal problem.

Example: 48 was separated into parts 11 and 37. 37 is 4 more than 3 times 11.

Let us solve the following problem using the 4 steps outlined above.

Problem: The length of a rectangle exceeds twice its width by 7. The perimeter of the rectangle is 50. Find the dimensions of the rectangle.

1. If y represents the width of the rectangle then $2y + 7$ represents its length.

2. Perimeter (2 lengths + 2 widths) = 50.

$$2(2y + 7) + 2y = 50$$

Sometimes, a diagram or some convenient arrangement of the data in the problem will help to set up the equation.

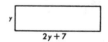

3. $2(2y + 7) + 2y = 50$
$$4y + 14 + 2y = 50$$
$$y = 6$$

4. The width of the rectangle is 6 and the length 19. The perimeter is $2(6) + 2(19) = 12 + 38 = 50$.

Number Problems

Example: Separate 96 into two parts such that the larger divided by the smaller yields a quotient of 18 and a remainder of 1.

► Let $x =$ the smaller number

Then $96 - x =$ the larger number

The relationship used to set up the equation is:

$$\frac{\text{Dividend}}{\text{divisor}} = \text{quotient} + \frac{\text{remainder}}{\text{divisor}}$$

$$\frac{96 - x}{x} = 18 + \frac{1}{x}$$

$$96 - x = 18x + 1$$

$$x = 5 \text{ and } 96 - x = 91$$

Check:

$$\frac{91}{5} = 18 + \frac{1}{5}$$

Example: The larger of two numbers is 1 less than twice the smaller. Three times the larger exceeds five times the smaller by 14.

► Let $x =$ the smaller number

Then $2x - 1 =$ the larger number

The relationship used to set up the equation is:

Three times the larger exceeds five times the smaller by 14.

$$3(2x - 1) = 5x + 14$$

$$6x - 3 = 5x + 14$$

$$x = 17$$

The numbers are 17 and 33, and $3(33) = 5(17) + 14$.

Example: Find three consecutive odd integers such that if 4 times the second be added to 3 times the third the result will be 1 less than 10 times the first.

► Let $x =$ the first odd number

Then $x + 2 =$ the second odd number

And $x + 4 =$ the third odd number

The relationship used to set up the equation is:
4 times the second + 3 times the third equals 10 times the first minus 1.

$$4(x + 2) + 3(x + 4) = 10x - 1$$
$$4x + 8 + 3x + 12 = 10x - 1$$
$$x = 7$$

The numbers are 7, 9, and 11.

Check: $4(9) + 3(11) = 10(7) - 1$
$$36 + 33 \quad\;\; = 70 - 1$$

NOTE: In representing consecutive numbers—

For consecutive numbers—x, $x + 1$, $x + 2$, etc.

For consecutive even numbers—x (first even number), $x + 2$, $x + 4$, etc.

For consecutive odd numbers—x (first odd number), $x + 2$, $x + 4$, etc.

EXERCISE 3

Solve the following problems:

1. Separate 39 into two parts such that 5 times the smaller diminished by the larger is 3.

2. The larger of two numbers exceeds twice the smaller by 3. Four times the larger exceeds 9 times the smaller by 5. Find the two numbers.

3. Separate 79 into two parts such that the larger divided by the smaller yields a quotient of 5 and a remainder of 7.

4. One number is 2 less than 3 times another number. One-half the larger exceeds the smaller by 2. Find the numbers.

5. Find 3 consecutive even integers such that the sum of twice the first and 5 times the second is 30 more than 3 times the third.

6. The difference of the squares of two consecutive numbers is 23. Find the numbers.

7. There are three numbers whose sum is 58. The second number exceeds twice the first number by 5, and the third number is one less than 3 times the first number. Find the three numbers.

8. Separate 93 into two parts such that the larger number divided by the smaller yields a quotient of 4 and a remainder of 8.

Age Problems

Example: A man is now 6 times as old as his son. In two years, the man will be 5 times as old as his son. Find the present ages of the man and his son.

▶ Let x = the present age of the son.
Then let $6x$ = the present age of the father.
$x + 2$ = the son's age in 2 years.
$6x + 2$ = the father's age in 2 years.

The relationship used in setting up the equation is:
In two years the father will be five times as old as the son.

$$6x + 2 = 5(x + 2)$$
$$6x + 2 = 5x + 10$$
$$x = 8$$

Check: The son is now 8 and the father is now 48.
In two years, the son will be 10 and the father will be 50.
At this time, the father will be 5 times as old as the son.

Example: John is now 18 years old and his brother, Charles, is 14 years old. How many years ago was John twice as old as Charles?

▶ Let x = the number of years ago John was twice as old as Charles.
Then $18 - x$ = John's age x years ago
And $14 - x$ = Charles' age x years ago

The relationship used in setting up the equation is:
x years ago, John was twice as old as Charles

$$18 - x = 2(14 - x); \quad x = 10$$

Check: 10 years ago, John was 8 and Charles was 4. At this time, John was twice as old as Charles.

Example: A mother is now 24 years older than her daughter. In 4 years, the mother will be 3 times as old as the daughter. What is the present age of each?

▶ Let x = the age of the daughter
Then $x + 24$ = the age of the mother
$x + 4$ = the age of the daughter in 4 years
$x + 24 + 4 = x + 28$ = the age of the mother in 4 years

The relationship used in setting up the equation is:
In 4 years, the mother will be 3 times as old as the daughter.

$$x + 28 = 3(x + 4); \quad x = 8$$

Check: The daughter is now 8 years old and the mother is 32 years old. In 4 years, the daughter will be 12 years old and the mother will be 36 years old. At this time, the mother will be 3 times as old as the daughter.

EXERCISE 4

Solve and check the following problems:

1. A man is now 3 times as old as his son. Six years ago, the man was 4 times as old as his son. Find the present ages of both father and son.

2. Bill is now 12 years older than Frank. Five years from now Bill will be twice as old as Frank. Find the present age of each.

3. A woman is 28 years older than her daughter. Eight years ago the woman was 3 times as old as her daughter. Find the present age of each.

4. Ten years ago A was 5 times as old as B. At present, A is 3 times as old as B. Find the present age of each.

5. Edward is now 30 years old and Robert is 12 years old. In how many years will Edward be twice as old as Robert?

Coin Problems

Example: A purse contains 19 coins worth $3.40. If the purse contains only dimes and quarters, how many of each coin are in the purse?

▶ Let x = the number of dimes in the purse
Then $19 - x$ = the number of quarters in the purse
$10x$ = the value of the dimes
$25(19 - x)$ = the value of quarters
The relationship used in setting up the equation is:
The value of the dimes + the value of the quarters = $3.40

$$10x + 25(19 - x) = 340$$
$$10x + 475 - 25x = 340$$
$$x = 9$$

There are 9 dimes and 10 quarters in the purse.

Check: The dimes are worth $.90 and the quarters are worth $2.50,
making a total of $3.40.

Example: A toy savings bank contains $17.30 consisting of
nickels, dimes, and quarters.   The number of dimes exceeds twice
the number of nickels by 3 and the number of quarters is 4 less
than 5 times the number of nickels.   How many of each coin are
in the bank?

▶   Let x = the number of nickels
Then $2x + 3$ = the number of dimes
And $5x - 4$ = the number of quarters

The relationship used in setting up the equation is:

Value of nickels + value of dimes + value of quarters = 1730.

$$5x + 10(2x + 3) + 25(5x - 4) = 1730 \quad x = 12$$

There are 12 nickels, 27 dimes and 56 quarters in the bank.

Check: The nickels are worth $.60, the dimes are worth $2.70, and
the quarters are worth $14.00, making a total of $17.30.

EXERCISE 5

Solve the following problems and check:

1. A coin box has $3.25 consisting of nickels and dimes.   The
number of dimes is 5 less than twice the number of nickels.   How
many of each kind are in the box?

2. A purse has 23 coins worth $4.70.   If the purse contains only
dimes and quarters, how many of each type of coin is in the purse?

3. A merchant deposited $11.50 in a bank.   The number of quar-
ters exceeded the number of half-dollars by 3 and the number of one
dollar bills was 5 less than the number of half-dollars.   Find the
number of one dollar bills the merchant deposited.

4. A man bought 80 stamps for $5.25, of which some were five-
cent stamps and some ten-cent stamps.   How many of each kind
did he buy?

5. The admission price at a movie house was $.60 in the balcony
and $.85 in the orchestra.   One evening a cashier's receipts were
$340.90.   If the cashier sold 479 tickets,' how many of each kind
did he sell?

6. A coin bank has a total of $10.30, consisting of nickels, dimes, and quarters. The number of dimes is two less than three times the number of nickels and the number of quarters exceeds the number of nickels by 6. How many of each coin were in the bank?

Mixture Problems

Methods of analyzing and solving mixture problems are illustrated in the following examples.

Example: A storekeeper has two kinds of cookies, one worth $.75 a pound and the other worth $.50 a pound. How many pounds of each should he use to make a mixture of 60 pounds worth $.55 a pound?

▶ Let x = the number of pounds of cookies at 75 cents
   Then $(60 - x)$ = the number of cookies at 50 cents a
   pound
The relationship used to set up the equation is:
Value of 75 cent cookies + Value of 50 cent cookies = Value of mixture.

$$75x + 50(60 - x) = 60(55)$$
$$75x + 3000 - 50x = 3300$$
$$25x = 300$$
$$x = 12 \text{ pounds of 75 cent cookies}$$
$$60 - x = 48 \text{ pounds of 50 cent cookies}$$

Check: The value of 12 pounds of cookies at 75 cents a pound is
         $9.00. The value of 48 pounds of cookies at 50 cents a
         pound is $24.00. The resulting mixture contains 60
         pounds worth $33.00. If 60 pounds of cookies are sold
         for $33.00 then each pound is sold for $.55.

Example: A chemist has 24 ounces of a 25% solution of argyrol. How much water must he add to reduce the strength of the argyrol to 20%?

▶ Let x = number of ounces of water to be added
The relationship used to set up the equation is (since only water is added)

Amount of argyrol in new mixture = amount of argyrol in old mixture.

$$.20(24 + x) = .25(24)$$
$$4.8 + .2x = 6$$
$$2x = 12, x = 6 \text{ ounces of water}$$

Check: The new mixture has 30 ounces. Of this, 6 ounces or 20% of the mixture is argyrol.

Example: How many quarts of pure alcohol must be added to 40 quarts of a mixture that is 35% alcohol to make a mixture that will be 48% alcohol.

▶ Let x = number of quarts of pure alcohol to be added
The relationship used to set up the equation is
Amount of alcohol in new mixture = Amount of alcohol in old mixture + Amount of alcohol added

$$.48(40 + x) = (.35)(40) + x$$
$$19.2 + .48x = 14 + x$$
$$x = 10 \text{ quarts of alcohol}$$

Check: Amount of alcohol in new mixture = 14 quarts + 10 quarts = 24 quarts. New mixture contains a total of 40 + 10 = 50 quarts. $\frac{24}{50} = 48\%$

Example: How much water must be evaporated from 120 pounds of solution which is 3% salt to make a solution of 5% salt?

▶ Let x = number of pounds of salt to be evaporated
The relationship used to set up the equation is
Amount of salt in new mixture = Amount of salt in old mixture.

$$.05(120 - x) = .03(120)$$
$$5(120 - x) = 3(120)$$
$$600 - 5x = 360$$
$$x = 48 \text{ pounds of water}$$

Check: Water in new mixture is 120 − 48 = 72 pounds. Of this

3% of 120 or 3.6 pounds is salt. $\frac{3.6}{72} = 5\%$.

Example: A truck radiator contains 32 quarts of a 20% solution of anti-freeze. How much of the original solution must be drawn off and replaced by pure anti-freeze to obtain a solution of 45% anti-freeze?

▶ Let $x =$ number of quarts of solution drawn off.

The relationship used to set up the equation is

Volume of anti-freeze left in 20% solution after drawing off x quarts $+$ Volume of anti-freeze added $=$ Volume of anti-freeze in 45% solution.

$$.20(32 - x) + x = .45(32)$$
$$6.4 - .20x + x = 14,40$$
$$.80x = 8$$
$$x = 10 \text{ quarts of anti-freeze added}$$

Check: The original mixture contained 32 quarts. From this 10 quarts were drawn off. Of this 20% or 2 quarts were anti-freeze. Since the original mixture contained 20% of 32 quarts or 6.4 quarts of anti-freeze, this left 4.4 quarts of anti-freeze. To this was added 10 quarts of anti-freeze making a total of 14.4 quarts of anti-freeze. 14.4 quarts is 45% of 32.

Example: Of 24 quarts of a mixture, 8% is iodine. Of another mixture, 4% is iodine. How many quarts of the second mixture should be added to the first mixture to obtain a mixture that is 5% iodine?

▶ Let $x =$ the number of quarts to be taken from second mixture.

The relationship used to set up the equation is:

Volume of iodine in the first mixture $+$ Volume of iodine in the second mixture $=$ Volume of iodine in the resulting mixture.

$$.08(24) + .04x = .05(24 + x)$$
$$1.92 + .04x = 1.20 + .05x$$
$$.01x = .72$$
$$x = 72 \text{ quarts of iodine}$$

Check: Volume of iodine in the first mixture is $(.08)(24)$ or 1.92 quarts. Volume of iodine in the second mixture is $(.04)(72)$ or 2.88 quarts. Total volume of iodine is 4.80 quarts is equal to 5% of 96 quarts.

Example: How many ounces of silver alloy which is 28% silver must be mixed with 24 ounces of silver which is 8% silver to produce a new alloy which is 20% silver?

▶ Let x = number of ounces of 28% silver to be used

The relationship used to set up the equation is

Volume of 28% silver + Volume of 8% silver = Volume of silver in mixture.

$$.28x + .08(24) = .20(x + 24)$$
$$28x + 8(24) = 20(x + 24)$$
$$8x = 288$$
$$x = 36 \text{ ounces of silver}$$

Check: Volume of 28% silver = $(.28)(36) = 10.08$

Volume of 8% silver = $(.08)(24) = 1.92$

Total amount of silver = 12 ounces

The total mixture contains $24 + 36 = 60$ ounces, and 12 ounces is 20% of 60 ounces.

EXERCISE 6

Solve and check the following problems:

1. How much water must be added to 20 ounces of a 10% solution of boric acid to reduce it to a 4% solution?

2. How much pure acid must be added to 120 c.c. of a 15% solution of the acid to strengthen it to a 25% solution?

3. A grocer wishes to blend two qualities of coffee. One quality sells for $.96 a pound and the other for $1.12 a pound. How many pounds of each quality must he use to obtain a blend of 20 pounds selling for $1.00 a pound?

4. A chemist has an 18% solution and a 45% solution of a disinfectant. How many ounces of each should be used to make 12 ounces of a 36% solution?

5. How many pounds of candy worth 45 cents per pound must be mixed with 20 pounds of candy worth 60 cents per pound to make a mixture which can be sold at 50 cents per pound?

6. A storekeeper has two kinds of nuts, one worth $.45 a pound and the other worth $.65 a pound. How many pounds of each should he use to make a mixture of 60 pounds selling for $.57 a pound?

7. How much water must be evaporated from 40 pounds of a solution which is 5% salt to make a solution which is 8% salt?

8. A car radiator contains 16 quarts of a 25% solution of anti-freeze. How much of the original solution must be drawn off and replaced by pure anti-freeze in order to obtain a solution of 40% anti-freeze?

9. A tank contains 60 pints of 15% acid. How much water must be evaporated to obtain a solution of 25% acid?

10. A car radiator contains 20 pints of a 20% solution of alcohol. How much must be drawn out and replaced with pure alcohol to obtain a solution of 60% alcohol?

11. How many quarts of cream which contains 76% butter fat and how many quarts of cream which contains 12% butter fat are needed to make a mixture of 54 quarts containing 60% butter fat?

12. One alloy contains 30% zinc and another alloy contains 60% zinc. How many pounds of each are needed to produce 90 pounds of zinc alloy that will be 50% zinc?

13. A chemist has 20 gallons of 30% acid. How much 80% acid must be added to produce a mixture that is 40% acid?

14. A chemist has 40 gallons of 20% acid. How much of the solution must be drawn out and replaced by a solution of 80% acid to obtain a solution which is 55% acid?

15. An alloy contains 18 pounds of tin and 6 pounds of copper. How much copper must be added to produce an alloy which will be 70% copper?

Business and Investment Problems

Interest is computed by multiplying the principal by the rate by the time, i.e. by using the formula $I = PRT$. For example, the interest on $400.00 at 3% for 2 years is $400.00 $\times \frac{3}{100} \times$ 2 = $24.00.

Example: A man invests $9,000, part at 3% and rest at 4%. If his total income from the two investments is $295, how much did he invest at each rate?

▶   Let x = amount invested at 3%
    Then ($9,000 - x$) = amount invested at 4%

The relationship used to set up this equation is

Income from 3% invested + Income from 4% investment = $295

$$.03x + .04(9,000 - x) = 295$$
$$3x + 4(9,000 - x) = 29,500$$
$$3x + 36,000 - 4x = 29,500$$
$$x = 6,500$$
$$9,000 - x = 2,500$$

Check: $6,500 at 3% for 1 year = $195
$2,500 at 4% for 1 year = $100
Total income = $295

Example: A man wishes to invest a part of $4200 in stocks earning 4% dividends and the remainder in bonds paying $2\frac{1}{2}$%. How much must he invest in stocks to receive an average return of 3% on the whole amount of money?

▶   Let x = amount invested at 4%
Then $4200 - x$ = amount invested at $2\frac{1}{2}$%
The relationship used to set up the equation is

Income from 4% investment + Income from $2\frac{1}{2}$% investment = 3% of $4200, or $126.

$$.04x + .025(4200 - x) = 126$$
$$40x + 25(4200 - x) = 126,000$$
$$40x + 105,000 - 25x = 126,000$$
$$15x = 21,000$$
$$x = 1,400$$
$$4200 - x = 2,800$$

Check:

4% of $1,400 = $56
$2\frac{1}{2}$% of $2,800 = $70
Total income = $126 which is 3% of $4,200

Example: A television set cost a dealer $102. At what price should he mark the set so that he can give a discount of 15% from the marked price and still make a profit of 20% on the selling price?

▶   Let x = The marked price
Then $.85x$ = The selling price

The relationship used to set up the equation is

Selling price = Cost + Profit

$$.85x = 102 + (.20)(.85x)$$
$$.85x = 102 + .17x$$
$$.68x = 102$$
$$x = \$150$$

Check: The marked price = $150. The selling price is 15% less or $127.50. Since the cost is $102, the profit is $25.50. The profit ($25.50) is 20% of the selling price ($127.50).

Example: At a movie showing there were 356 paid admissions. The total receipts were $287.40. If orchestra seats sold for $.90 and balcony seats for $.65, how many of each kind were sold?

▶   Let x = The number of $.90 seats sold
     Then $(356 - x)$ = The number of $.65 seats sold
The relationship used to set up the equation is

Orchestra receipts + Balcony receipts = Total receipts or $287.40

$$.90x + .65(356 - x) = 287.40$$
$$90x + 65(356 - x) = 287.40$$
$$90x + 23140 - 65x = 287.40$$
$$25x = 5600$$
$$x = 224 \text{ orchestra seats sold}$$
$$356 - x = 132 \text{ balcony seats sold}$$

Check: 224 at $.90 = $201.60
           132 at $.65 = $85.80
           356 tickets for $287.40

EXERCISE 7

Solve and check the following problems:

1. A man invested $9,200, part at 5% and the balance at 4%. If the total net income was $423, how much was invested at each rate?

2. A man invested $5,900, part at 5% and the balance at $3\frac{1}{2}$%. If the total net income was $244, how much was invested at each rate?

3. A man invested a part of $8,400 in bonds paying 3% interest and the balance in stocks paying 6% dividends. How much must he invest at each rate to obtain an average return of 4%?

4. A man invested a total of \$5,600, part at $5\frac{1}{2}\%$ and part at 4%. If the income on the $5\frac{1}{2}\%$ investment exceeded the income on the 4% investment by \$80, find the amount invested at each rate.

5. A suit cost a dealer \$42. At what price should he mark the suit so that he can give a discount of 20% and still make a profit of 25% on the selling price?

6. A camera cost a dealer \$54.00. At what price should he mark the camera so that he can give a discount of 10% and still make a profit of 25% on the selling price?

7. On an examination, a student received 5 credits for each correct answer, and lost two credits for each question unanswered or answered incorrectly. If the examination consisted of 80 questions and the student scored 274 credits, how many did he answer correctly?

8. During a special sale a merchant grouped all the men's shirts into two grades, one grade selling for \$3.25 and the other grade selling for \$2.49. If he sold 145 shirts during the sale and collected receipts of \$396.77, how many shirts of each grade did he sell?

Geometry Problems

Example: The length of a rectangle is 5 inches less than 3 times its width. If the perimeter of the rectangle is 54 inches, find the dimensions of the rectangle.

▶ Let $x =$ the width of the rectangle
Then $3x - 5 =$ the length of the rectangle
The relationship used to set up the equation is

Twice the length + Twice the width = Perimeter
$$2(3x - 5) + 2x = 54$$
$$6x - 10 + 2x = 54$$
$$8x = 64$$
$$x = 8 \text{ inches, width}$$
$$3x - 5 = 19 \text{ inches, length}$$

Check:
$$2(8) + 2(19) = 54$$
$$16 + 38 = 54$$
$$54 = 54$$

Example: The length of a rectangle is 3 inches less than twice the width.   If each dimension is increased by 4 inches, the area is increased by 88 square inches.   Find the dimensions of the original rectangle.

▶   Let x = the width of the rectangle
Then $2x - 3$ = the length of the rectangle

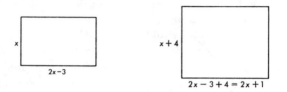

The dimensions of the new rectangle are $(x + 4)$ and $(2x + 1)$
The relationship used to set up the equation is

The area of the original rectangle + 88 square inches = The area of the new rectangle.

$$x(2x - 3) + 88 = (x + 4)(2x + 1)$$
$$2x^2 - 3x + 88 = 2x^2 + 9x + 4$$
$$12x = 84$$
$$x = 7 \text{ inches, width}$$
$$2x - 3 = 11 \text{ inches, length}$$

Check: Dimensions of the original rectangle are 7 inches and 11 inches.   The area of the original rectangle is 77 square inches.

Dimensions of the new rectangle are 11 inches and 15 inches.   The area of the new rectangle is 165 square inches.
$$77 + 88 = 165$$

Example: The length of a rectangular swimming pool is twice its width.   The pool is surrounded by a cement walk 4 feet wide.   The area of the walk is 784 square feet.   Find the dimensions of the pool.

▶   Let x = width of the pool
Then $2x$ = the length of the pool

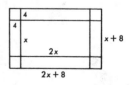

The dimensions of the outside walk are $(x + 8)$ and $(2x + 8)$
The relationship used in setting up the equation is
Total area — Area of pool = Area of walk
$$(2x + 8)(x + 8) - 2x^2 = 784$$
$$2x^2 + 24x + 64 - 2x^2 = 784$$
$$24x = 720$$
$$x = 30 \text{ feet, width of pool}$$
$$2x = 60 \text{ feet, length of pool}$$

Check: Outside dimensions are 38 by 68 and area is 2,584 sq. ft.
Dimensions of pool are 30 by 60 and area is 1,800 sq. ft.
Difference in areas is $2,584 - 1,800 = 784$ sq. ft., which is
the area of the walk.

Example: The width of a rectangle is 8 inches and the diagonal
of the rectangle is 2 inches greater than its length.   Find the length
of the rectangle.

▶   Let $x =$ the length of the rectangle
Then $x + 2 =$ the diagonal of the rectangle
The relationship used in setting up the equation is
Square of the width + Square of the length = Square of the
diagonal.
$$8^2 + x^2 = (x + 2)^2$$
$$64 + x^2 = x^2 + 4x + 4$$
$$4x = 60$$
$$x = 15 \text{ inches, length of the rectangle}$$

Check: $(\text{Width})^2 + (\text{Length})^2 = (\text{Diagonal})^2$
$$8^2 + 15^2 = 17^2$$
$$64 + 225 = 289$$
$$289 = 289$$

EXERCISE 8

Solve and check the following problems:
   1. The length of a rectangle is 3 inches more than twice its width.
Its perimeter is 48 inches.   Find its dimensions.
   2. If one side of a square is increased by 3 feet and its adjacent

side is decreased by 2 feet, then the area of the square is increased by 3 square feet. Find the side of the square.

3. The length of a rectangle is 5 greater than its width. If the length is decreased by 2 and the width increased by 1, the area remains unchanged. Find the dimensions of the rectangle.

4. The length of a rectangle is 1 less than twice its width. If the length is increased by 1 and the width is increased by 3, the area is increased by 35 square units. Find the dimensions of the rectangle.

5. The length of a rectangular swimming pool is three times its width. The pool is surrounded by a cement walk 3 feet wide. The area of the walk is 756 square feet. Find the dimensions of the pool.

6. A leg of a right triangle is 10 and the hypotenuse exceeds the other leg by 2. Find the other leg.

7. The width of a rectangle is 9 inches and the diagonal of the rectangle is 3 inches greater than the length of the rectangle. Find the length of the rectangle.

Motion Problems

The fundamental relationship used in writing equations for motion problems is Rate $\times$ Time $=$ Distance or $RT = D$. At times, it is more convenient to apply the formula in the forms $R = \dfrac{D}{T}$ or $T = \dfrac{D}{R}$. For example, the distance covered by a car traveling at a uniform rate of 40 miles per hour for 8 hours is $40 \times 8 = 320$ miles.

Example: At 2:00 P.M. two automobiles leave Chicago, one traveling at a uniform speed of 46 miles per hour and the other traveling west at a uniform speed of 42 miles per hour. At what time will the two automobiles be 462 miles apart?

▶   Let $x =$ The number of hours traveled by each car
     Then $46x =$ The distance covered by the first car
      And $42x =$ The distance covered by the second car

The relationship used in setting up the equation is

Distance covered by first car + Distance covered by second car = 462 miles.

$$46x + 42x = 462$$
$$88x = 462$$
$$x = 5\tfrac{1}{4} \text{ hours}$$

Therefore, the cars were 462 miles apart at 7:15 P.M.

Check: The first car covered $46 \times 5\tfrac{1}{4} = 241\tfrac{1}{2}$ miles

The second car covered $42 \times 5\tfrac{1}{4} = 220\tfrac{1}{2}$ miles

The sum of the distances covered by the two cars is 462 miles

Example: A plane left an air base at 9:00 A.M. flying at a uniform rate of 320 miles per hour. At 10:30 A.M. a second plane left the same base flying in pursuit at the uniform rate of 520 miles per hour. At what time will the second plane overtake the first?

▶ Let x = The number of hours flown by the pursuing plane

Then $x + \dfrac{3}{2}$ = The number of hours flown by the first plane

The relationship used in setting up the equation is

Distance covered by second car = Distance covered by first car

$$520x = 320\left(x + \frac{3}{2}\right)$$
$$520x = 320x + 480$$
$$x = 2\tfrac{2}{5} \text{ hours}$$

Therefore, the second plane will overtake the first at $10:30 + 2\tfrac{2}{5}$ hours = 10:30 + 2 hrs. and 24 minutes = 12:54 P.M.

Check: By 12:54 the first plane had flown 3 hours and 54 minutes or $3\tfrac{9}{10}$ hours at a rate of 320 miles per hour. The first plane had covered 1,248 miles.

By 12:54 the second plane had flown $2\tfrac{2}{5}$ hours at a rate of 520 miles per hour. The second plane had also covered 1,248 miles.

Example: A man took his family for an automobile ride driving at a uniform rate of 40 miles per hour. He estimated that his average rate on the return trip would be 32 miles per hour. How far did he drive if the round trip took $4\tfrac{1}{2}$ hours?

▶ Let x = The number of miles the man drove

Then $\dfrac{x}{40}$ = Time consumed on the trip out

And $\dfrac{x}{32}$ = Time consumed on the trip back

The relationship used in setting up the equation is

Time on trip out + Time on trip back = Round trip time

$$\frac{x}{40} + \frac{x}{32} = \frac{9}{2}$$

$$4x + 5x = 720, \; x = 80 \text{ miles}$$

The complete distance covered is 160 miles

Check: Time on trip out = $\dfrac{80}{40} = 2$ hours

Time on trip back = $\dfrac{80}{32} = 2\tfrac{1}{2}$ hours

Round trip = $4\tfrac{1}{2}$ hours

Example: A race is conducted on a circular track.   The faster of two runners can cover a lap in 30 seconds and the slower runner can cover a lap in 40 seconds.   If the runners start from the same place at the same time, how long will it take the faster runner to gain a lap on the second runner?

▶ Let x = Time required to gain a lap

Rate of faster runner = $\dfrac{1}{30}$ lap per second

Rate of slower runner = $\dfrac{1}{40}$ lap per second

The relationship used in setting up the equation is

Distance covered by faster runner − Distance covered by slower runner = 1 lap

$$\frac{x}{30} - \frac{x}{40} = 1$$

$$x = 120 \text{ seconds}$$

Check: In 120 seconds, faster runner covers $\dfrac{120}{30} = 4$ laps. In 120 seconds, slower runner covers $\dfrac{120}{40} = 3$ laps.

Example: A boat can travel 8 miles an hour in still water. If it can travel 15 miles down a stream in the same time that it can travel 9 miles up the stream, what is the rate of the stream?

▶ Let $x =$ Rate of the stream

Then $\dfrac{15}{8 + x} =$ Time spent going downstream (with current)

And $\dfrac{9}{8 - x} =$ Time spent going upstream (against current)

The relationship used in setting up the equation is

Time spent going downstream = Time spent going upstream

$$\frac{15}{8 + x} = \frac{9}{8 - x}$$
$$9(8 + x) = 15(8 - x)$$
$$x = 2 \text{ miles per hour, rate of current}$$

Check: Rate going upstream $= 8 + 2 = 10$ m.p.h. It took $1\frac{1}{2}$ hours to travel 15 miles.

Rate going downstream $= 8 - 2 = 6$ m.p.h. It took $1\frac{1}{2}$ hours to travel 9 miles.

Example: On a trip, a motorist traveled at a certain average speed for the first $2\frac{1}{2}$ hours. He then ran into heavy traffic and reduced his speed by 10 miles per hour for the next $1\frac{1}{4}$ hours. He traveled a total of 175 miles. What were the two speeds?

▶ Let $x =$ Rate of speed on first part of trip

Then $x - 10 =$ Rate of speed on second part of trip

The relationship used in setting up the equation is

Distance on first part of trip + Distance on second part of trip $= 175$ miles.

$$\frac{5x}{2} + \frac{5}{4}(x - 10) = 175$$
$$10x + 5x - 50 = 700$$
$$x = 50 \text{ miles per hour on first part of trip}$$
$$x - 10 = 40 \text{ miles per hour on second part of trip}$$

Check: $(50)(2\frac{1}{2}) + 40(1\frac{1}{4}) = 175$
$$125 + 50 = 175$$

EXERCISE 9

Solve and check the following problems:

1. Two automobiles start at the same time and place and travel in opposite directions. The first car averages 32 miles an hour and the second 36 miles an hour. In how many hours will the cars be 306 miles apart?

2. Two planes start at 10:00 A.M. from two airports which are 1,705 miles apart and fly toward each other at average rates of 280 miles per hour and 340 miles per hour. At what time will the planes pass each other?

3. Two trains starting at the same time from stations 396 miles apart, meet in $4\frac{1}{2}$ hours. How far has the faster train traveled when it meets the slower one, if the difference in their rates is 8 miles per hour.

4. At 1:00 P.M., a local train leaves Central Station traveling at an average speed of 36 miles per hour. At 2:30 P.M., an express train leaves the same station traveling at the average speed of 48 miles per hour. At what time will the express train overtake the local train?

5. A plane left an airport at 9:00 A.M. A second plane left the same airport at 11:30 A.M. traveling at the rate of 360 miles per hour. If the second plane overtook the first plane in 2 hours, at what rate of speed was the first plane traveling?

6. A car left New York traveling south at the average rate of 45 miles per hour. An hour and a half later, a second car left New York in pursuit of the first. If the second car is to overtake the first car in $4\frac{1}{2}$ hours, how fast must the second car travel?

7. A man had 5 hours to drive from his home to Springfield and back. If his average speed was 30 miles per hour, on the trip out and 45 miles per hour on the trip back, how far is the man's home from Springfield?

8. A man took a bicycle trip traveling at the rate of 12 miles per hour. On the return trip, he traveled in a car at the average rate of 28 miles per hour. If the round trip took 5 hours, how far did the man travel?

9. Don Jackson can run around a circular track in 35 seconds and Charles Edwards can run around the same track in 40 seconds. If the two boys start from the same place at the same time, how long will it take the faster runner to gain a lap on the slower runner?

10. A plane travels 240 miles per hour in still air. If it can travel 640 miles with the wind in the same time that it can travel 560 miles against the wind, find the speed of the wind.

11. A boat can travel 18 miles per hour in still water. If the boat can travel 85 miles with the current in the same time that it takes to travel 68 miles against the current, find the rate of the current.

12. A traveler goes part of a trip by bus at the average rate of 42 miles per hour and part of the trip by train at the average rate of 56 miles per hour. The entire trip covered 511 miles and took $10\frac{3}{4}$ hours. How many miles were traveled by bus?

13. A and B start from two towns 140 miles apart. A starts at 9:00 A.M., B starts at 10:00 A.M., and they meet at noon. If B travels 5 miles per hour faster than A, find the rate of each.

14. A plane flew from one city to another at 240 miles per hour. The plane returned at the rate of 270 miles per hour, and made the trip in $\frac{3}{4}$ hour less time. How far apart were the cities?

15. A river flows at the rate of 3 miles per hour. If a boat takes as long to go 75 miles downstream as it does to go 45 miles upstream, how fast would it travel in still water?

Work Problems

In solving work problems it should be borne in mind that, if a man can do a job in 5 days, he completes $\frac{1}{5}$ in one day, $\frac{2}{5}$ in 2 days, and $\frac{x}{5}$ in x days. In general, if it takes b days to complete a job, the part of it that can be done in a days is represented by the fraction $\frac{a}{b}$.

Example: If A can do a job in 8 days and B can do the same job in 12 days, how long would it take the two men working together?

► Let $x =$ The number of days it would take the two men working together

Then $\dfrac{x}{8}$ = The part of the job done by A

And $\dfrac{x}{12}$ = The part of the job done by B

The relationship used in setting up the equation is

Part of job done by A + Part of job done by B = 1 job

$$\frac{x}{8} + \frac{x}{12} = 1$$
$$3x + 2x = 24$$
$$x = 4\tfrac{4}{5} \text{ days}$$

Check:
$$\frac{4\tfrac{4}{5}}{8} + \frac{4\tfrac{4}{5}}{12} = 1$$
$$\frac{\tfrac{24}{5}}{8} + \frac{\tfrac{24}{5}}{12} = 1$$
$$\frac{3}{5} + \frac{2}{5} = 1$$

Example: A tank can be filled in 9 hours by one pipe, in 12 hours by a second pipe, and can be drained when full, by a third pipe, in 15 hours. How long would it take to fill the tank if it is empty, and if all pipes are in operation?

▶   Let x = The number of hours the pipes are in operation

Then $\dfrac{x}{9}$ = Part of tank filled by first pipe

And $\dfrac{x}{12}$ = Part of tank filled by second pipe

And $\dfrac{x}{15}$ = Part of tank emptied by third pipe

The relationship used in setting up the equation is

Part of tank filled by first pipe + Part of tank filled by second pipe − Part of tank emptied by third pipe = 1 Full tank.

$$\frac{x}{9} + \frac{x}{12} - \frac{x}{15} = 1$$
$$20x + 15x - 12x = 180$$
$$x = 7\tfrac{19}{23} \text{ hours}$$

Check: $\dfrac{7\frac{19}{23}}{9} + \dfrac{7\frac{19}{23}}{12} - \dfrac{7\frac{19}{23}}{15} = 1$

$$\left(\dfrac{180}{23} \cdot \dfrac{1}{9} \right) + \left(\dfrac{180}{23} \cdot \dfrac{1}{12} \right) - \left(\dfrac{180}{23} \cdot \dfrac{1}{15} \right) = 1$$

$$\dfrac{20}{23} + \dfrac{15}{23} - \dfrac{12}{23} = 1$$

Example: A man can do a job in 9 days and his son can do the same job in 16 days. They start working together. After 4 days the son leaves and the father finishes the job alone. How many days did the man take to finish the job alone?

▶ Let x = The number of days it takes the man to finish the job.

Note that the man actually works $(x + 4)$ days, and the son actually works 4 days.

The relationship used to set up the equation is

Part of job done by man + Part of job done by boy = 1 job

$$\dfrac{x + 4}{9} + \dfrac{4}{16} = 1$$

$$16(x + 4) + 4(9) = 144$$

$$x = 2\tfrac{3}{4} \text{ days}$$

Check: $\dfrac{2\frac{3}{4} + 4}{9} + \dfrac{4}{16} = 1$

$$\dfrac{3}{4} + \dfrac{1}{4} = 1$$

Example: A mechanic and his helper can repair a car in 8 hours. The mechanic works 3 times as fast as his helper. How long would it take the helper to make the repair, working alone?

▶ Let x = Number of hours it would take the mechanic working alone.

Then $3x$ = Number of hours it would take the helper working alone.

The relationship used in setting up the equation is

Part of job done by mechanic + Part of job done by helper = 1 job.

$$\frac{8}{x} + \frac{8}{3x} = 1$$

$$x = 10\tfrac{2}{3} \text{ hours by mechanic}$$

$$3x = 32 \text{ hours by helper}$$

Check:
$$\frac{8}{10\tfrac{2}{3}} + \frac{8}{32} = 1$$

$$\frac{24}{32} + \frac{1}{4} = 1$$

$$\frac{3}{4} + \frac{1}{4} = 1$$

EXERCISE 10

Solve and check the following problems:

1. A can do a job in 6 days and B can do the same job in 10 days. How long would it take both men working together?

2. Three pipes can fill a tank. The first pipe can fill it in 10 minutes, the second pipe can fill it in 12 minutes, and the third pipe can fill it in 15 minutes. How long would it take the three pipes to fill it?

3. A pool can be filled by one pipe in 6 hours, by a second pipe in 9 hours, and can be drained when full, by a third pipe in 8 hours. If the pool is empty and all three pipes are in operation, how long would it take to fill the pool?

4. A carpenter can do a job in 12 hours and his helper can do the same job in 16 hours. They start working together. After 4 days, the carpenter is called to another job and the helper completes the work. How long did it take the helper to finish up?

5. A man can paint his house in 6 days and his son can paint the house in 10 days. The man works alone for 4 days and then his son finishes the job. How many days did it take the son to complete the job?

6. A man can mow his lawn in 40 minutes and his son can mow the lawn in 50 minutes. The man starts to mow alone and is joined by his son after 10 minutes and they complete the job together. How long does the son work?

7. Two printing machines can complete a job in 12 hours. One machine works twice as fast as the other. How long would it take the slower machine to complete the job, working by itself?

8. A cistern can be filled by one pipe in 9 hours, and emptied by a second pipe in 12 hours. How long would it take to fill the cistern if both pipes are open?

9. A master plumber can complete a job in 6 working days. After he has worked for 2 days, he is called away and another plumber completes the job in 10 days. How long would it take the second plumber to do the job, working alone?

Ratio Problems

Example: The numerator and denominator of a fraction are in the ratio $5 : 4$. If 5 is added to the numerator and 12 is subtracted from the denominator, the value of the resulting fraction is $\frac{5}{2}$. Find the original fraction.

▶ Let $5x =$ Numerator of the fraction

Then $4x =$ Denominator of the fraction

The relationship used in setting up the equation is

$$\frac{\text{Original numerator} + 5}{\text{Original denominator} - 12} = \frac{5}{2}$$

$$\frac{5x + 5}{4x - 12} = \frac{5}{2}$$

$$2(5x + 5) = 5(4x - 12)$$

$$x = 7$$

Thus, the original fraction is $\dfrac{7 \times 5}{7 \times 4} = \dfrac{35}{28}$

Check: $\dfrac{\text{Original numerator} + 5}{\text{Original denominator} - 12} = \dfrac{35 + 5}{28 - 12} = \dfrac{40}{16} = \dfrac{5}{2}$

Example: A and B have sums of money in the ratio $4 : 3$. If A gives B \$12, the sums of money they have will be in the ratio 5:9. How much did each man possess originally?

▶ Let $4x =$ A's original amount

Then $3x =$ B's original amount

The relationship used in setting up the equation is

$$\frac{\text{A's original amount} - \$12}{\text{B's original amount} + \$12} = \frac{5}{9}$$

$$\frac{4x - 12}{3x + 12} = \frac{5}{9}$$

$$9(4x - 12) = 5(3x + 12)$$

$$x = 8$$

Thus, the original amounts were $8 \cdot 4 = \$32$ for A and $8 \cdot 3 = \$24$ for B.

Check: $\dfrac{\text{A's original amount} - \$12}{\text{B's original amount} + \$12} = \dfrac{32 - 12}{24 + 12} = \dfrac{20}{36} = \dfrac{5}{9}$

Example: Three numbers are in the ratio $3:5:7$. The sum of $\frac{1}{5}$ of the first number and $\frac{1}{2}$ of the last number is 9 less than the middle number. Find the numbers.

▶   Let $3x$ = The first number

Then $5x$ = The second number

And $7x$ = The third number

The relationship used in setting up the equation is

$\frac{1}{5}$ of the first number $+ \frac{1}{2}$ of the third number = the second number $- 9$

$$\frac{3x}{5} + \frac{7x}{2} = 5x - 9$$

$$x = 10$$

Thus, the three numbers are $3 \cdot 10 = 30$, $5 \cdot 10 = 50$, and $7 \cdot 10 = 70$.

Check: $\frac{1}{5}$ of the first number $= 6$, $\frac{1}{2}$ of the last number $= 35$

$$6 + 35 = 50 - 9$$

$$41 = 41$$

EXERCISE 11

Solve and check the following problems:

1. Two numbers are in the ratio $7:4$. Their difference is 27. Find the numbers.

2. Three partners are in business and agree to divide their annual profits in the ratio $3:2:4$. One year the firm made a profit of $34,884. What was each man's share of the profit?

3. The numerator and denominator of a fraction are in the ratio 3 : 8. If 3 is added to both numerator and denominator, the value of the resulting fraction is $\frac{3}{7}$. Find the original fraction.

4. A and B have sums of money in the ratio 9 : 5. If A gives B $12 the sums of money they have will be in the ratio 3 : 4. How much did each man possess originally?

5. Two numbers are in the ratio 2 : 3. If one-third of the smaller be added to one-fourth of the larger, the result will be 14 less than the smaller. Find the numbers.

6. Two cars are 300 miles apart. They travel toward each other at rates which are in the ratio of 7 : 8. If they meet after 5 hours, what are their rates?

Clock Problems

Example: At what time between 1 and 2 are the hands of a clock in opposite directions?

▶   At one o'clock there are 5 one-minute spaces between the hands of a clock. The hands of a clock will be in opposite directions when the distance between them will be 30 minute spaces. Thus, the minute hand will move 35 more minute spaces than the hour hand between one o'clock and the time the hands are in opposite directions.

Let x = The number of spaces the minute hand moves

Then $x - 35$ = The number of spaces the hour hand moves

Since the minute hand moves twelve times as fast as the hour hand:

$$x = 12(x - 35)$$
$$x = 38\tfrac{2}{11}$$

Thus, the hands will be in opposite directions at $1{:}38\tfrac{2}{11}$

Example: At what time between 10 and 10:30 are the hands of a watch at right angles?

▶   At 10 o'clock there are 10 one-minute spaces between the hands of a clock. The hands of a clock will be at right angles when the distance between them will be 15 minute spaces. Thus, the minute hand will move 5 more spaces than the hour hand between 10 o'clock and the time the hands will be at right angles.

Let x = The number of spaces the minute hand moves
Then $x - 5$ = The number of spaces the hour hand moves
$$x = 12(x - 5)$$
$$x = 5\tfrac{5}{11}$$

Thus, the hands will be at right angles at $10:05\tfrac{5}{11}$.

Example: At what time between 6 and 7 will the hands of a clock be together?

▶   At 6 o'clock there are 30 one-minute spaces between the hands of a clock. The hands of a clock will be together when there are no minute spaces between them. Thus, the minute hand will move 30 more minute spaces than the hour hand between 6 o'clock and the time the hands will be together.

Let x = The number of spaces the minute hand moves
Then $x - 30$ = The number of spaces the hour hand moves
Since the minute hand moves twelve times as fast as the hour hand
$$x = 12(x - 30)$$
$$x = 32\tfrac{8}{11}$$

Thus, the hands will be together at $6:32\tfrac{8}{11}$.

EXERCISE 12

Solve the following problems:

1. At what time between 4 and 5 are the hands of a clock pointing in opposite directions?

2. At what time between 7 and 7:30 are the hands of a watch at right angles?

3. At what time between 10 and 11 are the hands of a clock together?

ANSWERS

EXERCISE 1—Page 156

1. a. Identity
   b. Conditional equation
   c. Conditional equation
   d. Identity
   e. Conditional equation

2. a. Redundant
   b. Defective
   c. Redundant
   d. Redundant

3. a. 3
b. 5
c. 4
d. 8
e. 7
f. -2
g. -2
h. 10
i. 5

j. $\dfrac{60}{17}$

k. $-\dfrac{47}{4}$

l. $\dfrac{17}{3}$

m. no solution

n. $\dfrac{1}{2}$

o. $-\dfrac{1}{4}$

p. -7

q. $\dfrac{1}{7}$

r. -1

s. $\dfrac{1}{3}$

t. $\dfrac{1}{2}$

EXERCISE 2—Page 160

1. a. $\dfrac{ac - c}{a}$

b. $\dfrac{p}{a + b}$

c. 1

d. $\dfrac{8a}{3}$

e. 1

f. $\dfrac{a - b}{2}$

g. $d + c$

h. $\dfrac{2mn}{m + n}$

i. $-11a$

j. $\dfrac{a}{3}$

2. 20

3. 2

4. 6

5. 8

6. 21

7. $\dfrac{1}{25}$

8. a. $W = \dfrac{V}{lh}$

b. $r = \dfrac{s - a}{s}$

c. $n = \dfrac{2s}{a + l}$

d. $b = \dfrac{2a - b^1h}{h}$

e. $r = \dfrac{S - a}{S - l}$

f. $n = \dfrac{Cr}{E - Cr}$

g. $l = \dfrac{gt^2}{\pi^2}$

h. $q = \dfrac{fp}{p - f}$

i. $r = D - dq$

j. $R = \dfrac{Fs - Edp}{Ed}$

k. $t = \sqrt{\dfrac{2s}{g}}$

l. $V = \sqrt{\dfrac{2aE}{w}}$

m. $l = \dfrac{A - 2wh}{2h + 2w}$

n. $q = \dfrac{Ru^2}{k^2d^2}$

EXERCISE 3—Page 164

1. 7, 32
2. 7, 17
3. 12, 67
4. 6, 16

5. 8, 10, 12
6. 11, 12
7. 9, 23, 26
8. 76, 17

EXERCISE 4—Page 166

1. 54, 18
2. 7, 19
3. 22, 50
4. 20, 60
5. 6

EXERCISE 5—Page 167

1. 15 nickels, 25 dimes
2. 7 dimes, 16 quarters
3. 4 dollar bills
4. 55 5-cent stamps, 25 10-cent stamps
5. 265 balcony, 214 orchestra
6. 15 nickels, 43 dimes, 21 quarters

EXERCISE 6—Page 171

1. 30 ounces
2. 16 c.c.
3. 15 pounds at $.96 a pound, and 5 pounds at $1.12 a pound
4. 4 ounces of the 18% solution, and 8 ounces of the 45% solution
5. 40 pounds
6. 24 pounds of $.45 nuts and 36 pounds of $.65 nuts
7. 15 pounds
8. 3.2 quarts
9. 24 pints
10. 10 pints
11. 40.5 quarts of 76% butter fat, and 13.5 quarts of 12% butter fat
12. 30 pounds of 30% zinc, and 60 pounds of 60% zinc
13. 5 gallons
14. $23\frac{1}{3}$ gallons
15. 36 pounds

EXERCISE 7—Page 174

1. $5,500 at 5%, 3,700 at 4%
2. $2,500 at 5%, $3,400 at $3\frac{1}{2}$%
3. $5,600 at 3%, $2,800 at 6%
4. $3,200 at $5\frac{1}{2}$%, $2,400 at 4%
5. $70
6. $80
7. 62 right, 18 wrong
8. 47 at $3.25, 98 at $2.49

EXERCISE 8—Page 177

1. 7 inches by 17 inches
2. 9 inches
3. 8 inches by 3 inches
4. 9 feet by 5 feet
5. 90 feet by 30 feet
6. 24
7. 12

EXERCISE 9—Page 182

1. $4\frac{1}{2}$ hours
2. 12:45 P.M.
3. 216 miles
4. 7:00 P.M.
5. 160 m.p.h.
6. 60 m.p.h.
7. 90 miles
8. 84 miles
9. 280 seconds
10. 16 m.p.h.
11. 2 m.p.h.
12. 273 miles
13. 26 m.p.h., 31 m.p.h.
14. 1,620 miles
15. 12 m.p.h.

EXERCISE 10—Page 186

1. $3\frac{3}{4}$ days
2. 4 minutes
3. $6\frac{6}{11}$ hours
4. $6\frac{2}{3}$ hours
5. $3\frac{1}{3}$ days
6. $16\frac{2}{3}$ minutes
7. 36 hours
8. 36 hours
9. 15 days

EXERCISE 11—Page 188

1. 63, 36
2. $11,628, $7,752, $15,504
3. $\frac{17}{32}$
4. $36, $20
5. 48, 72
6. 28 m.p.h., 32 m.p.h.

EXERCISE 12—Page 190

1. $4:54\frac{6}{11}$　　　2. $7:21\frac{9}{11}$　　　3. $10:54\frac{6}{11}$

CHAPTER **8**

Systems of Linear Equations

DEFINITION: An equation of the form $ax + by + c = 0$ is called a linear equation in two variables, provided neither a nor b is equal to zero.

Example: $3x - 2y = 12$

This equation is satisfied by an infinite number of sets of values for x and y. For example, the following sets will satisfy the above equation.

x	6	5	4	3	2	1	0	-1	-2
y	3	$1\frac{1}{2}$	0	$-1\frac{1}{2}$	-3	$-4\frac{1}{2}$	-6	$-7\frac{1}{2}$	-9

If there is another linear equation in two variables, $2x + y = 1$, this equation is also satisfied by an infinite number of sets of values for x and y. The following sets will satisfy this second equation.

x	6	5	4	3	2	1	0	-1	-2
y	-11	-9	-7	-5	-3	-1	1	3	5

It will be noted that the values $x = 2$, $y = -3$, will satisfy both equations. The equations $3x - 2y = 12$, and $2x + y = 1$ are examples of *simultaneous linear equations* and the pair of values $x = 2$, $y = -3$ is called a *simultaneous solution* of the equations.

194

GRAPHICAL SOLUTION OF SETS OF SIMULTANEOUS LINEAR EQUATIONS

Let us consider the equations $3x + 2y = 4$ and $x - 2y = 8$. It is required to obtain the simultaneous solution of these equations. If we set up tables of sets of values we may not find a common set of values since this common set may include mixed numbers which will not appear in the tables.

It will be recalled that a graph of an equation contains all the number pairs and only those number pairs which satisfy the equation. In order to draw the graph, of the linear equation $3x + 2y = 4$ we may use the following number pairs. Although only two pairs are necessary a third pair is taken as a check.

x	0	2	4
y	2	-1	-4

For the graph of $x - 2y = 8$ we may use the following number pairs.

x	0	4	6
y	-4	-2	-1

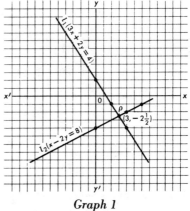

Graph 1

l_1 contains all the points whose number pairs satisfy the equation $3x + 2y = 4$. l_2 contains all the points whose number pairs satisfy the equation $x - 2y = 8$.

Point P, the intersection of the lines l_1 and l_2, therefore, represents the point whose number pair satisfies both equations simultaneously. The coordinates of point P are $x = 3$, $y = -2\frac{1}{2}$. If these values are substituted in the equations the equations are satisfied. One limitation of the graphical method is that it is very difficult to find fractional roots with accuracy.

CONSISTENT, INCONSISTENT, AND DEPENDENT SYSTEMS OF EQUATIONS

DEFINITION: If a set of linear equations has one common solution then the set is said to be *consistent*.

For example, the equations $3x + 2y = 4$ and $x - 2y = 8$, considered above have the common solution $x = 3$, $y = -2\frac{1}{2}$. The graphs of a system of two consistent linear equations consist of two intersecting straight lines.

DEFINITION: If a set of linear equations has no common solution then the set is said to be *inconsistent*.

For example, the equations $x + 2y = 5$ and $2x + 4y = 15$, have no common solution. The graphs of two inconsistent linear equations consist of two *parallel* straight lines. The graphs of the equations $x + 2y = 5$ and $2x + 4y = 15$ are shown below.

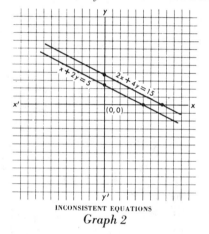

INCONSISTENT EQUATIONS
Graph 2

DEFINITION: If a set of linear equations has an infinite number of solutions then the set is said to be *dependent*.

For example, the equations $x + 2y = 5$ and $2x + 4y = 10$ have an infinite number of solutions. If the first of these equations is multiplied by 2 then the second of the two equations is obtained. The graphs of two dependent linear equations are represented by coincident straight lines. It may be considered that these lines intersect at all points and that each point represents a common solution. The graphs are shown on page 197.

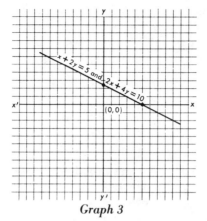

Graph 3

ELIMINATION BY ADDITION AND SUBTRACTION

In solving two linear equations in two unknowns we eliminate one of the unknowns by applying the addition or subtraction axiom.

Example: Solve and check the equations
$$x + 2y = 7$$
$$3x - 2y = 5$$

▶ $$x + 2y = 7$$
$$3x - 2y = 5$$

Adding these equations, we have $4x = 12$
$$x = 3$$
Let $x = 3$ in the first equation $3 + 2y = 7$
$$y = 2$$

Check: $3 + 2(2) = 7, \ 3 + 4 = 7$
$$3(3) - 2(2) = 5, \ 9 - 4 = 5$$

Example: Solve and check the equations
$$x + 4y = -7$$
$$2x - 3y = 8$$

▶ $$x + 4y = -7$$
$$2x - 3y = 8$$

Multiply the members of the first equation by 2, $2x + 8y = -14$. Now subtract corresponding members of the resulting equation from the second equation, we have $-11y = 22 \quad y = -2$

Let $y = -2$ in the first equation, $x - 8 = -7$

$$x = 1$$

Check: $1 + 4(-2) = -7,\ 1 - 8 = -7$

$2(1) - 3(-2) = 8,\quad 2 + 6 = 8$

Example: Solve and check the equations

$$2x - 5y = 4$$
$$3x - 2y = -16$$

▶ $2x - 5y = 4$
$3x - 2y = -16$

Multiply members of the first equation by 3 and members of the second equation by 2 to obtain

$$6x - 15y = 12$$
$$6x - \ 4y = -32$$

Subtracting corresponding members of the second resulting equation from members of the first resulting equations, we have

$$-11y = 44$$
$$y = -4$$

Let $y = -4$ in the first equation, $2x - 5(-4) = 4$

$$2x + 20 = 4$$
$$2x = -16$$
$$x = -8$$

Check: $2(-8) - 5(-4) = 4,\quad -16 + 20 = 4$

$3(-8) - 2(-4) = -16,\quad -24 + 8 = -16$

ELIMINATION BY SUBSTITUTION

This method consists of expressing one variable in terms of the other in one equation and substituting the result in the second equation:

Example: Solve and check the equations

$$x - 2y = 7$$
$$3x + 5y = 10$$

From the first equation, we obtain $x = 7 + 2y$. Substituting this result in the second equation, $3(7 + 2y) + 5y = 10$

$$21 + 6y + 5y = 10$$
$$11y = -11$$
$$y = -1$$

Let $y = -1$ in the first equation, $x - 2(-1) = 7$
$$x + 2 = 7$$
$$x = 5$$

Check: $\quad 5 - 2(-1) = 7, \quad 5 + 2 = 7$
$$3(5) + 5(-1) = 10, \; 15 - 5 = 10$$

Example: Solve by the substitution method and check the equations

$$2x - 3y = 4$$
$$4x + 2y = -16$$

▶
$$2x - 3y = 4$$
$$4x + 2y = -16$$

Solving for x in terms of y in the first equation, we have
$$x = \frac{4 + 3y}{2}$$

Substituting this result in the second equation, we have
$$4\left(\frac{4 + 3y}{2}\right) + 2y = -16$$

$$2(4 + 3y) + 2y = -16$$
$$8 + 6y + 2y = -16$$
$$y = -3$$

Substituting this result in the first equation $2x - 3(-3) = 4$
$$2x + 9 = 4$$
$$2x = -5$$
$$x = -\frac{5}{2}$$

Check: $\quad 2\left(-\frac{5}{2}\right) - 3(-3) = 4, \qquad -5 + 9 = 4$

$$4\left(-\frac{5}{2}\right) + 2(-3) = -16, \; -10 - 6 = -16$$

Example: Solve and check the equations
$$2x - 3y = 18$$
$$5x + 2y = 26$$

▶
$$2x - 3y = 18$$
$$5x + 2y = 26$$

Solving for x in the first equation, $x = \dfrac{18 + 3y}{2}$

Substituting this result in the second equation,

$$5\left(\frac{18 + 3y}{2}\right) + 2y = 26$$

$$\frac{90 + 15y}{2} + 2y = 26$$

Multiplying both members of the equation by 2, we have

$$90 + 15y + 4y = 52$$
$$19y = -38$$
$$y = -2$$

Since $x = \dfrac{18 + 3y}{2}$, $x = \dfrac{18 + 3(-2)}{2}$, $x = \dfrac{18 - 6}{2}$, $x = 6$

Check: $2(6) - 3(-2) = 18$, $12 + 6 = 18$
 $5(6) + 2(-2) = 26$, $30 - 4 = 26$

In general, either method may be used to solve a set of simultaneous linear equations. The student may select the method that appears to be more convenient. Below are solved more difficult sets of simultaneous linear equations.

Example: Solve and check the equations

$$\frac{4x + 3}{3} + 3y = 3 - y$$

$$\frac{3y - 6x}{4} + x = \frac{19}{4} - 6y$$

Multiplying both members of the first equation by 3

$$4x + 3 + 9y = 9 - 3y$$
$$4x + 12y = 6$$

Multiplying both members of the second equation by 4

$$3y - 6x + 4x = 19 - 24y$$
$$-2x + 27y = 19$$

Multiplying the last equation by 2, we obtain $-4x + 54y = 38$
$$ 4x + 12y = 6$$

Adding the corresponding members of the
two equations we obtain $ 66y = 44$
$$y = \frac{2}{3}$$

Substituting $y = \dfrac{2}{3}$ in the equation $4x + 12y = 6$, we obtain

$$4x + 12\left(\frac{2}{3}\right) = 6$$
$$4x + 8 = 6$$
$$4x = -2$$
$$x = -\frac{1}{2}$$

Check: The check is made in the first two equations.   A check in other equations, which might be simpler, would not give a check that is valid for the original equations.

$$\frac{4\left(-\dfrac{1}{2}\right) + 3}{3} + 3\left(\frac{2}{3}\right) = 3 - \frac{2}{3}$$

$$\frac{-2 + 3}{3} + 2 = 2\frac{1}{3}$$

$$\frac{1}{3} + 2 = 2\frac{1}{3}$$

$$\frac{3\left(\dfrac{2}{3}\right) - 6\left(-\dfrac{1}{2}\right)}{4}$$

$$-\frac{1}{2} = \frac{19}{4} - 6\left(\frac{2}{3}\right)$$

$$\frac{2 + 3}{4} - \frac{1}{2} = \frac{19}{4} - 4$$

$$\frac{5}{4} - \frac{2}{4} = \frac{19}{4} - \frac{16}{4}$$

$$\frac{3}{4} = \frac{3}{4}$$

Example: Solve and check the equations
$$a^2x - b^2y = a - b$$
$$2ab^2x - a^2by = -a^2 + 2b^2$$

▶

$$a^2x - b^2y = a - b$$
$$2ab^2x - a^2by = -a^2 + 2b^2$$

The method of solution selected is elimination by addition and subtraction.

Multiply both members of the first equation by $2b^2$, to obtain
$$2a^2b^2x - 2b^4y = 2ab^2 - 2b^3$$

Multiply both members of the second equation by a, to obtain
$$2a^2b^2x - a^3by = -a^3 + 2ab^2$$

Subtracting the second resulting equation from the first resulting equation, to obtain

$$-2b^4y + a^3by = -2b^3 + a^3$$
$$y = \frac{a^3 - 2b^3}{a^3b - 2b^4} = \frac{a^3 - 2b^3}{b(a^3 - 2b^3)} = \frac{1}{b}$$

Substitute this value of y in the first equation, to obtain

$$a^2x - b^2\left(\frac{1}{b}\right) = a - b$$
$$a^2x - b = a - b$$
$$a^2x = a$$
$$x = \frac{1}{a}$$

Check: $a^2\left(\dfrac{1}{a}\right) - b^2\left(\dfrac{1}{b}\right) = a - b, \quad a - b = a - b$

$2ab^2\left(\dfrac{1}{a}\right) - a^2b\left(\dfrac{1}{b}\right) = -a^2 + 2b^2, \ 2b^2 - a^2 = -a^2 + 2b^2$

Example: Solve and check the equations:
$$\frac{5}{x} + \frac{7}{y} = 31; \ \frac{3}{x} - \frac{11}{y} = -27$$

▶

$$\frac{5}{x} + \frac{7}{y} = 31$$
$$\frac{3}{x} - \frac{11}{y} = -27$$

In solving this set of equations it is convenient to regard $\dfrac{1}{x}$ and $\dfrac{1}{y}$ as the unknowns.

Multiply both members of the first equation by 3, to obtain
$$\frac{15}{x} + \frac{21}{y} = 93$$

Multiply both members of the second equation by 5, to obtain

$$\frac{15}{x} - \frac{55}{y} = -135$$

Subtract the second resulting equation from the first, to obtain

$$\frac{76}{y} = 228$$
$$228y = 76$$
$$y = \frac{1}{3}$$

Substitute this value for y in the first equation, to obtain

$$\frac{5}{x} + \frac{7}{\frac{1}{3}} = 31$$
$$\frac{5}{x} + 21 = 31$$
$$\frac{5}{x} = 10$$
$$10x = 5$$
$$x = \frac{1}{2}$$

Check: $\quad \dfrac{5}{\frac{1}{2}} + \dfrac{7}{\frac{1}{3}} = 31, 10 + 21 = 31$

$$\frac{3}{\frac{1}{2}} - \frac{11}{\frac{1}{3}} = -27, 6 - 33 = -27$$

EXERCISE 1

Solve the following sets of equations graphically:

1. $x + y = 7$
   $2x - y = 8$

2. $2x + y = 2$
   $x - 3y = 15$

3. $3x + y = -1$
   $x - 2y = -5$

4. $x + 2y = -1$
   $2x - 3y = 12$

5. $5x + 2y = -5$
   $2x - y = 7$

Solve and check the following sets of equations by addition and subtraction or by substitution:

6. $x - y = -1$
$3x - 2y = 3$

7. $3x - 2y = -8$
$5x + y = 4$

8. $3x + 2y = 1$
$2x - 3y = -8$

9. $5x - 3y = 9$
$7x + 5y = 8$

10. $3x + 10y = -6$
$2x - 8y = 7$

11. $7x + 8y = 19$
$5x + 6y = 13\frac{1}{2}$

12. $6x + 5y = 8$
$4x - 3y = 18$

13. $2x + 3y = 17$
$3x + 5y = 27$

14. $4x - 6y = 15$
$3x - 2y = 5$

15. $9x + 8y = 6$
$5x - 2y = -16$

16. $7x + 3y = 32$
$3x - 5y = 20$

17. $7x + 4y = 8$
$4x - 11y = 9$

18. $x - 5 = \dfrac{y - 2}{7}$

$4y - 3 = \dfrac{x + 10}{3}$

19. $\dfrac{3x + 4y}{2} - \dfrac{x}{4} = -\dfrac{3}{2}$

$\dfrac{2y - x}{4} + \dfrac{3y}{2} = \dfrac{3}{2}$

20. $\dfrac{2x + y}{3} - \dfrac{x + 2y}{2} = \dfrac{23}{6}$

$x - \dfrac{3x - 4y}{5} = 2$

21. $2x + 3y = 2a + 3b$
$3x - 2y = 3a - 2b$

22. $3x - 2y = 8c$
$5x + 4y = 6c$

23. $ax + by = 2ab$
$bx - ay = b^2 - a^2$

24. $\dfrac{x}{2a + b} + \dfrac{y}{2a - b} = 2$

$x - y = 2b$

25. $\dfrac{3}{x} + \dfrac{5}{y} = 2$

$\dfrac{4}{x} + \dfrac{7}{y} = 3$

26. $\dfrac{2}{x} - \dfrac{6}{y} = 2$

$\dfrac{7}{x} - \dfrac{1}{y} = 3\frac{2}{3}$

27. $\dfrac{3}{x} - \dfrac{2}{y} = -5$

$\dfrac{2}{x} - \dfrac{5}{y} = -7$

28. Classify each of the following systems of equations as consistent, inconsistent, or dependent.

a. $x + 2y = 7$ b. $3x + 2y = 9$ c. $x - 5y = 8$
$2x - y = 5$ $6x + 4y = 18$ $3x - 15y = 10$

d. $5x = 2y - 11$ e. $3x - 4y = 9$ f. $y = 3x - 2$
    $4y = 10x + 17$ $x + 7y = 6$ $12x - 8 = 4y$

SOLUTION OF SYSTEMS OF THREE LINEAR EQUATIONS IN THREE VARIABLES

In general, a system of three linear equations in three variables may be solved by methods of elimination. First, one unknown quantity is eliminated from any pair of equations and the same unknown quantity from another pair. Then, the resulting two equations in two unknowns may be solved by methods explained in the previous section.

   Example: Solve and check the following system of equations.
$$3x + 4y - 2z = 8$$
$$5x + 3y + 7z = -14$$
$$2x - 5y - 3z = 18$$

▶ In order to help the student follow the explanation the equations will be numbered.
$$(1) \ \ 3x + 4y - 2z = 8$$
$$(2) \ \ 5x + 3y + 7z = -14$$
$$(3) \ \ 2x - 5y - 3z = 18$$
In order to eliminate z from equations (1) and (2) we shall multiply equations (1) by 7 and equation (2) by 2.
$$(4) \ \ 21x + 28y - 14z = 56$$
$$(5) \ \ \ \ 10x + 6y + 14z = -28$$
Adding equations (4) and (5) we have (6) $31x + 34y = 28$.

We shall now eliminate z from equations (2) and (3) by multiplying equation (2) by 3 and equation (3) by 7.
$$(7) \ \ \ \ 15x + 9y + 21z = -42$$
$$(8) \ \ 14x - 35y - 21z = 126$$
Adding equations (7) and (8) we have (9) $29x - 26y = 84$.

We can now solve equations (6) and (9) simultaneously. We multiply equation (6) by 13 and equation (9) by 17, obtaining
$$(10) \ \ 403x + 442y = 364$$
$$(11) \ \ 493x - 442y = 1428$$
Adding equations (10) and (11) we have $896x = 1792$
$$(12) \hspace{3cm} \text{and } x = 2$$

(13) Substituting this value in equation (9) we have
$$29(-2) - 26y = 84$$
$$59 - 26y = 84$$
(14) $$y = -1$$
(15) Substituting $x = 2$ and $y = -1$ in equation (1) we have
$$3(2) + 4(-1) - 2z = 8$$
$$6 - 4 - 2z = 8$$
(16) $$z = -3$$
Check: $$3(2) + 4(-1) - 2(-3) = 8$$
$$6 - 4 + 6 = 8$$
$$5(2) + 3(-1) + 7(-3) = -14$$
$$10 - 3 - 21 = -14$$
$$2(2) - 5(-1) - 3(-3) = 18$$
$$4 + 5 + 9 = 18$$

Example: Solve and check the following set of equations
$$\frac{2}{x} + \frac{3}{y} - \frac{4}{z} = 29$$
$$\frac{3}{x} - \frac{2}{y} + \frac{5}{z} = -20$$
$$\frac{4}{x} + \frac{5}{y} - \frac{3}{z} = 35$$

▶ (1) $\frac{2}{x} + \frac{3}{y} - \frac{4}{z} = 29$

(2) $\frac{3}{x} - \frac{2}{y} + \frac{5}{z} = -20$

(3) $\frac{4}{x} + \frac{5}{y} - \frac{3}{z} = 35$

By multiplying equation (1) by 2 and equation (2) by 3 and adding the results we may eliminate the y-term.

(4) $\frac{4}{x} + \frac{6}{y} - \frac{8}{z} = 58$

(5) $\frac{9}{x} - \frac{6}{y} + \frac{15}{z} = -60$

(6) $\frac{13}{x} + \frac{7}{z} = -2$

By multiplying equation (2) by 5 and equation (3) by 2 and adding the results we may again eliminate the y-term.

(7) $\dfrac{15}{x} - \dfrac{10}{y} + \dfrac{25}{z} = -100$

(8) $\dfrac{8}{x} + \dfrac{10}{y} - \dfrac{6}{z} = 70$

(9) $\dfrac{23}{x} + \dfrac{19}{z} = -30$

By multiplying equation (6) by 19 and equation (9) by 7 and subtracting the results we may eliminate the z-terms and obtain an equation in x only.

(10) $\dfrac{247}{x} + \dfrac{133}{z} = -38$

(11) $\dfrac{161}{x} - \dfrac{133}{z} = -210$

(12) $\dfrac{86}{x} = 172$

(13) $172x = 86$

(14) $x = \dfrac{1}{2}$

(15) Substituting this value of x in equation 6, we have

$\dfrac{13}{\dfrac{1}{2}} + \dfrac{7}{z} = -2$

(16) $26 + \dfrac{7}{z} = -2$

(17) $\qquad z = -\dfrac{1}{4}$

(18) Substituting $x = \frac{1}{2}$ and $z = -\frac{1}{4}$ in equation (1) we have

$\dfrac{2}{\dfrac{1}{2}} + \dfrac{3}{y} - \dfrac{4}{-\dfrac{1}{4}} = 29$

(19) $4 + \dfrac{3}{y} + 16 = 29$

(20) $\qquad y = \dfrac{1}{3}$

Check: (1) $\dfrac{2}{\frac{1}{2}} + \dfrac{3}{\frac{1}{3}} - \dfrac{4}{-\frac{1}{4}} = 29$

$$4 + 9 + 16 = 29$$

(2) $\dfrac{3}{\frac{1}{2}} - \dfrac{2}{\frac{1}{3}} + \dfrac{5}{-\frac{1}{4}} = -20$

$$6 - 6 - 20 = -20$$

(3) $\dfrac{4}{\frac{1}{2}} + \dfrac{5}{\frac{1}{3}} - \dfrac{3}{-\frac{1}{4}} = 35$

$$8 + 15 + 12 = 35$$

EXERCISE 2

Solve and check the following sets of equations:

1. $3x + y + 2z = 5$
$x - 3y + 3z = 2$
$2x + 3y - z = 1$

2. $x - y + 5z = 3$
$2x + 3y - 2z = 21$
$x - 4y + z = -6$

3. $-x - 2y + 4z = -1$
$3x + 5y + 2z = 8$
$2x - 7y - 6z = -19$

4. $5x + 3y + 2z = -9$
$-3x - y + 5z = -17$
$4x - 2y + z = -18$

5. $3x - 2y - 3z = -1$
$6x + y + 2z = 7$
$9x + 3y + 4z = 9$

6. $2x - y + z = 5$
$4x - 3y = 5$
$6x + 2y + 2z = 7$

7. $\dfrac{4}{x} + \dfrac{6}{y} - \dfrac{8}{z} = 6$
$\dfrac{5}{x} + \dfrac{3}{y} + \dfrac{6}{z} = 2$
$\dfrac{3}{x} - \dfrac{9}{y} + \dfrac{10}{z} = -4$

8. $\dfrac{5}{x} - \dfrac{2}{y} + \dfrac{4}{z} = 2$
$\dfrac{7}{x} + \dfrac{4}{y} + \dfrac{6}{z} = 6$
$\dfrac{3}{x} - \dfrac{6}{y} + \dfrac{2}{z} = -1$

9. $\dfrac{2}{x} - \dfrac{9}{y} + \dfrac{8}{z} = -3$
$\dfrac{3}{x} + \dfrac{6}{y} - \dfrac{12}{z} = -1$
$\dfrac{5}{x} - \dfrac{12}{y} + \dfrac{4}{z} = -5$

10. $\dfrac{5}{x} + \dfrac{4}{y} - \dfrac{7}{z} = -27$
$\dfrac{3}{x} + \dfrac{6}{y} - \dfrac{4}{z} = -14$
$\dfrac{7}{x} - \dfrac{10}{y} + \dfrac{6}{z} = -14$

VERBAL PROBLEMS

Some verbal problems are solved more simply by using sets of linear equations, as illustrated below.

Number Problems

Example: Separate 120 into two parts such that the larger exceeds three times the smaller by 12.

▶ Let x = the larger number
And y = the smaller number
Then $x + y = 120$
$x = 3y + 12$

When these equations are solved simultaneously the larger number is found to be 93 and the smaller 27.

Example: Find two numbers such that the sum of twice the larger and the smaller is 64. But, if 5 times the smaller be subtracted from four times the larger the result is 16.

▶ Let x = the larger number
And y = the smaller number
Then $2x + y = 64$
$4x - 5y = 16$

When these equations are solved simultaneously the larger number is found to be 24 and the smaller number 16.

Example: If 3 be subtracted from the numerator of a certain fraction, the value of the fraction becomes $\frac{3}{5}$. If 1 be subtracted from the denominator of the same fraction then the value of the fraction becomes $\frac{2}{3}$. Find the original fraction.

▶ Let x = the numerator of the fraction
And y = the denominator of the fraction
Then (1) $\dfrac{x-3}{y} = \dfrac{3}{5}$, $5x - 15 = 3y$
(2) $\dfrac{x}{y-1} = \dfrac{2}{3}$, $3x = 2y - 2$
(3) $5x - 3y = 15$
(4) $3x - 2y = -2$

When these equations are solved simultaneously x is found to be 36 and y is found to be 55. Therefore, the original fraction is $\frac{36}{55}$.

EXERCISE 3

Solve the following problems:

1. Find two numbers whose sum is 27 and whose difference is 11.

2. Separate 86 into two parts such that when the larger is subtracted from 3 times the smaller the result is 10.

3. Find two numbers such that the sum of three times the larger and twice the smaller is 69. But, if five times the smaller be subtracted from twice the larger the result is 27.

4. The sum of two numbers is 53. The larger number exceeds 3 times the smaller by 5. Find the numbers.

5. If 4 is added to the numerator of a fraction the value of the fraction becomes $\frac{3}{4}$. If 2 is added to the denominator of the fraction the value of the fraction becomes $\frac{1}{2}$. Find the fraction.

6. The sum of two numbers is 82. If the larger number be divided by the smaller number the quotient is 8 and the remainder is 1. Find the numbers.

Digit Problems

Digit problems often involve the position value of the digits composing a number. For example, in the number 639, the number 6 represents 600 because of its position. It is called the hundreds' digit. For similar reasons, the 3 is called the tens' digit and the 9 the units' digit. The following note will be found helpful in solving digit problems.

Let t = tens' digit (for example, 7)

And u = units' digit (for example, 3)

Then $10t + u$ = the two-digit number ($10 \cdot 7 + 3 = 70 + 3 = 73$)

$10u + t$ = the two-digit number with the digits reversed
($10 \cdot 3 + 7 = 30 + 7 = 37$)

$t + u$ = the sum of the digits ($7 + 3 = 10$)

Example: The sum of the digits of a two-digit number is 9. The number is equal to 9 times the units' digit. Find the number.

▶ (1) $t + u = 9$
  (2) $10t + u = 9u$

If these two equations are solved simultaneously, $t = 4$ and $u = 5$ and the number is 45.

Example: The units' digit of a two-digit number is 4 less than 3 times the tens' digit. If the digits are reversed, a new number is formed which is 12 less than twice the original number. Find the number.

▶ (1) $u = 3t - 4$
  (2) $10u + t = 2(10t + u) - 12$

If these equations are solved simultaneously $t = 4$ and $u = 8$ and the number is 48.

EXERCISE 4

Solve the following problems:

1. The tens' digit of a two-digit number is 3 less than twice the units' digit. The sum of the digits is 12. Find the number.

2. The units' digit of a two-digit number exceeds the tens' digit by 4. The number with the digits reversed exceeds the sum of the digits by 45. Find the number.

3. The sum of the digits of a two-digit number is 12. If the digits are reversed, a new number is formed which is 12 less than twice the original number. Find the number.

4. The units' digit of a two-digit number is 3 less than the tens' digit. If 18 is added to the number, the sum is 7 less than 9 times the sum of the digits. Find the number.

5. The units' digit of a two-digit number is 5 less than the tens' digit. If the digits are reversed, a new number is formed which is $\frac{3}{8}$ of the original number. Find the original number.

6. When a certain two-digit number is divided by the sum of its digits, the quotient is 7. If the digits are reversed, the resulting number is 18 less than the original number. Find the original number.

7. A certain two-digit number is equal to 8 times the sum of the digits. The tens' digit exceeds 3 times the units' digit by 1. Find the number.

8. A two-digit number exceeds 4 times the number with the digits reversed by 9. The sum of the digits is equal to $\frac{1}{2}$ the number with the digits reversed. Find the number.

Mixture Problems

Example: In a chemical laboratory one carboy contains 12 gallons of acid and 18 gallons of water. Another carboy contains 9 gallons of acid and 3 gallons of water. How many gallons must be drawn from each carboy and combined to form a solution that is 7 gallons acid and 7 gallons water?

▶ Let x = Number of gallons taken from first carboy.
And y = Number of gallons taken from second carboy

(1) $x + y = 14$

(2) $\dfrac{12x}{30} + \dfrac{9y}{12} = 7$

In forming the second equation, it should be observed that $\frac{12}{30}$ of the liquid drawn from the first carboy is acid and $\frac{9}{12}$ of the liquid drawn from the second carboy is acid. The two quantities of liquid drawn from the two carboys yield 7 gallons of acid in the mixture.

When the equations are solved, it is found that

$$x = 10 \text{ and } y = 4$$

Example: A chemist has an 18% solution and a 45% solution of a disinfectant. How many ounces of each should be used to make 12 ounces of a 36% solution?

▶ Let x = Number of ounces from the 18% solution
And y = Number of ounces from the 45% solution

(1) $x + y = 12$

(2) $.18x + .45y = .36(12) = 4.32$

Note that .18 of the first solution is pure disinfectant and that .45 of the second solution is pure disinfectant. When the proper quantities are drawn from each mixture the result is 12 gallons of mixture which is .36 pure disinfectant, i.e., the resulting mixture contains 4.32 ounces of pure disinfectant.

When the equations are solved, it is found that $x = 4$ and $y = 8$.

Example: A man wants to obtain 15 gallons of a 24% alcohol solution by combining a quantity of 20% alcohol solution, a quantity of 30% alcohol solution and 1 gallon of pure water. How many gallons of each of the alcohol solutions must be used?

▶ Let x = The number of gallons of the 20% solution used
And y = The number of gallons of the 30% solution used
(1) $x + y + 1 = 15$
(2) $.20x + .30y = .24(15) = 3.6$

When these equations are solved it is found that 8 gallons of the 20% solution and 6 gallons of the 35% solution are used.

Example: Sand and gravel have been mixed in two separate piles. In the first pile the ratio of sand to gravel is 1 : 1, and in the second pile the ratio of sand to gravel is 1 : 4. A third pile, in which the ratio of sand to gravel is 1 : 3, is to be formed from the first two piles. If the third pile is to contain 15 cubic yards, how many cubic yards must be taken from each of the first two piles?

▶ Let x = Number of cubic yards to be taken from first pile
And y = Number of cubic yards to be taken from second pile
(1) $x + y = 15$
(2) $\frac{1}{2}x + \frac{1}{5}y = \frac{1}{4}(15)$

In forming the second equation it should be observed that when the ratio of sand to gravel in the first pile is 1 : 1 then $\frac{1}{2}$ of the material taken from the first pile is sand. Likewise, the ratio of sand to gravel in the second pile is 1 : 4 and therefore $\frac{1}{5}$ of the material taken from the second pile is sand. Since the ratio of sand to gravel in the third pile is 1 : 3 then $\frac{1}{4}$ of the 15 cubic yards in the third pile is sand.

When the equations are solved it is found that $x = 2\frac{1}{2}$ and $y = 12\frac{1}{2}$.

EXERCISE 5

Solve the following problems:

1. A builder has a mixture of 4 parts gravel to 3 parts sand and another mixture of 2 parts gravel to 1 part sand. How many

cubic feet of each of these mixtures should be used to form 25 cubic
feet of a mixture of 3 parts gravel to 2 parts sand?

2. Two tanks contain a mixture of water and insect spray. The
first tank has 15 gallons of water and 3 gallons of spray and the second has 6 gallons of water and 3 gallons of spray. How many gallons must be drawn from each tank to obtain 6 gallons of mixture
that is 20% spray?

3. A park superintendent has a sack of grass seed containing
70% permanent grasses and another sack of grass seed containing
52% permanent grasses. How many pounds should he take from
each sack to obtain 90 pounds of seed containing 60% permanent
grasses?

4. An alloy of tin and copper contains 14 pounds of tin and
26 pounds of copper. A second alloy of tin and copper contains
8 pounds of tin and 24 pounds of copper. How many pounds of
each alloy must be taken to form a third alloy containing 15 pounds
of tin and 35 pounds of copper?

5. A goldsmith has an alloy that is 65% pure gold and another
alloy that is 35% pure gold. How many ounces should he take from
each alloy to secure a third alloy of 60 ounces that will be 45% pure?

6. A farmer has cream testing 21% butter fat and milk testing 3%
butter fat. How many quarts of each must he take to obtain 60
quarts of milk testing 4% butter fat?

Business and Investment Problems

Example: A haberdasher sold 3 shirts and 4 ties to one cus-
tomer for $18.70. Another customer bought 4 shirts and 7 ties of
the same quality for $27.75. What was the price per shirt and price
per tie?

▶   Let x = Cost of one shirt
And y = Cost of one tie
(1) $3x + 4y = 1870$
(2) $4x + 7y = 2775$

When the equations are solved simultaneously it is found that
$x = 398$, i.e. the price of a shirt is $3.98 and $y = 169$, i.e., the
price of a tie is $1.69.

Example: A man derives an income of $309 from some money invested at 3% and some at $4\frac{1}{2}$%. If the amounts of the respective investments were interchanged, he would receive $336. How much did he originally invest at each rate?

▶ Let x = Amount originally invested at 3%
And y = Amount originally invested at $4\frac{1}{2}$%
(1) $.03x + .045y = 309$
(2) $.045x + .03y = 336$

When the equations are solved simultaneously it is found that $x = \$5,200$ and $y = \$3,400$.

Example: A man invested $7,800, part at 6% and part at 4%. If the income of the 4% investment exceeded the investment at 6% by $92, how much was invested at each rate?

▶ Let x = Amount invested at 4%
And y = Amount invested at 6%
(1) $x + y = 7800$
(2) $.04x = .06y + 92$

When the equations are solved it is found that $x = \$5,600$, i.e. the amount invested at 4%, and $y = \$2,200$, i.e. the amount invested at 6%.

EXERCISE 6

Solve the following problems:

1. One salesman sold 8 tables and 12 chairs for a total of $489.72. A second salesman sold 7 tables and 15 chairs for $462.21. Find the selling price of one table and of one chair.

2. One shopper bought 3 pounds of butter and 5 pounds of cheese for $5.59. A second shopper paid $4.85 for 4 pounds of butter and 3 pounds of cheese. What was the cost of one pound of butter and of one pound of cheese?

3. A man invested $7,500, part at 5% and the rest at 4%. The return on the 5% investment exceeded the income on the 4% investment by $15. How much was invested at each rate?

4. A man invested $15,200, part at $3\frac{1}{2}$% and the balance at 4%. The income on the 4% investment was $22 less than the income on the $3\frac{1}{2}$% investment. How much did he invest at each rate?

5. A man derives an income of $298 from some money invested at 5% and some at $4\frac{1}{2}$%. If the amounts of the respective investments were interchanged, he would receive an income of $291. How much did he invest at each rate?

6. A man invested part of his capital at 6% and part at $5\frac{1}{2}$%. He receives an annual income of $398. Had he interchanged the amounts of the respective investments his income would be $407. How much did he invest at each rate?

Motion Problems

Simultaneous equations are very useful in solving motion problems involving the influence of current and wind speeds.

Example: A man rows 18 miles downstream in 2 hours. He finds that it takes 6 hours to row back. What is the rate of rowing in still water and what is the rate of the stream?

▶   Let x = Rate of rowing in still water
     And y = Rate of stream

Note that the rate of motion downstream (i.e. with the current) is $x + y$. The rate of motion upstream (i.e. against the current) is $x - y$.

(1) $2(x + y) = 18$
(2) $6(x - y) = 18$

When the equations are solved it is discovered that $x = 6$ miles per hour and $y = 3$ miles per hour.

Example: Flying with the wind, an airplane can travel 1620 miles in 6 hours. Flying against the wind, the airplane can fly only $\frac{7}{9}$ of this distance in the same time. Find the speed of the plane in still air, and the speed of the wind.

▶   Let x = Speed of plane in still air
                     And y = Speed of wind
               Then $x + y$ = Speed of plane flying with the wind
                      $x - y$ = Speed of plane flying against the wind

(1) $6(x + y) = 1620$

(2) $6(x - y) = \dfrac{7}{9} \times 1620 = 1260$

When the equations are solved it is discovered that $x = 240$ miles per hour and $y = 30$ miles per hour.

Example: Two points, A and B, are 45 miles apart. A cabin cruiser leaves A for B at the same time that another cabin cruiser leaves B for A, going over the same route. The cruisers travel at uniform rates and meet at the end of 2 hours. Upon reaching its destination, each cruiser starts its return trip without delay, and they meet for the second time 15 miles from B. Find the rate of each cruiser.

▶ Let $x =$ Rate of cruiser leaving from A
And $y =$ Rate of cruiser leaving from B

Since the cruisers meet at the end of 2 hours the sum of the distances covered by both is 45 miles. This relationship yields the equation

(1) $2x + 2y = 45$

On the return trip the cruisers meet 15 miles from B. Therefore, the cruiser from A travels $45 + 15 = 60$ miles in the same time that the cruiser from B travels $45 + 30 = 75$ miles. This relationship yields the second equation.

(2) $\dfrac{60}{x} = \dfrac{75}{y}$ $\left(\text{Note that Time} = \dfrac{\text{Distance}}{\text{Rate}} \right)$

When these equations are solved simultaneously it is discovered that $x = 10$ and $y = 12\frac{1}{2}$.

EXERCISE 7

Solve the following problems:

1. A man rows 30 miles downstream in 3 hours. He finds that it takes 5 hours to row back. What is the rate of rowing in still water and what is the rate of the stream?

2. A pilot's destination is 1,080 miles from his starting point. On the trip out he flew against the wind and took 4 hours. On the trip back he flew with the wind and completed the trip in 3 hours. What is the rate of the plane in still air and what is the speed of the wind?

3. A man rows 12 miles upstream in 4 hours. He rows 20 miles downstream in 2 hours and 30 minutes. Find the man's rate of rowing in still water and the speed of the current.

4. Flying with the wind a pilot flew 1,600 miles in 4 hours. Flying against the wind on the return trip the pilot covered $\frac{4}{5}$ of this distance in the same time. What is the rate of the plane in still air and what is the speed of the wind?

Problems Involving Equations with Three Unknowns

Example: A, B, and C working together can complete a job in $2\frac{2}{3}$ days. B and C working together take 4 days to complete the job and A and C working together take $4\frac{4}{5}$ days to complete the job. How many days would it take each man working alone to complete the job?

▶   Let x = Number of days it takes A to do the job
And y = Number of days it takes B to do the job
And z = Number of days it takes C to do the job

$$(1)\ \ \frac{1}{x} + \frac{1}{y} + \frac{1}{z} = \frac{1}{2\frac{2}{3}} = \frac{3}{8}$$

$$(2)\ \ \ \ \ \ \ \frac{1}{y} + \frac{1}{z} = \frac{1}{4}$$

$$(3)\ \ \ \ \ \ \ \frac{1}{x} + \frac{1}{z} = \frac{1}{4\frac{4}{5}} = \frac{5}{24}$$

(4) If equation (2) be subtracted from equation (1) we have

$$\frac{1}{x} = \frac{3}{8} - \frac{1}{4} = \frac{1}{8} \text{ or } x = 8$$

(5) If equation (3) be subtracted from equation (1) we have

$$\frac{1}{y} = \frac{3}{8} - \frac{5}{24} = \frac{4}{24} \text{ or } y = 6$$

(6) If 6 is substituted for y in equation (2) we have

$$\frac{1}{6} + \frac{1}{z} = \frac{1}{4}$$

$$\frac{1}{z} = \frac{1}{4} - \frac{1}{6} \text{ or } z = 12$$

Example: The sum of the three digits of a number is 16. The sum of the first two digits exceeds the last digit by 2. If 27 is added to the number, the last two digits are interchanged. Find the number.

▶ Let h = the hundreds' digit
And t = the tens' digit
And u = the units' digit
(1) $h + t + u = 16$
(2) $h + t = u + 2$
(3) $100h + 10t + u + 27 = 100h + 10u + t$

Rewriting equation (2) we have (4) $h + t - u = 2$
Subtracting equation (4) from equation (1) we have (5) $2u = 14$
    and $u = 7$
Rewriting equation (3) we have (6) $9t - 9u = -27$
Since $u = 7$ we find (7) $t = 4$
Substituting $u = 7$ and $t = 4$ in equation (1) we find that $h = 5$
Therefore, the number is 547.

EXERCISE 8

Solve the following problems:

1. *A*, *B*, and *C* working together can complete a job in 3 days. *A* and *B* working together take $3\frac{3}{4}$ days to complete the job and *B* and *C* working together take 6 days to complete the job. How many days would it take each man working alone to complete the job?

2. The sum of three numbers is 45. The sum of twice the first number, three times the second number, and four times the third number is 142. The difference between 5 times the first number and twice the second number exceeds the third number by 13. Find the three numbers.

3. The sum of the three digits of a number is 15. The three-digit number exceeds the number with the digits reversed by 297. The sum of the first two digits is equal to twice the last digit. Find the number.

4. The sum of the first two digits of a three-digit number exceeds the third digit by 8. If the first two digits are interchanged the new number exceeds the original number by 270. If the number be divided by the sum of the digits the quotient is 31 and the remainder 15. Find the original number.

5. A man invested \$25,000 in three different investments. The first investment yielded $3\frac{1}{2}\%$, the second 4%, and the third yielded 5%. The total income was \$1,080. If the first two amounts had been interchanged the income would have been \$1,070. Find the amount invested at each rate.

ANSWERS

EXERCISE 1—Page 203

1. $x = 5, y = 2$

2. $x = 3, y = -4$

3. $x = -1, y = 2$

4. $x = 3, y = -2$

5. $x = 1, y = -5$

6. $x = 5, y = 6$

7. $x = 0, y = 4$

8. $x = -1, y = 2$

9. $x = \frac{3}{2}, y = -\frac{1}{2}$

10. $x = \frac{1}{2}, y = -\frac{3}{4}$

11. $x = 3, y = -\frac{1}{4}$

12. $x = 3, y = -2$

13. $x = 4, y = 3$

14. $x = 0, y = -\frac{5}{2}$

15. $x = -2, y = 3$

16. $x = 5, y = -1$

17. $x = \frac{4}{3}, y = -\frac{1}{3}$

18. $x = 5, y = 2$

19. $x = -2, y = \frac{1}{2}$

20. $x = 11, y = -3$

21. $x = a, y = b$

22. $x = 2c, y = -c$

23. $x = b, y = a$

24. $x = 2a + b, y = 2a - b$

25. $x = -1, y = 1$

26. $x = 2, y = -6$

27. $x = -1, y = 1$

28. a. consistent

b. dependent

c. inconsistent

d. inconsistent

e. consistent

f. dependent

EXERCISE 2—Page 208

1. $x = -1, y = 2, z = 3$

2. $x = 6, y = 3, z = 0$

3. $x = -1, y = 2, z = \dfrac{1}{2}$

4. $x = -2, y = 3, z = -4$

5. $x = \dfrac{2}{3}, y = -3, z = 3$

6. $x = \dfrac{1}{2}, y = -1, z = 3$

7. $x = 2, y = 3, z = -4$

8. $x = 1, y = 2, z = -2$

9. $x = -1, y = -15, z = -5$

10. $x = -\dfrac{1}{3}, y = 2, z = \dfrac{1}{2}$

EXERCISE 3—Page 210

1. 8, 19
2. 62, 24
3. 21, 3
4. 41, 12
5. $\frac{11}{20}$
6. 73, 9

EXERCISE 4 —Page 211

1. 75
2. 15
3. 48
4. 74
5. 72
6. 42
7. 72
8. 81

EXERCISE 5—Page 213

1. $17\frac{1}{2}$, $7\frac{1}{2}$
2. $4\frac{4}{5}$, $1\frac{1}{5}$
3. 40, 50
4. 25, 25
5. 20, 40
6. $3\frac{1}{3}$, $56\frac{2}{3}$

EXERCISE 6—Page 215

1. $49.98, $7.49
2. $.68, $.71
3. $3,500, $4,000
4. $8,400, $6,800
5. $3,800, $2,400
6. $2,600, $4,400

EXERCISE 7—Page 217

1. 8, 2   2. 315, 45   3. $5\frac{1}{2}$, $2\frac{1}{2}$ 4. 360, 40

EXERCISE 8—Page 219

1. *A*—6 days, *B*—10 days, *C*—15 days
2. 12, 14, 19
3. 825
4. 697
5. $6,000, $8,000, $11,000

CHAPTER **9**

Complex Numbers

DEFINITION: A square root of a negative number is called a pure imaginary.

Examples: $\sqrt{-5},\ \sqrt{-2},\ \sqrt{-16}$

DEFINITION: The unit in the system of imaginary numbers is $\sqrt{-1}$, frequently written as i.

DEFINITION: A number of the form $a + bi$ where a and b are real numbers is called a *complex number*.

Examples: $3 - 2i,\ -5 + \sqrt{7}i$

It follows that complex numbers $(a + bi)$ include all real numbers when $b = 0$. It also follows that a complex number in which $a = 0$ is a pure imaginary number.

Simplifying Imaginary Numbers

Imaginary numbers are in simplest form when expressed in terms of the unit i. The following examples will show how this is done.

Example: Express each of the following in terms of the unit *i*.

(a) $\sqrt{-16}$

(b) $\sqrt{-49}$

(c) $3\sqrt{-25}$

(d) $\frac{1}{2}\sqrt{-4}$

(e) $\sqrt{-8}$

(f) $2\sqrt{-75}$

▶ (a) $\sqrt{-16} = \sqrt{16} \cdot \sqrt{-1} = 4\sqrt{-1} = 4i$

(b) $\sqrt{-49} = \sqrt{49} \cdot \sqrt{-1} = 7\sqrt{-1} = 7i$

(c) $3\sqrt{-25} = 3\sqrt{25} \cdot \sqrt{-1} = 15\sqrt{-1} = 15i$

(d) $\frac{1}{2}\sqrt{-4} = \frac{1}{2}\sqrt{4} \cdot \sqrt{-1} = 1\sqrt{-1} = i$

(e) $\sqrt{-8} = \sqrt{8} \cdot \sqrt{-1} = 2\sqrt{2} \cdot \sqrt{-1} = 2\sqrt{2}i$

(f) $2\sqrt{-75} = 2\sqrt{75} \cdot \sqrt{-1} = 10\sqrt{3} \cdot \sqrt{-1} = 10\sqrt{3}i$

Example: Simplify $\sqrt{-9} + \sqrt{-36}$

▶ $\sqrt{-9} = \sqrt{9} \cdot \sqrt{-1} = 3i$

$\sqrt{-36} = \sqrt{36} \cdot \sqrt{-1} = 6i$

Therefore, $\sqrt{-9} + \sqrt{-36} = 9i$

Example: Simplify $3\sqrt{-100} - 2\sqrt{-49}$

▶ $3\sqrt{-100} = 3\sqrt{100} \cdot \sqrt{-1} = 30i$

$2\sqrt{-49} = 2\sqrt{49} \cdot \sqrt{-1} = 14i$

Therefore, $3\sqrt{-100} - 2\sqrt{-49} = 16i$

Example: Simplify $5\sqrt{-72} + 3\sqrt{-50}$

▶ $5\sqrt{-72} = 5\sqrt{72} \cdot \sqrt{-1} = 5\sqrt{36} \cdot \sqrt{2}i = 30\sqrt{2}i$

$3\sqrt{-50} = 3\sqrt{50} \cdot \sqrt{-1} = 3\sqrt{25} \cdot \sqrt{2}i = 15\sqrt{2}i$

Therefore, $5\sqrt{-72} + 3\sqrt{-50} = 45\sqrt{2}i$

Example: Simplify $8\sqrt{-\dfrac{1}{2}} - 4\sqrt{-\dfrac{9}{2}}$

▶ $8\sqrt{-\dfrac{1}{2}} - 4\sqrt{-\dfrac{9}{2}}$

$8\sqrt{-\dfrac{1}{2}} = 8\sqrt{\dfrac{1}{2}}\,i = 8\sqrt{\dfrac{1}{2}\cdot\dfrac{2}{2}}\,i = 4\sqrt{2}\,i$

$-4\sqrt{-\dfrac{9}{2}} = -4\sqrt{\dfrac{9}{2}}\,i = -4\sqrt{\dfrac{9}{2}\cdot\dfrac{2}{2}}\,i = -6\sqrt{2}\,i$

$4\sqrt{2}\,i - 6\sqrt{2}\,i = -2\sqrt{2}\,i$

Example: Simplify $\sqrt{-\dfrac{16}{27}} + \dfrac{1}{3}\sqrt{-\dfrac{25}{3}}$

▶ $\sqrt{-\dfrac{16}{27}} = \sqrt{\dfrac{16}{27}}\,i = \dfrac{4}{3}\sqrt{\dfrac{1}{3}}\,i = \dfrac{4}{9}\sqrt{3}\,i$

$\dfrac{1}{3}\sqrt{-\dfrac{25}{3}} = \dfrac{1}{3}\sqrt{\dfrac{25}{3}}\,i = \dfrac{5}{3}\sqrt{\dfrac{1}{3}}\,i = \dfrac{5}{9}\sqrt{3}\,i$

$\dfrac{4}{9}\sqrt{3}\,i + \dfrac{5}{9}\sqrt{3}\,i = \dfrac{9}{9}\sqrt{3}\,i = \sqrt{3}\,i$

EXERCISE 1

Simplify and express in terms of the unit i:

1. $\sqrt{-81} + \sqrt{-16}$
2. $\sqrt{-64} - \sqrt{-25}$
3. $\sqrt{-144} + \sqrt{-49}$
4. $3\sqrt{-9} + 2\sqrt{-4}$
5. $7\sqrt{-100} - 2\sqrt{-36}$
6. $8\sqrt{-81} - 4\sqrt{-25}$
7. $\dfrac{3}{4}\sqrt{-16} - \dfrac{2}{5}\sqrt{-100}$
8. $5\sqrt{-8} + 3\sqrt{-18}$

9. $9\sqrt{-27} - 2\sqrt{-12}$
10. $3\sqrt{-20} + 5\sqrt{-45}$
11. $\dfrac{1}{2}\sqrt{-32} - \dfrac{3}{5}\sqrt{-50}$
12. $\dfrac{1}{4}\sqrt{-48} + \dfrac{5}{6}\sqrt{-108}$
13. $12\sqrt{-\dfrac{1}{2}} - 6\sqrt{-\dfrac{25}{2}}$
14. $10\sqrt{-\dfrac{9}{5}} + 2\sqrt{-\dfrac{49}{5}}$

Addition and Subtraction of Complex Numbers

In adding or subtracting complex numbers we find the algebraic sum or difference of the real parts and the algebraic sum or difference of the imaginary parts separately as illustrated in the following examples.

Example: Add $5 - 3i$ and $-2 + i$

▶ $(5 - 3i) + (-2 + i) = 5 - 3i - 2 + i = 3 - 2i$

Example: Subtract $(5 - 2i)$ from $(8 - 6i)$

▶ $(8 - 6i) - (5 - 2i) = 8 - 6i - 5 + 2i = 3 - 4i$

Example: Add $6 + \sqrt{-16}$ and $5 - \sqrt{-9}$

▶ $6 + \sqrt{-16} = 6 + \sqrt{16}i = 6 + 4i$
   $5 - \sqrt{-9} = 5 - \sqrt{9}i = 5 - 3i$
   The sum is $(11 + i)$

Example: From $8 - \sqrt{-25}$ subtract $2 + \sqrt{-36}$

▶ $8 - \sqrt{-25} = 8 - \sqrt{25}i = 8 - 5i$
   $24 + \sqrt{-36} = 2 + \sqrt{36}i = 2 + 6i$
   The difference is $(6 - 11i)$

EXERCISE 2

Perform the indicated operations and express the results in the form $a + bi$:

1. $(6 + 5i) + (3 - 2i)$
2. $(8 - 3i) + (2 + 4i)$
3. $(7 + 2i) - (5 + 4i)$
4. $(9 - 5i) - (6 - 3i)$
5. $(-2 - 3i) + (-3 - 4i)$
6. $(-6 + 7i) - (-1 - i)$
7. $(5 - \sqrt{-100}) + (2 + \sqrt{-16})$
8. $(8 + \sqrt{-36}) + (3 - \sqrt{-49})$
9. $(10 + \sqrt{-9}) - (5 + \sqrt{-25})$
10. $(1 - \sqrt{-16}) - (3 - \sqrt{-81})$

Powers of i:

   The powers of i form a cyclical pattern.   We note that:

$$i = i \qquad\qquad\qquad i^3 = -i$$
$$i^2 = -1 \qquad\qquad\qquad i^4 = +1$$

This cycle repeats itself since $i^5 = i$, $i^6 = -1$, etc.

Example: Simplify $i^{16} + i^9$

▶ $i^{16} = (i^4)^4 = (+1)^4 = +1$
              $i^9 = (i^4)^2 \cdot i = (+1)^2 \cdot i = +i$

Therefore, $i^{16} + i^9 = 1 + i$

Example: Simplify $i^{31} + i^{26}$

▶ $i^{31} = (i^4)^7 \cdot i^3 = (+1) \cdot (-i) = -i$
              $i^{26} = (i^4)^6 \cdot i^2 = (+1) \cdot (-1) = -1$

Therefore, $i^{31} + i^{26} = -1 - i$

EXERCISE 3

Express the following in the form $a + bi$:

1. $i^5 + i^6$ 3. $i^{22} - i^{37}$
2. $i^8 + i^{15}$ 4. $i^{19} - i^{36}$

Multiplication of Complex Numbers

   The product of two imaginary numbers is obtained by writing these numbers as multiples of i and then performing the multiplication as shown below.

Example: Multiply $2\sqrt{-3}$ by $3\sqrt{-5}$

▶ $2\sqrt{-3} \times 3\sqrt{-5} = 2\sqrt{3}i \times 3\sqrt{5}i$
      $= 6\sqrt{15i^2} = -6\sqrt{15}$

Example: Multiply $5\sqrt{-3}(\sqrt{-6} + 2\sqrt{-12})$

▶ $5\sqrt{-3}(\sqrt{-6} + 2\sqrt{-12})$
          $5\sqrt{3}i(\sqrt{6}i + 2\sqrt{12}i)$
          $5\sqrt{18i^2} + 10\sqrt{36i^2}$
          $15\sqrt{2}(-1) + 10(6)(-1)$
          $-15\sqrt{2} - 60$

EXERCISE 4

Perform the indicated multiplications:

1. $\sqrt{-7} \cdot \sqrt{-3}$
2. $\sqrt{-6} \cdot \sqrt{-2}$
3. $3\sqrt{-8} \cdot 4\sqrt{-6}$
4. $5\sqrt{-3} \cdot \sqrt{-54}$
5. $2\sqrt{-3}(3\sqrt{-6} - 5\sqrt{-12})$
6. $\sqrt{-15}(2\sqrt{-3} + 4\sqrt{-5})$
7. $3\sqrt{-2}(\sqrt{-4} - 7\sqrt{-8})$
8. $6\sqrt{-9}(\sqrt{-1} + 2\sqrt{-4})$

The product of two complex numbers is obtained by multiplying them as algebraic binomials are multiplied and simplifying the results.

Example: Multiply $(5 + 3i)$ by $(4 - 6i)$

▶
$$
\begin{array}{r}
5 + 3i \\
4 - 6i \\
\hline
20 + 12i \\
\ \ \ \ \ - 30i - 18i^2 \\
\hline
20 - 18i - 18i^2
\end{array}
$$

Since $18i^2 = -18$ the simplified result is $(38 - 18i)$

Example: Multiply $(6 + \sqrt{-27})$ by $(5 - \sqrt{-3})$

▶
$$
\begin{array}{l}
6 + \sqrt{-27} = 6 + \sqrt{27}i = \ 6 + \ 3\sqrt{3}i \\
5 - \sqrt{-3} = 5 - \sqrt{3}i = \ 5 - \ \ \ \sqrt{3}i \\
\hline
30 + 15\sqrt{3}i \\
\ \ \ \ - \ 6\sqrt{3}i - 9i^2 \\
\hline
30 + \ 9\sqrt{3}i - 9i^2
\end{array}
$$

Since $-9i^2 = +9$ the simplified result is $39 + 9\sqrt{3}i$

Example: Simplify $(3 + 2i)^2$

▶ $(3 + 2i)^2 = 9 + 12i + 4i^2 = 9 + 12i - 4 = 5 + 12i$

Example: Simplify $(2 - 3i)^3$

▶ $(2 - 3i)^3 = 2^3 - 3(2)^2(3i) + 3(2)(3i)^2 - (3i)^3$
$= 8 - 12(3i) + 6(9i^2) - (27i^3)$
$= 8 - 36i - 54 + 27i$
$= -46 - 9i$

EXERCISE 5

Perform the indicated operation and simplify the result:

1. $(3 + 2i)(5 + i)$
2. $(6 - 5i)(4 + 3i)$
3. $(7 + 2i)(1 - 4i)$
4. $(-2 + 6i)(4 - i)$
5. $(-3 - 5i)(-2 + 3i)$
6. $(4 + \sqrt{-8})(3 + \sqrt{-2})$
7. $(5 + \sqrt{-3})(2 - \sqrt{-27})$
8. $(-2 + \sqrt{-3})(-3 - \sqrt{-12})$
9. $(5 + 4i)^2$
10. $(7 - 3i)^2$
11. $(-6 + 5i)^2$
12. $(-2 - 6i)^2$
13. $(1 + i)^3$
14. $(3 - 2i)^3$

Conjugate Complex Numbers

DEFINITION: The numbers $(a + bi)$ and $(a - bi)$ are called *conjugate complex numbers.*   (a and b are real numbers.) Examples are $(3 + 2i)$ and $(3 - 2i)$, $-4 + 5i$ and $-4 - 5i$. Note that the two conjugate complex numbers differ only in the sign of the imaginary part.

Example: Find the sum of the conjugate complex numbers $(5 + 12i)$ and $(5 - 12i)$

▶
$$5 + 12i$$
$$\underline{5 - 12i}$$

$$\text{Sum} = 10$$

Note: The sum of two conjugate complex numbers is always a real number.

Example: Find the product of the conjugate complex numbers $(3 + 5i)$ and $(3 - 5i)$.

▶
$$3 + 5i$$
$$3 - 5i$$
$$\overline{9 + 15i}$$
$$-15i - 25i^2$$
$$\overline{9 - 25i^2 = 9 - 25(-1) = 9 + 25 = 34}$$

Note: The product of two conjugate complex numbers is always a real number.

EXERCISE 6

Find the sums and products of the following pairs of conjugate complex numbers.
1. $(4 + i)$ and $(4 - i)$
2. $(-2 + 3i)$ and $(-2 - 3i)$
3. $(6 - 5i)$ and $(6 + 5i)$
4. $(-3 - 7i)$ and $(-3 + 7i)$

Division of Complex Numbers

The division of one complex number by another is accomplished by transforming the fraction into a fraction with a real denominator.

Example: Divide $3 + 4i$ by $2 - 5i$. Express the result in the form $a + bi$.

▶
$$(3 + 4i) \div (2 - 5i) = \frac{3 + 4i}{2 - 5i}$$

In order to transform the denominator into a real number, we multiply the numerator and the denominator of the above fraction by the conjugate of the denominator.

$$\frac{3 + 4i}{2 - 5i} \times \frac{2 + 5i}{2 + 5i} = \frac{6 + 23i + 20i^2}{4 - 25i^2} = \frac{6 + 23i - 20}{4 + 25}$$
$$\frac{-14 + 23i}{29} = \frac{-14}{29} + \frac{23i}{29}$$

Example: Express in the form $a + bi$, $\dfrac{7 - 4i}{3 + 2i}$

▶ $\dfrac{7 - 4i}{3 + 2i} \times \dfrac{3 - 2i}{3 - 2i} = \dfrac{21 - 26i + 8i^2}{9 - 4i} = \dfrac{21 - 26i - 8}{9 + 4}$

$$= \dfrac{13 - 26i}{13} = 1 - 2i$$

Example: Express in the form $a + bi$, $\dfrac{3 - \sqrt{3}i}{5 - \sqrt{3}i}$

▶ $$\dfrac{3 - \sqrt{3}i}{5 - \sqrt{3}i} = \dfrac{3 - \sqrt{3}i}{5 - \sqrt{3}i} \cdot \dfrac{5 + \sqrt{3}i}{5 + \sqrt{3}i}$$

$$= \dfrac{15 - 2\sqrt{3}i - 3i^2}{25 - 3i^2} = \dfrac{18 - 2\sqrt{3}i}{28}$$

$$= \dfrac{9 - \sqrt{3}i}{14} = \dfrac{9}{14} - \dfrac{\sqrt{3}}{14}i$$

EXERCISE 7

Perform the following divisions.   Express the results in the form $a + bi$.

1. $\dfrac{2 + i}{1 - i}$

2. $\dfrac{7 - 6i}{4 + i}$

3. $\dfrac{3 + 2i}{5 + 3i}$

4. $\dfrac{1 - 6i}{3 - 4i}$

5. $\dfrac{8}{3 + 2i}$

6. $\dfrac{5 - 3i}{1 + 3i}$

7. $\dfrac{3 - i}{2 - i}$

8. $\dfrac{5 + 4i}{6 - 5i}$

9. $\dfrac{2 + \sqrt{3}i}{2 - \sqrt{3}i}$

10. $\dfrac{5 - \sqrt{2}i}{4 + \sqrt{2}i}$

Graphic Representation of Complex Numbers

In plotting complex numbers we use a rectangular system, as indicated in the figure. The real number part of the complex number is located along the horizontal axis (marked the axis of real numbers on the diagram). The imaginary part of the complex number is located along the vertical axis (marked the axis of imaginary numbers on the diagram). Points in the *real plane* are located by designating number pairs, for example (3, 2).

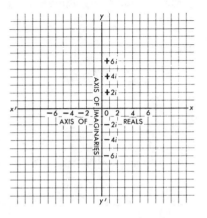

Points in the complex plane are located by designating *single* complex numbers, as shown below. Thus, every complex number may be represented by one and only one point in the plane. The plane is then called the *complex plane*.

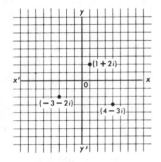

Example: Plot the number $1 + 2i$

▶ Count off one space to the right on the axis of real numbers. Then count 2 spaces up on the axis of imaginary numbers.

Example: Plot the number $4 - 3i$

▶ Count off 4 spaces to the right on the axis of real numbers. Then count 3 spaces down on the axis of imaginary numbers.

Example: Plot the number $-3 - 2i$

▶ Count off 3 spaces to the left on the axis of real numbers. Then count 2 spaces down on the axis of imaginary numbers.

EXERCISE 8

Plot the following numbers:

1. $3 + i$
2. $5 - 2i$
3. $-1 + 4i$
4. $1 - 2i$
5. $-3 + 4i$
6. $-2 - 7i$
7. $-4 + i$
8. $2 - 5i$

Graphical Addition and Subtraction of Complex Numbers

Let point A represent one complex number and let point B represent a second complex number. Now draw OA and OB. Through B draw a line parallel to OA and through A draw a line parallel to OB. These lines will intersect at a point C. Then C represents the sum of the two complex numbers A and B.

In this drawing, $OACB$ is a parallelogram because the opposite sides are parallel. The actual construction of the parallelogram may be accomplished by the use of compasses, as follows. With B as a center and with a radius equal to OA inscribe an arc. With A as a center and with a radius equal to OB inscribe a second arc. These arcs will intersect in the fourth vertex of the parallelogram.

In problems in physics, the lines OA and OB are interpreted as *vectors* (having both magnitude and direction). The diagonal OC is then the resultant of the two vectors OA and OB.

Example: Add graphically the complex numbers $(3 + 2i)$ and $(2 - 5i)$

▶ Locate point A $(3 + 2i)$. Locate point B $(5 - 3i)$. Draw OA and OB and complete the parallelogram thus determining point C. Point C $(5 - 3i)$ represents the required sum.

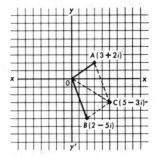

The graphical subtraction of two complex numbers can be performed in two ways, as shown below:

Example: From $(4 + 3i)$ subtract $(2 - i)$

Solution by Method 1: According to the rule for algebraic subtraction we change the sign of the subtrahend and then add. In this example, we must add $(4 + 3i)$ and $(-2 + i)$. Locate point A $(4 + 3i)$. Locate point B $(-2 + i)$. Draw OA and OB and then complete the parallelogram determining point C. Point C $(2 + 4i)$ represents the difference between $(4 + 3i)$ and $(2 - i)$.

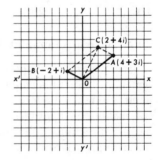

Solution by Method 2: Locate point A $(4 + 3i)$. Locate point B $(2 - i)$ and draw BO. Then extend BO backward through O determining B' so that $BO = OB'$. Now complete the parallelogram with OA and OB' as sides. Point C $(2 + 4i)$, the fourth vertex of the parallelogram, represents the difference between $(4 + 3i)$ and $(2 - i)$.

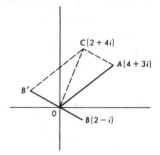

EXERCISE 9

Add the following complex numbers graphically:
1. $(1 + i)$ and $(2 + 3i)$
2. $(3 + 2i)$ and $(1 + 3i)$
3. $(5 - 2i)$ and $(3 + 3i)$
4. $(-2 + 3i)$ and $(6 - 2i)$
5. $(4 - 3i)$ and $(-5 + 2i)$
6. $(-1 - 2i)$ and $(1 + 3i)$

Subtract the following complex numbers graphically:
7. $(4 + 3i) - (3 + i)$
8. $(5 - 2i) - (2 - 3i)$
9. $(1 - 2i) - (-3 - 4i)$
10. $(-6 - i) - (2 + 3i)$

Polar Form of a Complex Number

A point A in the complex plane can be represented by the complex number $(x + iy)$. The
point A can also be located by drawing a line
OA making an appropriate angle θ with the
positive direction of line OB and then measur-
ing an appropriate distance $OA(\rho)$ along this
line. The angle θ and the distance ρ are called
the polar coordinates of the complex number.
The angle is called the *amplitude* and the distance ρ is called the
modulus of the complex number.

In the diagram, point A represents the complex number $x + iy$, written in rectangular
form. $OB = x, AB = y, \angle AOB = \theta, OA = \rho$.
Then, in the right triangle OAB, $\rho^2 = x^2 + y^2$
and $\rho = \sqrt{x^2 + y^2}$.

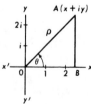

$$\sin \theta = \frac{y}{\rho} \text{ or } y = \rho \sin \theta$$

$$\cos \theta = \frac{x}{\rho} \text{ or } x = \rho \cos \theta$$

$x + iy = \rho \cos \theta + i\rho \sin \theta = \rho (\cos \theta + i \sin \theta)$.

The form $\rho(\cos \theta + i \sin \theta)$ is called the *polar* form of the complex number.

Writing a Complex Number in Polar Form

There are distinct advantages in writing complex numbers in the polar form in the solution of certain types of problems. These advantages will appear below as the applications of the polar form are developed.

The method of converting complex numbers into the polar form is given in the following examples.

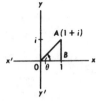

Example: Convert the complex number $(1 + i)$ from the rectangular form to the polar form.

▶ Plot $(1 + i)$ and designate the point by A. Draw $AB \perp$ to the real axis.

$$AB = 1 \text{ and } OB = 1.$$

$$\rho = \sqrt{x^2 + y^2} = \sqrt{1^2 + 1^2} = \sqrt{1 + 1}$$
$$= \sqrt{2}$$

$$\tan \theta = \frac{1}{1}, \theta = 45°$$

Since $x + iy = \rho(\cos \theta + i \sin \theta)$
$$1 + i \Rightarrow \sqrt{2}(\cos 45° + i \sin 45°)$$

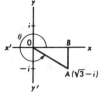

Example: Write the complex number $(\sqrt{3} - i)$ in polar form.

▶ Plot $(\sqrt{3} - i)$ and designate the point by A. Draw $AB \perp$ to the real axis.

$$\rho = \sqrt{x^2 + y^2} = \sqrt{3 + 1} = \sqrt{4} = 2$$

$$\tan \theta = -\frac{1}{\sqrt{3}}$$

$\therefore \theta = 330°$ (Note that θ is in the fourth quadrant since the point A is in the fourth quadrant.)

Since $x + iy = \rho(\cos 330° + i \sin 330°)$
$$\sqrt{3} - i = 2(\cos 330° + i \sin 330°)$$

Example: Write the complex number $(-2 + 3i)$ in polar form.

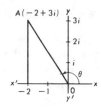

▶ Plot $(-2 + 3i)$ and designate the point by A. Draw $AB \perp$ to the real axis.

$$\rho = \sqrt{x^2 + y^2} = \sqrt{9 + 4} = \sqrt{13}$$

$$\tan \theta = -\frac{3}{2} = -1.5000$$

$\theta = 124°$, correct to the nearest degree.
(θ is an angle in the second quadrant since point A is in the second quadrant)
Since $x + iy = \rho (\cos \theta + i \sin \theta)$
$-2 + 3i = \sqrt{13} (\cos 124° + i \sin 124°)$

Example: Write the complex number $(2 - 2\sqrt{3i})$ in polar form.

▶ Plot $(2 - 2\sqrt{3}i)$ and designate the point by A. Draw $AB \perp$ to the real axis.

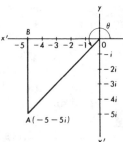

$$\rho = \sqrt{x^2 + y^2} = \sqrt{4 + 12} = \sqrt{16}$$
$$= 4$$

$$\tan \theta = -\frac{2\sqrt{3}}{2} = -\sqrt{3}$$

$\therefore \theta = 300°$ (θ is an angle in the fourth quadrant)
Since $x + iy = \rho (\cos \theta + i \sin \theta)$
$2 - 2\sqrt{3}i = 4 (\cos 300° + i \sin 300°)$

Example: Write the complex number $(-5 - 5i)$ in polar form.

▶ Plot $(-5 - 5i)$ and designate the point by A. Draw $AB \perp$ to the real axis.

$$\rho = \sqrt{x^2 + y^2} = \sqrt{25 + 25} = \sqrt{50} = 5\sqrt{2}$$
$$\tan \theta = \frac{-5}{-5} = 1$$

$\therefore \theta = 225°$ (θ is an angle in the third quadrant)
Since $x + iy = \rho (\cos \theta + i \sin \theta)$
$-5 - 5i = 5\sqrt{2}(\cos 225° + i \sin 225°)$

Example: Write the complex number $-3i$
in polar form.

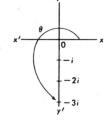

▶   Plot $-3i$ and designate the point by A.
As the diagram indicates, $\rho = 3$ and $\theta = 270°$.

$\therefore\ -3i = 3(\cos 270° + i \sin 270°)$

EXERCISE 10

Convert the following complex numbers from the rectangular
form to the polar form:

1. $2 + 2i$	**6.** $2 + 3i$
2. $1 + \sqrt{3}i$	**7.** $-3 - 4i$
3. $-3 + 3i$	**8.** $-2 - 2\sqrt{3}i$
4. $-1 - i$	**9.** $4 + 4\sqrt{3}i$
5. $\sqrt{3} - i$	**10.** $3 - 5i$

Changing from Polar Form to the Rectangular Form

The change from polar form to the rectangular form may be performed by simple substitution.   However, it is sometimes advisable
to work from a graph, as shown below.

Example: Write $4(\cos 150° + i \sin 150°)$ in the rectangular form

▶   Method 1: (By Direct Substitution)

$\cos 150° = -\dfrac{\sqrt{3}}{2}, \sin 150° = \dfrac{1}{2}$

Therefore $4 (\cos 150° + i \sin 150°)$

$= 4 \left(-\dfrac{\sqrt{3}}{2} + \dfrac{1}{2}i\right) = -2\sqrt{3} + 2i$

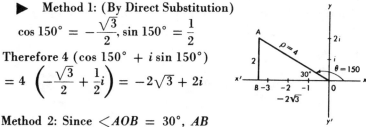

Method 2: Since $\angle AOB = 30°$, AB
$= 2$, and $OB = -2\sqrt{3}$
Therefore, $4 (\cos 150° + i \sin 150°)$
$= -2\sqrt{3} + 2i$

Example: Write $2(\cos 225° + i \sin 225°)$ in the form $a + bi$.
Method 1: (By Direct Substitution)
$$\cos 225° = -\frac{\sqrt{2}}{2}, \sin 225° = -\frac{\sqrt{2}}{2}$$
Therefore, $2 (\cos 225° + i \sin 225°)$
$$= 2 \left(-\frac{\sqrt{2}}{2} - \frac{\sqrt{2}}{2}i\right) = -\sqrt{2} - \sqrt{2}i$$

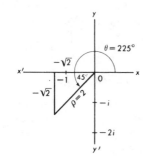

Method 2: Since $<AOB = 45°$, AB
$= -\sqrt{2}$ and $OB = -\sqrt{2}$
Therefore, $2 (\cos 225° + i \sin 225°)$
$= -\sqrt{2} - \sqrt{2}i$

EXERCISE 11

Write the following complex numbers in rectangular form:
1. $6 (\cos 45° + i \sin 45°)$ 6. $2 (\cos 330° + i \sin 330°)$
2. $3 (\cos 90° + i \sin 90°)$ 7. $2 (\cos 180° + i \sin 180°)$
3. $8 (\cos 240° + i \sin 240°)$ 8. $4 (\cos 315° + i \sin 315°)$
4. $10 (\cos 30° + i \sin 30°)$ 9. $7 (\cos 210° + i \sin 210°)$
5. $5 (\cos 135° + i \sin 135°)$ 10. $\frac{1}{2} (\cos 270° + i \sin 270°)$

APPLICATIONS OF THE POLAR FORM OF COMPLEX NUMBERS

Multiplication and Division of Complex Numbers in Polar Form

Let p and q be two complex numbers.   Suppose that their polar forms are:
$$p = r_1 (\cos \theta_1 + i \sin \theta_1)$$
$$q = r_2 (\cos \theta_2 + i \sin \theta_2)$$
Then $pq = r_1 r_2 [\cos (\theta_1 + \theta_2) + i \sin (\theta_1 + \theta_2)]$

And $\dfrac{p}{q} = \dfrac{r_1}{r_2} [\cos (\theta_1 - \theta_2) + i \sin (\theta_1 - \theta_2)], r_2 \neq 0$

These results may be applied as follows:

Example: Multiply the complex numbers $3 (\cos 20° + i \sin 20°)$
and $4 (\cos 70° + i \sin 70°)$.

▶ In multiplying two complex numbers in polar form, we multiply the moduli and add the amplitudes.

Therefore, the result is $3 \cdot 4 \: [\cos (20° + 70°) + i \sin (20° + 70°)]$

or $12 \: (\cos 90° + i \sin 90°)$

This result, written in the form $a + bi = 12(0 + i) = 12i$

Example: Divide the complex number 12 $(\cos 180° + i \sin 180°)$ by 4 $(\cos 45° + i \sin 45°)$

▶ In dividing two complex numbers in polar form, we divide the moduli and subtract the amplitudes.

Therefore, the result is $\frac{12}{4} \: [\cos (180° - 45°) + i \sin (180° - 45°)]$

or $3 \: (\cos 135° + i \sin 135°)$

This result written in the form $a + bi$

$$= 3 \left(-\frac{\sqrt{2}}{2} + \frac{\sqrt{2}}{2} i \right) = -\frac{3}{2}\sqrt{2} + \frac{3}{2}\sqrt{2} i$$

Example: Multiply the complex numbers $5(\cos 100° + i \sin 100°)$ and $2(\cos 50° + i \sin 50°)$.

▶ In multiplying two complex numbers in polar form we multiply the moduli and add the amplitudes.

Therefore, the result is $5 \cdot 2[\cos(100° + 50°) + i \sin(100° + 50°)]$ or $10(\cos 150° + i \sin 150°)$.

The result written in the form $a + bi = 10 \left(\frac{-\sqrt{3}}{2} + i \cdot \frac{1}{2} \right)$

$$= -5\sqrt{3} + 5i.$$

Example: Divide the complex number $6(\cos 170° + i \sin 170°)$ by $3(\cos 80° + i \sin 80°)$

▶ In dividing two complex numbers in polar form we divide the moduli and subtract the amplitudes.

Therefore, the result is $\frac{6}{3} \: [\cos(170° - 80°) + i \sin(170° - 80°)]$

$$= 2(\cos 90° + i \sin 90°).$$

The result written in the form $a + bi = 2(0 + i) = 2i$.

EXERCISE 12

Perform the indicated operations. Write the results in both the polar form and the $a + bi$ form.

1. $4(\cos 60° + i \sin 60°) \times 3(\cos 30° + i \sin 30°)$
2. $5(\cos 80° + i \sin 80°) \times 2(\cos 70° + i \sin 70°)$
3. $2(\cos 200° + i \sin 200°) \times 4(\cos 100° + i \sin 100°)$
4. $3(\cos 170° + i \sin 170°) \times 2(\cos 55° + i \sin 55°)$
5. $6(\cos 130° + i \sin 130°) \div 2(\cos 40° + i \sin 40°)$
6. $12(\cos 210° + i \sin 210°) \div 3(\cos 75° + i \sin 75°)$
7. $15(\cos 220° + i \sin 220°) \div 5(\cos 70° + i \sin 70°)$
8. $8(\cos 300° + i \sin 300°) \div 4(\cos 60° + i \sin 60°)$

De Moivre's Theorem

If n is a positive integer, then
$$[r(\cos \theta + i \sin \theta)]^n = r^n(\cos n\theta + i \sin n\theta)$$
This will be accepted without proof.

Applications of De Moivre's Theorem

De Moivre's Theorem is useful in raising a complex number to a desired power, as explained in the following examples.

Example: Find the value of $[\sqrt{2}(\cos 60° + i \sin 60°)]^4$

▶ $[\sqrt{2}(\cos 60° + i \sin 60°)]^4 = (\sqrt{2})^4(\cos 4 \cdot 60° + i \sin 4 \cdot 60°)$
$= 4(\cos 240° + i \sin 240°)$
This result written in the form $a + bi$ is $4 \left(-\dfrac{1}{2} - \dfrac{\sqrt{3}}{2}i \right)$
$= -2 - 2\sqrt{3}i$

Example: Find the value of $[3(\cos 135° + i \sin 135°)]^5$

▶ $[3(\cos 135° + i \sin 135°)]^5 = 3^5(\cos 5 \cdot 135° + i \sin 5 \cdot 135°)$
$= 3^5(\cos 675° + i \sin 675°) = 243(\cos 315° + i \sin 315°)$

This result written in the form $a + bi$ is $243 \left(\dfrac{\sqrt{2}}{2} - \dfrac{\sqrt{2}}{2} i \right)$

$$= \frac{243\sqrt{2}}{2} - \frac{243\sqrt{2}}{2} i$$

Example: Find the value of $(\sqrt{3} + i)^8$

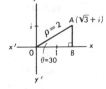

▶ Plot the point A representing the complex number $(\sqrt{3} + i)$. We find that the modulus $OA = 2$ and that the amplitude $\theta = 30°$. ∴ $\sqrt{3} + i = 2(\cos 30° + i \sin 30°)$ in polar form. By De Moivre's Theorem $[2(\cos 30° + i \sin 30°)]^8 = 2^8[\cos 8 \cdot 30° + i \sin 8 \cdot 30°] = 256(\cos 240° + i \sin 240°)$.
The result in polar form may now be changed to the $a + bi$ form.

$$256(\cos 240° + i \sin 240°) = 256 \left(-\frac{1}{2} - \frac{\sqrt{3}}{2} i \right)$$

$$= -128 - 128\sqrt{3}i$$

The solution of the problem is simplified considerably by using De Moivre's Theorem rather than by expanding by the Binomial Theorem.

Example: Find the value of $(-2 + 2i)^6$

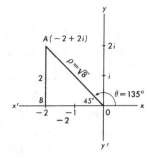

▶ Plot the point A representing the complex number $-2 + 2i$. We find that the modulus $OA = \sqrt{8}$ and that the amplitude $\theta = 135$.
∴ $-2 + 2i = \sqrt{8}(\cos 135° + i \sin 135°)$
By De Moivre's Theorem
$$[\sqrt{8}(\cos 135° + i \sin 135°)]^6$$
$$= (\sqrt{8})^6[\cos 6 \cdot 135° + i \sin 6 \cdot 135°]$$
$$= 512(\cos 810° + i \sin 810°) = 512(\cos 90° + i \sin 90°)$$
$$= 512(0 + i) = 512i$$

Example: Find the value of

$$\left(-\frac{1}{2} - \frac{\sqrt{3}}{2}i\right)^{25}$$

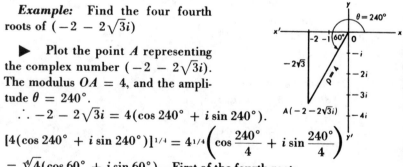

▶ Plot the point A representing the complex number $-\dfrac{1}{2} - \dfrac{\sqrt{3}}{2}i$.

The modulus is 1 and the amplitude is 240°.

$$\therefore -\frac{1}{2} - \frac{\sqrt{3}}{2}i = 1(\cos 240°$$
$$+ i \sin 240°)$$

By De Moivre's Theorem $[1(\cos 240° + i \sin 240°)]^{25}$
$$= (1)^{25}[\cos 25 \cdot 240° + i \sin 25 \cdot 240°]$$
$$= (\cos 6{,}000° + i \sin 6{,}000°) = (\cos 240° + i \sin 240°)$$
$$= -\frac{1}{2} - \frac{\sqrt{3}}{2}i$$

Extracting Roots of Complex Numbers

If n is a positive integer then, by an alternate form of De Moivre's Theorem

$$[r(\cos\theta + i \sin\theta)]^{\frac{1}{n}} = r^{\frac{1}{n}}\left(\cos\frac{\theta}{n} + i \sin\frac{\theta}{n}\right)$$

This form of De Moivre's Theorem can be used to extract roots of complex numbers, as explained in the examples below.

Example: Find the four fourth roots of $(-2 - 2\sqrt{3}i)$

▶ Plot the point A representing the complex number $(-2 - 2\sqrt{3}i)$. The modulus $OA = 4$, and the amplitude $\theta = 240°$.

$$\therefore -2 - 2\sqrt{3}i = 4(\cos 240° + i \sin 240°).$$

$$[4(\cos 240° + i \sin 240°)]^{1/4} = 4^{1/4}\left(\cos\frac{240°}{4} + i \sin\frac{240°}{4}\right)$$
$$= \sqrt[4]{4}(\cos 60° + i \sin 60°)\text{—First of the fourth roots.}$$

The complex number $(-2 - 2\sqrt{3}i)$ has four fourth roots. The other three fourth roots are obtained as follows:

$-2 - 2\sqrt{3}i = 4(\cos 240° + i \sin 240°)$
$= 4[\cos(240° + 360°) + i \sin(240° + 360°)]$
$= 4(\cos 600° + i \sin 600°)$

This result may be used to find the second fourth root.

$$[4(\cos 600° + i \sin 600°)]^{1/4} = 4^{1/4}\left(\cos \frac{600°}{4} + i \sin \frac{600°}{4}\right)$$

$= \sqrt[4]{4}(\cos 150° + i \sin 150°)$ Second of the fourth roots.

$-2 - 2\sqrt{3}i = 4[\cos(240° + 720°) + i \sin (240° + 720°)]$
$= 4(\cos 960° + i \sin 960°)$

$[4(\cos 960° + i \sin 960°)]^{1/4}$

$$= 4^{1/4}\left(\cos \frac{960°}{4} + i \sin \frac{960°}{4}\right)$$

$= \sqrt[4]{4}(\cos 240° + i \sin 240°)$ Third of the fourth roots.

$-2 - 2\sqrt{3}i = 4[\cos(240° + 1080°) + i \sin (240° + 1080°)]$
$= 4(\cos 1320° + i \sin 1320°)$

$[4(\cos 1320° + i \sin 1320°]^{1/4}$

$$= 4^{1/4}\left(\cos \frac{1320°}{4} + i \sin \frac{1320°}{4}\right)$$

$= \sqrt[4]{4}(\cos 330° + i \sin 330°)$ Fourth of the fourth roots.

$$\sqrt[4]{4} = (2^2)^{1/4} = 2^{1/2} = \sqrt{2}$$

$\therefore$ The four fourth roots of $(-2 - 2\sqrt{3}i)$ are:

$$\sqrt{2}(\cos \ 60° + i \sin \ 60°) = 2\left(\frac{1}{2} + \frac{\sqrt{3}}{2}i\right)$$

$$\sqrt{2}(\cos 150° + i \sin 150°) = \sqrt{2}\left(-\frac{\sqrt{3}}{2} + \frac{1}{2}i\right)$$

$$\sqrt{2}(\cos 240° + i \sin 240°) = \sqrt{2}\left(-\frac{1}{2} - \frac{\sqrt{3}}{2}i\right)$$

$$\sqrt{2}(\cos 330° + i \sin 330°) = \sqrt{2}\left(\frac{\sqrt{3}}{2} - \frac{1}{2}i\right)$$

Note: If the next multiple of 360° (i.e. 1440°) be added to 240° and the above process repeated then the result would be $\sqrt{2}(\cos 60° + i \sin 60°)$. In the same way, the other fourth roots of $(-2 - 2\sqrt{3}i)$ would be obtained again if other multiples of 360° are added to 240°.

The general method for finding the *n* *n*th roots of a complex number may be indicated as follows:

$[r(\cos \theta + i \sin \theta)]-$

$$= \sqrt[n]{r}\left[\cos \frac{\theta + k \cdot 360°}{n} + i \sin \frac{\theta + k \cdot 360°}{n}\right]$$

$$\text{for } k = 0, 1, \ldots, (n - 1)$$

In the above example we let $k = 0, 1, 2, 3$ to find the four fourth roots.

Example: Find the three cube roots of $-i$

▶   If $-i$ is plotted the modulus $= 1$
and the amplitude $= 270°$. ∴ $-i = 1(\cos 270° + i \sin 270°)$
$[1(\cos 270° + i \sin 270°)]^{1/3} = 1^{1/3}(\cos 90° + i \sin 90°)$
$= 1(0 + i) = i$

First cube root of $-i$

$1(\cos 270° + i \sin 270°) = 1(\cos 630° + i \sin 630°)$
$[1(\cos 630° + i \sin 630°)]^{1/3} = 1(\cos 210° + i \sin 210°)$
$$= -\frac{\sqrt{3}}{2} - \frac{1}{2}i$$

Second cube root of $-i$

$1(\cos 270° + i \sin 270°) = 1(\cos 990° + i \sin 990°)$
$[1(\cos 990° + i \sin 990°)]^{1/3} = 1(\cos 330° + i \sin 330°)$
$$= \frac{\sqrt{3}}{2} - \frac{1}{2}i$$

Third cube root of $-i$

Example: Find the six sixth roots of 1

▶ If 1 is plotted the modulus = 1 and
the amplitude = 0°.

∴ $1 = 1(\cos 0° + i \sin 60°)$

The six sixth roots of 1 are:

$$[1(\cos 0° + i \sin 0°)]^{1/6} = 1(\cos 0° + i \sin 0°) \qquad = 1$$

$$[1(\cos 360° + i \sin 360°)]^{1/6} = 1(\cos 60° + i \sin 60°) \qquad = \frac{1}{2} + \frac{\sqrt{3}}{2} i$$

$$[1(\cos 720° + i \sin 720°)]^{1/6} = 1(\cos 120° + i \sin 120°) = -\frac{1}{2} + \frac{\sqrt{3}}{2} i$$

$$[1(\cos 1080° + i \sin 1080°)]^{1/6} = 1(\cos 180° + i \sin 180°) = -1$$

$$[1(\cos 1440° + i \sin 1440°)]^{1/6} = 1(\cos 240° + i \sin 240°) = -\frac{1}{2} - \frac{\sqrt{3}}{2} i$$

$$[1(\cos 1800° + i \sin 1800°)]^{1/6} = 1(\cos 300° + i \sin 300°) = \frac{1}{2} - \frac{\sqrt{3}}{2} i$$

Example: Find the cube roots of $8i$

▶ If $8i$ is plotted the modulus = 8 and
the amplitude = 90°.

∴ $8i = 8(\cos 90° + i \sin 90°)$.

The three cube roots of $8i$ are

$$(8i)^{1/3} = [8(\cos 90° + i \sin 90°)]^{1/3} = 8^{1/3}\left(\cos\frac{90°}{3} + i \sin\frac{90°}{3}\right)$$

$$= \sqrt{3} + i \text{—First of the cube roots}$$

$$8i = 8(\cos 90° + i \sin 90°)$$

$$= 8[\cos(90° + 360°) + i \sin(90° + 360°)]$$

$$= 8(\cos 450° + i \sin 450°)$$

Using this result to find the second cube root

$$[8(\cos 450° + i \sin 450°)]^{1/3} = 8^{1/3}\left(\cos\frac{450°}{3} + i \sin\frac{450°}{3}\right)$$

$$= 2(\cos 150° + i \sin 150°)$$

$$= 2 \left(\frac{-\sqrt{3}}{2} + \frac{i}{2} \right) = -\sqrt{3} + i \text{—Second of the cube roots}$$

$$8i = 8(\cos 90° + i \sin 90°)$$
$$= 8[\cos(90° + 720°) + i \sin (90° + 720°)]$$
$$= 8(\cos 810° + i \sin 810°)$$

Using this result to find the third cube root

$$[8(\cos 810° + i \sin 810°)]^{1/3} = 8^{1/3} \left(\cos \frac{810°}{3} + i \sin \frac{810°}{3} \right)$$

$$= 2(\cos 270° + i \sin 270°)$$
$$= 2(0 - i) = -2i \text{—Third of the cube roots}$$

Example: Solve the equation $x^3 + 1 = 0$

▶
$$x^3 + 1 = 0$$
$$x^3 = -1$$
$$x = \sqrt[3]{-1}$$

It is therefore required to find the three cube roots of -1.
If -1 is plotted the modulus $= 1$ and the amplitude $= 180°$.
The three cube roots of -1 are
$$[1(\cos 180° + i \sin 180°)]^{1/3}$$

$$= 1(\cos 60° + i \sin 60°) = \frac{1}{2} + \frac{\sqrt{3}}{2} i$$

$$[1(\cos 540° + i \sin 540°)]^{1/3}$$

$$= 1 (\cos 180° + i \sin 180°) = -1$$

$$[1(\cos 900° + i \sin 900°)]^{1/3}$$

$$= 1(\cos 300° + i \sin 300°) = \frac{1}{2} - \frac{\sqrt{3}}{2} i$$

EXERCISE 13

Raise to the indicated power after first expressing the number in polar form:

1. $(1 + \sqrt{3}\, i)^3$
2. $(-1 - i)^5$

3. $\left(\dfrac{\sqrt{2}}{2} + \dfrac{\sqrt{2}}{2} i \right)^6$
4. $(\sqrt{3} - i)^8$
5. $(2 - 2i)^6$
6. $(-1 - \sqrt{3}\,i)^4$

Find all the roots after first expressing the number in polar form:

7. Square roots of $(1 - \sqrt{3}i)$
8. Cube roots of i
9. Fourth roots of $(-1 - \sqrt{3}i)$
10. Sixth roots -1
11. Fourth roots of -1
12. Fourth roots of $-1 + \sqrt{3}i$

Find all the roots of the following equations.

13. $x^3 - 1 = 0$
14. $x^3 + 8i = 0$
15. $x^2 + i = 0$
16. $x^3 - i = 0$

ANSWERS

EXERCISE 1—Page 224

1. $13i$
2. $3i$
3. $19i$
4. $13i$
5. $58i$
6. $52i$
7. $-i$
8. $19\sqrt{2}i$
9. $23\sqrt{3}i$
10. $21\sqrt{5}i$
11. $-\sqrt{2}i$
12. $6\sqrt{3}i$
13. $-9\sqrt{2}i$
14. $\frac{44}{5}\sqrt{5}i$

EXERCISE 2—Page 225

1. $9 + 3i$
2. $10 + i$
3. $2 - 2i$
4. $3 - 2i$
5. $-5 - 7i$
6. $-5 + 8i$
7. $7 - 6i$
8. $11 - i$
9. $5 - 2i$
10. $-2 + 5i$

EXERCISE 3—Page 226   *EXERCISE* 4—Page 227

1. $-1 + i$ **1.** $-\sqrt{21}$ **5.** $-18\sqrt{2} + 60$
2. $1 + i$ **2.** $-2\sqrt{3}$ **6.** $-6\sqrt{5} - 20\sqrt{3}$
3. $-1 - i$ **3.** $-48\sqrt{3}$ **7.** $-6\sqrt{2} + 84$
4. $-1 + i$ **4.** $-45\sqrt{2}$ **8.** -90

EXERCISE 5—Page 228

1. $13 + 13i$ **6.** $8 + 10\sqrt{2}i$ **11.** $11 - 60i$
2. $39 - 2i$ **7.** $19 - 13\sqrt{3}i$ **12.** $-32 + 24i$
3. $15 - 26i$ **8.** $12 + \sqrt{3}i$ **13.** $-2 + 2i$
4. $-2 + 26i$ **9.** $9 + 40i$ **14.** $-9 - 46i$
5. $21 + i$ **10.** $40 - 42i$

EXERCISE 6—Page 229

1. sum $= 8$, product $= 17$ **3.** sum $= 12$, product $= 61$
2. sum $= -4$, product $= 13$ **4.** sum $= -6$, product $= 58$

EXERCISE 7—Page 230

1. $\dfrac{1}{2} + \dfrac{3}{2} i$ **6.** $\dfrac{-2}{5} - \dfrac{9}{5} i$

2. $\dfrac{22}{17} - \dfrac{3}{17} i$ **7.** $\dfrac{7}{5} + \dfrac{i}{5}$

3. $\dfrac{21}{34} + \dfrac{1}{34} i$ **8.** $\dfrac{10}{61} + \dfrac{49}{61} i$

4. $\dfrac{27}{25} - \dfrac{14}{25} i$ **9.** $\dfrac{1}{7} + \dfrac{4\sqrt{3}}{7} i$

5. $\dfrac{24}{13} - \dfrac{16}{13} i$ **10.** $1 - \dfrac{\sqrt{2}}{2} i$

EXERCISE 9—Page 234

1. $3 + 4i$ **4.** $4 + i$ **7.** $1 + 2i$
2. $4 + 5i$ **5.** $-1 - i$ **8.** $3 + i$
3. $8 + i$ **6.** i **9.** $4 + 2i$
                                                10. $-4 - 4i$

EXERCISE 10—Page 237

1. $2\sqrt{2}(\cos 45° + i \sin 45°)$
2. $2(\cos 60° + i \sin 60°)$
3. $3\sqrt{2}(\cos 135° + i \sin 135°)$
4. $\sqrt{2}(\cos 225° + i \sin 225°)$
5. $2(\cos 330° + i \sin 330°)$
6. $\sqrt{13}(\cos 56° + i \sin 56°)$
7. $5(\cos 233° + i \sin 233°)$
8. $4(\cos 240° + i \sin 240°)$
9. $8(\cos 60° + i \sin 60°)$
10. $\sqrt{34}(\cos 301° + i \sin 301°)$

EXERCISE 11—Page 238

1. $3\sqrt{2} + 3\sqrt{2}i$
2. $3i$
3. $-4 - 4\sqrt{3}i$
4. $5\sqrt{3} + 5i$
5. $-\frac{5}{2}\sqrt{2} + \frac{5}{2}\sqrt{2}i$
6. $\sqrt{3} - i$
7. -2
8. $2\sqrt{2} - 2\sqrt{2}i$
9. $-\frac{7}{2}\sqrt{3} - \frac{7}{2}i$
10. $-\frac{i}{2}$

EXERCISE 12—Page 240

1. $12(\cos 90° + i \sin 90°), 12i$
2. $10(\cos 150° + i \sin 150°), -5\sqrt{3} + 5i$
3. $8(\cos 300° + i \sin 300°), 4 - 4\sqrt{3}i$
4. $6(\cos 225° + i \sin 225°), -3\sqrt{2} - 3\sqrt{2}i$
5. $3(\cos 90° + i \sin 90°), 3i$
6. $4(\cos 135° + i \sin 135°), -2\sqrt{2} + 2\sqrt{2}i$
7. $3(\cos 150° + i \sin 150°), -\frac{3}{2}\sqrt{3} + \frac{3}{2}i$
8. $2(\cos 240° + i \sin 240°), -1 - \sqrt{3}i$

EXERCISE 13—Page 246

1. -8

2. $4 + 4i$

3. $-i$

4. $-128 + 128\sqrt{3}i$

5. $512i$

6. $-8 - 8\sqrt{3}i$

7. $-\dfrac{\sqrt{6}}{2} + \dfrac{\sqrt{2}}{2}i, \dfrac{\sqrt{6}}{2} - \dfrac{\sqrt{2}}{2}i$

8. $\dfrac{\sqrt{3}}{2} + \dfrac{1}{2}i, \dfrac{-\sqrt{3}}{2} + \dfrac{1}{2}i, -i$

9. $\sqrt[4]{2}\left(\dfrac{1}{2} + \dfrac{\sqrt{3}}{2}i\right), \sqrt[4]{2}\left(-\dfrac{\sqrt{3}}{2} + \dfrac{i}{2}\right),$

   $\sqrt[4]{2}\left(-\dfrac{1}{2} - \dfrac{\sqrt{3}}{2}i\right), \sqrt[4]{2}\left(\dfrac{\sqrt{3}}{2} - \dfrac{i}{2}\right)$

10. $\dfrac{\sqrt{3}}{2} + \dfrac{1}{2}i, i, \dfrac{-\sqrt{3}}{2} + \dfrac{1}{2}i, \dfrac{-\sqrt{3}}{2} - \dfrac{1}{2}i, -i, \dfrac{\sqrt{3}}{2} - \dfrac{1}{2}i$

11. $\dfrac{\sqrt{2}}{2} + \dfrac{\sqrt{2}}{2}i, \dfrac{-\sqrt{2}}{2} + \dfrac{\sqrt{2}}{2}i, \dfrac{-\sqrt{2}}{2} - \dfrac{\sqrt{2}}{2}i, \dfrac{\sqrt{2}}{2} - \dfrac{\sqrt{2}}{2}i$

12. $\dfrac{\sqrt[4]{2}}{2}(\sqrt{3} + i), \dfrac{\sqrt[4]{2}}{2}(-1 + \sqrt{3}i), \dfrac{\sqrt[4]{2}}{2}(-\sqrt{3} - i),$

    $\dfrac{\sqrt[4]{2}}{2}(1 - \sqrt{3}i)$

13. $1, \dfrac{-1}{2} + \dfrac{\sqrt{3}i}{2}, -\dfrac{1}{2} - \dfrac{\sqrt{3}}{2}i$

14. $2i, -\sqrt{3} - i, \sqrt{3} - i$

15. $\dfrac{-\sqrt{2}}{2} + \dfrac{\sqrt{2}}{2}i, \dfrac{\sqrt{2}}{2} - \dfrac{\sqrt{2}}{2}i$

16. $\dfrac{\sqrt{3}}{2} + \dfrac{1}{2}i, \dfrac{-\sqrt{3}}{2} + \dfrac{1}{2}i, -i$

CHAPTER 10

The Quadratic Function

DEFINITION: A quadratic function of one variable is an expression of the form $ax^2 + bx + c$. In this expression a, b, and c are constants, $(a \neq 0)$.

Examples: 1. $3x^2 - 5x + 2$ $(a = +3, b = -5, c = +2)$
2. $-x^2 + 4x$ $(a = -1, b = +4, c = 0)$
3. $2x^2 - 7$ $(a = +2, b = 0, c = -7)$

DEFINITION: A *quadratic equation* is an equation obtained by equating a quadratic function to zero.

Examples: 1. $x^2 - 3x + 2 = 0$
2. $4x^2 - 5x = 0$

DEFINITION: An *incomplete quadratic equation* is a quadratic equation in which $b = 0$ or $c = 0$, or both $b = 0$ and $c = 0$.

Examples: 1. $x^2 - 5x = 0$
2. $x^2 - 7 = 0$

SOLUTION OF INCOMPLETE QUADRATIC EQUATIONS

Example: Solve $2x^2 - 6x = 0$

▶ $2x^2 - 6x = 0$

Factoring the left member of the equation we obtain
$$2x(x - 3) = 0$$
Since the product $2x(x - 3) = 0$ at least one of the factors must be equal to zero.

| If $2x = 0$ | If $x - 3 = 0$ |
| Then $x = 0$ | Then $x = 3$ |

If substituted in the equation both roots (0 and 3) will satisfy it.

Example: Solve the equation $x^2 - 25 = 0$

▶ 1: $x^2 - 25 = 0$
Transposing the constant $x^2 = 25$.
By taking the square root of each member of the equation we have $x = +5$ or -5.

▶ 2: $x^2 - 25 = 0$
Factoring $(x + 5)(x - 5) = 0$
Since the product $(x + 5)(x - 5) = 0$ then at least one of the factors must be equal to zero.

| If $x + 5 = 0$ | If $x - 5 = 0$ |
| Then $x = -5$ | Then $x = +5$ |

Both roots can be checked by substitution.

Example: Solve the equation $4a^2 - 9 = 0$
Transposing the constant, $4a^2 = 9$
Dividing both members of the equation by 4,
$$a^2 = \frac{9}{4}$$
Taking the square root of both members of the equation,
$$a = \pm \frac{3}{2}$$

EXERCISE 1

Solve the following equations:

1. $x^2 - 8x = 0$
2. $y^2 - 36 = 0$
3. $a^2 = 4a$
4. $3x^2 - 18x = 0$

5. $b^2 = 49$
6. $2x^2 - 72 = 0$
7. $48 = 3y^2$
8. $4a^2 = 28a$

SOLUTION OF QUADRATIC EQUATIONS BY FACTORING

In general, the simplest method of solving a quadratic equation is by factoring provided that the quadratic function involved is readily factorable. The following examples will illustrate the method.

Example: Solve the equation $4y^2 - y - 5 = 0$

▶ $4y^2 - y - 5 = 0$

Factoring the left member of the equation we obtain
$$(4y - 5)(y + 1) = 0$$
Since the product $(4y - 5)(y + 1) = 0$, at least one of the factors must be zero.

If $4y - 5 = 0$	If $y + 1 = 0$
Then $4y = 5$	
And $y = \dfrac{5}{4}$ or $1\dfrac{1}{4}$	Then $y = -1$

The roots $1\frac{1}{4}$ and -1 can be checked by substitution.

Example: Solve the equation $2x^2 - 7x + 6 = 0$

▶ $2x^2 - 7x + 6 = 0$

Factoring the left member of the equation we obtain
$$(2x - 3)(x - 2) = 0$$
Since the product $(2x - 3)(x - 2) = 0$ at least one of the factors must be zero.

If $2x - 3 = 0$	If $x - 2 = 0$
Then $2x = 3$	
And $x = 1\dfrac{1}{2}$	Then $x = 2$

The roots $1\frac{1}{2}$ and 2 can be checked by substitution.

Example: Solve the equation
$$x^2 - 2ax + a^2 - b^2 + 2bc - c^2 = 0$$

▶ $x^2 - 2ax + a^2 - b^2 + 2bc - c^2 = 0$

The equation may be written as,
$$(x - a)^2 - (b^2 - 2bc + c^2) = 0$$

$$(x - a)^2 - (b - c)^2 = 0$$
$$(x - a)^2 = (b - c)^2$$

$x - a = +(b - c)$ $\qquad\qquad$ $x - a = -(b - c)$

$x = a + b - c$ $\qquad\qquad$ $x = a - b + c$

EXERCISE 2

Solve and check each of the following equations:

1. $x^2 - 7x + 10 = 0$ $\qquad$ 11. $4a^2 - 3a - 1 = 0$
2. $a^2 + 6a + 8 = 0$ $\qquad$ 12. $2y^2 - 5y - 3 = 0$
3. $y^2 + 5y + 4 = 0$ $\qquad$ 13. $5a^2 - 9a - 2 = 0$
4. $x^2 - x - 12 = 0$ $\qquad$ 14. $2x^2 - 5x - 7 = 0$
5. $y^2 - 2y - 15 = 0$ $\qquad$ 15. $4a^2 + 4a - 3 = 0$
6. $x^2 + x - 42 = 0$ $\qquad$ 16. $6y^2 - 5y + 1 = 0$
7. $2x^2 + 3x + 1 = 0$ $\qquad$ 17. $12x^2 - 5x - 3 = 0$
8. $2x^2 + 5x - 3 = 0$ $\qquad$ 18. $6a^2 + 7a - 3 = 0$
9. $2y^2 - 7y - 15 = 0$ $\qquad$ 19. $12x^2 - 7x - 12 = 0$
10. $3a^2 - 11a + 6 = 0$ $\qquad$ 20. $10a^2 + 11a - 6 = 0$

SOLUTION OF QUADRATIC EQUATIONS BY COMPLETING THE SQUARE

Not all quadratic functions can be easily factored. For example, let us consider the equation $x^2 - 6x + 2 = 0$. The function $x^2 - 6x + 2$ cannot be factored. The following method will yield a solution.

Example: Solve the equation $x^2 - 6x + 2 = 0$

▶ $\quad x^2 - 6x + 2 = 0$

Transpose the constant, $x^2 - 6x = -2$.

The left member of the equation can be made a perfect square by adding 9. The 9 is obtained by taking $\frac{1}{2}$ the coefficient of x (in this case -6) and squaring the result.

Since 9 has been added to the left member of the equation, it is necessary to add 9 to the right member to maintain the equality of both members of the equation.

$$x^2 - 6x + 9 = -2 + 9$$
$$x^2 - 6x + 9 = 7$$

The left member of the equation is now a perfect square and may be written as follows:

$$(x - 3)^2 = 7$$

Now taking the square root of both members of the equation

$$x - 3 = +\sqrt{7} \qquad x - 3 = -\sqrt{7}$$
$$x = 3 + \sqrt{7} \qquad x = 3 - \sqrt{7}$$

This method of solving quadratic equations is called the method of *completing the square* since the left member of the equation is made a perfect square by adding the appropriate constant.

Example: Solve $2y^2 - 3y - 4 = 0$

▶ $2y^2 - 3y - 4 = 0$

Transpose the constant, $2y^2 - 3y = 4$
Divide both members of the equation by 2.

$$y^2 - \frac{3}{2}y = 2$$

Now, make the left member of the equation a perfect square by squaring $\frac{1}{2}$ of the coefficient of y, i.e. $(-\frac{3}{4})^2 = \frac{9}{16}$ and adding this quantity to both members of the equation:

$$y^2 - \frac{3}{2}y + \frac{9}{16} = 2 + \frac{9}{16} = \frac{41}{16}$$

$$\left(y - \frac{3}{4}\right)^2 = \frac{41}{16}$$

$$y - \frac{3}{4} = +\frac{\sqrt{41}}{4} \qquad\qquad y - \frac{3}{4} = -\frac{\sqrt{41}}{4}$$

$$y = \frac{3}{4} + \frac{\sqrt{41}}{4} \qquad\qquad y = \frac{3}{4} - \frac{\sqrt{41}}{4}$$

$$y = \frac{3 + \sqrt{41}}{4} \qquad\qquad y = \frac{3 - \sqrt{41}}{4}$$

Note: The method of completing the square may be used in cases where the quadratic function is factorable.

Example: Solve $y^2 - 2y - 3 = 0$

▶ 1: (By factoring) $y^2 - 2y - 3 = 0$
      $(y - 3)(y + 1) = 0$
      $y - 3 = 0$ $y + 1 = 0$
      $y = 3$ $y = -1$

▶ 2: (By completing the square) $y^2 - 2y - 3 = 0$
                                  $y^2 - 2y = 3$
Add 1 to both members of the equation to obtain
   $y^2 - 2y + 1 = 3 + 1$
   $(y - 1)^2 = 4$
   $y - 1 = +2$ $y - 1 = -2$
   $y = 3$ $y = -1$

EXERCISE 3

Solve each of the following equations by completing the square:
1. $x^2 - 5x - 6 = 0$ 6. $2x^2 - 8x + 1 = 0$
2. $2x^2 - 3x - 2 = 0$ 7. $2x^2 - 2x - 3 = 0$
3. $3y^2 + 2y - 1 = 0$ 8. $3x^2 + 5x - 1 = 0$
4. $5a^2 + 3a - 2 = 0$ 9. $3x^2 - 2x - 6 = 0$
5. $x^2 + 4x - 16 = 0$ 10. $2x^2 - 5x + 1 = 0$

SOLUTION OF QUADRATIC EQUATIONS BY THE FORMULA

Every quadratic equation can be written in the form
$$ax^2 + bx + c = 0 (a \neq 0)$$
This equation can be solved by the method of completing the square. The results of the solution of the equation $ax^2 + bx + c = 0$ are very useful as will be shown later in this chapter.
$$ax^2 + bx + c = 0$$
Since $a \neq 0$ we may divide both members of the equation by a to obtain $x^2 + \dfrac{bx}{a} + \dfrac{c}{a} = 0$

Transpose the constant, $x^2 + \dfrac{bx}{a} = -\dfrac{c}{a}$

Complete the square by adding $\left(\dfrac{b}{2a}\right)^2$ or $\dfrac{b^2}{4a^2}$ to both members of the equation.

$$x^2 + \frac{b}{a}x + \frac{b^2}{4a^2} = \frac{b^2}{4a^2} - \frac{c}{a} = \frac{b^2 - 4ac}{4a^2}$$

$$\left(x + \frac{b}{2a}\right)^2 = \frac{b^2 - 4ac}{4a^2}$$

Take the square of both members of the equation

$$x + \frac{b}{2a} = \frac{\pm\sqrt{b^2 - 4ac}}{2a}$$

$$x = -\frac{b}{2a} + \frac{\sqrt{b^2 - 4ac}}{2a} \qquad x = -\frac{b}{2a} - \frac{\sqrt{b^2 - 4ac}}{2a}$$

$$x = \frac{-b + \sqrt{b^2 - 4ac}}{2a} \qquad x = \frac{-b - \sqrt{b^2 - 4ac}}{2a}$$

The roots are usually written as follows:

$$x = \frac{-b \pm \sqrt{b^2 - 4ac}}{2a}$$

This is called the *quadratic formula.*

The quadratic formula may be used to solve any quadratic equation. The following examples illustrate the method.

Example: Solve $12x^2 - 5x = 3$

▶ $12x^2 - 5x = 3$

Transpose the constant, $12x^2 - 5x - 3 = 0$

Compare with $ax^2 + bx + c = 0$, $a = 12$, $b = -5$, $c = -3$.

The roots are $x = \dfrac{-b \pm \sqrt{b^2 - 4ac}}{2a}$

Substitute in this formula, $x = \dfrac{5 \pm \sqrt{25 + 144}}{24}$

$$x = \frac{5 \pm \sqrt{169}}{24}$$

Since $\sqrt{169} = 13$, $x = \dfrac{5 \pm 13}{24}$

$$x = \frac{5 + 13}{24} \qquad\qquad x = \frac{5 - 13}{24}$$

$$x = \frac{3}{4} \qquad\qquad\qquad x = -\frac{1}{3}$$

The roots are $\frac{3}{4}$ and $-\frac{1}{3}$.

Example: Solve $2x^2 - 7x - 1 = 0$

▶ $2x^2 - 7x - 1 = 0$
$a = 2, b = -7, c = -1$
$$x = \frac{-b \pm \sqrt{b^2 - 4ac}}{2a}$$
$$x = \frac{7 \pm \sqrt{49 + 8}}{4}$$
$$x = \frac{7 \pm \sqrt{57}}{4}$$

The roots are $\dfrac{7 + \sqrt{54}}{4}$ and $\dfrac{7 - \sqrt{54}}{4}$

By computing $\sqrt{57}$ we may write the roots to the nearest tenth, the nearest hundredth, etc.

Note: If the quantity under the radical sign is negative, the roots will be complex numbers.

Example: Solve the equation
$$(a + c - b)x^2 + 2cx + (b + c - a) = 0$$

▶ $(a + c - b)x^2 + 2cx + (b + c - a) = 0$

Use the formula, $x = \dfrac{-2c \pm \sqrt{4c^2 - 4(a + c - b)(b + c - a)}}{2(a + c - b)}$

$$x = \frac{-2c \pm \sqrt{4c^2 - 4(-a^2 + 2ab - b^2 + c^2)}}{2(a + c - b)}$$

$$x = \frac{-2c \pm \sqrt{\cancel{4c^2} + 4a^2 - 8ab + 4b^2 - \cancel{4c^2}}}{2(a + c - b)}$$

$$x = \frac{-2c \pm \sqrt{4(a^2 - 2ab + b^2)}}{2(a + c - b)}$$

$$x = \frac{-2c \pm \sqrt{4(a - b)^2}}{2(a + c - b)}$$

$$x = \frac{-2c + 2(a - b)}{2(a + c - b)} \qquad x = \frac{-2c - 2(a - b)}{2(a + c - b)}$$

$$x = \frac{-2c + 2a - 2b}{2(a + c - b)} \qquad x = \frac{-2c - 2a + 2b}{2(a + c - b)}$$

$$x = \frac{\cancel{2}(a - b - c)}{\cancel{2}(a + c - b)} \qquad x = \frac{\cancel{2}(-a + b - c)}{\cancel{2}(a + c - b)} = \frac{-(\cancel{a - b + c})}{(\cancel{a - b + c})}$$

$$x = \frac{a - b - c}{a - b + c} \qquad\qquad x = -1$$

EXERCISE 4

Solve the following quadratic equations by the formula:

1. $x^2 - 3x - 10 = 0$ 6. $y^2 - 3y - 7 = 0$
2. $y^2 + 3y - 18 = 0$ 7. $2a^2 + a - 5 = 0$
3. $2y^2 + 5y - 3 = 0$ 8. $3x^2 + 4x - 2 = 0$
4. $3x^2 + 7x - 6 = 0$ 9. $2y^2 - 5y + 6 = 0$
5. $x^2 - 2x - 1 = 0$ 10. $4x^2 + 2x + 5 = 0$

THEORY OF QUADRATIC EQUATIONS

Roots

In terms of the formula, the roots of the quadratic equation $ax^2 + bx + c = 0$ are

$$r_1 = \frac{-b + \sqrt{b^2 - 4ac}}{2a} \qquad r_2 = \frac{-b - \sqrt{b^2 - 4ac}}{2a}$$

Discriminant

The expression $b^2 - 4ac$ (i.e. the expression under the radical sign) is called the discriminant.

The discriminant is useful in describing the *nature of the roots*. The following table indicates how this is done.

Discriminant ($b^2 - 4ac$) *is*	*Roots are*
1. Positive or zero	Real
Negative	Complex
2. Perfect square	Rational
Not a perfect square	Irrational
3. Zero	Equal—(Since both roots are equal to $\dfrac{-b}{2a}$
4. Not zero	Unequal

Example: Determine the nature of the roots of the equation $2x^2 - 7x + 4 = 0$.

▶ $2x^2 - 7x + 4 = 0$
    Discriminant $= 49 - 32 = 17$
    Therefore, the roots are real, irrational, and unequal.

Example: Determine the nature of the roots of the equation $2x^2 - 3x + 3 = 0$.

▶ $2x^2 - 3x + 3 = 0$
Discriminant $= 9 - 24 = -15$
Therefore, the roots are complex.   (It is unnecessary to mention the other categories since complex roots of quadratic equations cannot be rational or equal).

Example: Determine the nature of the roots of the equation $4x^2 - 4x + 1 = 0$.

▶ Discriminant $= 16 - 16 = 0$
Therefore, the roots are real, rational, and equal.

EXERCISE 5

Determine the nature of the roots of the following equations:

1. $x^2 - 4x - 5 = 0$
2. $2x^2 + 11x - 6 = 0$
3. $x^2 - x + 5 = 0$
4. $3y^2 - 2y - 7 = 0$
5. $x^2 - 2x + 1 = 0$
6. $4y^2 - 2y + 3 = 0$
7. $3x^2 - 11x - 4 = 0$
8. $4y^2 - 12y + 9 = 0$
9. $2x^2 - 11x + 3 = 0$
10. $3a^2 - 2a + 1 = 0$

Sum and Product of the Roots

The roots of the equation $ax^2 + bx + c = 0$ are
$$r_1 = \frac{-b + \sqrt{b^2 - 4ac}}{2a} \quad \text{and} \quad r_2 = \frac{-b - \sqrt{b^2 - 4ac}}{2a}$$
Since these fractional expressions have the same denominator, we can find the sum by adding the numerators and placing the result over the common denominator.

Therefore, $r_1 + r_2 = -\dfrac{2b}{2a} = -\dfrac{b}{a}$

The product of the roots
$$r_1 r_2 = \left(\frac{-b + \sqrt{b^2 - 4ac}}{2a} \right) \left(\frac{-b - \sqrt{b^2 - 4ac}}{2a} \right)$$
In performing the actual multiplication, we note that the numerators may be written as $(-b + \text{Radical})(-b - \text{Radical})$.   The result of multiplying the numerators is $b^2 - \text{Radical}^2$.   This becomes $b^2 - (b^2 - 4ac) = b^2 - b^2 + 4ac = 4ac$.

The final result becomes $\dfrac{4ac}{4a^2} = \dfrac{c}{a}$

Therefore, $r_1 r_2 = \dfrac{c}{a}$

Example: Find the sum and product of the roots of the equation $7x = 2 - 3x^2$.

▶ Writing the equation in standard form we have
$$3x^2 + 7x - 2 = 0 \quad (a = 3, b = 7, c = -2)$$
$$\text{Sum of roots} = -\frac{b}{a} = -\frac{7}{3}$$
$$\text{Product of roots} = \frac{c}{a} = -\frac{2}{3}$$

Example: Find the sum and product of the roots of the equation $(p^2 - q^2)x^2 + (p + q)x + (p - q) = 0$.

▶ $\text{Sum of roots} = -\dfrac{b}{a} = -\dfrac{p + q}{p^2 - q^2} = -\dfrac{\cancel{p + q}}{\cancel{(p + q)}(p - q)}$
$= -\dfrac{1}{p - q}$

$\text{Product of roots} = \dfrac{c}{a} = \dfrac{p - q}{p^2 - q^2} = \dfrac{\cancel{p - q}}{(p + q)\cancel{(p - q)}} = \dfrac{1}{p + q}$

EXERCISE 6

Find the sum and product of the roots of each of the following equations:

1. $x^2 - 2x + 3 = 0$
2. $2y^2 - 7y = 6$
3. $5 + 2a^2 = 3a$
4. $4x^2 - 2x = 9$
5. $3x^2 = 17$
6. $5y^2 - y - 2 = 0$
7. $2x^2 - 9 = 0$
8. $3a^2 - 6 = 2a$

FORMING QUADRATIC EQUATIONS

Method 1: This method is based upon the process of reversing the steps of a solution by factoring.

Example: Write the quadratic equation, with integral coefficients, whose roots are $\frac{1}{4}$ and $-\frac{1}{2}$.

▶ $x = \dfrac{1}{4}$ $x = -\dfrac{1}{2}$

$x - \dfrac{1}{4} = 0$ $x + \dfrac{1}{2} = 0$

$\left(x - \dfrac{1}{4}\right)\left(x + \dfrac{1}{2}\ = 0\right)$

$x^2 + \dfrac{1}{4}x - \dfrac{1}{8} = 0$

To obtain integral coefficients multiply both members of the equation by 8.

$$8x^2 + 2x - 1 = 0$$

Method 2:   Consider the equation $ax^2 + bx + c = 0$.   If we divide both members of the equation by a, we obtain the result $x^2 + \dfrac{b}{a}x + \dfrac{c}{a} = 0$.   Note that the coefficient of $x\left(\text{i.e. } \dfrac{b}{a}\right)$ is identical with the sum of the roots with the sign changed.   Also, the constant term $\left(\dfrac{c}{a}\right)$ is identical with the product of the roots.

We can therefore write the equation $ax^2 + bx + c = 0$ as follows:

$x^2 +$ (sum of roots with sign changed)$x +$ product of roots $= 0$.

Example: Write the quadratic equation, with integral coefficients, whose roots are 2 and $-\frac{1}{4}$.

▶   The desired equation is

$x^2 +$ (sum of roots with sign changed)$x +$ product of roots $= 0$.

$x^2 - \dfrac{7}{4}x - \dfrac{1}{2} = 0$

$4x^2 - 7x - 2 = 0$

Example: Write the quadratic equation, with integral coefficients, whose roots are $\dfrac{1}{2} + \dfrac{\sqrt{-3}}{2}$ and $\dfrac{1}{2} - \dfrac{\sqrt{-3}}{2}$.

▶ The desired equation is

$x^2 +$ (sum of roots with sign changed)$x +$ product of roots $= 0$.

$$x^2 - \left(\frac{1}{2} + \frac{\sqrt{-3}}{2} + \frac{1}{2} - \frac{\sqrt{-3}}{2} \right) x + \left(\frac{1}{2} + \frac{\sqrt{-3}}{2} \right)$$

$$\left(\frac{1}{2} - \frac{\sqrt{-3}}{2} \right) = 0$$

$$x^2 - x + 1 = 0$$

EXERCISE 7

Form the quadratic equation, with integral coefficients, whose roots are:

1. $7, 2$
2. $3, 5$
3. $-2, 4$
4. $-6, -1$
5. $\frac{1}{2}, 8$
6. $-\frac{1}{4}, 6$

7. $\frac{1}{3}, \frac{1}{2}$
8. $-\frac{1}{5}, \frac{3}{2}$
9. $-\frac{3}{4}, -\frac{1}{2}$
10. $3 + \sqrt{2}, 3 - \sqrt{2}$
11. $-4 - \sqrt{3}, -4 + \sqrt{3}$
12. $1 + \sqrt{5}, 1 - \sqrt{5}$

Applications of the Theory of Quadratic Equations

The applications will be developed by the solution of a series of illustrative examples.

Example: Determine the value of k for which the roots of the equation $3x^2 - 7x - k + 3 = 0$ are equal.

▶ Recall that the roots of a quadratic equation are equal when $b^2 - 4ac = 0$.

$$b^2 - 4ac = 49 - 12(-k + 3) = 49 + 12k - 36 = 13 + 2k$$

When $13 + 12k = 0, 12k = -13, k = -\frac{13}{12}$

Example: Determine the value of k so that the equation $3x^2 + 5x - kx + 4 - 3k = 0$ has one root equal to 0.

▶ $3x^2 + 5x - kx + 4 - 3k = 0$

Rewrite the equation in standard form $3x^2 + (5 - k)x + 4 - 3k = 0$. One root of a quadratic equation will be zero if $c = 0$. Since $c = 4 - 3k$, $c = 0$ when $4 - 3k = 0$. When $4 - 3k = 0$, $k = \dfrac{4}{3}$.

Example: Determine the value of k so that the equation $5x^2 - kx + 2x - 3k + 8 = 0$ will have roots that are equal numerically but opposite in sign.

▶ The roots will be equal numerically but opposite in sign when $b = 0$, since the sum of the roots is zero under this condition.

Rewrite the equation in standard form $5x^2 + (-k + 2)x - 3k + 8 = 0$. Since $b = -k + 2$, $b = 0$ when $-k + 2 = 0$. When $-k + 2 = 0$, $k = 2$.

Example: Determine the value of k so that the equation $2x^2 + (k + 2)x + 16 = 0$ has one root equal to twice the other.

▶ Represent the roots by r and $2r$. The product of the roots $= 2r^2 = \dfrac{c}{a} = \dfrac{16}{2} = 8$. Thus, $2r^2 = 8$, $r^2 = 4$, $r = \pm 2$.

Since $r = 2$ is a root we may substitute 2 for x in the original equation and find the value of k.

$$2(2)^2 + (k + 2)2 + 16 = 0$$
$$8 + 2k + 4 + 16 = 0, k = -14$$

Since $r = -2$ is a root we may substitute -2 for x in the original equation and find the value of k.

$$2(-2)^2 + (k + 2)(-2) + 16 = 0$$
$$8 - 2k - 4 + 16 = 0, k = 10$$

Therefore, the values of k are -14 and $+10$.

Example: Determine the value of k so that the equation $2x^2 - 14x + k + 5 = 0$ has one root exceeding the other by 3.

▶ Represent the roots by r and $r + 3$.

The sum of the roots $= 2r + 3 = -\dfrac{b}{a} = 7$.

Thus $2r + 3 = 7$, $r = 2$.

Since $r = 2$ is a root we may substitute 2 for x in the original equation and find the value of k.

$$2(2)^2 - 14(2) + k + 5 = 0$$
$$8 - 28 + k + 5 = 0$$
$$k = 15$$

EXERCISE 8

Determine the value or values of k for which each of the following equations will have equal roots.

1. $x^2 - 3x + k = 0$
2. $2x^2 + 5x - k = 0$
3. $2y^2 + ky + 8 = 0$
4. $kx^2 + 2x - 2kx + k - 1 = 0$
5. $kx^2 + 4kx - 3x + k = 0$

Determine the value or values of k for which each of the following equations will have one root equal to 0.

6. $3x^2 + 2x + k - 5 = 0$
7. $5kx^2 - 3x + kx + 2k + 7 = 0$
8. $3y^2 + 5ky + 9 - 4k = 0$
9. $6x^2 - 3kx + 2x + k^2 - 2k - 8 = 0$
10. $2kx^2 + x - 2k^2 - 3k + 5 = 0$

Determine the value or values of k for which each of the following equations will have roots that are equal numerically but opposite in sign.

11. $3x^2 + kx + 5 = 0$
12. $2kx^2 + 2kx - 3x + 5 - k = 0$
13. $5x^2 + k^2x - 9x + 7 = 0$
14. $4x^2 + 3k^2x - 4kx - 15x + 2 = 0$

GRAPH OF THE QUADRATIC FUNCTION

In general, the graph of the quadratic function $y = ax^2 + bx + c$ is a parabola. The graph, for specific values of a, b, and c can be plotted by assigning convenient values to x, obtaining the corresponding values for y, and plotting the resulting number pairs.

Example: Plot the graph of $y = x^2 - 2x - 3$.

▶ In the following table, the values of x are chosen first, and then the corresponding values of y are computed.

x	4	3	2	1	0	-1	-2
y	5	0	-3	-4	-3	0	5

When these number pairs are plotted and joined by a smooth curve, the result is the graph below.

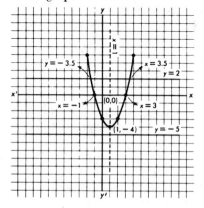

NOTES: 1. When $y = 0$ we can find the roots of the equation $x^2 - 2x - 3 = 0$. On the graph, $y = 0$ is the equation of the x-axis. Therefore, the roots of the equation $x^2 - 2x - 3 = 0$ can be read from the graph where the parabola cuts the x-axis. The roots are $x = 3$, and $x = -1$.

2. For the value $y = 2$ we can find the roots of the equation $x^2 - 2x - 3 = 2$, or $x^2 - 2x - 5 = 0$. On the graph, the line $y = 2$ cuts the parabola at the points where $x = 3.5$ and $x = -1.5$, approximately. The values $x = 3.5$ and $x = -1.5$ are the approximate roots of the equation $x^2 - 2x - 5 = 0$.

3. In general, quadratic equations which have real roots can be solved graphically (by the method described in note 2) but the roots will be determined approximately.

4. Consider the equation $x^2 - 2x - 3 = -5$. In this case, $y = -5$. On the graph, the line $y = -5$ does not cut the parabola. The reason for this is that the equation $x^2 - 2x - 3 = -5$ has no real roots.

Transpose the -5 and write the equation as $x^2 - 2x + 2 = 0$. The discriminant, $b^2 - 4ac = 4 - 8 = -4$. The negative discriminant is verification of the fact that the equation $x^2 - 2x - 3 = -5$ has complex roots.

5. The point $(1, -4)$ is the lowest point (minimum point) on the graph. The value -4 is the least value of the function $x^2 - 2x - 3$ for real values of x.

Example: a. Draw the graph of $y = -2x^2 + x + 3$.

b. From the graph find the roots of the equation $-2x^2 + x + 3 = 0$.

c. From the graph find the roots of the equation $-2x^2 + x + 3 = 2$.

d. Find the smallest integral value of y for which the roots of the equation $-2x^2 + x + 3 = y$ are complex.

e. Draw the axis of symmetry and write its equation.

f. From the graph find the maximum value of $-2x^2 + x + 3$.

▶   a. Make a table of values, selecting convenient values of x and then computing the corresponding values of y.

x	3	2	1	0	-1	-2	-3
y	-12	-3	2	3	0	-7	-18

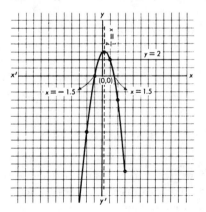

Note: In this case the parabola opens out downward. This will occur when the coefficient of the x^2 term is negative.

b. $x = -1.5$, $x = 1.5$

c. For the value $y = 2$ we can find the roots of the equation $-2x^2 + x + 3 = 2$ or $-2x^2 + x + 1 = 0$. On the graph, the line $y = 2$ cuts the parabola at the points where $x = 1$ and $x = -.5$ approximately. The values $x = 1$ and $x = -.5$ are the approximate roots of the equation $-2x^2 + x + 1 = 0$.

d. The line $y = 4$ on the graph yields the smallest integral value of y for which the line $y = k$ will not cut the graph of the parabola. Therefore, the roots of the equation $-2x^2 + x + 3 = 4$ are complex.

e. The equation of the axis of symmetry is $x = \frac{1}{4}$. $\left(x = -\frac{b}{2a} \right)$

f. The axis of symmetry passes through the maximum point. At this point $x = \frac{1}{4}$. If we substitute this value of x in the original equation $y = -2x^2 + x + 3$ we obtain the value $3\frac{1}{8}$ as the maximum value of the function $-2x^2 + x + 3$. Or this value may be read from the graph at the point where $x = \frac{1}{4}$.

EXERCISE 9

a. Draw the graph of the equation.

b. From the graph, find the roots of the equation when $y = 0$.

c. From the graph, find the roots of the equation when $y = -1$.

d. Draw the axis of symmetry and write its equation.

e. From the graph find the maximum or minimum value of the function.

1. $y = x^2 - 3x - 2$ 3. $y = 2x^2 - 4x - 3$
2. $y = -x^2 + 2x + 3$ 4. $y = -2x^2 - x + 5$

EQUATIONS IN QUADRATIC FORM

Some equations which have fractional exponents, are in radical form, or are of degree higher than the second can be solved by methods used to solve quadratic equations. The following examples will indicate how this is done.

Example: Solve the equation $x^4 - x^2 - 12 = 0$

▶ $x^4 - x^2 - 12 = 0$

Factor, $(x^2 - 4)(x^2 + 3) = 0$

Set each factor equal to zero, $x^2 - 4 = 0$, $x^2 + 3 = 0$.

$$x^2 = 4 \qquad\qquad x^2 = -3$$
$$x = \pm 2 \qquad\qquad x = \pm\sqrt{-3} = \pm i\sqrt{3}$$

These roots may be checked by substitution.

Example: Solve the equation $2y^{2/3} + 3y^{1/3} - 2 = 0$.

▶ $2y^{2/3} + 3y^{1/3} - 2 = 0$

Let $y^{1/3} = A$, Then $y^{2/3} = A^2$

The original equation becomes $2A^2 + 3A - 2 = 0$

Factor, $(2A - 1)(A + 2) = 0$

Set each factor equal to zero, $2A - 1 = 0 \qquad A + 2 = 0$

$$A = \frac{1}{2}, \qquad\qquad A = -2$$

Now, replace A by its equivalent $y^{1/3}$

$$y^{1/3} = \frac{1}{2} \qquad\qquad y^{1/3} = -2$$

Raise each member of the equation to the third power,

$$y = \frac{1}{8} \qquad\qquad y = -8$$

These roots may be checked by substitution.

Example: Solve the equation $3y^{-2} + y^{-1} - 10 = 0$.

▶ $3y^{-2} - y^{-1} - 10 = 0$

Factor, $(3y^{-1} - 5)(y^{-1} + 2) = 0$

Set each factor equal to zero, $3y^{-1} - 5 = 0 \qquad y^{-1} + 2 = 0$

$$3y^{-1} = 5 \qquad\qquad y^{-1} = -2$$
$$y^{-1} = \frac{5}{3} \qquad\qquad \frac{1}{y} = -2$$
$$\frac{1}{y} = \frac{5}{3} \qquad\qquad -2y = 1$$
$$y = \frac{3}{5} \qquad\qquad y = -\frac{1}{2}$$

These roots may be checked by substitution.

Example: Solve the equation

$$\left(x - \frac{2}{x}\right)^2 - 5\left(x - \frac{2}{x}\right) + 4 = 0$$

▶ $\left(x - \frac{2}{x}\right)^2 - 5\left(x - \frac{2}{x}\right) + 4 = 0$

Let $\left(x - \frac{2}{x}\right) = A$

Then, the original equation becomes $A^2 - 5A + 4 = 0$

Factor, $(A - 1)(A - 4) = 0$

Set each factor equal to zero

$A - 1 = 0$ $A - 4 = 0$

$A = 1$ $A = 4$

Replace A by its equivalent $x - \dfrac{2}{x}$

$x - \dfrac{2}{x} = 1$ $x - \dfrac{2}{x} = 4$

$x^2 - 2 = x$ $x^2 - 2 = 4x$

$x^2 - x - 2 = 0$ $x^2 - 4x - 2 = 0$

By factoring   By the formula

$(x - 2)(x + 1) = 0$ $x = \dfrac{4 \pm \sqrt{24}}{2} = 2 \pm \sqrt{6}$

$x = 2, x = -1$

The four roots may be check by substitution.

Example: Solve the equation $\dfrac{x^2 + 4}{4} + \dfrac{4x}{x^2 + 4} - 5 = 0$.

▶ $\dfrac{x^2 + 4}{x} + \dfrac{4x}{x^2 + 4} - 5 = 0$

Let $\dfrac{x^2 + 4}{x} = A$ Then $\dfrac{x}{x^2 + 4} = \dfrac{1}{A}$

Apply these substitutions in the original equation

$A + \dfrac{4}{A} - 5 = 0$

$A^2 + 4 - 5A = 0,$ $A^2 - 5A + 4 = 0$

Factor, $(A - 1)(A - 4) = 0$

Set each factor equal to zero

$A - 1 = 0$ $A - 4 = 0$

$A = 1$ $A = 4$

Replace A by $\dfrac{x^2 + 4}{x}$

$\dfrac{x^2 + 4}{x} = 1$ $\qquad\qquad$ $\dfrac{x^2 + 4}{x} = 4$

$x^2 + 4 = x$ $\qquad\qquad\qquad$ $x^2 + 4 = 4x$

$x^2 - x + 4 = 0$ $\qquad\qquad$ $x^2 - 4x + 4 = 0$

Use the formula $\qquad\qquad\qquad$ Factor

$x = \dfrac{1 \pm \sqrt{1 - 16}}{2}$ $\qquad\qquad$ $(x - 2)(x - 2) = 0$

$\qquad\qquad\qquad\qquad\qquad\qquad$ $x = 2, x = 2$

$x = \dfrac{1 \pm \sqrt{-15}}{2}$

The four roots are $\dfrac{1 \pm i\sqrt{15}}{2}, 2, 2.$

These roots may be checked by substitution.

EXERCISE 10

Solve the following equations:

1. $x^4 - 2x^2 - 15 = 0$
2. $x^4 - 11x^2 + 30 = 0$
3. $x^{2/3} - x^{1/3} - 6 = 0$
4. $2x^{2/3} + 7x^{1/3} - 15 = 0$
5. $x^{-2} + 2x^{-1} - 24 = 0$
6. $3x^{-2} + x^{-1} - 14 = 0$
7. $\left(x - \dfrac{3}{x} \right)^2 - \left(x - \dfrac{3}{x} \right) - 2 = 0$
8. $\left(y - \dfrac{6}{y} \right)^2 - 4\left(y - \dfrac{6}{y} \right) - 5 = 0$
9. $\dfrac{y^2 - 6}{y} + \dfrac{5y}{y^2 - 6} = 6$
10. $\dfrac{y^2 - 3}{y} + \dfrac{3y}{y^2 - 3} = \dfrac{13}{2}$

EQUATIONS CONTAINING RADICALS

In solving equations containing radicals, it is usually necessary to square both members of the equation. The process of squaring both members sometimes introduces *extraneous roots* i.e. roots which will not check when substituted in the original equation. For this reason, the roots of equations containing radicals must always be checked.

Example: Solve the equation $\sqrt{2x - 1} = x - 2$

▶ $\sqrt{2x - 1} = x - 2$

Square both members of the equation $2x - 1 = x^2 - 4x + 4$.
Transpose and rearrange, $x^2 - 6x + 5 = 0$
Factor $(x - 5)(x - 1) = 0$

$$x = 5 \qquad\qquad x = 1$$

Check for $x = 5$, $\sqrt{10 - 1} = 5 - 2$, $\sqrt{9} = 3, 3 = 3$
Check for $x = 1$, $\sqrt{2 - 1} = 1 - 2$, $1 = -1$

In checking, only the positive square root is used.
Therefore, the equation has only one root, $x = 5$.
The value $x = 1$, does *not* check; 1 is an extraneous root.

Example: Solve the equation $\sqrt{2x + 6} - \sqrt{x + 4} = \sqrt{x - 4}$

▶ $\sqrt{2x + 6} - \sqrt{x + 4} = \sqrt{x - 4}$

Transpose, $\sqrt{2x + 6} = \sqrt{x + 4} + \sqrt{x - 4}$
Square both members,

$$2x + 6 = x + 4 + 2\sqrt{(x + 4)(x - 4)} + x - 4$$
$$2x + 6 = 2x + 2\sqrt{(x + 4)(x - 4)}$$
$$6 = 2\sqrt{(x + 4)(x - 4)}$$

Divide both members by 2, $\sqrt{x^2 - 16} = 3$
Square both members, $x^2 - 16 = 9$

$$x^2 = 25$$
$$x = \pm 5$$

Only $x = 5$ satisfies the original equation. $x = -5$ is an extraneous root.

Example: Solve the equation
$$\sqrt{x + 3} + \sqrt{3x + 7} = \sqrt{12x - 8}$$

▶ $\sqrt{x + 3} + \sqrt{3x + 7} = \sqrt{12x - 8}$

Square both members,

$$x + 3 + 2\sqrt{(x + 3)(3x + 7)} + 3x + 7 = 12x - 8$$

Transpose and rearrange, $2\sqrt{(x + 3)(3x + 7)} = 8x - 18$

Divide both members by 2, $\sqrt{3x^2 + 16x + 21} = 4x - 9$
Square both members, $3x^2 + 16x + 21 = 16x^2 - 72x + 81$
Transpose and rearrange, $13x^2 - 88x + 60 = 0$
Factor, $(13x - 10)(x - 6) = 0$

$$13x - 10 = 0 \qquad\qquad x - 6 = 0$$
$$x = \frac{10}{13} \qquad\qquad\qquad x = 6$$

Only $x = 6$ satisfies the original equation. $x = \frac{10}{13}$ is an extraneous root.

Example: Solve the equation
$$2x^2 - 5x + \sqrt{2x^2 - 5x + 11} = 1$$

▶ $2x^2 - 5x + \sqrt{2x^2 - 5x + 11} = 1$
Add 11 to both members,
$$(2x^2 - 5x + 11) + \sqrt{2x^2 - 5x + 11} = 12$$
Let $\sqrt{2x^2 - 5x + 11} = A$, then $(2x^2 - 5x + 11) = A^2$
$$A^2 + A - 12 = 0$$
$$(A + 4)(A - 3) = 0$$
$$A + 4 = 0 \qquad\qquad A - 3 = 0$$
$$A = -4 \qquad\qquad A = 3$$
Replace A by $\sqrt{2x^2 - 5x + 11}$

$$\sqrt{2x^2 - 5x + 11} = -4 \qquad \sqrt{2x^2 - 5x + 11} = 3$$

Square both members, $2x^2 - 5x + 11 = 16$, $2x^2 - 5x + 11 = 9$

By formula, $x = \dfrac{5 \pm \sqrt{25 + 40}}{4}$ $\qquad 2x^2 - 5x + 2 = 0$

$$x = \frac{5 \pm \sqrt{65}}{4} \qquad (2x - 1)(x - 2) = 0$$
$$2x - 1 = 0, \, x - 2 = 0$$
$$x = \frac{1}{2}, \, x = 2$$

$x = \dfrac{1}{2}$ and $x = 2$ are roots

$x = \dfrac{5 \pm \sqrt{65}}{4}$ are extraneous roots

Example: Solve the equations $\quad \sqrt{3x + 1 + \sqrt{2x + 1}} = 4$

▶ $\quad \sqrt{3x + 1} + \sqrt{2x + 1} = 4$

Square both members, $3x + 1 + \sqrt{2x + 1} = 16$

Transpose, $\sqrt{2x + 1} = 15 - 3x$

Square both members, $2x + 1 = 225 - 90x + 9x^2$

Transpose, $9x^2 - 92x + 224 = 0$

Factor, $(9x - 56)(x - 4) = 0$

$9x - 56 = 0 \qquad\qquad\qquad x - 4 = 0$

$x = \dfrac{56}{9} \qquad\qquad\qquad\qquad x = 4$

$x = 4$ is a root

$x = \dfrac{56}{9}$ is an extraneous root

EXERCISE 11

Solve the following equations.   Check all roots and identify extraneous roots.

1. $\sqrt{3x + 7} = x - 1$
2. $11 + \sqrt{2x + 6} = 3x$
3. $\sqrt{2x + 3} = \sqrt{x + 1} + \sqrt{x - 2}$
4. $\sqrt{3x - 5} - \sqrt{2x - 5} = \sqrt{x - 6}$
5. $\sqrt{5x + 1} - \sqrt{x - 6} = \sqrt{3x + 4}$
6. $\sqrt{3x + 7} - \sqrt{x - 2} = \sqrt{5x - 6}$
7. $x^2 - 2x + 6 + \sqrt{x^2 - 2x + 6} = 12$
8. $3x^2 - x - \sqrt{3x^2 - x - 8} = 38$
9. $\sqrt{x + 1} + \sqrt{7x + 4} = 3$
10. $\sqrt{2x + 7} - \sqrt{x - 4} = 4$

MAXIMUM AND MINIMUM OF QUADRATIC FUNCTIONS

In the chapter on Functions and Graphs problems involving maximum and minimum values of functions were discussed.   The graph of the quadratic function $y = ax^2 + bx + c$ with real coefficients ($a \neq 0$) has a highest point if the graph opens out downward, or a lowest point if the graph opens out upward.   The coordinates of the maximum or minimum point may be found by completing the square as explained on page 275.

Example: Find the coordinates of the maximum or minimum point of $y = x^2 - 2x - 5$.

▶
$$y = x^2 - 2x - 5$$

To complete the square we add and subtract the square of half the coefficient of x. In this case, we add and subtract $[\frac{1}{2}(2)]^2$ or 1.

The equation now becomes
$$y = x^2 - 2x + 1 - 5 - 1 = x^2 - 2x + 1 - 6$$
This may be written as $y = (x - 1)^2 - 6$.

Since $(x - 1)^2$ is never negative for real values of x, the minimum value of $(x - 1)^2$ is zero. Therefore, the minimum value of y is -6 when $(x - 1)^2 = 0$, or $x = 1$.

The coordinates of the minimum point of $y = x^2 - 2x - 5$ are $(1, -6)$.

Example: Find the maximum value of the function
$$y = 2 - 18x - 3x^2$$

▶
$$y = 2 - 18x - 3x^2$$
Rewrite the function
$$y = -3x^2 - 18x + 2 = -3(x^2 + 6x) + 2$$

To complete the square we add and subtract the square of half the coefficient of x. In this case, we add and subtract $[\frac{1}{2}(6)]^2$ or 9.

The equation now becomes
$$y = -3(x^2 + 6x + 9) + 2 + 27 = -3(x + 3)^2 + 29$$

Note that the insertion of $+9$ in the parentheses was equivalent to subtracting 27. We therefore had to add 27 in order to avoid changing the value of the function.

Since $(x + 3)^2$ is never negative for real values of x, the maximum value of $y = 29$.

Note that the maximum value of y is obtained when $(x + 3) = 0$. Therefore, $x = -3$ when the function y has the maximum value 29.

Example: If a ball is thrown upward with an initial velocity of 120 feet per second, its height after t seconds is given by the formula $h = 120t - 16t^2$. Find the highest point the ball will reach.

$$h = 120t - 16t^2$$

Rewrite the function $h = -16(t^2 - \frac{15}{2} t)$.

To complete the square we add and subtract the square of half the coefficient of t. In this case, we add and subtract $[\frac{1}{2}(-\frac{15}{2})]^2$ or $\frac{225}{16}$.

The equation now becomes

$$h = -16\left(t^2 - \frac{15}{2}t + \frac{225}{16}\right) + 225 = -16\left(t - \frac{15}{4}\right)^2 + 225$$

Note that the insertion of $\frac{225}{16}$ in the parentheses was equivalent to subtracting 225. We therefore added 225 in order to avoid changing the value of the function.

Since $(t - \frac{15}{4})^2$ is never negative for any real value of t the maximum value of $h = 225$.

The maximum value of h is obtained when $(t - \frac{15}{4})^2 = 0$. Therefore, $t = \frac{15}{4}$ when h is a maximum, i.e. the ball reaches its greatest height $3\frac{3}{4}$ seconds after it is released.

Example: Find the area of the largest rectangle that can be inscribed in a triangle whose base is 10 and altitude 8.

▶ Let x be the length of the inscribed rectangle and let y be its width. In the similar triangles ABC and BDE

$$\frac{DE(x)}{AC(10)} = \frac{BF(8 - y)}{BG(8)}$$
$$8x = 80 - 10y$$
$$10y = 80 - 8x$$
$$y = \frac{80 - 8x}{10}$$

Area of the rectangle, $A = xy = \dfrac{x(80 - 8x)}{10}$

$$10A = 80x - 8x^2$$
$$A = 8x - \frac{4}{5}x^2$$

Rewrite the equation, $A = -\frac{4}{5}(x^2 - 10x)$.

To complete the square of the expression in the parentheses we add the square of half the coefficient of x. In this case, we add $[\frac{1}{2}(-10)]^2$ or 25 to the expression in the parentheses. The equation now becomes

$$A = -\tfrac{4}{5}(x^2 - 10x + 25) + 20 = -\tfrac{4}{5}(x - 5)^2 + 20.$$

Since we added -20 by inserting $+25$ in the parentheses we had to add $+20$ in order to avoid changing the value of the function.

Since $(x - 5)^2$ is never negative for any real value of x the maximum value of $A = 20$.

Note that the maximum value of A is obtained when $(x - 5)^2 = 0$. Therefore, $x = 5$ when A is a maximum and correspondingly $y = 4$.

Example: A clothing manufacturer can sell 500 suits when priced at $60 each. He can sell 50 additional suits for each $2 reduction in price per suit. At what price should he sell the suits in order to obtain the maximum gross return?

▶ Let $x = $ the number of $2 reductions to obtain the maximum gross return.

Then the maximum gross return, $M = (500 + 50x)(60 - 2x)$.

$$M = 30{,}000 + 2{,}000x - 100x^2$$

Rewrite the equation $M = -100(x^2 - 20x) + 30{,}000$.

Now, complete the square of the expression in the parentheses by adding the square of half the coefficient of x. In this case, we add $[\frac{1}{2}(-20)]^2$ or 100.

The equation now becomes

$$M = -100(x^2 - 20x + 100) + 30{,}000 + 10{,}000$$
$$= -100(x - 10)^2 + 40{,}000.$$

Since $(x - 10)^2$ is never negative for any real value of x, the maximum value of $M = \$40{,}000$.

The maximum value of M is obtained when $(x - 10)^2 = 0$. Therefore, $x = 10$ when M is a maximum. In other words, the manufacturer will obtain the maximum gross return when he sells the suits at $10(\$2)$ less than the original price, i.e. at $60 - $20 or at $40 per suit.

EXERCISE 12

1. Find the coordinates of the maximum or minimum points of each of the following:

a. $y = x^2 - 6x + 5$ c. $y = 4 - 6x - 3x^2$

b. $y = -2x^2 + 8x - 3$ d. $y = 2x + 4x^2 - 1$

2. Find two numbers whose sum is 16 and whose product is a maximum.

3. Find the dimensions of the rectangle with maximum area whose perimeter is 40 feet.

4. One side of a rectangular field is adjacent to a river. There are 240 yards of fencing available for the other three sides. What are the dimensions of the field when the area is a maximum?

5. Find the area of the largest rectangle that can be inscribed in a triangle whose base is 12 and altitude 6.

6. If a ball is thrown upward with an initial velocity of 192 feet per second, its height after t seconds is given by the formula $h = 192t - 16t^2$. Find the highest point the ball will reach.

VERBAL PROBLEMS

Number Problems

Example: Find the number which, increased by its reciprocal, is equal to $\frac{37}{6}$.

▶
$$\text{Let } x = \text{The number}$$
$$\text{Then } \frac{1}{x} = \text{The reciprocal}$$
$$x + \frac{1}{x} = \frac{37}{6}$$
$$6x^2 + 6 = 37x$$
$$6x^2 - 37x + 6 = 0$$
$$(6x - 1)(x - 6) = 0$$
$$6x - 1 = 0$$
$$x = \frac{1}{6}$$
$$x - 6 = 0$$
$$x = 6$$

The number may be taken as $\frac{1}{6}$ and the reciprocal as 6, or the number may be taken as 6 and the reciprocal as $\frac{1}{6}$.
Check: If the number is 6, the reciprocal is $\frac{1}{6}$.
   Then $6 + \frac{1}{6} = \frac{37}{6}$.

Example: Find 3 consecutive positive integers such that when 5 times the largest be subtracted from the square of the middle one the result exceeds three times the smallest by 7.

▶   Let x = The smallest number
                  Then $x + 1$ = The next larger number
                  And $x + 2$ = The largest number
$$(x + 1)^2 - 5(x + 2) = 3x + 7$$
$$x^2 + 2x + 1 - 5x - 10 = 3x + 7$$
$$x^2 - 6x - 16 = 0$$
$$(x - 8)(x + 2) = 0$$
   $x = 8$ $x = -2$ reject
$x + 1 = 9$
$x + 2 = 10$
Check: $(9)^2 - 5(10) = 3(8) + 7$
          $81 - 50 = 24 + 7, 31 = 31$

EXERCISE 13

Solve and check the following problems:
1. Find the number which, increased by its reciprocal, is equal to $\frac{26}{5}$.
2. Find 3 consecutive positive integers such that the sum of the first and 4 times the third is 147 less than the square of the second.
3. The larger of two numbers exceeds the smaller by 2. The sum of the squares of the two numbers is 340. Find the numbers.
4. The difference between two numbers is 2. The difference of their reciprocals is $\frac{1}{24}$. Find the numbers.
5. Separate 36 into two parts such that 5 times the square of the smaller diminished by twice the larger is equal to the square of the smaller.

Geometry Problems

Example: The perimeter of a rectangle is 56 inches and its area is 192 square inches. Find the dimensions of the rectangle.

▶ Let x = The number of inches in the length of the rectangle
Then $28 - x$ = The number of inches in the width of the rectangle

Since the perimeter is 56 inches, the sum of one length and one width is half the perimeter, or 28 inches.

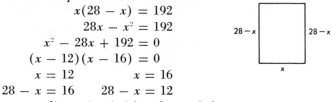

$$x(28 - x) = 192$$
$$28x - x^2 = 192$$
$$x^2 - 28x + 192 = 0$$
$$(x - 12)(x - 16) = 0$$
$$x = 12 \qquad x = 16$$
$$28 - x = 16 \qquad 28 - x = 12$$

Thus, one dimension is 16 inches and the other is 12 inches.

Check: If the length is 16 inches and the width is 12 inches, the perimeter is equal to $2(16) + 2(12) = 56$ inches. The area = $12 \times 16 = 192$ square inches.

Example: A piece of wire 56 inches long is bent into the form of a right triangle whose hypotenuse is 25 inches. Find the other two sides of the triangle.

▶ Let x = Length of one leg of the right triangle
Then $(31 - x)$ = Length of other leg of the right triangle
(Since the sum of the two legs is $56 - 25 = 31$)

$$x^2 + (31 - x)^2 = (25)^2$$
$$x^2 + 961 - 62x + x^2 = 625$$
$$2x^2 - 62x + 336 = 0$$
$$x^2 - 31x + 168 = 0$$
$$(x - 7)(x - 24) = 0$$
$$x = 7 \qquad x = 24$$
$$31 - x = 24 \qquad 31 - x = 7$$

Thus, the other two legs are 7 inches and 24 inches.

Check: The sides of the triangle are 7, 24, and 25.
$$(7)^2 + (24)^2 = (25)^2, 49 + 576 = 625$$

Example: A swimming pool, rectangular in shape, is 60 feet long and 40 feet wide. A walk of uniform width is built around the pool. If the area of the walk is 864 square feet, find the width of the walk.

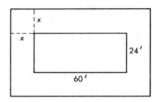

▶ Let x = Width of the walk

The outside dimensions of the pool are

length = $(60 + 2x)$,

width = $(40 + 2x)$

$$(60 + 2x)(40 + 2x) - 2,400 = 864$$
$$2,400 + 200x + 4x^2 - 2,400 = 864$$
$$4x^2 + 200x - 864 = 0$$
$$x^2 + 50x - 216 = 0$$
$$(x + 54)(x - 4) = 0$$
$$x = -54 \text{ reject}, \quad x = 4 \text{ feet}$$

The width of the walk is 4 feet.

Check: If the width of the walk is 4 feet, the outer dimensions are 68 feet by 48 feet. The outer area = $68 \times 48 = 3,264$ square feet. The area of the pool = $60 \times 40 = 2,400$ square feet. The area of the walk = $3,264 - 2,400 = 864$ square feet.

Example: A rectangular sheet of tin is twice as long as it is wide. From each corner a 3-inch square is cut out and the ends are turned up so as to form a box. If the volume of the box is 324 cubic inches, find the dimensions of the sheet of tin.

▶   Let $x = $ Width of the tin
          Then $2x = $ Length of the tin

W h e n   t h e
squares are cut
out and the ends
are turned up
the width of the
box is $(x - 6)$
and the length of
the box is $(2x -
6)$. The depth of
the box is 3 inches.   The relationship used to set up the equation is

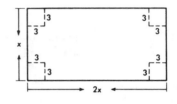

Depth $\times$ Width $\times$ Height $=$ Volume of the box
$$3(x - 6)(2x - 6) = 324$$
$$6x^2 - 54x + 108 = 324$$
$$x^2 - 9x - 36 = 0$$
$$(x - 12)(x + 3) = 0$$
$$x = 12, \quad x = -3 \text{ reject}$$

Thus, the dimensions of the sheet are 12″ by 24″.

Check: The dimensions of the box are 3″ by $(12'' - 6'')$ by $(24'' - 6'') = 3''$ by 6″ by 18″ $= 324$ cubic inches.

EXERCISE 14

Solve and check the following problems:

1. The perimeter of a rectangle is 42 inches and the area of the rectangle is 98 square inches.   Find the dimensions of the rectangle.

2. The hypotenuse of a right triangle is 30 inches.   The length of one leg exceeds the length of the other leg by 6 inches.   Find the lengths of the two legs.

3. A garden plot, rectangular in shape, is 80 feet long and 20 feet wide.   A walk of uniform width is built around the plot.   If the area of the walk is 1,344 square feet, find the width of the walk.

4. A rectangular sheet of metal is 3 times as long as it is wide. From each corner a 2-inch square is cut out and the ends are turned up so as to form a box.   If the volume of the box is 512 cubic inches, find the dimensions of the sheet of metal.

5. A picture is 12″ by 24″ and is surrounded by a frame of uniform width. If the area of the frame is 252 square feet, find the width of the frame.

Motion Problems

Example: An autoist traveling a distance of 60 miles would have reached his destination 30 minutes earlier if he had driven 6 miles an hour faster. Find the rate at which he traveled.

▶ There are two trips involved in this problem, an actual trip and a supposed trip.

Rate × Time = Distance

Actual	x	$\dfrac{60}{x}$	60
Supposed	$x + 6$	$\dfrac{60}{x + 6}$	60

Let x = Actual rate of speed

Then $x + 6$ = Supposed rate of speed

The data in this problem is collected in the box above.

The relationship used to set up the equation is

$$\text{Actual time} = \text{Supposed time} + \text{one-half hour}$$

$$\frac{60}{x} = \frac{60}{x + 6} + \frac{1}{2}$$

$$(60)(2)(x + 6) = 60(x)(2) + x(x + 6)$$

$$120x + 720 = 120x + x^2 + 6x$$

$$x^2 + 6x - 720 = 0$$

$$(x - 24)(x + 30) = 0$$

$$x = 24, \quad x = -30 \text{ reject}$$

Check: Actual speed was 24 miles per hour. Time was $60 \div 24$ $= 2\frac{1}{2}$ hours.

Supposed speed was 30 miles per hour. His time would have been $60 \div 30 = 2$ hours.

Thus, he would have saved one-half hour.

Example: A motorist drives a distance of 150 miles and then returns, the round trip taking 8 hours and 45 minutes. His speed returning is 10 miles per hour faster than his speed in going. What was his rate on the trip out?

▶   Let x = Rate on trip out
Then $x + 10$ = Rate on trip back

Rate × Time = Distance

Trip out	x	$\dfrac{150}{x}$	150
Trip back	$x + 10$	$\dfrac{150}{x + 10}$	150

The data in the problem is collected in the box above. The relation-ship used to set up the equation is

Time on trip out + Time on trip back = $8\frac{3}{4}$ hours.

$$\frac{150}{x} + \frac{150}{x + 10} = \frac{35}{4}$$
$$150(4)(x + 10) + 150(x)(4) = 35(x)(x + 10)$$
$$600x + 6{,}000 + 600x = 35x^2 + 350x$$
$$35x^2 - 850x - 6{,}000 = 0$$
$$7x^2 - 170x - 1{,}200 = 0$$
$$(7x + 40)(x - 30) = 0$$
$$x = -\frac{40}{7} \text{ reject}, \quad x = 30$$

Thus, the rate on the trip out was 30 miles per hour.

Check: Speed on trip out was 30 miles per hour. The time was $150 \div 30 = 5$ hours.

Speed on trip back was 40 miles per hour.
Time was $150 \div 40 = 3\frac{3}{4}$ hours.

Total time was $8\frac{3}{4}$ hours.

Example: A plane left an airfield to fly to a destination 1,860 miles away. After flying at a certain rate for 600 miles the wind changed, increasing the speed of the plane by 40 miles per hour, thus reducing the time of the trip by 45 minutes. What was the original rate of the plane?

▶ Let x = The original rate of the plane
And $x + 40$ = The increased rate of the plane

Rate × Time = Distance

First part of trip	x	$\dfrac{600}{x}$	600
Second part of trip	$x + 40$	$\dfrac{1,260}{x + 40}$	1,260

The relationship used to set up the equation is

Time on actual trip = Time on supposed trip $- \frac{3}{4}$ hour

$$\frac{600}{x} + \frac{1,260}{x + 40} = \frac{1,860}{x} - \frac{3}{4}$$

$$600(4)(x + 40) + 1,260(x)(4)$$
$$= 1,860(4)(x + 40) - 3(x)(x + 40)$$
$$2,400x + 96,000 + 5,040x = 7,440x + 297,600 - 3x^2 - 120x$$
$$3x^2 + 120x - 201,600 = 0$$
$$x^2 + 40x - 67,200 = 0$$
$$(x + 280)(x - 240) = 0$$
$$x = -280 \text{ reject}, \quad x = 240$$

Thus, the original rate of the plane was 240 miles per hour.

Check: Had the plane traveled 1,860 miles at 240 miles per hour, the trip would have taken $1,860 \div 240 = 7\frac{3}{4}$ hours.

The plane traveled 600 miles at 240 miles per hour taking $2\frac{1}{2}$ hours, then 1,260 miles at 280 miles per hour taking $4\frac{1}{2}$ hours.

EXERCISE 15

Solve and check the following problems:

1. A man traveled 240 miles. If he had gone 6 miles an hour faster he would have completed the trip in $1\frac{1}{3}$ hours less time. How fast did the man travel?

2. At his usual rate, a man can travel 60 miles by automobile in a certain time. If he were to increase his usual rate by 10 miles per hour he could travel the 60 miles in $\frac{1}{2}$ hour less time. Find the man's usual rate.

3. A man drove from his home to a destination 370 miles away. He drove at a certain rate for 80 miles. Then he finished the trip on a turnpike increasing his average speed by 18 miles per hour, thus reducing the time of the trip by $2\frac{1}{4}$ hours. What was the man's original rate of speed?

4. Two men travel between two cities by different routes, one route covering 240 miles and the other covering 255 miles. The man taking the shorter route travels 10 miles per hour faster than the other and completes the trip in $2\frac{1}{2}$ hours less time than the other. At what rates did the men travel?

5. Two men, A and B, start at the same time from a certain point and walk east and south respectively. At the end of 5 hours A has walked 5 miles farther than B and they are 25 miles apart. Find the rate of each.

Business Problems

Example: A merchant paid $1,800 for a group of men's suits. He sold all but 5 of the suits at $20 more per suit than he paid, thereby making a profit of $200 on the transaction. How many suits did the merchant buy?

▶   Let x = The number of suits the merchant bought

Then $\dfrac{1,800}{x}$ = The cost of each suit

The relationship used to set up the equation is

The number of suits sold $\times$ The selling price of each suit

$$= \$1,800 + \$200$$

$$(x - 5)\left(\frac{1,800}{x} + 20\right) = 2,000$$

$$1,800 + 20x - \frac{9,000}{x} - 100 = 2,000$$

$$20x - \frac{9,000}{x} = 300$$

$$20x^2 - 300x - 9,000 = 0$$

$$x^2 - 15x - 450 = 0$$

$$(x - 30)(x + 15) = 0$$

$$x = 30, \quad x = -15 \text{ reject}$$

The merchant bought 30 suits.

Check: The merchant bought 30 suits at $60 each.   He sold 25 suits at $80 each, thus taking in $2,000.

Example: A dealer can buy a certain number of ties for $30. If 5 more could be bought for the same money, the price would be $3.60 less per dozen.   What is the price per dozen?

▶   Let x = The number of ties the dealer can buy for $30

Then $\dfrac{30}{x}$ = The cost per tie

And $\dfrac{30}{x + 5}$ = The supposed cost per tie

The relationship used to set up the equation is

The supposed cost per dozen = The old cost per dozen minus $3.60

$$12\left(\frac{30}{x + 5}\right) = 12\left(\frac{30}{x}\right) - \frac{18}{5}$$
$$12(30)(5x) = 12(30)(5)(x + 5) - 18(x)(x + 5)$$
$$1{,}800x = 1{,}800x + 9{,}000 - 18x^2 - 90x$$
$$18x^2 + 90x - 9{,}000 = 0$$
$$x^2 + 5x - 500 = 0$$
$$(x + 25)(x - 20) = 0$$
$$x = -25 \text{ reject,} \quad x = 20$$

The man bought 20 ties for $30, which is at the rate of $18 per dozen.

Check: The man bought 20 ties for $30, paying $1.50 per tie or $18 per dozen.   Had he bought 25 ties for $30 he would have paid $1.20 per tie or $14.40 per dozen.   This would have been a saving of $3.60 per dozen.

EXERCISE 16

Solve and check the following problems:

1. A book dealer spent $100 for a number of copies of a certain book.   He sold all but two copies at a profit of $1 a book, thereby realizing a profit of $33 on the whole transaction.   How many copies of the book did he buy?

2. A number of boys agreed to contribute equally toward purchasing a basketball for $14. Later, 4 more boys joined the contributors, thus reducing the individual contribution by $.40. What was the amount of the original individual contribution per boy?

3. A dealer bought a number of radio sets for $576. He sold all but 3 at a profit of $9 per set, thereby making a total profit of $207. What was the original price per set?

4. A golfer can buy a certain number of balls for $21.60. If 6 more can be bought for the same money the price would be $2.16 less per dozen. What is the original price per dozen?

5. A motel owner rented a certain number of cabins for $288. When he reduced the price by $2 per cabin he rented 4 more cabins but the total receipts were $48 less. How many cabins did he rent at the higher rate?

Work Problems

Example: Working together, it takes a painter and his helper 4 days to complete a job. Working alone, it would take the helper 6 days more than the painter to do the job. How many days does it take the painter to complete the job alone?

▶ Let x = The number of days it takes the painter

Then $x + 6$ = The number of days it takes the helper

The relationship used in setting up the equation is

Part of job done by painter + Part of job done by helper = Total job

$$\frac{4}{x} + \frac{4}{x + 6} = 1$$
$$x^2 - 2x - 24 = 0$$
$$(x - 6)(x + 4) = 0$$
$$x = 6, \quad x = -4 \text{ reject}$$

It would take the painter 6 days to do the job.

Check: Part of job done by painter + Part of job done by helper = 1

$$\frac{4}{6} + \frac{4}{6 + 6} = 1$$
$$\frac{2}{3} + \frac{1}{3} = 1$$

Example: It takes Bill 4 hours longer to do a job than it takes John. They work together for 2 hours when John leaves. It then takes Bill 7 hours to complete the job. How long does it take John to complete the job, working alone?

▶ Let x = The number of hours it takes John to do the job

Then $x + 4$ = The number of hours it takes Bill to do the job

The relationship used to set up the equation is

Part of work done by John + Part of work done by Bill = 1 job

$$\frac{2}{x} + \frac{2 + 7}{x + 4} = 1$$
$$x^2 - 7x - 8 = 0$$
$$x = 8, \quad x = -1 \text{ reject}$$

It takes John 8 hours to complete the job.

Check: Part of job done by John + Part of job done by Bill = 1

$$\frac{2}{8} + \frac{2 + 7}{8 + 4} = 1$$
$$\frac{1}{4} + \frac{3}{4} = 1$$

EXERCISE 17

Solve and check the following problems:

1. A and B working together can do a job in 6 days. Working alone, it takes A 5 days longer than B to do the job. How long does it take B to do the job?

2. It takes Ed 10 more hours than it takes Frank to do a job. Working together, the two can do the job in 12 hours. How long would it take Frank to do the job, working alone?

3. A mechanic and his helper work on a job together for 3 days. The mechanic then leaves and the helper finishes the job by working 5 more days. If it takes the helper 3 more days than the mechanic to do the job when each works alone, how many days would it take the mechanic working alone?

4. It takes a slower machine 3 hours longer than it takes a faster machine to complete a job. The two machines work together for 4 hours when the faster machine breaks down and the slower machine completes the job in 6 more hours. How long would it take the slower machine to do the job, working alone?

Miscellaneous Problems

Example: The distance (S) in feet traveled by a certain falling body in t seconds is given by the formula $S = 16t^2 - 12t$. Find the time required for that body to fall 208 feet.

▶ In the formula, $S = 208$ and it is required to solve for t.

$$S = 16t^2 - 12t$$
$$208 = 16t^2 - 12t$$
$$4t^2 - 3t - 52 = 0$$
$$(t - 4)(4t + 13) = 0$$
$$t = 4, \quad \text{only} \quad t = -\frac{13}{4} \text{ reject}$$

It takes the body 4 seconds to fall 208 feet.

Example: A woman was told that it would take 1,440 small square tiles to cover the floor of her kitchenette, but that if she bought tiles 4 inches longer on each side, it would take only 160. Find the length of a side of each tile.

▶ Let x = Length of a side of the tile used
Then $x + 4$ = Length of a side of the tile she might have used
The relationship used in setting up the equation is
Area of kitchenette using small tiles

$$= \text{Area of kitchenette using larger tiles}$$

$$1440x^2 = 160(x + 4)^2$$

Divide both sides of the equation by 160, $9x^2 = (x + 4)^2$

$$9x^2 = x^2 + 8x + 16$$
$$x^2 - x - 2 = 0$$
$$(x + 1)(x - 2) = 0$$
$$x = -1 \text{ reject}, \quad x = 2$$

The smaller tile is 2″ by 2″ and the larger tile is 6″ by 6″.

Check: $1440(2)(2) = 160(6)(6)$

$$5,760 = 5,760$$

Example: A circular swimming pool is surrounded by a walk 8 feet wide. The area of the walk is $\dfrac{11}{25}$ of the area of the pool. Find the radius of the pool.

▶ Let x = The radius of the pool.

Then $x + 8$ = The outer radius of the pool.

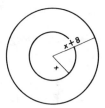

The relationship used in setting up the equation is

Outer area of the pool − Inner area of the pool
= Area of the walk

$$\pi(x + 8)^2 - \pi x^2 = \frac{11}{25}(\pi x^2)$$

$$25\pi(x + 8)^2 - 25\pi x^2 = 11\pi x^2$$

Divide both sides of the equation by π, $25(x + 8)^2 - 25x^2 = 11x^2$

$$25x^2 + 400x + 1,600 - 25x^2 = 11x^2$$
$$11x^2 - 400x - 1,600 = 0$$
$$(11x + 40)(x - 40) = 0$$
$$x = -\frac{40}{11} \text{ reject,} \quad x = 40$$

The radius of the pool is 40 feet.

Check: $\pi(48)^2 - \pi(40)^2 = \dfrac{11}{25}(\pi)(40)^2$

$$2,304\pi - 1,600\pi = 704\pi$$

ANSWERS

EXERCISE 1—Page 252

1. 0, 8   5. $+7, -7$
2. $+6, -6$ 6. $+6, -6$
3. 0, 4   7. $+4, -4$
4. 0, 6   8. 0, 7

EXERCISE 2—Page 254

1. 2, 5   12. $-\dfrac{1}{2}, 3$
2. $-2, -4$
3. $-1, -4$ 13. $-\dfrac{1}{5}, 2$
4. 4, -3
5. 5, -3 14. $\dfrac{7}{2}, -1$
6. $-7, 6$
7. $-\dfrac{1}{2}, -1$ 15. $-\dfrac{3}{2}, \dfrac{1}{2}$

8. $\dfrac{1}{2}, -3$ 16. $\dfrac{1}{3}, \dfrac{1}{2}$

9. $-\dfrac{3}{2}, 5$ 17. $\dfrac{3}{4}, -\dfrac{1}{3}$

10. $\dfrac{2}{3}, 3$ 18. $\dfrac{1}{3}, -\dfrac{3}{2}$

11. $-\dfrac{1}{4}, 1$ 19. $-\dfrac{3}{4}, \dfrac{4}{3}$

                           20. $\dfrac{2}{5}, -\dfrac{3}{2}$

EXERCISE 3—Page 256

1. 6, -1 3. $\dfrac{1}{3}, -1$

2. $-\dfrac{1}{2}, 2$ 4. $\dfrac{2}{5}, -1$

5. $-2 \pm 2\sqrt{5}$

6. $\dfrac{4 \pm \sqrt{14}}{2}$

7. $\dfrac{1 \pm \sqrt{7}}{2}$

8. $\dfrac{-5 \pm \sqrt{37}}{6}$

9. $\dfrac{1 \pm \sqrt{19}}{3}$

10. $\dfrac{5 \pm \sqrt{17}}{4}$

EXERCISE 4—Page 259

1. $5, -2$

2. $3, -6$

3. $\dfrac{1}{2}, -3$

4. $\dfrac{2}{3}, -3$

5. $1 \pm \sqrt{2}$

6. $\dfrac{3 \pm \sqrt{37}}{2}$

7. $\dfrac{-1 \pm \sqrt{41}}{4}$

8. $\dfrac{-2 \pm \sqrt{10}}{3}$

9. $\dfrac{5 \pm i\sqrt{23}}{4}$

10. $\dfrac{-1 \pm i\sqrt{19}}{4}$

EXERCISE 5—Page 260

1. Real, rational, unequal

2. Real, rational, unequal

3. Complex

4. Real, irrational, unequal

5. Real, rational, equal

6. Complex

7. Real, rational, unequal

8. Real, rational, equal

9. Real, irrational, unequal

10. Complex

EXERCISE 6—Page 261

1. $2, 3$

2. $3\dfrac{1}{2}, -3$

3. $1\dfrac{1}{2}, 2\dfrac{1}{2}$

4. $\dfrac{1}{2}, -2\dfrac{1}{4}$

5. $0, -\dfrac{17}{3}$

6. $\dfrac{1}{5}, -\dfrac{2}{5}$

7. $0, -\dfrac{9}{2}$

8. $\dfrac{2}{3}, -2$

EXERCISE 7—Page 263

1. $x^2 - 9x + 14 = 0$
2. $x^2 - 8x + 15 = 0$
3. $x^2 - 2x - 8 = 0$
4. $x^2 + 7x + 6 = 0$
5. $2x^2 - 17x + 8 = 0$
6. $4x^2 - 23x - 6 = 0$

7. $6x^2 - 5x + 1 = 0$
8. $10x^2 - 13x - 3 = 0$
9. $8x^2 + 10x + 3 = 0$
10. $x^2 - 6x + 7 = 0$
11. $x^2 + 8x + 13 = 0$
12. $x^2 - 2x - 4 = 0$

EXERCISE 8—Page 265

1. $\dfrac{9}{4}$
2. $-\dfrac{25}{8}$
3. $+8, -8$
4. 1
5. $\dfrac{3}{2}, \dfrac{1}{2}$
6. 5
7. $-\dfrac{7}{2}$

8. $\dfrac{9}{4}$
9. $4, -2$
10. $-\dfrac{5}{2}, 1$
11. 0
12. $\dfrac{3}{2}$
13. $+3, -3$
14. $-\dfrac{5}{3}, 3$

EXERCISE 9—Page 268

1. b. $3.6, -.6$
   c. $3.2, -.2$
   d. $x = \dfrac{3}{2}$
   e. min. $-\dfrac{17}{4}$
2. b. $3, -1$
   c. $3.2, -1.2$
   d. $x = 1$
   e. max. 4

3. b. $2.6, -.6$
   c. $2.4, -.4$
   d. $x = 1$
   e. min. -5
4. b. $1.4, -1.9$
   c. $1.5, -2$
   d. $x = -\dfrac{1}{4}$
   e. max. 5.1

EXERCISE **10**—Page 271

1. $\mp\sqrt{5}, \pm i\sqrt{3}$

2. $\pm\sqrt{5}, \pm\sqrt{6}$

3. $27, -8$

4. $\dfrac{27}{8}, -125$

5. $-\dfrac{1}{6}, \dfrac{1}{4}$

6. $-\dfrac{3}{7}, \dfrac{1}{2}$

7. $3, -1, \dfrac{-1 \pm \sqrt{13}}{2}$

8. $6, -1, -3, 2$

9. $6, -1, 3, -2$

10. $-\dfrac{3}{2}, 2, 3 \pm 2\sqrt{3}$

EXERCISE **11**—Page 274

1. $6, -1$ is extraneous

2. $5, \dfrac{23}{9}$ is extraneous

3. $3, -2$ is extraneous

4. $7, \dfrac{3}{2}$ is extraneous

5. $7, \dfrac{-15}{11}$ is extraneous

6. $3, \dfrac{-59}{11}$ is extraneous

7. $7.3, -1, 5$ and -3 are extraneous

8. $-\dfrac{11}{3}, 4, \dfrac{1 \pm \sqrt{397}}{6}$ is extraneous

9. $3, 20$ is extraneous

10. $5, \dfrac{17}{4}$ is extraneous

EXERCISE **12**—Page 278

1. a. $3, -4$

　　b. $2, 5$

　　c. $-1, 7$

　　d. $-\dfrac{1}{4}, -\dfrac{5}{4}$

2. $8, 8$

3. 10 by 10

4. 60 by 120

5. 18 feet

6. 576 feet

EXERCISE 13—Page 279

1. $\frac{1}{5}$
2. 14, 15, 16
3. 12, 14
4. 6, 8
5. 32, 4

EXERCISE 14—Page 282

1. 14 by 7
2. 18, 24
3. 6 feet
4. 36 by 12
5. 3 inches

EXERCISE 15—Page 285

1. 30 m.p.h.
2. 30 m.p.h.
3. 40 m.p.h.
4. 30 m.p.h.
5. A 40 m.p.h.
   B 30 m.p.h.

EXERCISE 16—Page 287

1. 40
2. $1.40
3. $18
4. $10.80
5. 36

EXERCISE 17—Page 289

1. 10 days
2. 20 hours
3. 9 days
4. 15 hours

CHAPTER **11**

Systems of Equations Involving Quadratics

The general form of the quadratic equation in two unknowns is $ax^2 + bxy + cy^2 + dx + ey + f = 0$. In this equation, a, b, c, d, e, and f are constants and at least one of the coefficients a, b, c is not equal to zero.

It is shown in analytic geometry that the graph of a quadratic equation in two unknowns is a conic section. A conic section is a section cut from a cone of two nappes by a plane. As shown in the diagram, a conic section may be a circle, a parabola, an ellipse, or a hyperbola. In special cases, the intersection may be two straight lines, a single line, or two intersecting lines.

GRAPHS OF CONIC SECTIONS

Example: Draw the graph of the equation $x^2 + y^2 = 9$

▶ This equation is of the form $x^2 + y^2 = r^2$. In general, the graph of this equation is a circle whose center is at the origin and whose radius is r.

Therefore, the graph of $x^2 + y^2 = 9$ is a circle whose center is at the origin and whose radius is 3.

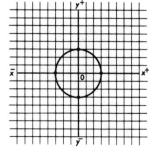

Example: Draw the graph of the equation $(x - 4)^2 + (y + 5)^2 = 49$

▶ This equation is the form
$$(x - h)^2 + (y - k)^2 = r^2$$
In general, the graph of this equation is a circle whose center is at the point (h, k) and whose radius is r.

Therefore, the graph of $(x - 4)^2 + (y + 5)^2 = 49$ is a circle whose center is at $(4, -5)$ and whose radius is 7.

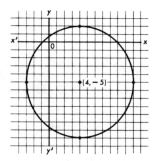

Example: Draw the graph of the equation
$$x^2 - 4x + y^2 + 6y = 12$$

▶ The equation $x^2 - 4x + y^2 + 6y = 12$ may easily be transformed into an equation of the form $(x - h)^2 + (y - k)^2 = r^2$ if we complete squares.

Adding 13 to both members of the equation we have

$$x^2 - 4x + 4 + y^2 + 6y + 9$$
$$= 12 + 13$$
$$(x - 2)^2 + (y + 3)^2 = 25$$

The center of this circle is at $(2, -3)$ and the radius is 5.

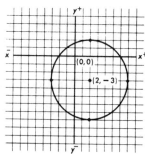

Example: **Draw the graph of the equation** $y = x^2 - 2x - 7$

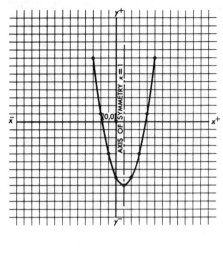

▶ The graph of an equation of the form $y = x^2 - 2x - 7$ is a parabola and its axis of symmetry is the line $x = \frac{-b}{2a}$. In this case, the axis of symmetry is the line $x = \frac{-(-2)}{2} = +1$.

We shall plot the graph by obtaining points on the graph. It will be convenient to take a series of values of x centering about $x = +1$. The following table shows the points used in plotting the graph:

x	-3	-2	-1	0	1	2	3	4	5
y	8	1	-4	-7	-8	-7	-4	1	8

Example: **Draw the graph of the equation**
$$y^2 - 4y + 2x - 14 = 0$$

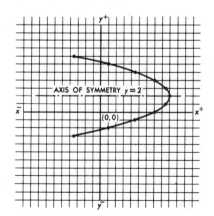

▶ The equation $y^2 - 4y + 2x - 14 = 0$ may be written as $x = \frac{-y^2}{2} + 2y + 7$. We recognize this as a parabola with axis of symmetry $y = \frac{-b}{2a}$, $y = \frac{-2}{2(-\frac{1}{2})} = \frac{-2}{-1} = +2$. Thus, in finding points on the graph it will be convenient to take a series of values of y centering about $y = +2$. The following table shows the points used in plotting the graph.

y	-3	-2	-1	0	1	2	3	4	5	6	7
x	$-3\frac{1}{2}$	1	$4\frac{1}{2}$	7	$8\frac{1}{2}$	9	$8\frac{1}{2}$	7	$4\frac{1}{2}$	1	$-3\frac{1}{2}$

Example: Draw the graph of the equation $4x^2 + 9y^2 = 36$

▶ $4x^2 + 9y^2 = 36$

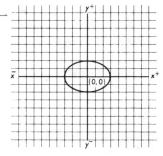

If we solve for y in terms of x we obtain the result $y = \pm\frac{1}{3}\sqrt{36 - 4x^2}$ $= \pm\frac{2}{3}\sqrt{9 - x^2}$. From this result we may conclude that x cannot be greater than $+3$ or less than -3. Values of x greater than $+3$ or less than -3 yield imaginary values for y. If we take valid values for x and compute the corresponding values for y we obtain the table below.

x	3	2	1	0	-1	-2	-3
y	0	± 1.5	± 1.9	± 2	± 1.9	± 1.5	0

Example: Draw the graph of the equation $9x^2 - 25y^2 = 225$

▶ $9x^2 - 25y^2 = 225$

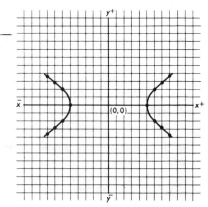

If we solve for y in terms of x we obtain the result $y = \pm\frac{1}{5}\sqrt{9x^2 - 225}$ $= \pm\frac{3}{5}\sqrt{x^2 - 25}$. From this result we may conclude that x cannot have values between -5 and $+5$. Values of x between -5 and $+5$ yield imaginary values for y. If we take other values for x and compute the corresponding values for y we obtain the table below:

x	5	6	7	8	-5	-6	-7	-8
y	0	± 2.0	± 2.9	± 3.8	0	± 2.0	± 2.9	± 3.8

Example: Draw the graph of the equation $xy = 12$.

▶ $xy = 12$

If we solve for y in terms of x we obtain the result $y = \dfrac{12}{x}$. If we take a series of values for x and compute the corresponding values for y we obtain the table below:

x	12	6	4	2	1	-6	-2	-1
y	1	2	3	6	12	-2	-6	-12

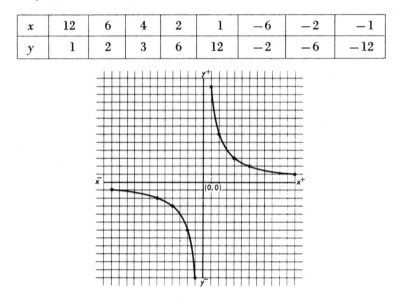

EXERCISE 1

Draw the graphs of the following equations.

1. $x^2 + y^2 = 16$
2. $(x - 1)^2 + (y + 2)^2 = 6$
3. $x^2 + y^2 - 2y = 5$
4. $x^2 + 8x + y^2 - 4y = 29$
5. $x^2 + 4y^2 = 16$

6. $36x^2 + 9y^2 = 324$
7. $y^2 = 16x$
8. $x^2 = -12y$
9. $x^2 + 4x - 2y - 5 = 0$
10. $3x = y^2 + 2y - 35$
11. $4x^2 - 9y^2 = 36$
12. $xy = 18$

Solution of Systems of Equations by Graphs

It will be recalled that systems of simultaneous linear equations have been solved by graphs. The method involved drawing the graphs of the linear equations and obtaining the coordinates of the points of intersection. The same general method is used in solving systems of equations involving quadratic equations.

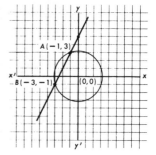

Example: Solve graphically
$$y = 2x + 5$$
$$x^2 + y^2 = 10$$

▶ The graph of the first equation is a straight line. The following table may be used to plot the line $y = 2x + 5$.

x	-2	0	1
y	1	5	7

The graph of the second equation is a circle with center at the origin and radius $= \sqrt{10} = 3.2$.

The points A and B are the intersections of the graphs of the two equations.

$$A(x = -1, y = 3)$$
$$B(x = -3, y = -1)$$

Example: Solve graphically

$$x + 2y = -4$$
$$y = 2x^2 - 3x - 5$$

▶   The graph of the first equation is a straight line. The following table may be used to plot the line $x + 2y = -4$.

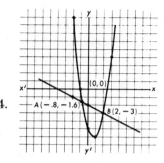

x	-2	0	1
y	-1	-2	$-2\frac{1}{2}$

The graph of the second equation is a parabola. The axis of symmetry of the parabola $x = \frac{-b}{2a} = \frac{3}{4}$. In plotting the graph it will be convenient to obtain points centering about $x = \frac{3}{4}$. The following table shows the points used in plotting the graph of $y = 2x^2 - 3x - 5$.

x	-2	-1	0	1	2	3	4
y	9	0	-5	-6	-3	4	15

The points A and B are the intersections of the graphs of the two equations. The coordinates of these points of intersection are the root of the two equations.

$A(x = -.8, y = -1.6)$ $B(x = 2, y = -3)$

The first two roots are estimated correct to the nearest tenth.

Example: Solve graphically.

$$(x - 3)^2 + y^2 = 16$$
$$x^2 - y^2 = 9$$

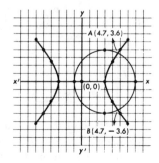

▶   The graph of the first equation is a circle, with center at point $(3,0)$ and radius $= 4$.

The graph of the second equation is a hyperbola. If we solve for y in terms of x we obtain the result

$y = \pm \sqrt{x^2 - 9}$. From this result we may conclude that x cannot have values between -3 and $+3$. Values of x between -3 and $+3$ yield imaginary values for y. The table below shows the points used in plotting the graph of $x^2 - y^2 = 9$.

x	-3	-4	-5	-6	$+3$	$+4$	$+5$	$+6$
y	0	±2.6	±4	±5.2	0	±2.6	±4	±5.2

The points A and B are the intersections of the graphs of the two equations. The coordinates of the points of intersection are

$A(x = 4.7, y = 3.6)$ $\qquad\qquad$ $B(x = 4.7, y = -3.6)$

These sets of roots are estimated correct to the nearest tenth.

In general, a circle and a hyperbola will intersect in 4 points and there will be 4 sets of roots. The algebraic solution of these two equations will show that the other two sets of roots are imaginary.

Example: Solve graphically.

$$x^2 + 4y^2 = 25$$
$$xy = -6$$

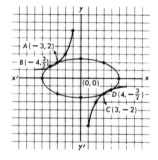

▶ The graph of the equation $x^2 + 4y^2 = 25$ is an ellipse. If we solve for y in terms of x we obtain the result $y = \pm \sqrt{25 - x^2}$. From this result we may conclude that x cannot be greater than $+5$ or less than -5. Values of x greater than $+5$ or less than -5 yield imaginary values for y. The table below shows the points used in plotting the graph of $x^2 + 4y^2 = 25$.

x	5	4	3	2	1	0	-1	-2	-3	-4	-5
y	0	±1.5	±2	±2.3	±2.4	±2.5	±2.4	±2.3	±2	±1.5	0

The graph of the second equation is a hyperbola. If we solve for y in terms of x we obtain the result $y = \frac{-6}{x}$. The table below shows the points used in plotting the graph of the equation $xy = -6$.

x	6	4	3	2	1	-1	-2	-3	-4	-6
y	-1	-1.5	-2	-3	-6	6	3	2	1.5	1

The points A, B, C, and D are the intersections of the graphs of the two equations. The coordinates of the points of intersection are:
$$A(x = -3, y = 2), \quad B(x = -4, y = \tfrac{3}{2}), \quad C(x = 3, y = -2),$$
$$D(x = 4, y = -\tfrac{3}{2})$$

These roots may be checked by direct substitution in the original equations.

Example: Solve graphically.
$$x^2 + y^2 = 10$$
$$4x^2 + y^2 = 13$$

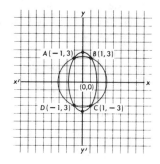

▶ The graph of the equation $x^2 + y^2 = 10$ is a circle with center at the origin and radius $= \sqrt{10} = 3.2$.

The graph of the second equation is an ellipse. If we solve this equation for y in terms of x we obtain the result $y = \pm\sqrt{13 - 4x^2}$. The quantity $13 - 4x^2$ cannot be negative. There-fore, we conclude that x must lie between -1.8 and $+1.8$. Values outside these numbers will make y imaginary. The table below shows the points used in plotting the graph:

x	-1.8	-1.5	-1.2	-1	0	1	1.2	1.5	1.8
y	0	± 2	± 2.7	± 3	± 3.6	± 3	± 2.7	± 2	0

The points A, B, C, and D are the intersections of the graphs of the two equations. The coordinates of these points of intersection are the roots of the two equations. $A(x = -1, y = +3)$, $B(x = +1, y = +3)$, $C(x = +1, y = -3)$, $D(x = -1, y = -3)$.

These solutions may be checked by direct substitution in the original equations.

Example: Solve graphically.

$$y = x^2 - 2x - 8$$
$$y^2 - x^2 = 24$$

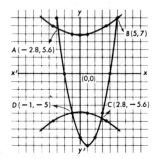

▶   The graph of the first equation is a parabola. The axis of symmetry is the line $x = 1$. The table below shows the points used in plotting the graph of $y = x^2 - 2x - 8$.

The graph of the second equation is a hyperbola. If we solve this equation for x in terms of y we obtain the result $x = \pm \sqrt{y^2 - 24}$. This result indicates that y cannot have values between -4.9 and $+4.9$. Values of y between -4.9 and $+4.9$ will yield imaginary values for x. The table below shows the points used in plotting the graph.

x	-3	-2	-1	0	1	2	3	4	5
y	7	0	-5	-8	-9	-8	-5	0	7

x	$+4.9$	$+5$	$+6$	$+7$	-4.9	-5	-6	-7
y	0	± 1	± 3.5	± 5	0	± 1	± 3.5	± 5

The points A, B, C, and D are the intersections of the graphs of the two equations. The coordinates of these points of intersection are the roots of the two equations. $A(x = -2.8, y = +5.6)$, $B(x = +5, y = +7)$, $C(x = +2.8, -5.6)$, $D(x = -1, y = -5)$

The values of the roots at points A and C are estimated correct to the nearest tenth.

EXERCISE 2

Solve the following sets of equations graphically:

1. $x^2 + y^2 = 20$
   $3x + y = 2$

2. $x^2 + y^2 = 34$
   $2x + 3y = 9$

3. $y = x^2 - 4x - 5$
   $2x + y = 3$

4. $y = x^2 + x - 3$
   $y = 2x - 1$

5. $x^2 + y^2 = 16$
   $y^2 = 2x - 8$

6. $(x + 1)^2 + y^2 = 9$
   $x^2 - y^2 = 4$

7. $(x - 1)^2 + y^2 = 9$
   $y = x^2 - 3x$

8. $(x - 3)^2 + (y - 5)^2 = 25$
   $xy = 6$

9. $2x^2 + 3y^2 = 29$
   $y^2 - x^2 = 8$

10. $x^2 + y^2 = 26$
    $10x^2 + y^2 = 35$

11. $y = x^2 - x + 1$
    $y^2 - x^2 = 5$

12. $2x^2 + y^2 = 54$
    $x = y^2 - y - 1$

ALGEBRAIC SOLUTION OF SYSTEMS OF EQUATIONS OF HIGHER DEGREE

DEFINITION: The *degree of an equation* containing several unknown quantities is equal to the greatest sum of the exponents of the unknown quantities contained in any term.

For example, the equation $x^2 y + 3x + 2y^2 = 5$ is of the third degree.

DEFINITION: A *homogeneous equation* is an equation all of whose terms are of the same degree.

Examples: $x^3 + x^2 y + xy^2 = y^3$
$y^4 + 3xy^3 + 5x^2 y^2 = 7x^3 y$

Note that homogeneous equations do not contain constant terms.

DEFINITION: A *symmetrical equation* is an equation that remains unchanged when the unknown quantities are interchanged.

Examples: $x^4 + 2x^2 y^2 + y^4 = 9$
$x + y + 7 = x^2 + y^2 + xy$

One Linear Equation and the Other Quadratic

A system of equations consisting of one linear and one quadratic
is solved by expressing one of the unknowns in the linear equation in
terms of the other and substituting the result in the quadratic equation.

Example: Solve the system of equations

$$x^2 + y^2 = 13$$
$$3x - y = 11$$

▶   1. $x^2 + y^2 = 13$
         2. $3x - y = 11$

In equation 2, solving for y,

         3. $y = 3x - 11$

Substituting this result in equation 1, we have

         4. $x^2 + (3x - 11)^2 = 13$
         5. $x^2 + 9x^2 - 66x + 121 = 13$
         6. $10x^2 - 66x + 108 = 0$

Dividing both members by 2 we have

         7. $5x^2 - 33x + 54 = 0$

Factoring   8. $(5x - 18)(x - 3) = 0$
              9. $5x - 18 = 0$ $x - 3 = 0$
             10. $x = \dfrac{18}{5}$ $x = 3$

Substituting in equation 3, we have

$$y = -\frac{1}{5} \qquad\qquad y = -2$$

Check for $x = \dfrac{18}{5}, y = -\dfrac{1}{5}$

         1. $\dfrac{324}{25} + \dfrac{1}{25} = 13, \dfrac{325}{25} = 13$
         2. $\dfrac{54}{5} + \dfrac{1}{5} = 11, \dfrac{55}{5} = 11$

Check for $x = 3, y = -2$

         1. $9 + 4 = 13$
         2. $9 + 2 = 11$

    Example: Solve the system of equations

$$3x^2 - xy = 3$$
$$6x - y = 10$$

▶   1. $3x^2 - xy = 3$
       2. $6x - y = 10$

In equation 2, solve for y in terms of x,

     3. $y = 6x - 10$

Substituting this result in equation 1, we have

     4. $3x^2 - x(6x - 10) = 3$
     5. $3x^2 - 6x^2 + 10x = 3$
     6. $3x^2 - 10x + 3 = 0$
     7. $(3x - 1)(x - 3) = 0$
     8. $3x - 1 = 0$ $x - 3 = 0$
     9. $x = \dfrac{1}{3}$ $x = 3$

Substituting in equation 3, we have

     10. $y = -8$ $y = 8$

Check for $x = \dfrac{1}{3}, y = -8$

     1. $3\left(\dfrac{1}{9}\right) - \left(\dfrac{1}{3}\right)(-8) = 3, \dfrac{1}{3} + \dfrac{8}{3} = 3$

     2. $6\left(\dfrac{1}{3}\right) - (-8) = 10, 2 + 8 = 10$

Check for $x = 3, y = 8$

     1. $3(3)^2 - 3(8) = 3, 27 - 24 = 3$
     2. $6(3) - 8 = 10, 18 - 8 = 10$

    Example: Solve the system

$$2x + 5y = 1$$
$$2x^2 + 3xy = 9$$

▶   1. $2x + 5y = 1$
       2. $2x^2 + 3xy = 9$

In equation 1, solve for y in terms of x,

     3. $y = \dfrac{1 - 2x}{5}$

Substituting this value in equation 2, we have

4. $2x^2 + 3x\left(\dfrac{1 - 2x}{5}\right) = 9$

5. $2x^2 + \dfrac{3x - 6x^2}{5} = 9$

6. $10x^2 + 3x - 6x^2 = 45$

7. $4x^2 + 3x - 45 = 0$

8. $(4x + 15)(x - 3) = 0$

9. $x = \dfrac{-15}{4}$ $x = 3$

Substituting in equation 3, we have

10. $y = \dfrac{1 - 2\left(\dfrac{-15}{4}\right)}{5} = \dfrac{17}{10}$

$y = \dfrac{1 - 6}{5} = -1$

The roots are

$x = -\dfrac{15}{4}$ $x = 3$

$y = \dfrac{17}{10}$ $y = -1$

Check for $x = \dfrac{-15}{4}, y = \dfrac{17}{10}$

1. $2\left(\dfrac{-15}{4}\right) + 5\left(\dfrac{17}{10}\right) = 1,\quad \dfrac{-15}{2} + \dfrac{17}{2} = 1$

2. $2\left(\dfrac{225}{16}\right) + 3\left(\dfrac{-15}{4}\right)\left(\dfrac{17}{10}\right) = 9$

$\dfrac{225}{8} - \dfrac{153}{8} = 9,\quad \dfrac{72}{8} = 9$

Check for $x = 3,\quad y = -1$

1. $6 - 5 = 1$

2. $18 - 9 = 9$

Example: Solve the system

$2x - y = 6$

$y^2 = 8x$

▶ 1. $2x - y = 6$
2. $y^2 = 8x$

Solving for x in terms of y in equation 1, we have

3. $x = \dfrac{6 + y}{2}$

Substituting this result in equation 2, we have

4. $y^2 = \dfrac{8(6 + y)}{2}$
5. $y^2 = 4y + 24$
6. $y^2 - 4y - 24 = 0$

Since the left member of equation 6 cannot be factored we may find y by using the quadratic formula

$$y = \frac{-b \pm \sqrt{b^2 - 4a^2}}{2a}$$

7. $y = \dfrac{4 \pm \sqrt{16 + 96}}{2} = \dfrac{4 \pm \sqrt{112}}{2}$

$\qquad = \dfrac{4 \pm 4\sqrt{7}}{2} = 2 \pm 2\sqrt{7}$

Substituting in equation 3, we have

8. $x = \dfrac{6 + (2 \pm 2\sqrt{7})}{2} = \dfrac{8 \pm 2\sqrt{7}}{2} = 4 \pm \sqrt{7}$

The roots are $x = 4 + \sqrt{7}$ and $x = 4 - \sqrt{7}$
$\qquad\qquad y = 2 + 2\sqrt{7} \qquad y = 2 - 2\sqrt{7}$

Check for $x = 4 + \sqrt{7}, y = 2 + 2\sqrt{7}$

1. $2(4 + \sqrt{7}) - (2 + 2\sqrt{7}) = 6,$
$8 + 2\sqrt{7} - 2 - 2\sqrt{7}, 6 = 6$
2. $(2 + 2\sqrt{7})^2 = 8(4 + \sqrt{7})$
$4 + 8\sqrt{7} + 28 = 32 + 8\sqrt{7}$
$32 + 8\sqrt{7} = 32 + 8\sqrt{7}$

BOTH EQUATIONS OF THE FORM $ax^2 + by^2 = c$

Example: Solve the equations
$2x^2 - 3y^2 = 20$
$5x^2 + 2y^2 = 88$

▶   1. $2x^2 - 3y^2 = 20$
      2. $5x^2 + 2y^2 = 88$

Multiplying both members of equation 1 by 2, we have

      3. $4x^2 - 6y^2 = 40$

Multiplying both members of equation 2 by 3, we have

      4. $15x^2 + 6y^2 = 264$

Adding corresponding members of equations 3 and 4, we have

      5. $19x^2 = 304$
      6. $x^2 = 16$
      7. $x = +4$ $x = -4$

Substituting in equation 1 we have

      8. $32 - 3y^2 = 20$ $y = \pm 2$

The four sets of root are shown in the following table:

x	$+4$	$+4$	-4	-4
y	$+2$	-2	$+2$	-2

The roots may be checked by substitution in the original equations.

Example: Solve the equations
      $4x^2 - 3y^2 = 36$
      $3x^2 + 5y^2 = 56$

▶   1. $4x^2 - 3y^2 = 36$
      2. $3x^2 + 5y^2 = 56$

Multiplying both members of equation 1 by 5, we have

      3. $20x^2 - 15y^2 = 180$

Multiplying both members of equation 2 by 3, we have

      4. $9x^2 + 15y^2 = 168$

Adding corresponding members of equations 3 and 4, we have

      5. $29x^2 = 348$
      6. $x^2 = 12$
      7. $x = +\sqrt{12} = +2\sqrt{3}$ $x = -\sqrt{12} = -2\sqrt{3}$

Substituting in equation 1 we have

      8. $48 - 3y^2 = 36$
      9. $y = \pm 2$

The four sets of roots are shown in the following table:

x	$+2\sqrt{3}$	$+2\sqrt{3}$	$-2\sqrt{3}$	$-2\sqrt{3}$
y	$+2$	-2	$+2$	-2

EQUATIONS OF THE FORM $ax^2 + bxy + cy^2 = d$

There are two general methods of solving systems of equations containing equations of the form $ax^2 + bxy + cy^2 = d$. The first method involves the elimination of the absolute term, factoring the resulting equation, and using the derived relationship between the variables to complete the solution. The second method involves the use of the substitution $y = mx$ in both equations.

Example: Solve the set of equations
$$x^2 - xy + y^2 = 21$$
$$y^2 - 2xy = -15$$

▶ 1. $x^2 - xy + y^2 = 21$
2. $y^2 - 2xy = -15$

In order to eliminate the constant terms we shall multiply both members of equation 1 by 5 and both members of equation 2 by 7.

3. $5x^2 - 5xy + 5y^2 = 105$
4. $-14xy + 7y^2 = -105$

Adding corresponding members of the two equations, we have

5. $5x^2 - 19xy + 12y^2 = 0$

Factoring 6. $(5x - 4y)(x - 3y) = 0$

7. $5x - 4y = 0 \qquad x - 3 = 0$

8. $x = \dfrac{4y}{5} \qquad x = 3y$

Substituting these values in equation 2, we have

9. $y^2 - \dfrac{8y^2}{5} = -15 \qquad y^2 - 6y^2 = -15$

10. $y^2 = 25 \qquad y^2 = 3$

11. $y = \pm 5 \qquad y = \pm\sqrt{3}$

12. $x = \dfrac{4y}{5} = \pm 4 \qquad x = 3y = \pm 3\sqrt{3}$

The roots arranged in tabular form are

x	$+4$	-4	$+3\sqrt{3}$	$-3\sqrt{3}$
y	$+5$	-5	$+\sqrt{3}$	$-\sqrt{3}$

The roots may be checked by substitution in the original equation.

Example: Solve the set of equations
$$2y^2 - 4xy + 3x^2 = 17, \quad y^2 - x^2 = 16$$

▶   1. $2y^2 - 4xy + 3x^2 = 17$
    2. $y^2 - x^2 = 16$

In solving this system of equations we shall use the second general method, i.e. we shall let $y = mx$.
If $y = mx$ equation 1 becomes
      3. $2m^2x^2 - 4mx^2 + 3x^2 = 17$
If $y = mx$ equation 2 becomes
      4. $m^2x^2 - x^2 = 16$
We shall now solve equations 3 and 4 for x^2 in terms of m.

   5. $x^2(2m^2 - 4m + 3) = 17$ or $x^2 = \dfrac{17}{2m^2 - 4m + 3}$

   6. $x^2(m^2 - 1) = 16$ or $x^2 = \dfrac{16}{m^2 - 1}$

   7. Therefore, $\dfrac{17}{2m^2 - 4m + 3} = \dfrac{16}{m^2 - 1}$

   8. $32m^2 - 64m + 48 = 17m^2 - 17$

   9. $15m^2 - 64m + 65 = 0$

   10. $(5m - 13)(3m - 5) = 0$

   11. $m = \dfrac{13}{5}$ $\qquad\qquad\qquad m = \dfrac{5}{3}$

   12. Since $y = mx$, $y = \dfrac{13x}{5} \qquad y = \dfrac{5x}{3}$

Substituting in equation 2, we have

   13. $\dfrac{169x^2}{25} - x^2 = 16 \qquad \dfrac{25x^2}{9} - x^2 = 16$

   14. $x = \pm\dfrac{5}{3} \qquad\qquad x = \pm 3$

   15. $y = \pm\dfrac{13}{3} \qquad\qquad y = \pm 5$

The roots arranged in tabular form are:

x	$+\frac{5}{3}$	$-\frac{5}{3}$	$+3$	-3
y	$+\frac{13}{3}$	$-\frac{13}{3}$	$+5$	-5

The roots may be checked by direct substitution in the original equation.

SPECIAL METHODS OF SOLUTION OF SYSTEMS OF QUADRATIC EQUATIONS

A. Symmetrical equations can often be solved by letting $x = u + v$, and $y = u - v$.

Example: Solve the set of equations

$$x^2 + 3xy + y^2 = -\frac{11}{4}$$

$$x^2 - xy + y^2 = \frac{49}{4}$$

▶   1. $x^2 + 3xy + y^2 = -\dfrac{11}{4}$

       2. $x^2 - xy + y^2 = \dfrac{49}{4}$

Let $x = u + v$, and $y = u - v$

       3. $(u + v)^2 + 3(u + v)(u - v) + (u - v)^2$
$$= -\frac{11}{4}$$

       4. $(u + v)^2 - (u + v)(u - v) + (u - v)^2 = \dfrac{49}{4}$

       5. $u^2 + 2uv + v^2 + 3u^2 - 3v^2 + u^2 - 2uv + v^2$
$$= -\frac{11}{4}$$

       6. $u^2 + 2uv + v^2 - u^2 + v^2 + u^2 - 2uv + v^2$
$$= \frac{49}{4}$$

       7. $5u^2 - v^2 = -\dfrac{11}{4}$

$$8. \ u^2 + 3v^2 = \frac{49}{4}$$

Multiply equation 7 by 3

$$9. \ 15u^2 - 3v^2 = -\frac{33}{4}$$

Add the corresponding members of equations 8 and 9

$$10. \ 16u^2 = \frac{16}{4} = 4$$

$$11. \ u^2 = \frac{1}{4}$$

$$12. \ u = +\frac{1}{2} \qquad\qquad u = -\frac{1}{2}$$

Substituting in equation 7, we have

$$13. \ v = \pm 2 \qquad\qquad v = \pm 2$$

$$14. \ x = u + v = \frac{5}{2}, \ -\frac{3}{2} \qquad x = u + v = \frac{3}{2}, \ -\frac{5}{2}$$

$$15. \ y = u - v = -\frac{3}{2}, \ \frac{5}{2} \qquad y = u - v = -\frac{5}{2}, \ \frac{3}{2}$$

The roots arranged in tabular form are:

x	$\frac{5}{2}$	$-\frac{3}{2}$	$\frac{3}{2}$	$-\frac{5}{2}$
y	$-\frac{3}{2}$	$\frac{5}{2}$	$-\frac{5}{2}$	$\frac{3}{2}$

The roots may be checked by direct substitution in the original equations.

B. Some systems of equations may be solved by dividing one equation by the other.

Example: Solve the set of equations
$$x^3 - y^3 = 37$$
$$x - y = 1$$

▶   1. $x^3 - y^3 = 37$
       2. $x - y = 1$

Divide the members of equation 1 by the corresponding members of equation 2.

$\qquad$ 3. $\dfrac{x^3 - y^3}{x - y} = 37$

$\qquad$ 4. $x^2 + xy + y^2 = 37$

From equation 2 we have

$\qquad$ 5. $x = y + 1$

Substituting in equation 4 we have

$\qquad$ 6. $y^2 + 2y + 1 + y^2 + y + y^2 = 37$

$\qquad$ 7. $3y^2 + 3y = 36$, or $y + y^2 - 12 = 0$

$\qquad$ 8. $(y + 4)(y - 3) = 0$

$\qquad$ 9. $y + 4 = 0 \qquad\qquad\qquad y - 3 = 0$

$\qquad$ 10. $y = -4 \qquad\qquad\qquad y = 3$

$\qquad$ 11. $x = y + 1 = -3 \qquad\qquad x = y + 1 = 4$

The roots arranged in tabular form are:

x	-3	4
y	-4	3

C. Some systems of equations may be solved by regarding quantities such as $\dfrac{1}{x}$, xy, $x + y$, etc. as the unknown quantities.

Example: Solve the set of equations

$$\frac{1}{x} + \frac{1}{y} = 2$$

$$\frac{1}{x^2} + \frac{1}{y^2} = 34$$

▶ $\qquad$ 1. $\dfrac{1}{x} + \dfrac{1}{y} = 2$

$\qquad$ 2. $\dfrac{1}{x^2} + \dfrac{1}{y^2} = 34$

Squaring both members of equation 1, we have

$\qquad$ 3. $\dfrac{1}{x^2} + \dfrac{2}{xy} + \dfrac{1}{y^2} = 4$

Subtracting members of equation 3 from corresponding members of equation 2, we have

$\qquad$ 4. $-\dfrac{2}{xy} = 30$

Adding equations 2 and 4, we have

5. $\dfrac{1}{x^2} - \dfrac{2}{xy} + \dfrac{1}{y^2} = 64$

6. $\left(\dfrac{1}{x} - \dfrac{1}{y} \right)^2 = 64$

7. $\dfrac{1}{x} - \dfrac{1}{y} = +8$ $\qquad\qquad\qquad$ $\dfrac{1}{x} - \dfrac{1}{y} = -8$

8. $\dfrac{1}{x} + \dfrac{1}{y} = 2$ $\qquad\qquad\qquad$ $\dfrac{1}{x} + \dfrac{1}{y} = 2$

9. $\dfrac{2}{x} = 10, \quad x = \dfrac{1}{5}$ $\qquad\qquad$ $\dfrac{2}{x} = -6, \quad x = -\dfrac{1}{3}$

Substitute these values in equation 1

10. $y = -\dfrac{1}{3}$ $\qquad\qquad\qquad\qquad$ $y = \dfrac{1}{5}$

The roots arranged in tabular form are:

x	$\frac{1}{5}$	$-\frac{1}{3}$
y	$-\frac{1}{3}$	$\frac{1}{5}$

Example: Solve the set of equations
$x^2y^2 + xy = 6$
$x + 2y = -5$

▶ 1. $x^2y^2 + xy = 6$
2. $x + 2y = -5$

Factoring equation 1, we have

3. $x^2y^2 + xy - 6 = 0$ or $(xy + 3)(xy - 2) = 0$

4. $xy + 3 = 0$ $\qquad\qquad\qquad$ $xy - 2 = 0$

5. $x + 2y = -5$ $\qquad\qquad\qquad$ $x + 2y = -5$

6. $x = -2y - 5$ $\qquad\qquad\qquad$ $x = -2y - 5$

Substituting in equation 4, we have

7. $y(-2y - 5) + 3 = 0$ $\qquad$ $y(-2y - 5) - 2 = 0$

8. $-2y^2 - 5y + 3 = 0$ $\qquad\quad$ $-2y^2 - 5y - 2 = 0$

9. $2y^2 + 5y - 3 = 0$ $\qquad\quad$ $2y^2 + 5y + 2 = 0$

10. $(2y - 1)(y + 3) = 0$ $\qquad$ $(2y + 1)(y + 2) = 0$

11. $y = \dfrac{1}{2}, \quad y = -3$ $\qquad$ $y = -\dfrac{1}{2}, \quad y = -2$

Substitute these results in equation 2

12. $x = -6, \quad x = 1$ $\qquad\qquad$ $x = -4, \quad x = -1$

The roots arranged in tabular form are:

x	-6	1	-4	-1
y	$\frac{1}{2}$	-3	$-\frac{1}{2}$	-2

D. In some cases, the quadratic term may be eliminated and the system of equations may be treated as one linear equation and one quadratic equation.

    Example: Solve the set of equations
$$3xy - 2x + y = -31$$
$$2xy - x - 3y = -47$$

▶   1. $3xy - 2x + y = -31$
      2. $2xy - x - 3y = -47$
Multiply equation 1 by 2 and equation 2 by 3
      3. $6xy - 4x + 2y = -62$
      4. $6xy - 3x - 9y = -141$
Subtract equation 3 from equation 4
      5. $x - 11y = -79$, or $x = 11y - 79$
Substitute for x in equation 1
      6. $3y(11y - 79) - 2(11y - 79) + y = -31$
      7. $33y^2 - 258y + 189 = 0$
      8. $11y^2 - 86y + 63 = 0$
      9. $(11y - 9)(y - 7) = 0$
     10. $11y - 9 = 0$ | $y - 7 = 0$
     11. $y = \dfrac{9}{11}$ | $y = 7$
     12. $x = 11y - 79 = -70$ | $x = 11y - 79 = -2$

The roots arranged in tabular form are:

x	-70	-2
y	$\frac{9}{11}$	7

EXERCISE 3

Solve and check the following systems of equations:

A. One equation linear and the other quadratic

1. $2x^2 + y^2 = 17$
  $2x + y = 1$

2. $x^2 + 3y^2 = 13$
  $x - y = 3$

3. $x^2 + y^2 = 10$
  $2x - y = 5$

4. $x^2 = 2y + 10$
  $3x - y = 9$

5. $x^2 - 3xy + y^2 = -1$
  $y - 2x = 1$

6. $2x - 9y = 7$
  $4x^2 + 9y^2 = 13$

7. $4x + 3y = 20$
  $8x^2 - 3xy = 8$

8. $3x + 4y = -6$
  $x^2 + y^2 = 13$

9. $2x - 3y = 7$
  $x^2 - 6y = 19$

10. $4x + 3y = 7$
  $2x^2 + 5xy + y^2 = 2$

B. Sets of equations of the form $ax^2 + by^2 = c$

11. $2x^2 + 3y^2 = 35$
  $x^2 - y^2 = -5$

12. $3x^2 - y^2 = 7$
  $2x^2 + 3y^2 = 23$

13. $3x^2 + 5y^2 = 93$
  $2x^2 - 7y^2 = -31$

14. $3x^2 + 2y^2 = 30$
  $5x^2 - 9y^2 = -61$

C. Equations of the form $ax^2 + bxy + cy^2 = d$
  Use the method of factoring or the $y = mx$ substitution

15. $x^2 + 3xy = -5$
  $y^2 - xy = 6$

16. $x^2 + 6xy = 28$
  $xy + 8y^2 = 4$

17. $x^2 + y^2 = 50$
  $x^2 + 3xy - 3y^2 = 25$

18. $3xy + 2y^2 = -10$
  $2x^2 + xy = 12$

19. $3x^2 - 4xy + y^2 = 8$
  $4x^2 + xy - y^2 = -16$

20. $x^2 + xy + 2y^2 = 44$
  $2x^2 - xy + y^2 = 16$

21. $x^2 + y^2 = 5$
  $3xy + 4y^2 = 10$

22. $3xy - y^2 + x^2 = 9$
  $x^2 + 2y^2 = 27$

23. $xy = -9$
  $x^2 + 3y^2 = 36$

24. $3xy - 8x^2 = -8$
  $3y^2 + 4xy = 40$

D. Special Methods

25. $x^2 + 4xy + y^2 = -11$
$5x^2 + 7xy + 5y^2 = 23$

26. $x^2 + xy + y^2 = 37$
$x^2 + y^2 = 25$

27. $x^3 + y^3 = -19$
$x + y = -1$

28. $x^3 - y^3 = 61$
$x - y = 1$

29. $x^4 - y^4 = 240$
$x^2 + y^2 = 20$

30. $x^3 - y^3 = 98$
$x - y = 2$

31. $\dfrac{1}{x} - \dfrac{1}{y} = 2$
$\dfrac{1}{x^2} - \dfrac{1}{y^2} = 28$

32. $5x^2 + xy = 4$
$x^2 y^2 + 6 + 7xy = 0$

33. $x + xy + y = 29$
$x^2 + xy + y^2 = 61$

34. $3x^2 - x + y = 42$
$4x^2 + 9y = 46$

VERBAL PROBLEMS

The solution of verbal problems involving systems of quadratic equations follows the general principles outlined above in connection with other types of problems.

Example: The sum of two numbers is 11 and the sum of their squares is 65. Fnd the numbers.

▶   Let x = one number
And y = the other number
Then 1. $x + y = 11$
     2. $x^2 + y^2 = 65$
     3. $x = 11 - y$
     4. $(11 - y)^2 + y^2 = 65$
     5. $121 - 22y + y^2 + y^2 = 65$
     6. $2y^2 - 22y + 56 = 0$
     7. $y^2 - 11y + 28 = 0$
     8. $(y - 7)(y - 4) = 0$
     9. $y = 7$ $y = 4$
    10. $x = 4$ $x = 7$
The numbers are 4 and 7.

Example: The hypotenuse of a right triangle is 68 inches. The area of the triangle is 960 square inches. Find the lengths of the legs.

▶ Let x = the number of inches in one
    leg
And y = the number of inches in the
    other leg
Then 1. $x^2 + y^2 = 68^2 = 4{,}624$

2. $\dfrac{xy}{2} = 960$, $xy = 1{,}920$

3. $2xy = 3{,}840$

Add corresponding members of equations 1 and 3

4. $x^2 + 2xy + y^2 = 8{,}464$
5. $(x + y)^2 = 8{,}464$
6. $x + y = +92$ $x + y = -92$
7. $y = 92 - x$ $y = -92 - x$

Substituting in equation 1, we have

8. $x^2 + (92 - x)^2 = 4{,}624$
              $x^2 + (-92 - x)^2 = 4{,}624$
9. $x^2 + 8{,}464 - 184x + x^2 = 4{,}624$
              $x^2 + 8{,}464x + 184x + x^2 = 4{,}624$
10. $x^2 - 92x + 1{,}920 = 0$
              $x^2 + 92x + 1{,}920 = 0$
11. $(x - 60)(x - 32) = 0$
              $(x + 60)(x + 32) = 0$
12. $x = 60, x = 32$
              $x = -60, x = -32$—rejected
13. $y = 92 - x = 32, y = 60$

The legs are 60 inches and 32 inches.

Example: The fence around a rectangular plot of ground is 128 yards long. The area of the lot is 960 square yards. Find the dimensions of the rectangular plot.

▶   Let $x =$ the number of yards in the length
    And $y =$ the number of yards in the width
    Then 1. $2x + 2y = 128$ or $x + y = 64$
         2. $xy = 960$
         3. $y = 64 - x$
         4. $x(64 - x) = 960$
         5. $64x - x^2 = 960$
         6. $x^2 - 64x + 960 = 0$
         7. $(x - 24)(x - 40) = 0$
         8. $x = 24$ $x = 40$
         9. $y = 64 - x = 40$ $y = 64 - x = 24$

The dimensions of the plot are 40 by 24.

Example: A dealer bought a shipment of shoes for $480. He sold all but 5 pairs at a profit of $6 per pair, thereby making a profit of $290 on the shipment. How many pairs of shoes were in the original shipment?

▶   Let $x =$ the number of pairs of shoes in the shipment
    And $y =$ the cost of a pair of shoes
    Then 1. $xy = 480$
         2. $(x - 5)(y + 6) = 770$
         3. $xy + 6x - 5y - 30 = 770$
         4. $xy + 6x - 5y = 800$
         5. Since $xy = 480, 6x - 5y = 800 - 480 = 320$
         6. $x = \dfrac{320 + 5y}{6}$
         7. $y\left(\dfrac{(320 + 5y)}{6}\right) = 480$
         8. $5y^2 + 320y - 2{,}880 = 0$
         9. $y^2 + 64y - 576 = 0$
        10. $(y - 8)(y + 72) = 0$
        11. $y = 8, \quad y = -72$ reject
        12. Since $xy = 480, x = 60$

There were 60 pairs of shoes in the original shipment.

Example: The product of a two-digit number and the sum of its digits is 324. The units' digit exceeds the tens' digit by 3. Find the number.

▶   Let t = tens' digit
And u = units' digit
Then 1. $(10t + u)(t + u) = 324$
        2. $u = t + 3$
Substituting $u = t + 3$ in equation 1, we have
        3. $(10t + t + 3)(t + t + 3) = 324$
        4. $(11t + 3)(2t + 3) = 324$
        5. $22t^2 + 39t - 315 = 0$
        6. $(22t + 105)(t - 3) = 0$
        7. $t = \dfrac{-105}{22}$ reject, $t = 3$
        8. Since $u = t + 3$, $u = 6$
The number is 36.

Example: Two stations are 300 miles apart. At a certain time a passenger train leaves A for B, and at the same time a freight train leaves B for A. The two trains meet at a point 100 miles from B. Had the speed of the passenger train been 10 miles an hour greater, it would have reached B 9 hours before the freight train reached A. Find the speed of each train.

▶   Let x = rate of passenger train
And y = rate of freight train
Since the trains meet 100 miles from B, the passenger train covers 200 miles in the same time that the freight train covers 100 miles. Therefore,

$$1.\ \frac{200}{x} = \frac{100}{y},\ 100x = 200y,\ x = 2y$$

At an increased speed, the passenger train covers the 300 miles in 9 hours less time than the freight train.
Therefore.

$$2.\ \frac{300}{x + 10} = \frac{300}{y} - 9$$
        3. $300y = 300(x + 10) - 9y(x + 10)$
        4. $300y = 300x + 3,000 - 9xy - 90y$

Substituting $2y$ for x

     5. $300y = 600y + 3{,}000 - 18y^2 - 90y$

     6. $18y^2 - 210y - 3{,}000 = 0$

     7. $3y^2 - 35y - 500 = 0$

     8. $(3y + 25)(y - 20) = 0$

     9. $y = \dfrac{-25}{3}$ reject, $y = 20$

Since $x = 2y$, $x = 2(20)$, $x = 40$.

The rate of the freight train is 20 miles per hour.

The rate of the passenger train is 40 miles per hour.

EXERCISE 4

Solve the following problems:

1. The sum of two numbers is 12 and the sum of their squares is 80. Find the numbers.

2. The hypotenuse of a right triangle is 50 inches. The area of the triangle is 336 square inches. Find the lengths of the legs.

3. The fence around a rectangular plot of ground is 168 yards long. The area of the lot is 1,728 square yards. Find the dimensions of the rectangular plot.

4. A man worked a certain number of days to earn $240. If he had received $2 less per day, he would have had to work 10 days longer to earn the same amount. How many days did he work?

5. The product of a two-digit number and the tens' digit is 384. The units' digit is 2 less than the tens' digit. Find the number.

6. Two towns on opposite sides of a lake are 33 miles apart by water. At 6:00 A.M. a boat starts from each town for the other town, traveling at uniform speed. The boats pass each other at 9:00 A.M. One boat arrives at its destination 1 hour and 6 minutes earlier than the other. Find the time it takes each boat to make the trip.

7. A number of young men purchased a camp, each paying the same amount. If there had been two more men in the company, each would have paid $12 less. If there had been three fewer men in the company, each would have paid $24 more. How many men were there and how much did each pay?

8. The distance between two cities is 270 miles. A motorist made a round-trip between the cities. On the return trip he met heavy traffic and reduced his average rate by 5 miles per hour. As a result, the return trip took him $\frac{3}{4}$ of an hour longer than the trip out. What was his rate of speed on the trip out?

9. The product of two numbers exceeds the larger of the two by 9 and the smaller of the two by 16. Find the two numbers.

10. The area of a small garden is 240 square feet. If the length of the garden be increased by 2 feet and the width of the garden be decreased by 2 feet, then the area of the garden will be decreased by 20 square feet. Find the dimensions of the garden.

11. The area of a triangle is 36 square inches. If the base is increased by 1 inch and the altitude is decreased by 2 inches, the area becomes 30 square inches. Find the base and the altitude of the original triangle.

12. A boat is anchored 3 miles from a straight shore. A camp C is located on the shore 10 miles from A, the point on the shore nearest the boat. A man walks a certain distance from C toward A at 3 miles an hour. He then rows straight to the boat at 4 miles an hour. If the entire trip took him $3\frac{1}{4}$ hours, how many hours did he walk?

ANSWERS

EXERCISE 2

1. $x = 2, y = -4$
  $x = -.8, y = 4.4$
2. $x = -3, y = 5$
  $x = 5.8, y = -.8$
3. $x = 4, y = -5$
  $x = -2, y = 7$
4. $x = 2, y = 3$
  $x = -1, y = -3$
5. $x = 4, y = 0$
6. $x = 2, y = 0$
  $x = -3, y = 2.2$
  $x = -3, y = -2.2$
7. $x = 3.5, y = 1.8$
  $x = -.7, y = 2.6$

8. $x = 6, y = 1$
  $x = .6, y = 9.5$
9. $x = 1, y = 3$
  $x = 1, y = -3$
  $x = -1, y = 3$
  $x = -1, y = -3$
10. $x = 1, y = 5$
  $x = 1, y = -5$
  $x = -1, y = 5$
  $x = -1, y = -5$
11. $x = 2, y = 3$
  $x = -.7, y = 2.3$
12. $x = 5, y = -2$
  $x = 4.8, y = 2.9$

EXERCISE 3

1. $x = -\dfrac{4}{3}, y = \dfrac{11}{3}$

   $x = 2, y = -3$

2. $x = \dfrac{7}{2}, y = \dfrac{1}{2}$

   $x = 1, y = -2$

3. $x = 3, y = 1$

   $x = 1, y = -3$

4. $x = 4, y = 3$

   $x = 2, y = -3$

5. $x = 2, y = 5$

   $x = -1, y = -1$

6. $x = -1, y = -1$

   $x = \dfrac{17}{10}, y = -\dfrac{2}{5}$

7. $x = -\dfrac{1}{3}, y = \dfrac{64}{9}$

   $x = 2, y = 4$

8. $x = 2, y = -3$

   $x = -\dfrac{86}{25}, y = \dfrac{27}{25}$

9. $x = -1, y = -3$

   $x = 5, y = 1$

10. $x = -\dfrac{1}{2}, y = 3$

    $x = \dfrac{31}{13}, y = -\dfrac{11}{13}$

11.

x	$+2$	$+2$	-2	-2
y	$+3$	-3	$+3$	-3

12.

x	-2	$+2$	$+2$	-2
y	$+\sqrt{5}$	$-\sqrt{5}$	$+\sqrt{5}$	$-\sqrt{5}$

13.

x	$+4$	$+4$	-4	-4
y	$+3$	-3	$+3$	-3

14.

x	$+2$	$+2$	-2	-2
y	$+3$	-3	$+3$	-3

15.

x	$-\frac{5}{2}$	$+\frac{5}{2}$	$+1$	-1
y	$+\frac{3}{2}$	$-\frac{3}{2}$	-2	$+2$

16.

x	$+4$	-4	$+14$	-14
y	$+\frac{1}{2}$	$-\frac{1}{2}$	-2	$+2$

17.

x	$+5$	-5	$+7$	-7
y	$+5$	-5	-1	$+1$

18.

x	$+4$	-4	$+3$	-3
y	-5	$+5$	-2	$+2$

19.

x	$+1$	-1	$2i\sqrt{2}$	$-2i\sqrt{2}$
y	$+5$	-5	$4i\sqrt{2}$	$-4i\sqrt{2}$

20.

x	$+2$	-2	$+\sqrt{2}$	$-\sqrt{2}$
y	$+4$	-4	$+3\sqrt{2}$	$-3\sqrt{2}$

21.

x	-1	$+1$	$+2$	-2
y	$+2$	-2	$+1$	-1

22.

x	$+\sqrt{3}$	$-\sqrt{3}$	$+5$	-5
y	$+2\sqrt{3}$	$-2\sqrt{3}$	-1	$+1$

23.

x	$+3$	-3	$+3\sqrt{3}$	$-3\sqrt{3}$
y	-3	$+3$	$-\sqrt{3}$	$+\sqrt{3}$

24.

x	$+\frac{1}{2}$	$-\frac{1}{2}$	$+\frac{2}{3}\sqrt{6}$	$-\frac{2}{3}\sqrt{6}$
y	-4	$+4$	$+\frac{10\sqrt{6}}{9}$	$-\frac{10}{9}\sqrt{6}$

25.

x	$+3$	-3	-2	$+2$
y	-2	$+2$	$+3$	-3

26.

x	$+4$	-4	$+3$	-3
y	$+3$	-3	$+4$	-4

27.

x	2	-3
y	-3	$+2$

28.

x	-4	$+5$
y	-5	$+4$

29.

x	$+4$	$+4$	-4	-4
y	$+2$	-2	$+2$	-2

30.

x	$+5$	-3
y	$+3$	-5

31.

x	$\frac{1}{8}$
y	$\frac{1}{6}$

32.

x	$+1$	-1	$+\sqrt{2}$	$-\sqrt{2}$
y	-1	$+1$	$-3\sqrt{2}$	$+3\sqrt{2}$

33.

x	$+5$	$+4$	$-5 - i\sqrt{14}$	$-5 + i\sqrt{14}$
y	$+4$	$+5$	$-5 + i\sqrt{14}$	$-5 - i\sqrt{14}$

34.

x	$+4$	$-\frac{83}{23}$
y	-2	$-\frac{358}{529}$

EXERCISE 4

1. 8, 4
2. 14, 48
3. 48 yds. by 36 yds.
4. 30 days
5. 64
6. $6\frac{3}{5}$ hours, $5\frac{1}{2}$ hours
7. 18 men, $120 each
8. 45 miles per hour
9. 9 and 2, or -8 and -1.
10. 20 feet by 12 feet
11. base = 9 inches
    altitude = 8 inches
12. 2 hours

CHAPTER **12**

Progressions

ARITHMETIC PROGRESSIONS

DEFINITION: A sequence is a set of numbers arranged in a definite order and formed according to a definite law.

DEFINITION: An arithmetic progression (A.P.) is a sequence of numbers each of which is obtained from the preceding one by adding a constant quantity to it. The constant quantity is called the *common difference*.

Examples: 1. 7, 9, 11, 13, 15 . . . . (common difference is 2)
2. 10, 6, 2, -2, -6, -10 . . . . (common difference is -4)
3. $2\frac{1}{2}$, 3, $3\frac{1}{2}$, 4 . . . . . . (common difference is $\frac{1}{2}$)

Formula for the *n*th Term of an A.P.

Designate the first term of an A.P. by a and the common difference by d. Then the terms of the series can be written as follows:

1	2	3	4 . . . . n
a	$a + d$	$a + 2d$	$a + 3d$ $a + (n - 1)d$

If the *n*th term is called l we have the formula $l = a + (n - 1)d$. The following examples will show how this formula is applied:

Example: Find the 15th term of the progression 2, $6\frac{1}{2}$, 11, ...

▶ Use the formula $l = a(n - 1)d$.
   In this formula $a = 2, d = 4\frac{1}{2}, n = 15, l = ?$
   $l = 2 + \frac{9}{2}(15 - 1) = 2 + \frac{9}{2} \times 14 = 2 + 63 = 65$.

Example: Which term of the series 7, 4, 1, -2, ... is -44?

▶ Use the formula $l = a + d(n - 1)$.
   In this formula $l = -44, a = 7, d = -3, n = ?$
   $-44 = 7 - 3(n - 1)$
   $-44 = 7 - 3n + 3, 3n = 54, n = 18$
   -44 is the 18th term of the series.

Example: In an A.P. of 20 terms the first term is $8\frac{1}{4}$ and the last term is $-\frac{5}{4}$. Find the common difference.

▶ Use the formula $l = a + d(n - 1)$.
   In this formula $a = 8\frac{1}{4}, 1 = 1\frac{1}{4}, n = 20$.
   $-\frac{5}{4} = \frac{33}{4} + d(20 - 1)$
   $-\frac{5}{4} = \frac{33}{4} + 19d$
   $19d = -\frac{38}{4}, d = -\frac{2}{4} = -\frac{1}{2}$.

EXERCISE 1

1. Find the 14th term of the progression 3, 7, 11, ...
2. Find the 40th term of the series $-26, -24, -22$, ...
3. Find the 35th term of the progression 3, $4\frac{1}{2}$, 6, ...
4. Find the 20th term of the progression $4\frac{1}{2}$, 3, $1\frac{1}{2}$, ...
5. Write the 30th term of the series $x - y, x, x + y$, ...
6. Which term of the series 7, 13, 19, ... is 151?
7. Which term of the series 6, 1, -4, ... is -74?
8. Which term of the series 4, $3\frac{1}{3}$, $2\frac{2}{3}$, ... is -8?
9. In an A.P. of 16 terms the first term is -3 and the last term is 27. Find the common difference.
10. In an A.P. of 37 terms the first term is 8 and the last term is -1. Find the common difference.

Arithmetic Means

In an A.P., the terms between any two other terms are called the arithmetic means between the two given terms.

Example: In the A.P. 2, 5, 8, 11, 14 the numbers 5, 8, and 11 are the arithmetic means between 2 and 14.

Example: When there is *one* arithmetic mean between two numbers it is called *the arithmetic mean* of the two numbers.

Let the numbers be p and q and the arithmetic mean be x. Then

$$q - x = x - p, 2x = p + q, \text{ and } x = \frac{p + q}{2}.$$

Thus, the arithmetic mean of two numbers is equal to one-half the sum of the two numbers. The arithmetic mean of two numbers is called their *average*.

Example: Insert 5 arithmetic means between 5 and -16.

▶ Use the formula $l = a + (n - 1)d$.
In this formula $a = 5, l = -16, n = 7$.
$-16 = 5 + (7 - 1)d, -16 = 5 + 6d, 6d = -21,$
$d = -3\frac{1}{2}$.
Therefore, the progression is 5, $1\frac{1}{2}$, -2, $-5\frac{1}{2}$, -9, $-12\frac{1}{2}$, -16.

Example: Insert 3 arithmetic means between x and y.

▶ Use the formula, $l = a + (n - 1)d$.
In this formula, $a = x, l = y,$ and $n = 5$.
$$y = x + (5 - 1)d$$
$$y = x + 4d$$
$$d = \frac{y - x}{4}$$

Therefore, the progression is x, $x + \dfrac{y - x}{4}$, $x + \dfrac{2(y - x)}{4}$,

$x + \dfrac{3(y - x)}{4}$, y.

When simplified, the progression is x, $\dfrac{3x + y}{4}$, $\dfrac{x + y}{2}$,

$\dfrac{x + 3y}{4}$, y.

EXERCISE 2

1. Insert 4 arithmetic means between 2 and 12.

2. Insert 5 arithmetic means between 5 and 13.

3. Insert 5 arithmetic means between 12 and −6.

4. Find the arithmetic means between (a) 4 and 6 (b) −3 and +5 (c) $6\frac{1}{2}$ and $9\frac{1}{2}$ (d) x and y (e) $3\sqrt{5}$ and $-5\sqrt{5}$.

5. Insert 3 arithmetic means between 6 and −3.

The Sum of an Arithmetic Progression

Let S_n represent the sum of n terms of an A.P.

Then $S_n = a + (a + d) + (a + 2d) + \ldots + (l - d) + l$.

If we write S_n in reverse order we have

$$S_n = l + (l - d) + (l - 2d) + \ldots + (a + d) + a$$

If we add these two equations we have

$$2S_n = (a + l) + (a + l) + (a + l) + \ldots$$
$$+ (a + l) \; (a + l)$$

In other words, we have an $(a + l)$ for each term.

Since there are n terms we have

$$2S_n = n(a + l) \qquad S_n = \frac{n}{2}(a + l)$$

Since $l = a + d(n - 1)$ we have by substitution

$$S_n = \frac{n}{2}[a + a + d(n - 1)] \qquad S_n = \frac{n}{2}[2a + d(n - 1)]$$

It will be observed that there are 5 quantities involved in an A.P. They are:

l = last term
a = first term
n = number of terms
d = common difference
S_n = sum

The student can readily recall these quantities by noting that they spell out the word *lands*. If any three of these quantities are given in a problem, then the other two may be found by using the appropriate formula. The examples below will show how this is done.

Example: Find the sum of 36 terms of the A.P. 7, 10, 13, 16, ...

▶ In this problem, $l = ?, a = 7, n = 36, d = 3, S_n = ?$

We may first find l, $l = a + d(n - 1)$, $l = 7 + 3(35)$, $l = 112$.

Now we may find S_n, $S_n = \dfrac{n}{2}(a + l)$, $S_n = \dfrac{36}{2}(7 + 112)$

$= 18(119) = 2,142$ or We may use the formula

$S_n = \dfrac{n}{a}[2a + d(n - 1)]$ $S_n = \dfrac{36}{2}[14 + 3(35)]$

$= 18[14 + 105] = 18(119) = 2,142$

Note that we may find S_n without finding l first by using the formula $S_n = \frac{n}{2}[2a + d(n - 1)]$. Either method is satisfactory. The choice is a matter of preference.

Example: In an A.P., $S_n = -102, a = 8,$ and $l = -25$. Find n and d.

▶ $S_n = \dfrac{n}{2}(a + l),$ $-102 = \dfrac{n}{2}(8 - 25)$

$-102 = -\dfrac{17n}{2},$ $\therefore n = 12$

$l = a + d(n - 1),$ $-25 = 8 + d(12 - 1),$
$-25 = 8 + 11d,$ $\therefore d = -3$

Example: The first row of a theatre has 36 seats and each succeeding row has 2 additional seats. The theatre has a total of 3,000 seats. How many rows of seats does the theatre have? How many seats are there in the last row?

▶ In this problem $a = 36, d = 2,$ and $S_n = 3,000$.

$S_n = \dfrac{n}{2}[2a + d(n - 1)]$

$3,000 = \dfrac{n}{2}[72 + 2(n - 1)]$

$3,000 = \dfrac{n}{2}[72 + 2n - 2] = \dfrac{n}{2}[70 + 2n]$

$3,000 = 35n + n^2$, $n^2 + 35n - 3,000 = 0$
$(n - 40)(n + 75) = 0$
$n = 40$ (The root $n = -75$ is rejected)
$l = a + d(n - 1)$
$l = 36 + 2(40 - 1) = 36 + 2(39) = 36 + 78 = 114$

Example: Find three numbers in arithmetic progression whose sum is 27 and whose product is 405.

▶   The three numbers may be represented as a, $a + d$, and $a + 2d$. However, the solution will be simplified if the numbers are represented as $a - d$, a, and $a + d$.

Since the sum of the numbers is 27, we have $a - d + a + a + d = 27$, $3a = 27$, and $a = 9$. Since $a = 9$, the three numbers may now be represented as $9 - d$, d, and $9 + d$.

Since the product of the numbers is 405, we have
$$(9 - d)(d)(9 + d) = 405, \quad \text{or} \quad 729 - 9d^2 = 405.$$
$$9d^2 = 324, d^2 = 36, d = \pm 6.$$

Hence, the progression is $9 - 6, 9, 9 + 6$, or 3, 9, 15. If we use -6 as the common difference we obtain the same progression in reverse order 15, 9, 3.

Example: A man pays off a debt of \$6,000 by \$500 annual installments and 6% interest on the unpaid balance. What is the total amount of interest that he pays?

▶   The first installment is \$500 plus interest on \$6,000 at 6%, or \$360. The second installment is \$500 plus interest on \$5,500 at 6%, or \$330. The third installment is \$500 plus interest on \$5,000 at 6%, or \$300. Thus, the interest payments \$360, \$330, \$300, etc. form an arithmetic progression.

In this progression, $a = 360$, $d = -30$, and $n = 12$.
Using the formula,

$$S_n = \frac{n}{2}[2a + d(n - 1)]$$
$$S_n = \frac{12}{2}[720 - 30(12 - 1)]$$
$$S_n = 6[720 - 330] = 6[390] = \$2,340.$$

Example: In an arithmetic progression whose first term is 5, the second term is to the sixth term as the fourth term is to the eleventh term. Find the fifteenth term.

▶   If the first term of the progression is 5 and the common difference is d then the second term is $5 + d$, the sixth term is $5 + 5d$, the fourth term is $5 + 3d$, and the eleventh term is $5 + 10d$.

$$\frac{5 + d}{5 + 5d} = \frac{5 + 3d}{5 + 10d}$$

$$(5 + d)(5 + 10d) = (5 + 3d)(5 + 5d)$$

$$25 + 55d + 10d^2 = 25 + 40d + 15d^2$$

$$5d^2 - 15d = 0$$

$$5d(d - 3) = 0$$

$$d = 0, \text{reject}, d = 3.$$

The fifteenth term is $5 + 14d = 5 + 3(14) = 47$.

Example: Find the sum of all the numbers between 70 and 200 that are divisible by 11.

▶ The numbers between 70 and 200 that are divisible by 11 form an arithmetic progression. The first number in this progression is 77, the common difference is 11, and the last number is 198.

Use the formula $l = a + d(n - 1)$ to find n.

$$198 = 77 + (n - 1)11$$

$$198 = 77 + 11n - 11$$

$$11n = 132$$

$$n = 12$$

Use the formula $S_n = \frac{n}{2}(a + l)$

$$S_n = \frac{12}{2}(77 + 198) = 6(275) = 1,650$$

EXERCISE 3

1. Find the sum of the first 18 terms of the A.P., 7, 11, 15, 19. . . . .

2. Find the sum of the first 15 terms of the A.P. whose first term is -8 and whose last term is 19.

3. How many terms of the progression $-7, -4, -1, \ldots$ are required to give a sum of 210?

4. Given $a = 32, n = 18, S_n = -36$. Find d and l.

5. A man started a savings fund by putting \$150 in the bank and increasing the amount by \$25 each year. In how many years did he save \$4,875 (disregarding interest)?

6. Given $l = 9, d = -\frac{3}{4}, n = 25$. Find a and S_n.

7. Given $S_n = -208, l = -8, n = 16$. Find a and d.

8. Given $a = 4, l = -22, S_n = -99$. Find n and d.

HARMONIC PROGRESSIONS

DEFINITION: A harmonic progression (H.P.) is a sequence of numbers whose reciprocals are in arithmetic progression.

Example: The numbers $-\frac{1}{4}$, -1, $\frac{1}{2}$, $\frac{1}{5}$, $\frac{1}{8}$, $\frac{1}{11}$ are in harmonic progression because their reciprocals -4, -1, 2, 5, 8, 11 are in arithmetic progression.

A harmonic progression cannot have 0 as one of its terms since the reciprocal of 0 is undefined.

Example: Show that the following set of numbers form a harmonic progression:

$$\frac{3}{7}, \frac{1}{4}, \frac{3}{17}, \frac{3}{22}, \frac{1}{9}$$

► $\dfrac{3}{7}, \dfrac{1}{4}, \dfrac{3}{17}, \dfrac{3}{22}, \dfrac{1}{9}$

Inverting each term we have

$$\frac{7}{3}, \frac{4}{1}, \frac{17}{3}, \frac{22}{3}, \frac{9}{1}$$

This set of numbers is in A.P. with a common difference of $1\frac{2}{3}$. Hence, $\frac{3}{7}, \frac{1}{4}, \frac{3}{17}, \frac{3}{22}, \frac{1}{9}$ are in H.P.

Example: Find the 8th term of the H.P. $\frac{7}{2}, \frac{7}{5}, \frac{7}{8}, \ldots\ldots\ldots$

► Write the reciprocals of the terms in the H.P. $\frac{2}{7}, \frac{5}{7}, \frac{8}{7}$.
In this A.P. the common difference is $\frac{3}{7}$.
$$l = a + d(n - 1)$$
$$l = \frac{2}{7} + \frac{3}{7}(8 - 1) = \frac{2}{7} + \frac{3}{7}(7) = \frac{2}{7} + 3 = 3\frac{2}{7} \text{ or } \frac{23}{7}.$$
The 8th term of the H.P. is $\frac{7}{23}$.

Example: Insert five harmonic means between $\frac{1}{6}$ and $-\frac{1}{18}$.

► The reciprocals of the given terms are 6 and -18.
First, we insert 5 arithmetic means between 6 and -18.
$$l = a + d(n - 1)$$
$$-18 = 6 + d(6), 6d = -24, d = -4.$$
The A.P. is $6, 2, -2, -6, -10, -14, -18$.
The H.P. is $\frac{1}{6}, \frac{1}{2}, -\frac{1}{2}, -\frac{1}{6}, -\frac{1}{10}, -\frac{1}{14}, -\frac{1}{18}$.
There is no general formula for finding the sum of the terms of an H.P.

EXERCISE 4

Show that the following sets of numbers are in H.P. and write two additional terms of each H.P.

1. $\frac{1}{2}, \frac{1}{4}, \frac{1}{6}, \frac{1}{8}, \ldots$

2. $5, 1\frac{2}{3}, 1, \frac{5}{7}, \ldots$

3. $\frac{3}{2}, \frac{1}{2}, \frac{3}{10}, \frac{3}{14}, \ldots$

4. $-6, 3, 1\frac{1}{5}, \frac{3}{4}, \ldots$

In the following problems find the indicated term:

5. $\frac{3}{4}, \frac{3}{7}, \frac{3}{10}, \frac{3}{13}, \ldots \ldots \ldots \ldots \ldots \ldots$ (7th term)

6. $\frac{2}{5}, \frac{4}{15}, \frac{1}{5}, \frac{4}{25}, \ldots \ldots \ldots \ldots \ldots \ldots$ (8th term)

7. $\frac{2}{3}, \frac{1}{2}, \frac{2}{5}, \frac{1}{3}, \ldots \ldots \ldots \ldots \ldots \ldots$ (10th term)

8. $-\frac{2}{7}, -\frac{1}{2}, -2, 1, \ldots \ldots \ldots \ldots \ldots$ (9th term)

9. Insert 3 harmonic means between $\frac{1}{5}$ and $\frac{1}{15}$.

10. Insert 4 harmonic means between $\frac{1}{2}$ and $\frac{1}{17}$.

11. Insert 4 harmonic means between $\frac{4}{3}$ and $\frac{2}{9}$.

12. Insert 6 harmonic means between $-1\frac{1}{5}$ and $\frac{2}{3}$.

GEOMETRIC PROGRESSIONS

A geometric progression (G.P.) is a sequence of numbers in which each term after the first is obtained by multiplying the preceding term by a fixed number.

Example: 2, 6, 18, 54, ...

The fixed number multiplier is called the *common ratio*. In the example above the common ratio is 3.

Formula for the Nth Term of a G.P.

Designate the first term of a G.P. by a and the common ratio by r. Then the terms of the series can be written as follows:

(1)	(2)	(3)	(4)	. . . . .	(n)
a	ar	ar^2	ar^3	. . . . .	ar^{n-1}

If the nth term is called l we have the formula

$$l = ar^{n-1}$$

The following examples will show how this formula is applied.

Example: Find the 8th term of the progression 4, −2, 1, ...

▶   Use the formula $l = ar^{n-1}$

In this formula $a = 4, r = -\frac{1}{2}, n = 8, l = ?$

In general, r can be found in a given series by forming an equation involving any two consecutive terms.   In this case, the equation

is $4r = -2, r = -\frac{1}{2}$.

$$l = (4) \left(-\frac{1}{2}\right)^{8-1} = 4 \left(-\frac{1}{2}\right)^{7} = 4 \left(-\frac{1}{128}\right) = -\frac{1}{32}$$

Example: Which term of the progression 36, 18, 9 . . . . is $\frac{9}{32}$?

▶   Use the formula $l = ar^{n-1}$
  In this formula $l = \frac{9}{32}, a = 36, r = \frac{1}{2}, n = ?$
  $\frac{9}{32} = 36(\frac{1}{2})^{n-1}$
  Dividing both sides of the equation by 36

$$\frac{1}{128} = \left(\frac{1}{2}\right)^{n-1} \quad \text{or} \quad \left(\frac{1}{2}\right)^{n-1} = \frac{1}{2^7} = \left(\frac{1}{2}\right)^{7}$$

$$\therefore \; n - 1 = 7 \quad \text{and} \quad n = 8$$

Example: A machine costs \$5,600.   It depreciates 20% of its value each year.   What is the value of the machine after 4 years?

▶   Use the formula $l = ar^{n-1}$
  In this formula, $a = 5,600, r = 80\%$ or $\frac{4}{5}, n = 4$.
  We take $r = \frac{4}{5}$ since the value of each year is $\frac{4}{5}$ of the value of the preceding year.

$$l = 5,600 \left(\frac{4}{5}\right)^{4-1}, \; l = 5,600 \left(\frac{4}{5}\right)^{3}, \; l = 5,600 \times \frac{64}{125}$$

$$l = \$2,867.20$$

EXERCISE 5

1. Find the fifth term of the series 1, 3, 9, ...
2. Find the sixth term of the series 1, $-\frac{1}{2}, \frac{1}{4}$, ...

3. Find the eighth term of the series $\frac{1}{9}$, $\frac{1}{3}$, 1, ...
4. Find the ninth term of the series $-\frac{3}{2}$, 3, -6, ...
5. Find the seventh term of the series 243, -162, 108, ...
6. Which term of the series 2, -6, 18, ... is 162?
7. Which term of the series $\frac{1}{9}$, $\frac{1}{3}$, 1, ... is 81?
8. Which term of the series 64, -16, 4, ... is $-\frac{1}{256}$?

Geometric Means

DEFINITION: In a G.P., the terms that lie between any two other terms are called the geometric means between those terms.

Example: In the series, 1, 2, 4, 8, 16, 32 the terms 2, 4, 8, and 16 are the geometric means between 1 and 32.

In the case where there is only one geometric mean between two numbers the geometric mean is the mean proportional between these numbers. If the numbers are a and b and the geometric mean is G then $a:G = G:b$, $G^2 = ab$, and $G = \pm\sqrt{ab}$.

Example: Insert 5 geometric means between $\frac{1}{8}$ and 8.

▶ Use the formula $l = ar^{n-1}$
In this formula, $l = 8$, $a = \frac{1}{8}$, $r = ?$, $n = 7$.

$$8 = \left(\frac{1}{8}\right)(r)^{7-1}, \ 8 = \frac{1}{8}r^6, \ r^6 = 64, \ r = \pm 2.$$

There are two solutions:

$\frac{1}{8}$, $\frac{1}{4}$, $\frac{1}{2}$, 1, 2, 4, 8

and

$\frac{1}{8}$, $-\frac{1}{4}$, $\frac{1}{2}$, -1, 2, -4, 8

Example: Find the geometric mean between $x^2 - y^2$ and $\dfrac{x + y}{x - y}$.

▶ The geometric mean (G) between two numbers is obtained by using the formula $G = \pm\sqrt{ab}$. In this case,

$$G = \pm \sqrt{(x^2 - y^2) \cdot \frac{(x + y)}{(x - y)}}$$

$$= \pm \sqrt{(x + y)(x - y) \cdot \frac{(x + y)}{(x - y)}}$$

$$= \pm \sqrt{(x + y)^2} = \pm (x + y)$$

Example: Insert 4 geometric means between $\dfrac{c^3}{3}$ and $\dfrac{81}{c^2}$.

▶   Use the formula, $l = ar^{n-1}$
In this formula,

$$l = \frac{81}{c^2}, \ a = \frac{c^3}{3}, \ n = 6, \ r = ? \quad \frac{81}{c^2} = \frac{c^3}{3} \cdot r^{6-1}$$

$$\frac{81}{c^2} = \frac{c^3 r^5}{3}$$

$$c^5 r^5 = 243$$

$$r^5 = \frac{3^5}{c^5}$$

$$r = \frac{3}{c}$$

The progression is

$$\frac{c^3}{3}, \ c^2, \ 3c, \ 9, \ \frac{27}{c}, \ \frac{81}{c^2}$$

EXERCISE 6

Insert between the given numbers the indicated number of geometric means:

1. 6 and 486	(3 means)
2. 3,645 and 15	(4 means)
3. $\frac{2}{3}$ and $-\frac{81}{16}$	(4 means)
4. -2 and $\frac{1}{16}$	(4 means)
5. 56 and $-\frac{7}{16}$	(6 means)

The Sum of a Geometric Progression

Let S_n represent the sum of n terms of a G.P.
Then $S_n = a + ar + ar^2 + \ldots \ldots \ldots \ldots + ar^{n-1}$.

Multiply both sides of this equation by r, obtaining
$$rS_n = ar + ar^2 + \ldots + ar^{n-1} \ldots + ar^n$$
Subtracting the second equation from the first
$$S_n - rS_n = a - ar^n$$
$$S_n (1 - r) = a - ar^n$$
$$S_n = \frac{a - ar^n}{1 - r}$$

Since $l = ar^{n-1}$ then $rl = ar^n$.

$\therefore$ The formula may be written $S_n = \dfrac{a - rl}{1 - r}$.

In working out problems in G.P. it should be observed that there are 5 quantities involved. They are l, a, n, r, S_n. If any 3 of these quantities are given then the others may be found by using the appropriate formula.

Example: Find the sum of the first 5 terms of the G.P. $\frac{1}{3}$, 2, 12 . . . .

▶ Use the formula $Sn = \dfrac{a - ar^n}{1 - r}$.

In this formula $a = \frac{1}{3}$, $r = 6$, $n = 5$, $S_n = ?$

$$S_n = \frac{\frac{1}{3} - \frac{1}{3}(6)^5}{1 - 6} = \frac{\frac{1}{3} - 2,592}{-5} = \frac{\frac{-7,775}{3}}{-5} = \frac{1,555}{3}$$

Example: For the G.P. 81, 27, 9 . . . . . , find the seventh term and the sum of seven terms.

▶ To find l use the formula $l = ar^{n-1}$.

In this formula, $a = 81$, $r = \frac{1}{3}$, $n = 7$, $l = ?$,

$$l = 81(\tfrac{1}{3})^{7-1} = 81(\tfrac{1}{3})^6 = \tfrac{1}{9}.$$

To find S_n we may use the formula

$$S_n = \frac{a - rl}{1 - r} = \frac{81 - \frac{1}{3}(\frac{1}{9})}{1 - \frac{1}{3}} = \frac{81 - \frac{1}{27}}{\frac{2}{3}} = \frac{\frac{2,186}{27}}{\frac{2}{3}} = \frac{1,093}{9}$$

Example: Given $a = 81$, $r = \frac{2}{3}$, $S_n = 228\frac{7}{9}$.
Find: n and l.

▶ Use the formula $S_n = \dfrac{a - ar^n}{1 - r}$.

$$228\tfrac{7}{9} = \frac{81 - 81(\tfrac{2}{3})^n}{1 - \tfrac{2}{3}}$$

$$\frac{2{,}059}{9} = \frac{81 - 81(\tfrac{2}{3})^n}{\tfrac{1}{3}}$$

$$\tfrac{1}{3}(2{,}059) = 9[81 - 81(\tfrac{2}{3})^n]$$

$$\frac{2{,}059}{3} = 729 - 729(\tfrac{2}{3})^n$$

$$729(\tfrac{2}{3})^n = 729 - \frac{2{,}059}{3} = \frac{128}{3}$$

$$(\tfrac{2}{3})^n = \frac{128}{3} \times \frac{1}{729} = \frac{128}{2{,}187} = \frac{2^7}{3^7} = (\tfrac{2}{3})^7$$

$$\therefore n = 7$$

To find l, use the formula $l = ar^{n-1}$

$$l = 81(\tfrac{2}{3})^{7-1} = 81(\tfrac{2}{3})^6 = 81\,\frac{64}{729} = \frac{64}{9}$$

Example: The seventh term of a G.P. is $\tfrac{729}{2}$ and the second term is 48. Find the sum of the first seven terms of the series.

▶ Use the formula $l = ar^{n-1}$.
In this formula, $l = \tfrac{729}{2}$, $a = 48$, $n = 6$.

$$\frac{729}{2} = 48r^5, \quad r^5 = \frac{243}{32}, \quad r = \frac{3}{2}$$

To find the first term of the original series use the relationship

$$\frac{3x}{2} = 48, \quad x = 32$$

To find the sum of the first 7 terms of the series use the formula

$$S_n = \frac{ar^n - a}{r - 1}$$

$$S_n = \frac{32\left(\dfrac{3}{2}\right)^7 - 32}{\dfrac{3}{2} - 1} = \frac{\dfrac{2{,}187}{4} - 32}{\dfrac{1}{2}} = \frac{2{,}059}{2}$$

Example: Given the G.P. 3, 6, 12, ... Prove that if $k(k \neq 0)$ is added to each term of this progression, the resulting terms do not form a G.P.

▶ If k is added to the first three terms of the G.P. we obtain $3 + k, 6 + k, 12 + k$. If these terms are in G.P. then

$$\frac{6 + k}{3 + k} = \frac{12 + k}{6 + k} \quad \text{or} \quad (6 + k)^2 = (3 + k)(12 + k)$$

Then $36 + 12k + k^2 = 36 + 15k + k^2$.

And $12k = 15k$.

But this is impossible. Hence, $3 + k$, $6 + k$, and $12 + k$ do not form a G.P.

Example: If $\frac{1}{b-a}$, $\frac{1}{2b}$, and $\frac{1}{b-c}$ are in A.P. prove that a, b, and c are in G.P.

▶ Since $\frac{1}{b-a}$, $\frac{1}{2b}$, and $\frac{1}{b-c}$ are in A.P.,

$$\frac{1}{b-c} - \frac{1}{2b} = \frac{1}{2b} - \frac{1}{b-a}$$

Combining fractions on both sides of the equation, we obtain

$$\frac{b + c}{2b(b - c)} = \frac{-a - b}{2b(b - a)}$$

Multiplying both sides of the equation by $2b$, we obtain

$$\frac{b + c}{b - c} = \frac{-a - b}{b - a}$$

$$(b + c)(b - a) = (b - c)(-a - b)$$

$$b^2 - ab + bc - ac = -ab - b^2 + ac + bc$$

$$2b^2 = 2ac$$

$$\text{or} \quad b^2 = ac$$

i.e., b is the mean proportional between a and c, or a, b, and c are in G.P.

Example: Find the seventh term of an A.P. whose first, second, and fifth terms are in G.P. and whose first term is 2.

▶ The first term of the A.P. is 2, the second term is $(2 + d)$ and the fifth term is $(2 + 4d)$. These terms are in G.P. or

$$2(2 + 4d) = (2 + d)^2$$

$$(4 + 8d) = 4 + 4d + d^2$$

$$d^2 - 4d = 0$$

$$d = 0, \text{ reject}, \quad d = 4$$

The seventh term is $2 + 6d$ or $2 + 6(4) = 26$.

Example: During the first year a business was in operation the net profit was \$10,000. Each succeeding year the profit was increased by 20% over the preceding year. What was the total profit for the first 5 years?

▶ Use the formula $S_n = \dfrac{ar^n - a}{r - 1}$

In this formula, $a = \$10,000$, $r = 120\%$ or 1.2, and $n = 5$

$$S_n = \frac{10,000(1.2)^5 - 10,000}{1.2 - 1} = \$74,400$$

correct to the nearest hundred dollars.

EXERCISE 7

1. Find the sum of the first 8 terms of the series $\frac{1}{4}$, $\frac{1}{2}$, 1, ...

2. Find the sum of the first 5 terms of the series 64, −48, 36, ...

3. For the series 243, −81, +27, ... find the seventh term and the sum of seven terms.

4. For the series 32, −48, 72, ... find the eighth term and the sum of eight terms.

5. For the series 27, 36, 48, ... find the sixth term and the sum of six terms.

6. In a G.P., $a = \frac{81}{2}$, $r = \frac{1}{3}$, $S_n = 60\frac{2}{3}$. Find n and l.

7. In a G.P., $a = \frac{64}{3}$, $r = \frac{1}{2}$, $S_n = 42\frac{1}{3}$. Find n and l.

INFINITE GEOMETRIC PROGRESSIONS

Let us consider the G.P. 1, $\frac{1}{2}$, $\frac{1}{4}$, $\frac{1}{8}$, $\frac{1}{16}$, $\frac{1}{32}$, $\frac{1}{64}$, ... In this progression $r = \frac{1}{2}$.

Now let us examine the sum of this G.P. ($s_3 =$ sum for 3 terms) as we increase the number of terms.

$S_1 = 1$
$S_2 = 1\frac{1}{2}$
$S_3 = 1\frac{3}{4}$
$S_4 = 1\frac{7}{8}$
$S_5 = 1\frac{15}{16}$
$S_6 = 1\frac{31}{32}$
$S_7 = 1\frac{63}{64}$
$S_8 = 1\frac{127}{128}$

This summation is shown on the diagram below:

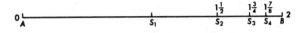

'If we take additional terms and find the sum, we note that the sum gets closer and closer to 2. However, the sum will never reach 2 no matter how many terms are taken.

This situation is described in symbols as follows:

$$\text{Limit } S_n = 2$$
$$n \to \infty$$

We read this symbolic statement "As n increases indefinitely, S_n approaches the limit 2." Notice that, in this case, S_n is not a sum in the ordinary sense but is the *limit of a sum.*

A G.P. in which the number of terms increases without bound is called an *infinite geometric progression.*

Formula for the Sum of an Infinite Geometric Progression

The formula for a G.P. is $S_n = \frac{a - ar^n}{1 - r}$. This may be written as $S_n = \frac{a}{1 - r} - \frac{ar^n}{1 - r}$. Assume that the numerical value of r is less than 1. Then the value of r^n decreases as n increases. In fact, r^n can be made as small as desired if n is taken sufficiently large. Thus, in finding the sum, the quantity $\frac{ar^n}{1 - r}$ which is subtracted from $\frac{a}{1 - r}$ can be made as small as desired. In other words, Sn can be made to approach $\frac{a}{1 - r}$ as a limit. In symbols, this is written

$$\text{Limit } S_n = \frac{a}{1 - r}$$
$$n \to \infty$$

This may be written

$$S_n = \frac{a}{1 - r}(\,|\,r\,|\, < 1)$$
$$n \to \infty$$

Example: Find the sum of the infinite geometric progression $6, 2, \frac{2}{3}, \ldots$.

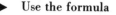 Use the formula

$$S_n = \frac{a}{1-r}$$
$$n \to \infty$$

In this formula $a = 6$, and $r = \frac{1}{3}$.

$$S_n = \frac{6}{1 - \frac{1}{3}} = \frac{6}{\frac{2}{3}} = 9$$
$$n \to \infty$$

Example: Find the sum of the infinite geometric progression $8, -6, \frac{9}{2}, \ldots$.

▶ Use the formula

$$S_n = \frac{a}{1-r}$$
$$n \to \infty$$

In this formula, $a = 8$, $r = -\frac{3}{4}$.

$$S_n = \frac{8}{1 - \left(-\frac{3}{4}\right)} = \frac{8}{\frac{7}{4}} = \frac{32}{7}$$
$$n \to \infty$$

Example: A rubber ball is dropped from the top of a tower 90 feet high. The ball hits the ground and rebounds to a height $\frac{2}{3}$ of the height from which it was dropped. It continues to do this until it comes to rest. How many feet does the ball travel before it comes to rest?

▶ The ball drops 90 feet and rebounds 60 feet. It then drops 60 feet and rebounds 40 feet. We must find the sum of the terms of the progression 90, 60, 60, 40, 40, etc.

The terms 60, 40, etc., form an infinite geometric progression. Use the formula

$$S_n = \frac{a}{1-r}$$
$$n \to \infty$$

$$S_n = \frac{60}{1 - \frac{2}{3}} = \frac{60}{\frac{1}{3}} = 180$$
$$n \to \infty$$

In order to obtain the final result we must double 180 and add 90. The final result is 450 feet.

EXERCISE 8

Find the sums of the following infinite geometric progressions:

1. $9, 6, 4, \ldots \ldots \ldots$
2. $10, 2, \frac{2}{5}, \ldots \ldots \ldots$
3. $8, -4, 2, \ldots \ldots \ldots$
4. $3, \frac{3}{2}, \frac{3}{4}, \ldots \ldots \ldots$
5. $8, -6, 4\frac{1}{2}, \ldots \ldots \ldots$
6. $18, -4, \frac{8}{9}, \ldots \ldots \ldots$

Repeating Decimals

It can be shown by advanced methods that all rational fractions can be written as decimals which terminate or repeat in groups of numbers. For example:

$\frac{1}{2} = .5$ (terminating decimal)

$\frac{1}{3} = .3333 \ldots$ (repeating decimal)

$\frac{1}{4} = .25$ (terminating decimal)

$\frac{1}{5} = .2$ (terminating decimal)

$\frac{1}{6} = .166666 \ldots$ (repeating decimal, 6 repeats after the first place)

$\frac{1}{7} = .142857142857 \ldots$ (repeating decimal, cycle is 142857)

Now, every terminating or repeating decimal can be written as a rational fraction. For terminating decimals this is a simple process, e.g. $.4 = \frac{4}{10}$. For repeating decimals, we use the formula

$$\underset{n \to \infty}{S_n} = \frac{a}{1 - r}$$

Example: Write $.272727 \ldots$ as a common fraction.

▶ $.272727 \ldots$ may be written as

$$.27 + .0027 + .000027 + \ldots$$

This is an infinite G.P. with $a = .27$ and $r = .01$.

$$\underset{n \to \infty}{S_n} = \frac{a}{1 - r} \quad , \quad \underset{n \to \infty}{S_n} = \frac{.27}{1 - .01} = \frac{.27}{.99} = \frac{3}{11}$$

Example: Write $.56666 \ldots$ as a common fraction.

▶ $.56666 \ldots$ may be written as $.5 + .06 + .006 + .0006$.
The series $.06 + .006 + .0006 \ldots$ is an infinite G.P. with $a = .06$ and $r = .1$.

$$S_n = \frac{a}{1 - r} \quad , \quad S_n = \frac{.06}{1 - .1} = \frac{.06}{.9} = \frac{6}{90} = \frac{1}{15}$$
$$\begin{array}{cc} n \to \infty & n \to \infty \end{array}$$

Now add .5 to $\frac{1}{15}$, $\frac{5}{10} + \frac{1}{15}$.

$$\frac{15 + 2}{30} = \frac{17}{30}$$

EXERCISE 9

Express each of the following as common fractions:

1. .11111 . . . .
2. .363636 . . . .
3. .212121 . . . .
4. .83333 . . . .
5. .53333 . . . .
6. .2181818 . . . .

ANSWERS

EXERCISE 1

1. 55
2. 52
3. 54
4. −24
5. $x + 28y$
6. 25
7. 17
8. 19
9. 2
10. $-\dfrac{1}{4}$

EXERCISE 2

1. 4, 6, 8, 10
2. $6\frac{1}{3}, 7\frac{2}{3}, 9, 10\frac{1}{3}, 11\frac{2}{3}$
3. 9, 6, 3, 0, −3
4. a. 5
   b. 1
   c. 8
   d. $\dfrac{x + y}{2}$
   e. $-\sqrt{5}$
5. $3\frac{3}{4}, 1\frac{1}{2}, -\frac{3}{4}$

EXERCISE 3

1. 738
2. $82\frac{1}{2}$
3. 15
4. $d = -4, l = -36$

5. 15
6. $a = 27, Sn = 450$
7. $a = -18, d = \frac{2}{3}$
8. $n = 11, d = -1\frac{3}{5}$

EXERCISE 4

1. $\frac{1}{10}, \frac{1}{12}$

2. $\frac{5}{9}, \frac{5}{11}$

3. $\frac{1}{6}, \frac{3}{22}$

4. $\frac{6}{11}, \frac{3}{7}$

5. $\frac{3}{22}$

6. $\frac{4}{45}$

7. $\frac{1}{6}$

8. $\frac{2}{17}$

9. $\frac{2}{15}, \frac{1}{10}, \frac{2}{25}$

10. $\frac{1}{5}, \frac{1}{8}, \frac{1}{11}, \frac{1}{14}$

11. $\frac{2}{3}, \frac{4}{9}, \frac{1}{3}, \frac{4}{15}$

12. $-2, -6, 6, 2, \frac{6}{5}, \frac{6}{7}$

EXERCISE 5

1. 81

2. $-\frac{1}{32}$

3. 243

4. -384

5. $\frac{64}{3}$

6. 5

7. 7

8. 8

EXERCISE 6

1. $18, 54, 162$ and $-18, 54, -162$

2. $1,215, 405, 135, 45$

3. $-1, \frac{3}{2}, -\frac{9}{4}, \frac{27}{8}$

4. $1, -\frac{1}{2}, \frac{1}{4}, -\frac{1}{8}$

5. $-28, 14, -7, \frac{7}{2}, -\frac{7}{4}, \frac{7}{8}$

EXERCISE 7

1. $2\frac{55}{4}$

2. $45\frac{1}{4}$

3. $\frac{1}{3}, 182\frac{1}{3}$

4. $-546\frac{3}{4}, -315\frac{1}{4}$

5. $113\frac{7}{9}, 374\frac{1}{9}$

6. $6, \frac{1}{6}$

7. $7, \frac{1}{3}$

EXERCISE 8

1. 27

2. $12\frac{1}{2}$

3. $5\frac{1}{3}$

4. 6

5. $4\frac{4}{7}$

6. $14\frac{8}{11}$

EXERCISE 9

1. $\frac{1}{9}$

2. $\frac{4}{11}$

3. $\frac{7}{33}$

4. $\frac{5}{6}$

5. $\frac{8}{15}$

6. $\frac{12}{55}$

CHAPTER 13

Ratio, Proportion, and Variation

DEFINITION: A ratio is the indicated quotient of two quantities. This quotient may be expressed as a fraction, or by the symbol : placed between the two quantities. Thus the ratio of a to b may be written as $\frac{a}{b}$ or $a : b$.

DEFINITION: A proportion is an expression of the equality of two ratios as $\frac{a}{b} = \frac{c}{d}$ or $a : b = c : d$.

Each quantity in a proportion is called a *term*. The first and last terms of the proportion are called the *extremes*. The second and third terms of the proportion are called the *means*.

In the proportion $a : b = c : d$, d is called the *fourth proportional* to a, b and c.

In the proportion $a : b = b : c$, c is called the *third proportional* to a and b.

A proportion in which the second and third terms are identical is called a *mean proportion* and the common term is called the *mean proportional* between the other two. Thus, $x : y = y : z$ is an example of a mean proportion and y is the mean proportional between x and z.

FUNDAMENTAL PROPERTIES OF PROPORTIONS

1. In a proportion, the product of the means is equal to the product of the extremes.

If $a : b = c : d$, then $ad = bc$

If $5 : 7 = 9 : x$ then $5x = 63$ and $x = 12\frac{3}{5}$

2. If the product of two quantities is equal to the product of two other quantities then either pair may be made the means and the other pair the extremes.

If $ab = cd$ then $a : c = d : b$ or $b : c = d : a$ or $d : a = b : c$, etc.

3. If the proportion $a : b = c : d$ is true then the following proportions are true:

$a : c = b : d$ (called alternation)

$b : a = d : c$ (called inversion)

$(a + b) : b = (c + d) : d$ (called addition)

$(a - b) : b = (c - d) : d$ (called subtraction)

$(a + b) : (a - b) = (c + d) : (c - d)$ (called addition and subtraction)

Example: Write the following as ratios reduced to simplest form:

a. 9 inches to 2 feet = 9 inches to 24 inches = $9 : 24 = 3 : 8$

b. 45 minutes to 3 hours = 45 minutes to 180 minutes = $45 : 180 = 1 : 4$

c. 660 yards to 1 mile = 660 yards to 1,760 yards = $660 : 1,760 = 3 : 8$

d. $(x^3 - x^2 - 6x) : (x^3 - 5x^2 + 6x)$

$$= \frac{x^3 - x^2 - 6x}{x^3 - 5x^2 + 6x} = \frac{x(x^2 - x - 6)}{x(x^2 - 5x + 6)} = \frac{x(x-3)(x + 2)}{x(x-3)(x - 2)}$$

$$= \frac{x + 2}{x - 2}$$

The result may be expressed as $(x + 2) : (x - 2)$.

Example: Find the fourth proportional to 3, 7 and 8.

▶ Since x is the fourth proportional,

$$3 : 7 = 8 : x, \quad 3x = 56, \quad x = 18\frac{2}{3}$$

Example: Find the third proportional to 11 and 5.

▶ $11 : 5 = 5 : x, \quad 11x = 25, \quad x = 2\frac{3}{11}$

Example: Find the mean proportional between 3 and 27.

▶   Since x is the mean proportional,
$$3 : x = x : 27, x^2 = 81, x = \pm 9$$

Example: Find the value of a in the proportion $(a + 6) :$ $(2a - 3) = 8 : (a - 12)$.

▶ $(a + 6) : (2a - 3) = 8 : (a - 12)$

Since the product of the means is equal to the product of the extremes
$$(a + 6)(a - 12) = 8(2a - 3)$$
$$a^2 - 6a - 72 = 16a - 24$$
$$a^2 - 22a - 48 = 0$$
$$(a - 24)(a + 2) = 0$$
$$a = +24, \quad a = -2$$

Example: If $6x = 7y$ find the ratio $x : y$.

▶ $6x = 7y$

We make use of the fundamental property of proportions.   If the product of two numbers is equal to the product of two other numbers then either pair may be made the means and the other pair the extremes.

Since $6x = 7y$, either the pair $6x$ or the pair $7y$ may be made the extremes.   Making $6x$ the extremes we have $x : y = 7 : 6$.

EXERCISE 1

1. Write the following as ratios reduced to lowest terms:
   a.   10 inches to 3 feet
   b.   1100 feet to 1 mile
   c.   2 quarts to 6 gallons
   d.   18 weeks to 1 year
   e.   12 ounces to 2 pounds
   f.   3 minutes to 40 seconds
   g. $(a^3 + 2a^2 - 24a) : (a^3 - 16a)$
   h. $(x^2 + 2x - 3) : (x^2 + 3x)$

2. In each case, find the fourth proportional to the three given quantities:

   a. 4, 9 and 12
   b. 6, 5 and 7
   c. $\frac{3}{2}$, 4 and 8
   d. a, $2a$ and $(a + 1)$

3. In each case, find the third proportional to the two given quantities:

   a. 5 and 10
   b. 7 and 9
   c. $\frac{2}{3}$ and 6
   d. y and $2y$

4. In each case, find the mean proportion between the two given quantities:

   a. 6 and 24
   b. 2 and 32
   c. $\frac{1}{4}$ and 9
   d. b and $16b$

5. Find the value of x in each of the following proportions:

   a. $x : 16 = 15 : 12$
   b. $32 : x = 75 : 50$
   c. $(x + 2) : (x - 3)$
      $= (x + 8) : (x - 1)$
   d. $(x + 5)(3x - 6) = 3 : 8$
   e. $(2x + 3) : 5 = (4x - 7) : 9$
   f. $(4x - 1) : (2x + 5) = (6x - 7) : (3x + 2)$
   g. $(x + 3) : (3x - 1) = (x + 7) : (2x + 11)$
   h. $x : (x - 5) = (2x - 9) : (3x - 7)$

6. In each case, find the ratio $x : y$ if

   a. $3x = 4y$
   b. $7x = 12y$
   c. $7x + 3y = 5x + 9y$
   d. $(3x - 2y) : (x + 2y)$
      $= 5 : 7$

Solving Problems by the Use of Proportions

In general, when dealing with two unknown quantities which are in the ratio $a : b$, it is advisable to represent these quantities as ax and bx.

Example: A and B are partners in business and their respective share of the profits of the business are in the ratio 5 : 7. The profits for a certain year were \$64,104,72. What was each man's share of the profits?

▶   Let $5x$ = A's share of the profits
                    And $7x$ = B's share of the profits
            Then $5x + 7x$ = \$64,104.72
                      $12x$ = \$64,104.72
                        x = \$5,342.06

A's share of the profits was $5(\$5,342.06)$ = \$26,710.30
B's share of the profits was $7(\$5,342.06)$ = \$37,394.42

Example: The three sides of a triangle are a, b, and c inches in length. Side c is divided in the ratio of the other two sides. How many inches are there in each segment of side c.

▶   Let ax = the length of one segment
                    And bx = the length of the other segment
            Then $ax + bx$ = c
                  $x(a + b)$ = c
                          $x = \dfrac{c}{a + b}$

Therefore, one segment is $a\left(\dfrac{c}{a + b}\right)$ or $\dfrac{ac}{a + b}$

The other segment is $b\left(\dfrac{c}{a + b}\right)$ or $\dfrac{bc}{a + b}$

Example: Separate the number n into two parts which have the ratio of $p : q$.

▶   Let px = one part
                    And qx = the second part
            Then $px + qx$ = n
                  $x(p + q)$ = n
                          $x = \dfrac{n}{p + q}$

$$\text{One part} = p\left(\frac{n}{p+q}\right) = \frac{pn}{p+q}$$

$$\text{Second part} = q\left(\frac{n}{p+q}\right) = \frac{qn}{p+q}$$

Example: Two numbers are in the ratio $5:7$. If 15 be added to both these numbers they will then be in the ratio $4:5$. Find the original numbers.

$$\text{Let } 5x = \text{the first number}$$
$$\text{And } 7x = \text{the second number}$$
$$\text{Then } \frac{5x+15}{7x+15} = \frac{4}{5}$$
$$4(7x+15) = 5(5x+15)$$
$$28x+60 = 25x+75$$
$$3x = 15$$
$$x = 5$$
$$\text{The first number} = 5 \times 5 = 25$$
$$\text{The second number} = 7 \times 5 = 35$$

Example: A wire is 162 inches long. Divide this wire into three parts which will have the ratio $2:3:7$.

▶

$$\text{Let } 2x = \text{The length of one part of the wire}$$
$$\text{And } 3x = \text{The length of the second part of the wire}$$
$$\text{And } 7x = \text{The length of the third part of the wire}$$
$$\text{Then } 2x + 3x + 7x = 162$$
$$12x = 162$$
$$x = 13\tfrac{1}{2}$$
$$2(13\tfrac{1}{2}) = 27 \text{ inches} = \text{The length of one part of the wire}$$
$$3(13\tfrac{1}{2}) = 40\tfrac{1}{2} \text{ inches} = \text{The length of the second part of the wire}$$
$$7(13\tfrac{1}{2}) = 94\tfrac{1}{2} \text{ inches} = \text{The length of the third part of the wire}$$

Example: A man traveled M miles in H hours. How many miles can he travel in S hours at the same rate?

▶   Let $x =$ The number of miles the man can travel in S
                hours

$$\text{Then } \frac{M}{H} = \frac{x}{S}$$
$$Hx = MS$$
$$x = \frac{MS}{H}$$

EXERCISE 2

Solve the following problems:

1. Brass is an alloy consisting of copper and zinc in the ratio $2:1$. How many ounces of copper and of zinc are there in 57 ounces of brass?

2. Two men agree to divide the profit on a business venture in the ratio $4:7$. If the total profit on the venture was \$6,197.62, what was each man's share?

3. Water is composed of hydrogen and oxygen in the ratio of $1:8$. How many grams of each element are contained in 250 grams of water?

4. A man died and left an estate of \$28,714. His will provided that his wife, daughter, and son receive shares in the ratio $3:2:2$. How much did each receive?

5. The perimeter of a rectangle is 198 inches. The length and width of the rectangle are in the ratio $5:4$. Find the length and width of the rectangle.

6. Separate 124 into two parts which will have the ratio $3:5$.

7. Two numbers are in the ratio $2:7$. If 4 be added to each number they will be in the ratio $1:3$. Find the original numbers.

8. A line segment is 50 inches long. Divide this line segment into two parts which have the ratio $a:b$.

9. A man spent \$60.00 for p shirts. At the same price, what would be the cost of the q shirts?

10. It took a driver a hours to complete a trip of b miles. How long would it take the driver to complete a trip of c miles, traveling at the same rate?

VARIATION

Direct Variation

A variable y is said to vary directly as a second variable x, if the ratio $\frac{y}{x}$ is a constant k. This is usually written $y = kx$ and is sometimes written $y \propto x$. The *constant of variation* is k. For example, the circumference (C) of a circle varies directly as the diameter (D). In this case, $C = \pi D$ and the constant of variation is π.

Example: The cost (c) of traveling on a toll highway varies directly as the mileage (m) covered. If a trip of 90 miles costs $1.35, what is the cost of a trip of 140 miles?

► 1: According to the conditions of the problem, $c = km$, and $c = \$1.35$ when $m = 90$.
Therefore, $\$1.35 = 90k$ and $k = \$.015$.
For a trip of 140 miles, $c = km$, $c = (.015)(140) = \$2.10$.

► 2: Variation problems may be conveniently solved by proportion, especially if it is not required to find the constant of variation k. In many problems, however, the constant k is very useful information.

$$\frac{c_1}{c_2} = \frac{m_1}{m_2}$$
$$\frac{1.35}{c_2} = \frac{90}{140}$$
$$c_2 = \frac{(140)(\$1.35)}{90} = \$2.10$$

NOTE: In the solution of variation problems involving the determination of the constant of variation (k) the following four steps are usually required.
1. The statement of the law of variation. In this case, $c = km$.
2. Substitution of given values to determine k.
   In this case, $k = \$.015$.
3. The restatement of the law of variation with k replaced by its numerical value. In this case, $c = \$.015m$.
4. The substitution of the value of one or more variables to determine the value of the required variable. In this case,
$$c = (\$.015)(140)$$

Inverse Variation

A variable y is said to vary inversely as a second variable x, if the product yx is equal to a constant k. This is usually written $xy = k$ and is sometimes written $y \propto \frac{1}{x}$ or $y = \frac{k}{x}$. For example, the electrical resistance (R) of a wire of fixed length varies inversely as the cross-section area (A) of the wire. In this case, $R = \frac{k}{A}$ and the constant of variation is expressed by k.

Example: The volume (V) of a gas in a container at a constant temperature varies inversely as the pressure (p). If the volume of a gas is 32 cubic centimeters when the pressure is 8 pounds, find the pressure when the volume is 12 cubic centimeters.

▶ 1: According to the conditions of the problem,
$$Vp = k \text{ and } V = 32, \text{ when } p = 8$$
Therefore, $(32)(8) = k$, and $k = 256$.
For a volume of 12 cubic centimeters,
$$12p = 256, \text{ and } p = 21\tfrac{1}{3} \text{ pounds}$$

▶ 2: Using a proportion, since the volume and pressure vary inversely, their relationship may be expressed by the inverse proportion.
$$\frac{p_1}{p_2} = \frac{V_2}{V_1}$$
$$\frac{8}{p_2} = \frac{12}{32}$$
$$12p_2 = 8(32), \; 12p_2 = 256, \; p_2 = 21\tfrac{1}{3} \text{ pounds}$$

Joint Variation

A variable z is said to vary jointly as the variables x and y, if z varies directly as the product of x and y. This is usually written $z = kxy$ and is sometimes written as $z \propto xy$. If z varies jointly as three variables x, y, and w then $z = kxyw$.

This definition may be extended to include more than three variables.

Example: The pressure (p) of a gas varies jointly as the absolute temperature (t) and the density (d). If the pressure is 30 pounds when the temperature is 280° and the density is 1.2, what is the pressure when the density is .75 and the temperature is 336°?

▶ 1: According to the conditions of the problem, $p = ktd$ and $p = 30$ when $t = 280$ and $d = 1.2$.

Therefore, $30 = k(280)(1.2)$, $336k = 30$, $k = \dfrac{5}{56}$.

Therefore, $p = \dfrac{5}{56}td$

$$p = \frac{5}{56} \cdot 336 \cdot \frac{3}{4} = 22\frac{1}{2} \text{ pounds}$$

▶ 2: Using a proportion
$$\frac{p_1}{p_2} = \frac{t_1 d_1}{t_2 d_2}$$
$$\frac{30}{p_2} = \frac{(280)(1.2)}{(336)(.75)}$$
$$p_2 = 22.5 \text{ pounds}$$

Direct Variation as the Square

A variable y is said to vary directly as the square of a second variable x if $y = kx^2$. This is sometimes written as $y \propto x^2$. For example, the surface area (A) of a sphere varies directly as the square of its radius (R).

In this case, $A = 4\pi R^2$ and the constant of variation is 4π.

Example: For a body falling freely from rest under gravity the number of feet (s) through which the body falls varies as the square of the time (t) in seconds. If an object falls 64 feet in 2 seconds, how many seconds will it take an object to reach the ground if it is dropped from a plane 3,600 feet above the ground?

▶ 1: According to the conditions of the problem,
$s = kt^2$
and $s = 64$ when $t = 2$

Therefore, $64 = k(2)^2$, $4k = 64$, $k = 16$, and $s = 16t^2$.

$$3600 = 16t^2, \quad t^2 = \frac{3600}{16} = 225, \quad t = 15 \text{ seconds.}$$

▶ 2: Using a proportion

$$\frac{s_1}{s_2} = \frac{t_1^2}{t_2^2}$$

$$\frac{64}{3600} = \frac{4}{t_2^2}, \quad t_2 = 15 \text{ seconds.}$$

Inverse Variation as the Square

A variable y is said to vary inversely as the square of a second variable x if $y = \dfrac{k}{x^2}$. This is sometimes written as $y \propto \dfrac{1}{x^2}$. For example, the electrical resistance (R) of a wire of given length varies inversely as the square of the diameter (D) of the wire. This is expressed as

$$R = \frac{k}{D^2}$$

Example: The intensity of illumination (I) received from a source of light varies inversely as the square of the distance (D) from the source of light. If the intensity at a distance of 15 feet is 24 candle-power, what is the intensity at 25 feet?

▶ 1: According to the conditions of the problem,

$$I = \frac{k}{D^2} \text{ and } I = 24 \text{ when } D = 15.$$

Therefore, $24 = \dfrac{k}{(15)^2}$, $k = 5,400$, and $I = \dfrac{5,400}{D^2}$

$$I = \frac{5,400}{(25)^2}, \quad I = \frac{216}{25} \text{ or } 8\frac{16}{25} \text{ candle-power.}$$

▶ 2: Using a proportion,

$$\frac{I}{I_2} = \frac{D_2^2}{D_1^2}$$

$$\frac{24}{I_2} = \frac{(25)^2}{(15)^2}$$

$$I_2 = \frac{(24)(15)^2}{(25)^2} = 8\frac{16}{25} \text{ candle-power.}$$

Summary of Types of Variation

Summary of Types of Variation 363

Combined Variation

When a variable y varies as other variables x, y, and z according to a combination of types of variation studied above we have a case of combined variation. For example, if y varies directly as x and inversely as the square of z we have $y = \dfrac{kx}{z^2}$. Or, if y varies jointly as the square of x and w, we have $y = kx^2w$.

Example: The electrical resistance (R) of a wire varies directly as its length (L), and inversely as the square of its diameter (D). If the resistance of 720 feet of copper wire $\frac{1}{4}$ inch in diameter is $1\frac{1}{2}$ ohms, what is the resistance of 960 feet of the same type of copper wire $\frac{1}{2}$ inch in diameter?

▶ 1: According to the conditions of the problem,
$$R = \frac{kl}{D^2} \text{ and } R = \frac{3}{2} \text{ when } L = 720 \text{ and } D = \frac{1}{4}.$$
Therefore, $\dfrac{3}{2} = k\dfrac{(720)}{(\frac{1}{4})^2}, \dfrac{3}{2} = k(720)(16), k = \dfrac{1}{7,680}$
$$R = \frac{1}{7,680}\left[\frac{960}{(\frac{1}{2})^2}\right], \quad R = \frac{1}{2} \text{ ohm.}$$

▶ 2: Using a proportion
$$\frac{R_1}{R_2} = \frac{L_1}{L_2} \cdot \frac{D_2{}^2}{D_1{}^2}$$
$$\frac{\frac{3}{2}}{R_2} = \frac{720}{960} \cdot \frac{\frac{1}{4}}{\frac{1}{16}}, \frac{\frac{3}{2}}{R_2} = \frac{180}{60} = \frac{3}{1}$$
$$R_2 = \tfrac{1}{2} \text{ ohm.}$$

Summary of Types of Variation

Direct variation—$y = kx$ or $\dfrac{y}{x} = k$

Inverse variation—$xy = k$ or $x = \dfrac{k}{y}$

Joint variation—$z = kxy$

Direct variation as the square—$y = kx^2$

Inverse variation as the square—$y = \dfrac{k}{x^2}$

Combined variation—$y = \dfrac{kx}{z}$, $y = \dfrac{kx}{z^2}$, $y = \dfrac{kx^2}{z}$, etc.

Example: Express each of the following relations as an equation, using k as the constant of variation.

a. The surface area (A) of a cube varies directly as the square of an edge (e). Answer, $A = ke^2$

b. The volume (V) of a cone varies jointly as the product of its base (b) and altitude (h). Answer, $V = kbh$

c. The volume (V) of a gas varies directly as the absolute temperature (T) and inversely as the pressure (P). Answer, $V = \dfrac{kT}{P}$

d. The number of units of heat (H) generated by an electric current in a circuit varies directly as the product of the resistance (R), the time (t), and the square of the current (A).

Answer, $H = kRtA^2$.

e. The distance (D) required for an automobile to stop varies directly as the square of its speed (S) and inversely as the coefficient of friction (c) of its tires on the road. Answer, $D = \dfrac{ks^2}{c}$

EXERCISE 3

1. Express each of the following relations as an equation, using k as the constant of variation.

a. The cost (c) of building a home varies directly as the floor area (A) of the house.

b. The number of men (N) needed to complete a job varies inversely as the number of hours (H) each man works.

c. The lateral area (L) of a cylinder varies jointly as the radius (R) of its base and its altitude (H).

d. The mechanical advantage (A) of a lever varies directly as the length (L) of the lever arm and inversely as the length of the resistance arm (R).

    e. The attraction (A) between two bodies varies inversely as the square of the distance (D) between them.

    f. The available power (P) in a jet of water varies jointly as the cube of its velocity (V) and the area (A) of the jet.

    g. The volume of a sphere (V) varies directly as the cube of its radius (R).

    h. The kinetic energy (E) of a moving body varies jointly as its weight (W) and the square of its speed (S).

    i. The height (H) of a circular cylinder varies directly as the volume (V) and inversely as the square of the radius (R) of the base.

    j. The velocity (V) of sound in air varies directly as the square root of the absolute temperature (t) of the air.

    2. The pressure of wind on a sail varies jointly as the area of the sail and the square of the wind's velocity. When the wind is 20 miles per hour, the pressure on 2 square feet is 4 pounds. What is the velocity of the wind when the pressure on 9 square feet is 32 pounds?

    3. The time (t) of oscillation of a simple pendulum varies directly as the square root of the length (l). If a pendulum 1 foot long has a time of 1.2 seconds, what is the time of a pendulum 2.25 feet long?

    4. An automobile uses 9 gallons of gasoline in traveling 150 miles at 25 miles per hour. If gasoline consumption varies jointly as the distance traveled and the square root of the speed, find how many gallons would be used on a like trip of 144 miles at a speed of 36 miles per hour.

    5. The amount of electrical current required to melt a fuse wire varies directly as the three-halves power of the diameter. If the current required to melt a wire of 0.09 inches is 27 amperes, what current will melt a wire of diameter 0.04 inches?

    6. The electrical resistance of a cable varies directly as its length and inversely as the square of its diameter. If a cable 1,000 feet long and $\frac{1}{2}$ inch in diameter has a resistance of 0.08 ohms, find the resistance in a cable 1 inch in diameter and 500 feet long.

    7. The force of attraction (F) between two bodies varies directly as the product of their masses (m) and (M) and inversely as the square of the distance, (d) between them.

If $F = 6$ when $m = 64$, $M = 108$, and $d = 24$, find m if $F = 96$, $M = 256$, and $d = 18$.

8. The weight (W) of a body above the surface of the earth varies inversely as the square of the distance (D) from the center of the earth. If a body weighs 15 pounds when it is 4,000 miles from the center of the earth, how much will it weigh at a distance of 4,500 miles from the center?

9. The safe load of a horizontal beam supported at both ends varies jointly as the breadth and the square of the depth, and inversely as the length between supports. If 2″ x 6″ white pine joist, 8 feet long between supports, safely supports 600 pounds, what is the safe load of a 4″ x 8″ beam of the same material 9 feet long?

10. The crushing load of a solid square oak pillar varies directly as the fourth power of its thickness and inversely as the square of its length. If a 4-inch pillar, 6 feet high, is crushed by a weight of 98 tons, what weight will crush a pillar of the same wood 8 inches thick and 10 feet high?

11. The pressure of gas in a tank varies jointly as its density and its absolute temperature. When the density is 1.5 and the temperature is 280°, the pressure is 12 pounds per square inch. What is the temperature when the density is 1.6 and the pressure is 32 pounds per square inch?

12. If a principal (P) be invested at simple interest, the interest (I) varies jointly as the principal and the number of years (N). If an investment of \$12,000 yields \$1,080 in 2 years, how large a principal will yield \$2,520 in $3\frac{1}{2}$ years?

ANSWERS

EXERCISE 1

1. a. $5 : 18$
   b. $5 : 24$
   c. $1 : 12$
   d. $9 : 26$
   e. $3 : 8$
   f. $9 : 2$
   g. $(a + 6) : (a + 4)$
   h. $(x - 1) : x$

2. a. 27

b. $5\frac{5}{6}$

c. $21\frac{1}{3}$

d. $2(a + 1)$

3. a. 20

b. $11\frac{4}{7}$

c. 54

d. $4y$

4. a. ± 12

b. ± 8

c. $\pm \frac{1}{2}$

d. $\pm 4b$

5. a. 20

b. $21\frac{1}{3}$

c. $5\frac{1}{2}$

d. 58

e. 31

f. 3

g. $-8, +5$

h. $-15, +3$

6. a. $x : y = 4 : 3$

b. $x : y = 12 : 7$

c. $x : y = 3 : 1$

d. $x : y = 3 : 2$

EXERCISE 2

1. 38 ounces of copper,
19 ounces of zinc

2. $2,253.68, $3,943.94

3. $27\frac{7}{9}$ grams of hydrogen
$222\frac{2}{9}$ grams of oxygen

4. Wife received $12,306.
Each child received $8,204

5. Length 55 inches,
width 44 inches

6. $46\frac{1}{2}, 77\frac{1}{2}$

7. 16 and 56

8. $\dfrac{50a}{a + b}, \dfrac{50b}{a + b}$

9. $\dfrac{\$60q}{p}$

10. $\dfrac{ac}{b}$ hours

EXERCISE 3

1. a. $c = ka$

b. $N = \dfrac{k}{h}$

c. $L = kRH$

d. $A = \dfrac{KL}{R}$

e. $A = \dfrac{k}{D^2}$

f. $P = kV^3A$

g. $V = kR^3$

h. $E = kWs^2$

i. $H = \dfrac{kV}{R^2}$

j. $V = k\sqrt{t}$

2. $26\frac{2}{3}$ miles per hour

3. 1.8 seconds

4. $10\frac{46}{125}$ gallons

5. 8 amperes

6. .01 ohms

7. 243

8. $\frac{320}{27}$ pounds

9. $1,896\frac{8}{27}$ pounds

10. 564.48 tons

11. $700°$

12. $16,000

CHAPTER **14**

Inequalities

DEFINITION: An inequality is a statement that two quantities are unequal. All numbers referred to in this chapter are real numbers.

SYMBOLS OF INEQUALITY

$a \neq b$ means that a and b are unequal without indicating which is the greater.

$a > b$ means that a is greater than b [in this case $(a - b)$ is positive].

$a < b$ means that a is less than b [in this case $(a - b)$ is negative].

$a \geq b$ means that a is greater than b or that a is equal to b.

$a \leq b$ means that a is less than b or that a is equal to b.

$a < x < b$ means that x lies between a and b [x is greater than a but less than b].

Sense of Inequalities

If two inequalities are expressed by the same sign they are said to be in the same sense. If two inequalities are expressed by different signs, they are said to be in the opposite sense.

Examples: 7 > 5 and 3 > 1 are two inequalities in the same sense. 9 > 4 and 2 < 8 are two inequalities in the opposite sense.

PROPERTIES OF INEQUALITIES

1. An inequality will be unchanged in sense if the same quantity is added to or subtracted from each member.

Examples: Given the inequality 9 > 7. If 3 is added to both members we have the inequality 12 > 10 which has the same sense as the original.

Likewise, if 3 is subtracted from both members we have the inequality 6 > 4 which again has the same sense as the original.

2. Because of property 1, a term may be transposed from one member of an inequality to the other without changing the sense of the inequality.

Example: Given the inequality 11 > 6. Subtracting 4 from both members yields 7 + 4 > 6, or 7 > 6 − 4 or 7 > 2.

3. An inequality will be unchanged in sense if all its terms are multiplied or divided by the same positive number.

Example: Given the inequality 15 > 12. Multiplying both members by the positive number 2 yields the inequality 30 > 24. Dividing both members by the positive number 3 yields the inequality 5 > 4.

4. An inequality will be unchanged in sense if both members are positive and both are raised to the same positive power.

Example: Given the inequality 5 > 2. Raising both members to the third power yields the inequality $5^3 > 2^3$ or 125 > 8.

5. If the corresponding members of two inequalities in the same sense are added to each other or multiplied by each other the results will be inequalities in the same sense.

Examples: Given the inequalities 12 > 3 and 7 > 2. Then 12 + 7 > 3 + 2 or 19 > 5, and 12 × 7 > 3 × 2 or 84 > 6.

6. If both members of an inequality are multiplied or divided by the same negative number then the sense of the inequality will be reversed.

Examples: Given the inequality $15 > 12$. If both members are multiplied by -2 then $-30 < -24$.

Given an inequality $15 > 12$. If both members are divided by -3 then $-5 < -4$.

7. If the members of an inequality are subtracted from equals the result will be an inequality in the opposite sense.

Example: If $18 = 18$
                 And $10 > 4$
                 Then $8 < 14$

8. If two equal quantities (except zero) are divided by unequals in one sense the results will be unequal in the opposite sense.

Example: Given the equality $36 = 36$. If the members are divided by $9 > 2$ in that order the result will be the inequality $4 < 18$.

ABSOLUTE AND CONDITIONAL INEQUALITIES

In connection with the study of equalities, it has been observed that equalities are of two general types, identities and conditional equations. Similarly, there are two types of inequalities, *absolute inequalities* and *conditional inequalities*.

An *absolute inequality* is an inequality which is true for all permissible values of the letters involved.

Example: $a^2 + b^2 \leqq 2ab$

A *conditional inequality* is an inequality which is true only for particular values of the letters involved.

Example: The inequality $x + 3 > 4$ is true only for positive values of x greater than 1.

Proofs of Absolute Inequalities

Absolute inequalities represent relationships which are generally true and are given general proofs. The following examples will illustrate the method.

Example: Prove that if $a > b$ and $b > c$ then $a > c$.

Proof: Since $a > b$, let $a = b + x$ (x is positive)

Since $b > c$, let $b = c + y$ (y is positive)

Adding, we have $a + b = b + c + x + y$

Subtracting b from both numbers, we have $a = c + x + y$

Therefore $a > c$

Example: Prove that $a^2 + b^2 > 2ab$ if $a \neq b$.

Proof: In analyzing this proof

If $a^2 + b^2 > 2ab$

Then $a^2 - 2ab + b^2 > 0$

And $(a - b)^2 > 0$

Since $a \neq b$, $(a - b)^2$ is a positive number and is therefore greater than zero.

We may now write the proof with the above steps reversed.

Since $a \neq b$, let $a > b$, and $a - b > 0$

$(a - b)^2 > 0$ If a number is greater than zero, then its square is greater than zero

or $a^2 - 2ab + b^2 > 0$

$a^2 + b^2 > 2ab$ Transposing

This is a fundamental inequality relationship and may be used as a theorem.

Example: If $a \neq b \neq c$, prove that
$$a^2 + b^2 + c^2 > ab + bc + ac$$

Proof: $a^2 + b^2 > 2ab$ using the result of example 2

$b^2 + c^2 > 2bc$

$a^2 + c^2 > 2ac$

$2a^2 + 2b^2 + 2c^2 > 2ab + 2bc + 2ac$ If unequals are added to unequals in a certain order the results are unequal in the same order

$a^2 + b^2 + c^2 > ab + bc + ac$ Dividing both sides by 2

Example: Prove that if $a \neq b$, then $\dfrac{a}{b} + \dfrac{b}{a} > 2$. ($a$ and b are positive).

Proof: If $\dfrac{a}{b} + \dfrac{b}{a} > 2$

Then $\dfrac{a^2 + b^2}{ab} > 2$ Combining the fractions in the left member

and $a^2 + b^2 > 2ab$ An inequality will be unchanged in
sense if both members are multiplied
by the same quantity.

But the last step is true.   Therefore, the original statement is true
and the proof may be given by reversing the above steps.

Note: This exercise proves the general theorem—The sum of a
positive rational number (other than 1) and its reciprocal is greater
than 2.

Example: Prove that $a(b^2 + ab + c^2) + c(b^2 + bc + a^2) >$
$6\,abc$ (a, b, and c are positive).

Proof: $a^2 + b^2 > 2ab, c(a^2 + b^2) > 2abc$ Using the result of
example 2

$b^2 + c^2 > 2bc, a(b^2 + c^2) > 2abc$

$a^2 + c^2 > 2ac, b(a^2 + c^2) > 2abc$

$c(a^2 + b^2) + a(b^2 + c^2) + b(a^2 + c^2) > 6abc$

If unequals are added to unequals in a certain sense the
results are unequal in the same sense

$a(b^2 + c^2) + c(a^2 + b^2) + a^2b + c^2b > 6abc$

$a(b^2 + c^2) + a^2b + c(a^2 + b^2) + c^2b > 6abc$

$a(b^2 + c^2 + ab) + c(a^2 + b^2 + cb) > 6abc$

Example: Prove that $(ab + xy)(ax + by) > 4abxy$.
(a, b, x, and y are positive).

Proof: In analyzing this problem, we multiply the two binomials
on the left obtaining

$a^2bx + ab^2y + ax^2y + bxy^2 > 4abxy$

$bx(a^2 + y^2) + ay(b^2 + x^2) > 4abxy$

This suggests the following proof:

$a^2 + y^2 > 2ay, bx(a^2 + y^2) > 2abxy$

$b^2 + x^2 > 2bx, ay(b^2 + x^2) > 2abxy$

$bx(a^2 + y^2) + ay(b^2 + x^2) > 4abxy$

$a^2bx + bxy^2 + ab^2y + ax^2y > 4abxy$

$a^2bx + ab^2y + bxy^2 + ax^2y > 4abxy$

$ab(ax + by) + xy(by + ax) > 4abxy$

This is equivalent to $(ab + xy)(ax + by) > 4abxy$

Example: If $a^2 + b^2 + c^2 = 1$ and $x^2 + y^2 + z^2 = 1$, prove
that $ax + by + cz < 1$.

Proof: $a^2 + x^2 > 2ax$
$b^2 + y^2 > 2by$
$c^2 + z^2 > 2cz$
$a^2 + b^2 + c^2 + x^2 + y^2 + z^2 > 2ax + 2by + 2cy$

> If unequals are added to unequals in a certain sense, the results are unequal in the same sense

$2 > 2ax + 2by + 2cz$ Substitution
$1 > ax + by + cz$ or $ax + by + cz < 1$

Example: If a and b are positive and unequal prove (a) that their arithmetic mean is greater than their geometric mean, (b) that their geometric mean is greater than their harmonic mean.

Proof: (a) We must prove that $\dfrac{a + b}{2} > \sqrt{ab}$

> or $a + b > 2\sqrt{ab}$
> or $(a + b)^2 > (2\sqrt{ab})^2$
> or $a^2 + 2ab + b^2 > 4ab$
> or $a^2 - 2ab + b^2 > 0$ Transposing $4ab$
> or $(a - b)^2 > 0$

But the last step is true. Therefore, the original statement may be proved by reversing the steps.

> (b) We must prove that $\sqrt{ab} > \dfrac{2ab}{(a + b)}$

> or $ab > \dfrac{4a^2b^2}{(a + b)^2}$
> or $ab(a + b)^2 > 4a^2b^2$
> or $(a + b)^2 > 4ab$
> or $a^2 - 2ab + b^2 > 0$
> or $(a - b)^2 > 0$

But the last step is true. Therefore, the original statement may be proved by reversing the steps.

Example: Prove that $b^2c^2 + c^2a^2 + a^2b^2 > abc(a + b + c)$

Proof: $a^2 + b^2 > 2ab,\ c^2(a^2 + b^2) > 2abc^2,\ a^2c^2 + b^2c^2 > 2abc^2$
$a^2 + c^2 > 2ac,\ b^2(a^2 + c^2) > 2ab^2c,\ a^2b^2 + b^2c^2 > 2ab^2c$
$b^2 + c^2 > 2bc,\ a^2(b^2 + c^2) > 2a^2bc,\ a^2b^2 + a^2c^2 > 2a^2bc$
$2a^2b^2 + 2a^2c^2 + 2b^2c^2 > 2abc^2 + 2ab^2c + 2a^2bc$

If $\neq$s are added to $\neq$s in a certain order then the results are $\neq$ in the same order

$$2(a^2b^2 + a^2c^2 + b^2c^2) > 2abc(c + b + a)$$
$$a^2b^2 + a^2c^2 + b^2c^2 > abc(a + b + c)$$

Example: Prove that $(a^4 + b^4)(a^2 + b^2) > (a^3 + b^3)^2$

Proof: $\quad (a^4 + b^4)(a^2 + b^2) > (a^3 + b^3)^2$

or $a^6 + a^4b^2 + a^2b^4 + b^6 > a^6 + 2a^3b^3 + b^6$

or $a^4b^2 + a^2b^4 > 2a^3b^3$

or $a^2b^2(a^2 + b^2) > 2a^3b^3$

or $(a^2 + b^2) > 2ab$

But the last step is true. Therefore the original statement may be proved by reversing the steps.

Example: If $a > b$ prove that

    (a) $a^a b^b > a^b b^a$

    (b) $\log \dfrac{b + 1}{a + 1} > \log \dfrac{b}{a}$

Proof: (a) $a^a b^b > a^b b^a$

or $\dfrac{a^a}{a^b} > \dfrac{b^a}{b^b}$ Dividing both members by $a^b b^b$

or $a^{a-b} > b^{a-b}$

Since $a > b$ the last step is true. Therefore, the original statement may be proved by reversing the steps.

    (b) $\log \dfrac{b + 1}{a + 1} > \log \dfrac{b}{a}$

or $\log (b + 1) - \log(a + 1) > \log b - \log a$

or $\log (b + 1) + \log a > \log b + \log(a + 1)$

    Transposing

or $\log a(b + 1) > \log b(a + 1)$

or $\log(ab + a) > \log(ab + b)$

Since $a > b$ the last step is true. Therefore, the original statement may be proved by reversing the steps.

EXERCISE 1

If the letters represent unequal, positive numbers, prove the following:

1. If $a > b$ and $c > d$ prove that $a - c > b - d$ if $a - b > c - d$
2. If $a > b$ and $c > 0$ prove that $ac > bc$
3. $2a^2 + b^2 > a^2 + 2ab$
4. $a^2 + 3b^2 > 2b(a + b)$
5. $\dfrac{(a + b)}{2} > \dfrac{2ab}{(a + b)}$
6. $a^3b + ab^3 > 2a^2b^2$

7. $\dfrac{a^2 + 1}{a} > 2$ if $a \neq 1$
8. $\dfrac{(a^2 + 5b^2)}{4} > \dfrac{ab}{2 + b^2}$
9. If $a > b$ then $\dfrac{a}{b} > \dfrac{(a + m)}{(b + m)}$
10. $a^3 - b^3 > 3a^2b - 3ab^2$
11. $a^2 + b^2 + c^2 > ab + ac + bc$
12. If $a^2 + b^2 = 1$ and $x^2 + y^2 = 1$ prove that $ax + by < 1$
13. $3(a^2 + b^2 + c^2) > (a + b + c)^2$
14. $(a + b)(b + c)(a + c) > 8abc$

Conditional Inequalities

Just as conditional equations are true for some values of the unknown and false for other values of the unknown, conditional inequalities are true under some conditions and false under others. Solving an inequality involves the determination of those values of the unknown for which the given inequality is true:

Example: Solve the inequality $2x + 1 > 9$.

▶ $2x + 1 > 9$
Transposing $2x > 8$
Dividing by 2, $x > 4$

▶ By graph: This inequality may be solved by graphs.
$$2x + 1 > 9$$
Transposing $2x - 8 > 0$
Let $y = 2x - 8$ and graph the straight line

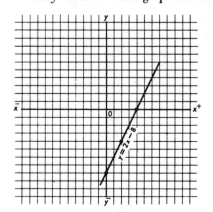

As the graph indicates, y is positive for all values of x greater than 4.

Example: Solve the inequality $\dfrac{8x - 9}{4} > \dfrac{x + 1}{2}$

▶ $\dfrac{8x - 9}{4} > \dfrac{x + 1}{2}$
Multiplying by 4, $8x - 9 > 2x + 2$
Transposing, $8x - 2x > 9 + 2$
$$6x > 11$$
Dividing by 6, $x > 1\frac{5}{6}$

Example: Solve the inequality $\dfrac{2 - x + 12}{2} > \dfrac{3x - 5}{3}$

▶ $\dfrac{2 - x + 12}{2} > \dfrac{3x - 5}{3}$
Multiplying both members by 6, $6 - 3x + 36 > 6x - 10$
Transposing, $-9x > 52$

Dividing both members by -9, $x < \dfrac{52}{9}$

If both members of an inequality are multiplied or divided by the same negative number, then the sense of the inequality will be reversed.

Example: Solve the inequality $\dfrac{2}{x-1} > \dfrac{5}{3x-7}$

▶ $\dfrac{2}{x-1} > \dfrac{5}{3x-7}$

Multiplying both members of the inequality by $(x-1)$ $(3x-7)$, we have

$$2(3x-7) > 5(x-1)$$
$$6x - 14 > 5x - 5$$

Transposing, $6x - 5x > 14 - 5$
$$x > 9$$

Example: Solve the inequality $x^2 - 4x + 3 > 0$.

▶ $x^2 - 4x + 3 > 0$

Factoring $(x-3)(x-1) > 0$

In order that the product of the two binomials be greater than 0 it is necessary that both binomials be positive or that both binomials be negative.

If both binomials are positive then $x - 3 > 0$ and $x - 1 > 0$ simultaneously, i.e. $x > 3$ and $x > 1$. Both conditions are satisfied simultaneously if $x > 3$.

If both binomials are negative then $x - 3 < 0$ and $x - 1 < 0$ simultaneously, i.e. $x < 3$ and $x < 1$. Both conditions are satisfied simultaneously if $x < 1$.

Therefore $x^2 - 4x + 3 > 0$ for all values of $x < 1$ and $x > 3$.

▶ **By graphs:** It is desired that we determine all values of x for which the function $x^2 - 4x + 3 > 0$. Let $y = x^2 - 4x + 3$ and plot the graph.

x	-1	0	1	2	3	4	5
y	8	3	0	-1	0	3	8

From the graph, it can be seen that the function $x^2 - 4x + 3 > 0$ when $x > 3$ and when $x < 1$.

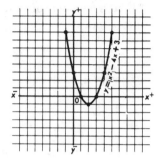

Example: Solve the inequality
$$x^2 - 3x - 10 > 0$$
Factoring $(x - 5)(x + 2) > 0$
In order that the product of these two binomials be greater than zero it is necessary that both binomials be positive or that both be negative.

If both binomials are positive then $x - 5 > 0$ and $x + 2 > 0$ simultaneously, i.e. $x > 5$ and $x > -2$. Both conditions are satisfied simultaneously if $x > 5$.

If both binomials are negative then $x - 5 < 0$ and $x + 2 < 0$ simultaneously, i.e. $x < 5$ and $x < -2$. Both conditions are satisfied simultaneously if $x < -2$.

Therefore $x^2 - 3x - 10 > 0$ for all values of $x < -2$ and $x > 5$.

Example: Solve the inequality $x^2 - 2x - 15 < 0$.

▶ $x^2 - 2x - 15 < 0$
      Factoring $(x - 5)(x + 3) < 0$
In order that the product of these two binomials be less than zero it is necessary that one binomial be positive and that the other be negative.

If $x - 5 > 0$ and $x + 3 < 0$, then $x > 5$ and $x < -3$ simultaneously. Both these conditions *cannot* be satisfied simultaneously.

If $x - 5 < 0$ and $x + 3 > 0$ then $x < 5$ and $x > -3$ simultaneously. Therefore $x^2 - 2x - 15 < 0$ for all values of $x > -3$ and $x < 5$, or $5 > x > -3$.

EXERCISE 2

Solve the following inequalities:
1. $3x + 2 > 17$
2. $4x - 1 > 25$
3. $\dfrac{6x - 1}{5} > \dfrac{x}{3} + 5$

4. $\dfrac{2x}{3} - \dfrac{x-2}{4} > 3$

5. $\dfrac{3-(x-8)}{4} > \dfrac{2x-5}{6}$

6. $\dfrac{5-(3x-2)}{3} > \dfrac{3-(x-1)}{7}$

7. $\dfrac{7}{3x+1} > \dfrac{3}{2x+4}$

8. $\dfrac{5}{x-1} > \dfrac{8}{4x-7}$

9. $x^2 - 5x + 6 > 0$

10. $x^2 - 7x + 12 > 0$

11. $x^2 - 2x - 8 > 0$

12. $x^2 - 3x - 4 > 0$

13. $x^2 - x - 6 < 0 \cdot$

14. $x^2 + 4x - 21 < 0$

ANSWERS

EXERCISE 2

1. $x > 5$
2. $x > 6\frac{1}{2}$
3. $x > 6$
4. $x > 6$
5. $x < 6\frac{1}{7}$
6. $x < \frac{37}{18}$
7. $x > -5$
8. $x > \frac{9}{4}$
9. $x > 3$ or $x < 2$
10. $x > 4$ or $x < 3$
11. $x > 4$ or $x < -2$
12. $x > 4$ or $x < -1$
13. $3 > x > -2$
14. $3 > x > -7$

CHAPTER 15

Theory of Equations

In a previous chapter, methods of solving quadratic equations have been developed. In this chapter, the theory of the polynomial will be studied. This will lead to methods of solving higher degree equations.

Polynomials of Degree n

The general polynomial of degree n in a single variable x may be written in the form $f(x) = a_0x^n + a_1x^{n-1} + a_2x^{n-2} + \ldots \ldots$ $a_{n-1}x + a_n$, where n is a positive integer and $a_0, a_1, a_2 \ldots \ldots, a_n$ are constants with $a_0 \neq 0$.

If the polynomial above is set equal to zero we have the *polynomial equation of degree* n, $a_0x^n + a_1x^{n-1} + a_2x^{n-2} + \ldots \ldots \ldots$ $+ a_{n-1}x + a_n = 0$. For convenience, we frequently write this equation as $f(x) = 0$. If $f(x) = 2x^3 - 4x^2 + 3x - 1$, then $f(a) = 2a^3 - 4a^2 + 3a - 1$.

Example: If $f(x) = 3x^3 - 5x + 12$, find $f(-2)$.

▶ $f(-2)$ is obtained by substituting -2 for x in the original expression.

$$f(-2) = 3(-2)^3 - 5(-2) + 12 = 3(-8) + 10 + 12$$
$$= -24 + 22 = -2$$

Example: If $f(x) = x^2 + 2x - 2$, find $f(a - 3)$.

▶ $f(a - 3)$ is obtained by substituting $(a - 3)$ for x in the original expression.

$(a - 3)^2 + 2(a - 3) - 2 = a^2 - 6a + 9 + 2a - 6 - 2$

$= a^2 - 4a + 1$

Example: If $f(x,y) = 2x^2 + 5y^2 - 3xy - 4$ find (a) $f(1,3)$, (b) $f(-2,1)$.

▶ (a) In $f(x,y) = 2x^2 + 5y^2 - 3xy - 4$ substitute $x = 1$, and $y = 3$. $f(1,3) = 2(1)^2 + 5(3)^2 - 3(1)(3) - 4$

$= 2 + 45 - 9 - 4 = 34$

(b) In $f(x,y) = 2x^2 + 5y^2 - 3xy - 4$ substitute $x = -2$, $y = 1$. $f(-2,1) = 2(-2)^2 + 5(1)^2 - 3(-2)(1) - 4$

$= 8 + 5 + 6 - 4 = 15.$

EXERCISE 1

1. If $f(x) = 2x^3 - 3x^2 + 7x - 4$, find (a) $f(0)$, (b) $f(-2)$.
2. If $f(y) = 3y - 2$, find $f(2 - a)$.
3. If $f(x) = x^2 - 3x$, find $f(a - 1)$.
4. If $f(x) = 4x^3 - 5x - 7$, find (a) $f(-x)$, (b) $f(-1)$, (c) $f(2)$.

Remainder Theorem

If a polynomial $f(x)$ is divided by a binomial of the form $(x-a)$ until a remainder free of x is obtained then the remainder is equal to $f(a)$.

Proof: If $f(x)$ be divided by $(x-a)$ the following identity is true:

$$f(x) = Q(x-a) + R$$
$$\text{In this identity } f(a) = Q(a-a) + R$$
$$f(a) = Q(0) + R$$
$$f(a) = R$$

Application: If $f(x) = 2x^3 - 5x^2 + x + 7$ be divided by $x - 2$ the quotient is $2x^2 - x - 1$, and the remainder is 5.

i.e. $\dfrac{2x^3 - 5x^2 + x + 7}{x-2} = 2x^2 - x - 1 + \dfrac{5}{x-2}.$

According to the Remainder Theorem the remainder is 5, and may be obtained from the original polynomial $2x^3 - 5x^2 + x + 7$ by finding $f(2)$.

$$f(2) = 2(2)^3 - 5(2)^2 + 2 + 7 = 16 - 20 + 2 + 7 = 5$$

Example: Find the remainder when $2x^3 - 4x^2 + 3$ is divided by $(x+1)$.

▶ If $f(x)$ be divided by $(x-a)$ the remainder is equal to $f(a)$. In this example $(x-a)$ is represented by $x+1$, or $a = -1$.

∴ The remainder is equal to $f(-1)$.

$$f(-1) = 2(-1)^3 - 4(-1)^2 + 3 = 2(-1) - 4(1) + 3$$
$$= -2 - 4 + 3 = -3.$$

Example: Find the remainder when $x^4 - x^3 + 2x^2 - 5x + 4$ is divided by $(x-2)$.

▶ The remainder is equal to $f(2)$.

$$f(2) = (2)^4 - (2)^3 + 2(2)^2 - 5(2) + 4 = 16 - 8 + 8$$
$$- 10 + 4 = 10$$

Example: Find the remainder when $y^{25} - y^7 + 5$ is divided by $(y + 1)$.

▶ The remainder is equal to $f(-1)$.

$$f(-1) = (-1)^{25} - (-1)^7 + 5 = (-1) - (-1) + 5$$
$$= -1 + 1 + 5 = 5.$$

EXERCISE 2

Find the remainder after each of the following divisions are performed.

1. $x^3 + 2x^2 - 3x + 4$ by $(x-3)$
2. $2y^3 - 3y^2 + 2y - 5$ by $(y+1)$
3. $x^4 - 5x^3 - 2x^2 + 3x - 6$ by $(x-2)$
4. $y^{15} - y^5 + 1$ by $(y-1)$
5. $3x^3 - 4x + 7$ by $(x+2)$

Factor Theorem

If a is a root of the equation $f(x) = 0$, then $(x-a)$ is a factor of $f(x)$.

Proof: If $f(x)$ be divided by $(x-a)$ the following identity is true:
$$f(x) = Q(x-a) + R$$
In this identity $f(a) = Q(a-a) + R$
$$f(a) = R$$
Since a is a root of $f(x) = 0$, $f(a) = 0$. $\therefore$ $R = 0$. i e. When $f(x)$ is divided by $(x-a)$ the remainder is zero. This is equivalent to the statement that $(x-a)$ is a factor of $f(x)$.

Application: Consider the polynomial $x^3 - 7x^2 + 7x + 15$. $f(3) = (3)^3 - 7(3)^2 + 7(3) + 15 = 27 - 63 + 21 + 15 = 0$. Since $f(3) = 0$, $(x-3)$ is a factor of $f(x)$. $f(x) = (x-3)(x^2 - 4x - 5)$.

Converse of Factor Theorem

If $(x - a)$ is a factor of $f(x)$, then a is a root of $f(x) = 0$.

Application: $(x-2)$ is a factor of $x^3 - 3x^2 - 10x + 24$. Therefore, 2 is a root of the equation $x^3 - 3x^2 - 10x + 24 = 0$.

Example: Show that $(x-2)$ is a factor of $x^3 + x^2 - 11x + 10$.

▶ $(x-2)$ is a factor of $x^3 + x^2 - 11x + 10$, if $f(2) = 0$.
$f(2) = (2)^3 + (2)^2 - 11(2) + 10 = 8 + 4 - 22 + 10 = 0$.
$\therefore$ $(x - 2)$ is a factor of $x^3 + x^2 - 11x + 10$.

Example: Show that $(x + 4)$ is a factor of
$$x^3 - 3x^2 - 18x + 40.$$

▶ $(x+4)$ or $[x - (-4)]$ is a factor of $x^3 - 3x^2 - 18x + 40$ if $f(-4) = 0$.
$f(-4) = (-4)^3 - 3(-4)^2 - 18(-4) + 40$
$= -64 - 48 + 72 + 40 = 0$.
$\therefore$ $(x+4)$ is a factor of $x^3 - 3x^2 - 18x + 40$.

Example: For what value of K is $(x-1)$ a factor of $2x^{17} + Kx^{11} - 4$?

▶ $(x - 1)$ is a factor of $2x^{17} + Kx^{11} - 4$ if $f(1) = 0$.
$f(1) = 2(1)^{17} + K(1)^{11} - 4 = 2 + K - 4$
$2 + K - 4 = 0$, if $K = 2$.
$\therefore$ $(x - 1)$ is a factor of $2x^{17} + Kx^{11} - 4$, if $K = 2$.

Example: Show that $x^{2n} - a^{2n}$ is exactly divisible by $x + a$ for every positive integer n.

▶ $(x + a)$ is a factor of $x^{2n} - a^{2n}$ if $f(-a) = 0$. $f(-a)$
   $= (-a)^{2n} - a^{2n}$. $a^{2n} - a^{2n} = 0$.

∴ $(x + a)$ is a factor of $x^{2n} - a^{2n}$.

Example: Show that $x^7 - a^7$ is exactly divisible by $(x - a)$.

▶ $(x - a)$ is a factor of $x^7 - a^7$ if $f(a) = 0$. $f(a)$
   $= a^7 - a^7 = 0$.

∴ $(x - a)$ is a factor of $(x^7 - a^7)$.

EXERCISE 3

1. Show that $(x - 2)$ is a factor of $x^3 + x^2 - 10x + 8$.

2. Show that $(x - 3)$ is a factor of $x^5 - 15x^3 + 35x^2 - 5x + 24$.

3. Show that $(x + 4)$ is a factor of $x^3 - 3x^2 - 18x + 40$.

4. Show that $(x + 5)$ is a factor of $x^3 - 2x^2 - 41x - 30$.

5. For what value of m is $(x + 2)$ a factor of $x^3 + x^2 + mx - 4$?

6. Find the value of K if $x^3 - 3x^2 + Kx - 9$ is exactly divisible by $(x - 3)$.

7. Show that $x^5 + a^5$ is exactly divisible by $(x + a)$.

8. Show that $x^n - a^n$ is divisible by $(x - a)$ for positive integer n.

Synthetic Division

It is often necessary to divide a polynomial $f(x)$ by a binomial of the form $(x - a)$. The method of synthetic division is a short method of performing such a division.

Example: Divide $2x^3 - 7x^2 + 5x - 1$ by $(x - 3)$.

▶
$$\begin{array}{r} 2 - 7 + 5 - 1 \quad \underline{|\,3} \\ \underline{+ 6 - 3 + 6} \\ 2 - 1 + 2\,\underline{|+ 5} \end{array}$$

1. Write the coefficients of $f(x)$ in descending powers of x, as given. If a term is missing its coefficient is zero.

2. The divisor $(x - 3)$ corresponds to the divisor $(x - a)$, or $a = 3$. Write 3 as indicated.

3. Bring the first coefficient 2 below the line.

4. Multiply the number 2 below the line by the divisor 3 and place the result $+6$ under the second-coefficient 7.

5. Add -7 and $+6$ and place the result -1 under the line.

6. Multiply the number -1 below the line by the divisor 3 and place the result -3 under the third coefficient $+5$.

7. Add $+5$ and -3 and place the result $+2$ under the line.

8. Multiply the number $+2$ below the line by the divisor 3 and place the result $+6$ under the fourth coefficient -1.

9. Add -1 and $+6$ and place the result $+5$ under the line.

Since $2x^3 - 7x^2 + 5x - 1$ was divided by $(x - 3)$ the quotient is a second degree expression. The quotient, $2x^2 - x + 2$ is obtained by taking the numbers below the line in order. The remainder is $+5$. The final result may be written as follows:

$$\frac{2x^3 - 7x^2 + 5x - 1}{x - 3} = 2x^2 - x + 2 + \frac{5}{x - 3}$$

Example: Divide $x^4 - 2x^2 + 7x - 5$ by $(x + 2)$.

▶ Since an x^3 term is missing we must fill the place of this term with the coefficient 0. Also, since $(x + 2)$ represents $(x - a)$, $a = -2$.

$$
\begin{array}{rrrrr|r}
1 & + 0 & - 2 & + 7 & - 5 & \underline{-2} \\
 & - 2 & + 4 & - 4 & - 6 & \\
\hline
1 & - 2 & + 2 & + 3 & \underline{- 11} &
\end{array}
$$

The result is $x^3 - 2x^2 + 2x + 3 - \dfrac{11}{x + 2}$.

Example: Divide $3x^4 - 16x^3 + 14x + 24x - 8$ by $(x - \frac{1}{3})$.

▶
$$
\begin{array}{rrrrr|r}
3 & - 16 & + 14 & + 24 & - 8 & \underline{\tfrac{1}{3}} \\
 & + 1 & - 5 & + 3 & + 9 & \\
\hline
3 & - 15 & + 9 & + 27 & \underline{+ 1} &
\end{array}
$$

∴ The result is $3x^3 - 15x^2 + 9x + 27 + \dfrac{1}{(x - \frac{1}{3})}$.

Example: If $f(x) = 2x^3 + 3x^2 - 4x + 7$ find $f(-4)$.

▶
$$
\begin{array}{rrrr|r}
2 & + 3 & - 4 & + 7 & \underline{-4} \\
 & - 8 & + 20 & - 64 & \\
\hline
2 & - 5 & + 16 & \underline{- 57} &
\end{array}
$$

∴ $f(-4) = -57$.

Notice that the Remainder Theorem may be used as an alternative to substitution.

▶ **Example:**   Determine whether 3 is a root of the equation
$2x^4 - 3x^3 - 6x - 8 - 3 = 0$.

▶   3 is a root of $f(x) = 0$, if $f(3) = 0$.

$$
\begin{array}{ccccccc}
2 & - & 3 & - 6 & - 8 & - 3 & \;\underline{\lvert 3} \\
  & + & 6 & + 9 & + 9 & + 3 & \\
\hline
2 & + & 3 & + 3 & + 1 & \underline{\;\lvert\; 0} &
\end{array}
$$

Since $f(3) = 0$, 3 is a root of $f(x) = 0$. Note also that if $f(3) = 0$
then $(x - 3)$ is a factor of $f(x)$.

EXERCISE 4

1. Perform by synthetic division:

　　a. $3x^3 - 2x^2 - 7x - 5 \div (x - 2)$

　　b. $2x^4 + 5x^3 - x^2 + 4x - 2 \div (x + 3)$

　　c. $2x^3 + 3x^2 + 4x - 8 \div (x - \frac{1}{2})$

　　d. $6x^4 + 5x^3 - 2x^2 + 8x + 1 \div (x + \frac{1}{3})$

　　e. $x^5 - 2x^3 - 7 \div (x - 3)$

2. If $f(x) = 5x^3 + 7x^2 - x - 5$ find (a) $f(2)$, (b) $f(-3)$.

3. If $f(x) = 2x^4 - x^3 + 3x^2 - 7x + 2$ find (a) $f(-1)$,
(b) $f(-2)$.

4. Determine whether the numbers in parentheses are roots of the
given equations:

　　a. $x^3 - 5x^2 + 15x - 75 = 0$ (5)

　　b. $y^3 - 3y^2 - 18y + 40 = 0$ (−4)

　　c. $x^4 - x^3 + 5x^2 + x - 6 = 0$ (−2)

　　d. $6x^3 + 11x^2 - x - 3 = 0$ ($\frac{1}{2}$)

Basic Theorems on Theory of Equations

1. Fundamental Theorem of Algebra—Every polynomial equation with real or complex coefficients has at least one root, which may be real or complex.

This theorem will be applied only to polynomial equations with real coefficients.

2. Every polynomial equation of degree n has exactly n roots. The roots need not be distinct.

Examples: $x^5 + 4x^3 - 7x^2 + 2x - 5 = 0$ has 5 roots.

$x^{12} - x + 2 = 0$ has 12 roots.

3. If $(a + bi)$ is a root of the equation $f(x) = 0$ with real co-efficients, then the conjugate complex number $(a - bi)$ is also a root.

Examples: If it is known that $(1 + 2i)$ is a root of the equation $2x^3 - 5x^2 + 12x - 5 = 0$ then $(1 - 2i)$ is also a root of this equation.

If it is known that $(-1 - i)$ is a root of the equation $x^4 + x^3 - 2x - 6x - 4 = 0$ then $(-1 + i)$ is also a root of this equation.

4. If $(a + \sqrt{b})$ is a root of the equation $f(x) = 0$ with rational coefficients, then $(a - \sqrt{b})$ is also a root.

Example: If it is known that $(3 - \sqrt{2})$ is a root of the equation $x^3 - 8x^2 + 19x - 14 = 0$, then $(3 + \sqrt{2})$ is also a root.

Depressed Equations

Consider the equation $x^3 - 4x^2 + x + 6 = 0$. If it is given that one root of this equation is 3, then $f(3) = 0$; also, $f(x) = 0$ is divisible by $(x - 3)$. If we perform this division

$$
\begin{array}{r}
1 - 4 + 1 + 6 \quad \underline{|3} \\
+ 3 - 3 - 6 \\
\hline
1 - 1 - 2 \quad \underline{|\ 0}
\end{array}
$$

the quotient is $x^2 - x - 2$ and the original equation may be written as $(x - 3)(x^2 - x - 2) = 0$. If we set $x^2 - x - 2 = 0$ we may find the other roots of $f(x) = 0$. The equation $x^2 - x - 2 = 0$ is called the depressed equation.

Example: One root of the equation $2x^3 + 3x^2 - 23x - 12 = 0$ is -4. Find the other roots.

$$
\begin{array}{r}
\blacktriangleright \qquad 2 + 3 - 23 - 12 \quad \underline{|-4} \\
- 8 + 20 + 12 \\
\hline
2 - 5 - 3 \quad |\ 0
\end{array}
$$

The depressed equation is $2x^2 - 5x - 3 = 0$.

$$(2x + 1)(x - 3) = 0$$
$$2x + 1 = 0, \quad x - 3 = 0$$

$$x = -\frac{1}{2}, \quad x = 3$$

Example: One root of the equation $2x^3 - 9x^2 + 2x + 1 = 0$ is $\frac{1}{2}$. Find the other roots.

$$
\begin{array}{rrrr|r}
2 & -\ 9 & +\ 2 & +\ 1 & \quad\underline{\frac{1}{2}} \\
  & +\ 1 & -\ 4 & -\ 1 & \\
\hline
2 & -\ 8 & -\ 2 & \underline{\ 0} &
\end{array}
$$

The depressed equation is $2x^2 - 8x - 2 = 0$, which can be written as $x^2 - 4x - 1 = 0$. Solution of this equation by the quadratic formula gives

$$x = \frac{4 \pm \sqrt{16 + 4}}{2}$$

$$x = \frac{4 \pm \sqrt{20}}{2} = 2 \pm \sqrt{5}.$$

Example: Find the roots of the equation

$$x^4 + x^3 - 5x^2 + x - 6 = 0$$

if two of its roots are 2 and -3.

$$
\begin{array}{rrrrr|r}
1 & +\ 1 & -\ 5 & +\ 1 & -\ 6 & \quad\underline{2} \\
  & +\ 2 & +\ 6 & +\ 2 & +\ 6 & \\
\hline
1 & +\ 3 & +\ 1 & +\ 3 & \underline{\ 0} &
\end{array}
$$

The first depressed equation is $x^3 + 3x^2 + x + 3 = 0$.

$$
\begin{array}{rrrr|r}
1 & +\ 3 & +\ 1 & +\ 3 & \quad\underline{-3} \\
  & -\ 3 & +\ 0 & -\ 3 & \\
\hline
1 & +\ 0 & +\ 1 & \underline{\ 0} &
\end{array}
$$

The second depressed equation is $x^2 + 1 = 0$.
Solution of this equation yields

$$x^2 = -1$$
$$x = \pm i$$

EXERCISE 5

In the following equations one or two roots are given in parentheses.   Find the other roots.

1. $x^3 - 9x^2 + 26x - 24 = 0$ (2)
2. $2x^3 + x^2 - 13x + 6 = 0$ (−3)
3. $x^3 - 2x^2 - 5x - 12 = 0$ (4)
4. $6x^3 + 3x^2 - 10x - 5 = 0$ $(-\frac{1}{2})$
5. $6x^4 - 43x^2 + 78x^2 - 5x - 12 = 0$ $(-\frac{1}{3})(4)$
6. $x^4 + 8x^3 + 22x^2 + 23x + 6 = 0$ $(-2)(-3)$
7. $8x^4 - 26x^3 + 45x^2 - 7x - 5 = 0$ $(\frac{1}{2})(-\frac{1}{4})$
8. $24x^4 + 22x^3 - 43x^2 + 17x - 2 = 0$ $(\frac{1}{3})(-2)$

Formation of Equations

If the roots of an equation are given the equation may be formed as shown below.

Example: Write the equation having only the roots 2, − 3, and 1.

▶   Since the roots of the equation are 2, − 3 and 1 the factors of the polynomial $f(x)$ are $(x - 2)(x + 3)(x - 1)$.   The equation is $(x - 2)(x + 3)(x - 1) = 0$, or $x^3 - 7x + 6 = 0$.

Example: Write the equation having − 1 as a double root and the other two roots 4 and 0.

▶   Since the roots of the equation are − 1, − 1, 4, and 0, the factors of the polynomial $f(x)$ are $(x + 1)(x + 1)(x - 4)(x - 0)$.
The equation is $(x + 1)(x + 1)(x - 4)(x) = 0$, or $x^4 - 2x^3 - 7x^2 - 4x = 0$.

Example: Write the equation, with rational coefficients and of lowest possible degree, three of whose roots are − 2, − 1, and 3 + i.

▶   Since 3 + i is a root of the equation with real coefficients, then 3 − i must also be a root.
Therefore, the equation is of the fourth degree, and its roots are − 2, − 1, and 3 + i, and 3 − i.

The equation is $(x + 2)(x + 1)[x - (3 + i)][x - (3 - i)]$
$= 0$, or $x^4 - 3x^3 - 6x^2 + 18x + 20 = 0$.

Example: Write the equation, with rational coefficients and of lowest possible degree, two of whose roots are $2 + i$ and $3 - \sqrt{2}$.

▶ Since $2 + i$ is a root of the equation with real coefficients then $2 - i$ must also be a root. Therefore, the equation is of the fourth degree, and the roots are $2 + i$, $2 - i$, $3 + \sqrt{2}$, and $3 - \sqrt{2}$.

The equation is $[x - (2 + i)][x - (2 - i)][x - (3 + \sqrt{2})]$ $[x - (3 - \sqrt{2})] = 0$, or $(x^2 - 4x + 5)(x^2 - 6x + 7)$, or $x^4 - 10x^3 + 36x^2 - 58x + 35 = 0$.

Example: Form the equation with integral coefficients having only the roots $\frac{1}{2}$, $-\frac{1}{3}$, $+3i$, $-3i$.

▶ The equation is $(x - \frac{1}{2})(x + \frac{1}{3})(x - 3i)(x + 3i) = 0$ or
$$\left(x^2 - \frac{x}{6} - \frac{1}{6}\right)(x^2 + 9) = 0$$

Multiply both sides of the equation by 6 to remove fractions:
$(6x^2 - x - 1)(x^2 + 9) = 0$, or $6x^4 - x^3 + 53x^2 - 9x - 9 = 0$

EXERCISE 6

1. Write the equations having only the following roots:

a. $3, -2, 1$ c. $-3, -3, -1$

b. $4, -5, 2$ d. $5, -4, -2, 0$

2. Write the equations, with rational coefficients and of lowest possible degree, three of whose roots are:

a. $3, 1, 1 + i$ d. $-1, 2, -3, +i$

b. $4, -1, -2i$ e. $\frac{1}{4}, -\frac{1}{3}, 2 - i$

c. $5, -1, 2, +\sqrt{3}$ f. $-\frac{3}{4}, \frac{1}{6}, \sqrt{2}$

3. Write the equations with rational coefficients of lowest possible degree, two of whose roots are:

a. $4 + i, 5 + \sqrt{2}$ c. $2 + 3i, 4 - \sqrt{3}$

b. $3 - \sqrt{5}, i$ d. $5 - 2i, 1 + 2\sqrt{3}$

4. Form the equation with integral coefficients having only the following roots:

a. $\frac{1}{2}, -3, 4$ c. $\frac{3}{4}, 1, \frac{1}{2}, -\frac{1}{6}$

b. $\frac{2}{3}, -\frac{1}{2}, 5$ d. $-\frac{1}{3}, \frac{5}{6}, 2, -\frac{1}{4}$

Relations Between Roots and Coefficients

Consider the equation $a_0 x^n + a_2 x^{n-1} + a_1 x^{n-2} + \ldots + a_{n-1} x + a_n = 0$. If both sides of this equation are divided by a_0, we have an equation of the form $x^n + b_1 x^{n-1} + b_2 x^{n-2} + \ldots + b_{n-1} x + b_n = 0$. For this equation it can be shown that

$-b =$ sum of the roots $(r_1 + r_2 + r_3 + \ldots\ldots\ldots\ldots\ldots\ldots r_n)$

$b_2 =$ sum of the roots taken 2 at a time $(r_1 r_2 + r_1 r_3 + r_2 r_3 + \ldots)$

$-b_3 =$ sum of the roots taken 3 at a time $(r_1 r_2 r_3 + r_2 r_3 r_4 \ldots\ldots r_n)$

$(-1)^n b_n =$ product of the roots $(r_1 r_2 r_3 \ldots\ldots\ldots\ldots\ldots\ldots r_n)$

$(-1)^n$ will be negative when n is an odd number

$(-1)^n$ will be positive when n is an even number.

Application to Third Degree Equations

The third degree equation may be written

$x^3 +$ (sum of roots with sign changed) x^2

     $+$ (sum of roots taken 2 at a time) x

     $+$ (product of roots with sign changed) $= 0$

In the equation $x^3 + 5x^2 - 7x - 6 = 0$, the sum of the roots is -5, the sum of the roots taken 2 at a time is -7, and the product of the roots is $+6$.

In the equation $2x^3 - 9x^2 + 5x + 1 = 0$ we divide both members by 2, obtaining $x^3 - \dfrac{9x^2}{2} + \dfrac{5x}{2} + \dfrac{1}{2} = 0$. The sum of the roots is $\dfrac{9}{2}$, the sum of the roots taken 2 at a time is $\frac{5}{2}$, and the product of the roots is $-\frac{1}{2}$.

Application to Fourth Degree Equations

The fourth degree equation may be written

$x^4 +$ (sum of roots with sign changed) x^3

     $+$ (sum of roots taken 2 at a time) x^2

     $+$ (sum of roots taken 3 at a time, with sign changed) x

     $+$ (product of roots) $= 0$.

In the equation $x^4 - 6x^3 + 4x - x - 2 = 0$, the sum of the roots is $+6$, the sum of the roots taken 2 at a time is $+4$, the sum of the roots taken 3 at a time is $+1$, and the product of the roots is -2.

In the equation $3x^4 + 5x^3 - 6x^2 + 4x - 2 = 0$ we divide both members by 3 obtaining $x^4 + \dfrac{5x^3}{3} - 2x^2 + \dfrac{4x}{3} - \dfrac{2}{3} = 0$. The sum of the roots is $-\frac{5}{3}$, the sum of the roots taken 2 at a time is -2, the sum of the roots taken 3 at a time is $-\frac{4}{3}$, and the product is $-\frac{2}{3}$.

Example: In the equation $x^3 + 5x^2 - 9x + K = 0$, find the value of K if the one root is the negative of another.

▶ Let the roots be a, $-a$, and b. Then the sum of the roots $a - a + b = -5$ or $b = -5$. If -5 is a root then $x = -5$. Substituting -5 for x we have $-125 + 125 + 45 + K = 0$, or $K = -45$.

Example: Solve the equation $x^3 - 6x^2 + 3x + 10 = 0$, if it is given that the roots are in arithmetic progression.

▶ Let the roots be $a - b$, a, and $a + b$. Then the sum of the roots, $3a = 6$, or $a = 2$.

$$
\begin{array}{r}
1 - 6 + 3 + 10 \quad \underline{\lvert 2} \\
+ 2 - 8 - 10 \\
\hline
1 - 4 - 5 \quad \underline{\lvert 0}
\end{array}
$$

The depressed equation is $x^2 - 4x - 5 = 0$
$$(x - 5)(x + 1) = 0 \quad \text{and} \quad x = 5, x = -1.$$
∴ The roots are 2, 5, and -1.

Example: The difference between two roots of the equation $x^3 + 3x^2 + K = 0$ is 3. Find the values of K.

▶ Let the roots be a, $a + 3$, and b. Then the sum of the roots $a + a + 3 + b = -3$ or $2a + b = -6$. The sum of the roots taken 2 at a time
$a(a + 3) + ab + (a + 3)b = 0$
$a^2 + 3a + ab + ab + 3b = 0 \quad \text{or} \quad a^2 + 3a + 2ab + 3b = 0.$
$b = -6 - 2a$
$a^2 + 3a + 2a(-6 - 2a) + 3(-6 - 2a) = 0$
$a^2 + 3a - 12a - 4a^2 - 18 - 6a = 0, \text{ or } -3a^2 - 15a - 18 = 0$

$a^2 + 5a + 6 = 0$
$(a + 3)(a + 2) = 0, a = -3, a = -2.$
$b = -6 - 2a$
$b = -6 + 6 = 0, b = -6 + 4 = -2$
$K = $ product of roots with sign changed
$\therefore \quad K = 0, -4$

Example: Solve the equation $2x^3 - 7x^2 + 7x - 2 = 0$ if the roots are in geometric progression.

▶ Let the roots be $\frac{a}{r}$, a, and ar. Then the product of the roots is a^3. Then $a^3 = 1$, and $a = 1$.

$$
\begin{array}{r}
2 - 7 + 7 - 2 \quad \underline{|1} \\
+ 2 - 5 + 2 \\
\hline
2 - 5 + 2 \quad \underline{|0}
\end{array}
$$

The depressed equation is $2x^2 - 5x + 2 = 0$.
$$(2x - 1)(x - 2) = 0$$
$$x = \frac{1}{2}, \ 2$$

$\therefore$ The roots are $1, \frac{1}{2}$, and 2.

Example: Let a, b, c, be the roots of the equation $x^3 + 3x^2 - 4x - 12 = 0$, find:

a. $a + b + c$

b. $ab + ac + bc$

c. abc

d. $a^2 + b^2 + c^2$

e. $\dfrac{1}{a} + \dfrac{1}{b} + \dfrac{1}{c}$

f. $\dfrac{1}{ab} + \dfrac{1}{ac} + \dfrac{1}{bc}$

(a) The sum of the roots $(a + b + c) = -3$.

(b) The sum of the roots taken 2 at a time
$(ab + ac + bc) = -4$.

(c) The product of the roots $+12$.

(d) $a^2 + b^2 + c^2 = (a + b + c)^2 - 2(ab + ac - bc)$
$= 9 + 8 = 17$.

(e) Note that $abc \left(\dfrac{1}{a} + \dfrac{1}{b} + \dfrac{1}{c} \right) = bc + ac + ab$

$$\text{or } 12 \left(\frac{1}{a} + \frac{1}{b} + \frac{1}{c} \right) = -4$$

$$\frac{1}{a} + \frac{1}{b} + \frac{1}{c} = -\frac{1}{3}$$

(f) Note that $abc \left(\dfrac{1}{ab} + \dfrac{1}{ac} + \dfrac{1}{bc} \right) = c + b + a$

$$12 \left(\frac{1}{ab} + \frac{1}{ac} + \frac{1}{bc} \right) = -3$$

$$\frac{1}{ab} + \frac{1}{ac} + \frac{1}{bc} = -\frac{1}{4}$$

Example: Form the equation whose roots are -2, -3, 4, and -1.

▶   The equation is of the form $x^4 + ax^3 + bx^2 + cx + d = 0$.
Where a = sum of roots with sign changed
$$= -(-2 - 3 + 4 - 1) = +2$$
b = sum of roots taken 2 at a time
$$= (-2)(-3) + (-2)(4) + (-2)(-1)$$
$$+ (-3)(4) + (-3)(-1) + (4)(-1)$$
$$= 6 - 8 + 2 - 12 + 3 - 4 = -13$$
c = sum of roots taken 3 at a time with sign changed
$$= -[(-2)(-3)(4) + (-2)(-3)(-1)$$
$$+ (-2)(4)(-1) + (-3)(4)(-1)]$$
$$= -(24 - 6 + 8 + 12) = -38$$
d = product of roots = $(-2)(-3)(4)(-1) = -24$
The equation is $x^4 + 2x^3 - 13x^2 - 38x - 24 = 0$.

EXERCISE 7

1. Let a, b, and c be the roots of the equation $x^3 - 5x^2 + 7x - 9 = 0$. Find (a) $a + b + c$ (b) $ab + ac + bc$ (c) abc.

2. The product of 2 of the roots of the equation $x^3 + px^2 + qx + 12 = 0$ is -6. What is the third root?

3. The difference between 2 roots of the equation $x^3 + 3x^2 - 13x + K = 0$ is 4.   Find the value of K.

4. Find the integral value of K such that one root of the equation $x^3 - 5x^2 - 2x + K = 0$ shall be 5 more than another.

5. Find the roots of the equation $4x^3 - 12x^2 + 11x - 3 = 0$ if they are in arithmetic progression.

6. Find the values of K if two roots of the equation $x^3 - 3x + K = 0$ are equal.

7. Find the value of k if the sum of two of the roots of the equation $24x^3 + kx^2 + 64x + 24 = 0$ is zero.

8. For the equation $3x^4 - 2x^3 + x^2 - 5x + 7 = 0$ find
   a. the sum of the roots
   b. the product of the roots

9. Let a, b, c be the roots of the equation $x^3 + 2x^2 + 5x - 12 = 0$.   Find
   a. $a + b + c$
   b. $ab + ac + bc$
   c. abc
   d. $a^2 + b^2 + c^2$
   e. $\dfrac{1}{a} + \dfrac{1}{b} + \dfrac{1}{c}$
   f. $\dfrac{1}{ab} + \dfrac{1}{ac} + \dfrac{1}{bc}$

10. Solve the equation $4x^3 + 16x^2 - 9x - 36 = 0$, the sum of two of the roots being zero.

11. For the equation $x^3 + 3x^2 - 5x + K = 0$, find the value of K if
   a. -2 is a root
   b. the sum of 2 roots is zero
   c. the roots are in arithmetic progression

12. Form the equation whose roots are
   a. 5, -1, and 3
   b. 2, -3, 1, -4
   c. $\tfrac{1}{2}$, $-\tfrac{3}{2}$, 4

Multiplying the Roots of an Equation by a Constant

Given the equation $x^n + b_1x^{n-1} + b_2x^{n-2} \ldots . b_n = 0$, it is sometimes useful to transform this equation into another one whose roots are equal to K times the roots of the given equation.   To accomplish this we multiply the coefficient b_1 by K, the coefficient b_2 by K^2, the coefficient b_3 by K^3, etc.

Example: Write an equation whose roots are twice the roots of the equation $x^4 - 2x^3 - 7x^2 + 8x + 12 = 0$.

▶ 1: $x^4 - 2x^3 - 7x^2 + 8x + 12 = 0$

$x^4 - (2)(2x^3) - (2)^2(7x^2) + (2)^3(8x)$
$+ (2)^4(12) = 0$

$x^4 - (2)(2x^3) - (4)(7x^2) + (8)(8x)$
$+ (16)(12) = 0$

$y^4 - 4y^3 - 28y^2 + 64y + 192 = 0$

Note: The roots of the original equation are 2, -2, 3, and -1. The corresponding roots of the new equation are 4, -4, 6, and -2.

▶ 2: $x^4 - 2x^3 - 7x^2 + 8x + 12 = 0$

Let $y = 2x$

Then $x = \dfrac{y}{2}$

$$\left(\frac{y}{2}\right)^4 - 2\left(\frac{y}{2}\right)^3 - 7\left(\frac{y}{2}\right)^2 + 8\left(\frac{y}{2}\right) + 12 = 0$$

$$\frac{y^4}{16} - 2\cdot\frac{y^3}{8} - \frac{7y^2}{4} + \frac{8y}{2} + 12 = 0$$

$$y^4 - 4y^3 - 28y^2 + 64y + 192 = 0$$

Example: Write an equation whose roots are the roots of the equation $2x^3 - x - 1 = 0$, multiplied by -3.

▶ Write the equation as $2x^3 + 0x^2 - x - 1 = 0$
$2x^3 + (-3)(0x^2) - (-3)^2(x) - (-3)^3(1) = 0$
$2y^3 - 9y + 27 = 0$

Example: Write an equation whose roots are the roots of the equation $x^4 - 6x^3 + 7x^2 - 5x - 8 = 0$, divided by 2.

▶ Division by 2 is equivalent to multiplying by $\frac{1}{2}$.

$$x^4 - \left(\frac{1}{2}\right)6x + \left(\frac{1}{2}\right)^2 7x - \left(\frac{1}{2}\right)^3 5x - \left(\frac{1}{2}\right)^4(8) = 0$$

$$x^4 - 3x^3 + \frac{7x^2}{4} - \frac{5x}{8} - \frac{1}{2} = 0$$

If we multiply both sides of this equation by 8, we have
$$8y^4 - 24y^3 + 14y^2 - 5y - 4 = 0$$

Example: Write an equation whose roots are the negatives of the roots of the equation $x^4 - 15x^2 + 10x + 24 = 0$.

▶   Write the above equation as
$$x^4 + 0x^3 - 15x^2 + 10x + 24 = 0$$
$$x^4 + (-1)(0x) - (-1)^2(15x^2) + (-1)^3(10x)$$
$$+ (-1)^4(24) = 0$$
$$y^4 - 15y^2 - 10y + 24 = 0$$
The roots of the original equation are $-1, +2, +3, -4$.
The roots of the new equation are $+1, -2, -3, +4$.

The same result may be obtained by substituting $(-x)$ for x in the original equation as follows:
$$(-x)^4 - 15(-x)^2 + 10(-x) + 24 = 0$$
$$y^4 - 15y^2 - 10y + 24 = 0$$

Example: Write an equation whose roots are the negatives of the roots of the equation $3x^3 - 5x^2 + 2x - 7 = 0$.

▶   1: $3x^3 - (-1)(5x^2) + (-1)^2(2x) - (-1)^3(7) = 0$
        $3y^3 + 5y^2 + 2y + 7 = 0$

▶   2: $3(-x)^3 - 5(-x)^2 + 2(-x) - 7 = 0$
        $-3y^3 - 5y^2 - 2y - 7 = 0$
       or $3y^3 + 5y^2 + 2y + 7 = 0$

EXERCISE 8

Write the equations whose roots are the roots of the given equation multiplied by the number in parentheses.

  1. $x^3 - 2x^2 - x + 1 = 0$ (3)

  2. $2x^3 + 2x^2 - 4x + 8 = 0$ $(\frac{1}{2})$

  3. $x^4 + 2x^3 - x^2 + 4 = 0$ (-2)

  4. $3x^3 - 2x^2 + 3x - 5 = 0$ (10)

  5. $x^3 - 6x^2 - 3x + 18 = 0$ $(\frac{1}{3})$

  6. $3x^5 + 4x^3 - 8x + 16 = 0$ $(-\frac{1}{2})$

  7. $2x^3 + 7x^2 + 3x - 5 = 0$ (-1)

  8. $3x^4 - 2x^3 - 7x^2 + 2x + 4 = 0$ (-1)

Rational Roots

The following theorems are very useful in finding rational roots
of polynomial equations.

Theorem 1:   If the polynomial equation
$$x^n + a_1 x^{n-1} + a_2 x^{n-2} + \ldots + a_{n-1} x + a_n = 0$$
(coefficient of the x^n term is 1) has a rational root, the root is an integer and a factor of a_n.

Example: If the equation $x^4 - 6x^2 + 4x - 8 = 0$ has rational
roots, these roots must be whole numbers and factors of -8.   Therefore, the possible rational roots are $\pm 1, \pm 2, \pm 4, \pm 8$, and no others.

Theorem 2:   If the equation $a_0 x^n + a_1 x^{n-1} + a_2 x^{n-2} + \ldots \ldots$
$+ a_{n-1} x + a_n = 0$ with integral coefficients has rational roots of the
form $\dfrac{p}{q}$, where $\dfrac{p}{q}$ is in its lowest terms, then p is a factor of a_n and q is
a factor of a_0.

Example: If the equation $2x^3 - 17x^2 + 11x + 15 = 0$ has rational roots of the form $\dfrac{p}{q}$ then p must be a factor of 15 and q a factor
2.   The possible rational roots are conveniently shown in the following table:

p	± 1	± 3	± 5	± 15
q	± 1	± 2		

For example, $\dfrac{p}{q}$ may be $\pm \dfrac{3}{2}, \pm \dfrac{5}{2}, \pm \dfrac{1}{2}$ etc.

Example: Solve the equation $x^3 + 2x^2 - 9x - 18 = 0$.

▶   If this equation has rational roots, they cannot be fractions
since the coefficient of the highest power of x is 1.   Also the rational
roots must be factors of 18.   Thus, the possible rational roots are
$\pm 1, \pm 2, \pm 6, \pm 9, \pm 18$.   If we test ± 1 and $+ 2$ we find that these
numbers are not roots.   (Testing -2)

$$1 + 2 - 9 - 18 \quad \lfloor -2$$
$$\underline{\quad - 2 + 0 + 18}$$
$$1 + 0 - 9 \quad \lfloor 0$$

$\therefore$ -2 is a root. The depressed equation is $x^2 - 9 = 0$. The other 2 roots are ± 3.

Example: Solve the equation $x^3 - 5x^2 - 9x + 45 = 0$.

▶ The possible rational roots are factors of 45, or ± 1, ± 3, ± 5, ± 9, ± 15, ± 45.
When tested, $+1$ and -1 are found to be not roots.

$$1 - 5 - 9 + 45 \quad \lfloor 3$$
$$\underline{\quad + 3 - 6 - 45}$$
$$1 - 2 - 15 \quad \lfloor 0$$

$\therefore$ 3 is a root.
The depressed equation is $x^2 - 2x - 15 = 0$.
$$(x - 5)(x + 3) = 0$$
$$x = 5 \quad \text{or} \quad -3.$$

$\therefore$ The three roots are 3, 5, and -3.

Example: Solve the equation $x^4 - 6x^3 + 10x^2 - 6x + 9 = 0$.

▶ The possible rational roots are ± 1, ± 3, ± 9. When tested $+1$, and -1 are found to be not roots.

$$1 - 6 + 10 - 6 + 9 \quad \lfloor 3$$
$$\underline{\quad + 3 - 9 + 3 - 9}$$
$$1 - 3 + 1 - 3 \quad \lfloor 0$$

$\therefore$ 3 is a root.
The depressed equation is $x^3 - 3x^2 + x - 3 = 0$. The possible rational roots are ± 1, ± 3. However, ± 1 cannot be roots of the depressed equation since they were eliminated as roots of the original equation.

$$1 - 3 + 1 - 3 \quad \lfloor 3$$
$$\underline{\quad + 3 + 0 + 3}$$
$$1 + 0 + 1 \quad \lfloor 0$$

$\therefore$ 3 is a root of the depressed equation. The second depressed equation is $x^2 + 1 = 0$. The roots of this equation are $\pm i$. The roots of the original equation are $+3$, $+3$, $+i$, $-i$.

Example: Solve the equation $x^3 + 3x^2 - 5x - 10 = 0$.

▶   The possible rational roots are ± 1, ± 2, ± 5, ± 10.   When tested, $+1$ and -1 are found to be not roots.

$$
\begin{array}{r}
1 + 3 - 5 - 10 \quad \underline{|2} \\
+ 2 + 10 + 10 \\
\hline
1 + 5 + 5 \quad \underline{|0}
\end{array}
$$

The depressed equation is $x^2 + 5x + 5 = 0$.   When solving by the quadratic formula the roots of the depressed equation are found to be

$$x = \frac{-5 \pm \sqrt{5}}{2}$$

$\therefore$ The roots of the original equation are $2, \dfrac{-5 \pm \sqrt{5}}{2}$.

EXERCISE 9

Solve the following equations:
1. $x^3 + 3x^2 - 4x - 12 = 0$
2. $x^3 + 4x^2 - 17x - 60 = 0$
3. $x^3 - 7x^2 + 8x + 4 = 0$
4. $x^3 - 2x^2 - 2x + 12 = 0$
5. $x^3 - 4x^2 + 2x + 4 = 0$
6. $x^4 + x^3 - x^2 - 7x - 6 = 0$
7. $x^4 + 2x^3 - 3x^2 - 4x + 4 = 0$
8. $x^4 - 5x^3 + 7x^2 + 3x - 10 = 0$
9. $x^4 - 2x^3 - 3x^2 + 4x - 12 = 0$
10. $x^4 - 6x^3 + 17x^2 - 28x + 20 = 0$

In solving polynomial equations having the coefficient of the highest power of $x \neq 1$ the two methods illustrated below may be used.

Example: Solve the equation $3x^3 + 2x^2 - 19x + 6 = 0$.

▶   1:   If this equation has rational roots they must be of the form $\dfrac{p}{q}$ with p a factor of 6 and q a factor of 3.   The possible rational roots are shown in the following table.

p	± 1	± 2	± 3	± 6
q	± 1	± 3		

The possibilities are ± 1, $\pm \frac{1}{3}$, ± 2, $\pm \frac{2}{3}$, ± 3, ± 6.
When tested, ± 1 are found to be not roots.

$$
\begin{array}{rrrrr}
3 & + \ 2 & - \ 19 & + \ 6 & \underline{\left|\frac{1}{3}\right.} \\
 & + \ 1 & + \ 1 & - \ 6 & \\
\hline
3 & + \ 3 & - \ 18 & \underline{\left|\ 0\right.} &
\end{array}
$$

$\therefore \quad \frac{1}{3}$ is a root.

The depressed equation is $3x^2 + 3x - 18 = 0$. Dividing by 3 we have $x^2 + x - 6 = 0$.

$$(x + 3)(x - 2) = 0$$
$$x = -3 \quad \text{or} \quad +2$$

The three roots are $\frac{1}{3}$, -3, $+2$.

▶ 2: $3x^3 + 2x^2 - 19x + 6 = 0$.

All possible fractional roots of this equation must have the denominator 3. If we transform this equation by multiplying the roots of the equation by 3, we will obtain a new equation whose rational roots are whole numbers.

$$3x^3 + 2x^2 - 19x + 6 = 0$$
$$x^3 + \frac{2x^2}{3} - \frac{19x}{3} + \frac{6}{3} = 0$$

Multiplying the roots by 3 we have

$$x^3 + 3\left(\frac{2}{3}\right)(x^2) - (3)^2\left(\frac{19}{3}\right)x + (3)^3\left(\frac{6}{3}\right) = 0$$

or $\quad x^3 + 2x^2 - 57x + 54 = 0$. In this new equation, the rational roots are integers and factors of 54. If we test 1 as a root:

$$
\begin{array}{rrrrr}
1 & + \ 2 & - \ 57 & + \ 54 & \underline{\left|\ 1\right.} \\
 & + \ 1 & + \ 3 & - \ 54 & \\
\hline
1 & + \ 3 & - \ 54 & \underline{\left|\ 0\right.} &
\end{array}
$$

$\therefore \quad 1$ is a root of this new equation. The depressed equation is $x^2 + 3x - 54 = 0$.

$$(x + 9)(x - 6) = 0$$
$$x = -9 \quad \text{or} \quad x = 6.$$

∴   The roots of the new equation are 1, −9, and 6.

The roots of the original equation are these roots divided by 3.

∴   The roots of the original equation are $\frac{1}{3}$, −3, and 2.

Example: Solve the equation
$$18x^4 + 51x^3 + 29x^2 - 4x - 4 = 0.$$

▶   1:   The rational roots are included in the table:

p	± 1	± 2	± 4			
q	± 1	± 2	± 3	± 6	± 9	± 18

The possibilities ± 1 and $+\frac{1}{2}$ can be eliminated by testing.

$$
\begin{array}{r}
18 + 51 + 29 - 4 - 4 \quad \underline{\left| -\frac{1}{2} \right.} \\
- \;\;\, 9 - 21 - 4 + 4 \\
\hline
18 + 42 + \;\; 8 - 8 \;\;\underline{\left| 0 \right.}
\end{array}
$$

∴ $-\frac{1}{2}$ is a root.

The depressed equation is $18x^3 + 42x^2 + 8x - 8 = 0$, or dividing by 2, $9x^3 + 21x^2 + 4x - 4 = 0$. The next possible root in order is $\frac{1}{3}$.

$$
\begin{array}{r}
9 + 21 + \;\; 4 - 4 \quad \underline{\left| \frac{1}{3} \right.} \\
+ \;\; 3 + \;\; 8 + 4 \\
\hline
9 + 24 + 12 \;\;\underline{\left| 0 \right.}
\end{array}
$$

∴ $\frac{1}{3}$ is a root and the depressed equation is $9x^2 + 24x + 12 = 0$. Dividing by 3,

$$3x^2 + 8x + 4 = 0$$
$$(3x + 2)(x + 2) = 0$$
$$x = -\frac{2}{3}, \quad x = -2.$$

∴   The roots are $-\frac{1}{2}$, $\frac{1}{3}$, $-\frac{2}{3}$, and -2.

▶   2: $18x^4 + 51x^3 + 29x^2 - 4x - 4 = 0.$

$$x^4 + \frac{51x^3}{18} + \frac{29x^2}{18} - \frac{4x}{18} - \frac{4}{18} = 0$$

$$x^4 + \frac{17x^3}{6} + \frac{29x^2}{18} - \frac{2x}{9} - \frac{2}{9} = 0$$

In order to eliminate fractional roots we must transform this equation by multiplying its roots by 6.

$$x^4 + (6)\frac{(17x)^3}{6} + (6)^2\left(\frac{29}{18}\right)x^2 - (6)^3\left(\frac{2}{9}\right)x - (6)^4\left(\frac{2}{9}\right) = 0$$

$$x^4 + 17x + 58x - 48x - 288 = 0.$$

Rational roots of this new equation must be integers and factors of 288.

$$
\begin{array}{r}
1 + 17 + 58 - 48 - 288 \quad \underline{|\,2} \\
 + 2 + 38 + 192 + 288 \\
\hline
1 + 19 + 96 + 144 \qquad \underline{|\,0}
\end{array}
$$

∴   2 is a root.

$$
\begin{array}{r}
1 + 19 + 96 + 144 \quad \underline{|-3} \\
- 3 - 48 - 144 \\
\hline
1 + 16 + 48 \qquad \underline{|\,0}
\end{array}
$$

∴ -3 is a root.

$$x^2 + 16x + 48 = 0$$
$$(x + 4)(x + 12) = 0$$
$$x = -4 \quad \text{or} \quad -12$$

∴   The roots of the transformed equation are $2, -3, -4, -12$.

∴   The roots of the original equation are $\frac{1}{3}, -\frac{1}{2}, -\frac{2}{3},$ and -2.

EXERCISE 10

Solve the following equations:

1. $2x^3 - x^2 - 22x - 24 = 0$
2. $2x^3 - 5x + 8x - 3 = 0$
3. $6x^3 + 13x^2 - 4x - 15 = 0$
4. $3x^4 + 8x^3 + 6x^2 + 3x - 2 = 0$
5. $2x^4 + 5x^3 + 3x^2 + x - 2 = 0$
6. $24x^3 - 10x^2 - 13x + 6 = 0$
7. $4x^4 + 9x^2 - 11x + 3 = 0$
8. $2x^4 + 9x^3 + 15x^2 + 14x - 12 = 0$
9. $3x^4 + 4x^3 + 2x^2 - x - 2 = 0$
10. $2x^4 - 3x^3 - 6x^2 - 8x - 3 = 0$
11. $6x^4 - 5x^3 + 7x^2 - 5x + 1 = 0$
12. $3x^4 - 16x^3 + 14x^2 + 24x - 9 = 0$
13. $8x^4 - 2x^3 - 41x^2 - 34x - 6 = 0$
14. $9x^4 - 18x^3 - 13x^2 + 8x + 4 = 0$
15. $4x^4 - 8x^3 - 5x^2 + 18x - 9 = 0$

Upper and Lower Bounds of Real Roots

It is often useful to know upper and lower bounds of the roots of an equation.

DEFINITION: If no real root of $f(x) = 0$ is greater than b_1, or less than b_2, then b_1 is called an upper bound and b_2 is called a lower bound for the real roots of $f(x) = 0$. Upper and lower bounds are usually given as integers.

Upper and lower bounds may be found by using the following theorem:

Theorem: If the coefficient of the highest power of x in $f(x) = 0$ is positive and if each number in the last line of the synthetic division of $f(x)$ by $(x - b_1)$ is positive or zero, then no real root of $f(x) = 0$ is greater than b_1.

To find a lower bound of b_2, the synthetic division is performed with $f(-x)$ being divided by $(x - b_2)$. This is true because the roots of $f(-x) = 0$ are the negatives of the roots of $f(x) = 0$.

This theorem does not state that an upper bound is the least upper bound or that a lower bound is the greatest lower bound.

Example: Find integers that are upper and lower bounds of the roots of the equation $x^3 + 2x^2 - 5x - 15 = 0$.

$$
\begin{array}{rrrr|r}
1 + 2 - 5 - 15 & \underline{1} \\
+ 1 + 3 - 2 \\
\hline
1 + 3 - 2 \,|\, -17
\end{array}
$$

$$
\begin{array}{rrrr|r}
1 + 2 - 5 - 15 & \underline{2} \\
+ 2 + 8 + 6 \\
\hline
1 + 4 + 3 \,|\, -9
\end{array}
$$

$$
\begin{array}{rrrr|r}
1 + 2 - 5 - 15 & \underline{3} \\
+ 3 + 15 + 30 \\
\hline
1 + 5 + 10 \,|\, +15
\end{array}
$$

Since the signs of all the terms in the bottom row of the synthetic division with 3 are positive *then 3 is an upper bound* of the positive roots of $f(x) = 0$. This conclusion makes it unnecessary to try $+5$ and $+15$ as possible real roots.

$$f(-x) = -x^3 + 2x^2 + 5x - 15$$

We can now use the equation $f(-x) = 0$ or $x^3 - 2x^2 - 2x + 15 = 0$ to find a lower bound.

$$
\begin{array}{rrrr|r}
1 & -\ 2 & -\ 5 & +\ 15 & \underline{\ 1} \\
 & +\ 1 & -\ 1 & -\ 6 & \\
\hline
1 & -\ 1 & -\ 6 & \underline{\ +\ 9} & \\
\end{array}
$$

$$
\begin{array}{rrrr|r}
1 & -\ 2 & -\ 5 & +\ 15 & \underline{\ 2} \\
 & +\ 2 & +\ 0 & -\ 10 & \\
\hline
1 & +\ 0 & -\ 5 & \underline{\ +\ 5} & \\
\end{array}
$$

$$
\begin{array}{rrrr|r}
1 & -\ 2 & -\ 5 & +\ 15 & \underline{\ 3} \\
 & +\ 3 & +\ 3 & -\ 6 & \\
\hline
1 & +\ 1 & -\ 2 & \underline{\ +\ 9} & \\
\end{array}
$$

$$
\begin{array}{rrrr|r}
1 & -\ 2 & -\ 5 & +\ 15 & \underline{\ 4} \\
 & +\ 4 & +\ 8 & +\ 12 & \\
\hline
1 & +\ 2 & +\ 3 & \underline{\ +\ 27} & \\
\end{array}
$$

Since the signs in the bottom row of the synthetic division with 4 are positive *then* -4 *is a lower bound* of the negative roots of $f(x) = 0$. Thus -5 and -15 do not have to be tested.

Example: Find integers that are upper and lower bounds of the roots of the equation $2x^4 - 7x^3 + x^2 - 5x - 20 = 0$.

▶ If the integers 1, 2, and 3, are tried negative signs will be found in the last line of the synthetic division.

$$
\begin{array}{rrrrr|r}
2 & -\ 7 & +\ 1 & -\ 5 & -\ 20 & \underline{\ 4} \\
 & +\ 8 & +\ 4 & +\ 20 & +\ 60 & \\
\hline
2 & +\ 1 & +\ 5 & +\ 15 & \underline{\ +\ 40} & \\
\end{array}
$$

Since 4 is the first integer that yields only positive signs in the last line of synthetic division, *4 is an upper bound* of the positive roots of $f(x) = 0$. $f(-x) = 2x^4 + 7x^3 + x^2 + 5x - 20 = 0$.

The integer 1 yields a negative sign in the last line of synthetic division.

$$
\begin{array}{rrrrr|r}
2 & +\ 7 & +\ 1 & +\ 5 & -\ 20 & \underline{\ 2} \\
 & +\ 4 & +\ 22 & +\ 46 & +\ 102 & \\
\hline
2 & +\ 11 & +\ 23 & +\ 51 & \underline{\ +\ 82} & \\
\end{array}
$$

-2 *is a lower bound* of the negative roots of the equation.

EXERCISE 11

In each case find integers that are upper and lower bounds of the roots of the equations:

1. $x^3 + 2x^2 + 8x - 12 = 0$
2. $2x^3 + x^2 - 6x - 18 = 0$
3. $6x^3 - 8x^2 - 3x - 12 = 0$
4. $x^4 - 5x^3 + 4x^2 + 2x - 30 = 0$
5. $2x^4 - 6x^3 - 8x^2 + 5x - 16 = 0$
6. $4x^4 + x^3 - 7x^2 - 18x - 20 = 0$

Descartes' Rule of Signs

In a polynomial $f(x)$ with real coefficients arranged in descending powers of x, if two successive terms differ in sign, then a variation in sign occurs.

Example: Consider the polynomial $x^4 - 3x^3 + 2x^2 + 3x - 5$. The signs of the terms of this polynomial arranged in order are $+ - + + -$. There are 3 variations in sign as shown by the loops above the signs.

Example: How many variations in sign are there in the polynomial $x^5 - 4x^4 + 3x^3 - 2x^2 + 3x + 5$.

▶ The signs in order are $+ - + - + +$. There are 4 variations in sign as shown above.

Descartes' Rule of Signs: The number of positive roots of an equation $f(x) = 0$ is either equal to the number of variations of sign in $f(x)$ or to that number decreased by an even integer. The number of negative roots of $f(x) = 0$ is equal to the number of variations of sign in $f(-x)$ or to that number decreased by an even integer.

Example: Determine the nature of the roots of the equation
$$2x^6 + 5x^2 - 3x + 7 = 0.$$

▶ For $f(x) = 0$ the signs are $+ + - +$ or 2 variations.
For $f(-x) = 0$ the signs are $+ + + +$ or no variations.
There are 2 positive roots or 0 positive roots.
There are no negative roots.

The possible combinations of roots are:

	Positive	Negative	Complex
(a)	2	0	4
(b)	0	0	6

Complex roots occur in conjugate pairs.

Example: Determine the nature of the roots of the equation
$$2x^3 - 3x^2 - 2x + 5 = 0.$$

▶ For $f(x) = 0$ the signs are $+ - - +$, or 2 variations.
For $f(-x) = 0$ the signs are $- - + +$, or 1 variation.
The possible combinations of roots are:

	Positive	Negative	Complex
(a)	2	1	0
(b)	0	1	2

Example: Determine the nature of the roots of the equation
$x^4 - 3x^3 - 12x^2 + 7x + 3 = 0.$

▶ For $f(x) = 0$ the signs are $+ - - + +$, or 2 variations.
For $f(-x) = 0$ the signs are $+ + - - +$, or 2 variations.
The possible combinations of roots are

	Positive	Negative	Complex
(a)	2	2	0
(b)	2	0	2
(c)	0	2	2
(d)	0	0	4

Example: Determine the nature of the roots of the equation
$x^5 + 3x^2 + 7 = 0.$

▶ For $f(x) = 0$ the signs are $+ + +$, or 0 variations.
For $f(-x) = 0$ the signs are $- + +$, or 1 variation.
∴ There is one negative root and there are 4 complex roots.

Example: Determine the nature of the roots of the equation
$x^6 + 4x^4 + 3x^2 + 16 = 0.$

▶ For $f(x) = 0$ the signs are $+ + + +$ or 0 variations.
   For $f(-x) = 0$ the signs are $+ + + +$ or 0 variations.
∴ There are no positive roots, no negative roots and 6 complex
   roots.

Example: Determine the nature of the roots of the equation
$x^3 - 3x^2 + 5x - 7 = 0$.

▶ For $f(x) = 0$ the signs are $+ - + -$ or 3 variations.
   For $f(-x) = 0$ the signs are $- - - -$ or 0 variations.
The possible combinations of roots are

	Positive	Negative	Complex
(a)	3	0	0
(b)	1	0	2

EXERCISE 12

By means of Descartes' Rule of Signs determine the nature of the
roots of the following equations:
1. $2x^3 - 7x - 5 = 0$
2. $x^5 + 3x^2 + 1 = 0$
3. $x^6 + 2x^4 + 8x^2 + 4 = 0$
4. $x^3 - 5x^2 + 2 = 0$
5. $x^6 + 4x^2 - 10 = 0$
6. $x^4 - 2x^3 + 3x^2 - 7x + 2 = 0$
7. $3x^5 - 2x^3 + 5x - 8 = 0$
8. $2x^4 + 3x^2 - 2x + 5 = 0$
9. $x^6 - 3x^5 + 2x^2 - 3x + 1 = 0$
10. $x^5 + 3x^3 - 5x - 16 = 0$

Graphs of Polynomials

A graph is very helpful in determining the irrational roots of a
polynomial equation. In practice, by the use of the remainder
theorem, synthetic division may be used to find the value of $f(x)$,
as illustrated in the following example.

Example: Plot the graph of $y = f(x) = x^3 - x^2 - 9x + 13$.

$$
\begin{array}{rrrr|r}
1 & -1 & -9 & +13 & \underline{\,3\,} \\
 & +3 & +6 & -9 & \\
\hline
1 & +2 & -3 & +4 &
\end{array}
$$

$$
\begin{array}{rrrr|r}
1 & -1 & -9 & +13 & \underline{\,-2\,} \\
 & -2 & +6 & +6 & \\
\hline
1 & -3 & -3 & +19 &
\end{array}
$$

$$
\begin{array}{rrrr|r}
1 & -1 & -9 & +13 & \underline{\,2\,} \\
 & +2 & +2 & -14 & \\
\hline
1 & +1 & -7 & -1 &
\end{array}
$$

$$
\begin{array}{rrrr|r}
1 & -1 & -9 & +13 & \underline{\,-3\,} \\
 & -3 & +12 & -9 & \\
\hline
1 & -4 & +3 & +4 &
\end{array}
$$

$$
\begin{array}{rrrr|r}
1 & -1 & -9 & +13 & \underline{\,-4\,} \\
 & -4 & +20 & -44 & \\
\hline
1 & -5 & +11 & -31 &
\end{array}
$$

x	$+3$	$+2$	$+1$	0	-1	-2	-3	-4
y	$+4$	-1	$+4$	$+13$	$+20$	$+19$	$+4$	-31

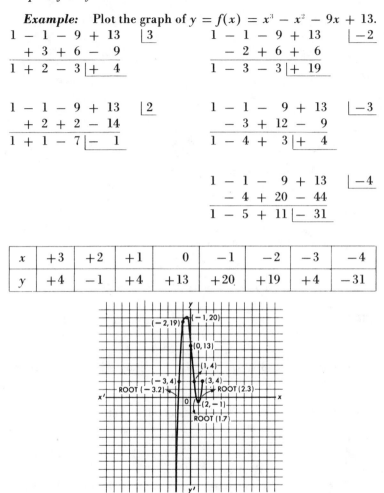

From the graph it appears that the function $f(x)$ is zero for a value between 1 and 2, between 2 and 3, and also for a value between -3 and -4. The abscissas of these points are real roots of the equation $f(x) = 0$. The roots may be estimated from the graph as 1.7, 2.3, and -3.2.

The graph of a polynomial will be a smooth curve. A closer approximation to the real roots may be obtained by using a larger scale.

When imaginary roots occur, they will not appear on the graph.

EXERCISE 13

By means of graphs find the approximate real roots of the following equations:

1. $x^3 - x - 3 = 0$
2. $x^3 - 3x - 4 = 0$
3. $x^3 - 5x - 3 = 0$

4. $x^3 + x^2 - 3x - 10 = 0$
5. $x^3 - 3x^2 + 5x - 9 = 0$
6. $x^3 + 3x^2 - 6x - 2 = 0$

Location of Real Roots

A root of $f(x) = 0$ will exist between the real numbers a and b when $f(a)$ and $f(b)$ have unlike signs.

Example: Locate the real roots in the equation
$$x^3 + x - 15 = 0$$

▶ $f(0) = -15, f(1) = -13, f(2) = -5, f(3) = +15$
     A real root lies between 2 and 3.

The application of Descartes' Rule of Signs indicates that this is the only real root.

Note: If we plot the points $(0, -15)$, $(1, -13)$, $(2, -5)$ and $(3, 15)$ we can readily see that the graph cuts the x-axis between 2 and 3 on the x-axis.

Example: Locate the real roots of the equation
$$x^3 + 3x^2 - 4x - 5 = 0$$

▶ $f(-4) = -5, f(-3) = +7, f(-2) = +7, f(-1) = +1,$
     $f(0) = -5, f(1) = -5, f(2) = +7.$

∴ There are real roots between -3 and -4, between 0 and -1, and between 1 and 2.

EXERCISE 14

Determine the location of the real roots in each of the following equations.

1. $x^3 - 3x^2 - 12 = 0$
2. $x^3 + 2x - 7 = 0$
3. $x^3 + 5x^2 - 3 = 0$
4. $x^3 - 5x^2 + 4x + 5 = 0$
5. $2x^4 - 3x^3 + x - 8 = 0$
6. $x^3 - 2x^2 - 6x - 1 = 0$

Finding Irrational Roots by Linear Interpolation

The method of finding irrational roots by linear interpolation is described in the following example.

Example: Find a real root, to two decimal places, of the equation, $x^3 + x - 5 = 0$.

▶ To find the nature of the roots apply Descartes' Rule of Signs. Since $f(x) = 0$ has one variation of sign, $f(x) = 0$ has no more than 1 positive root. Since $f(-x) = 0$ has no variation of sign, $f(x)$ has no negative roots. Therefore, $f(x) = 0$ has one positive and two complex roots.

To locate the positive root we note that $f(0) = -5, f(1) = -3,$ $f(2) = +5$. Therefore, the positive root lies between 1 and 2. The following graph of the equation will help to clarify the next steps.

x	-2	-1	0	1	2
y	-15	-7	-5	-3	5

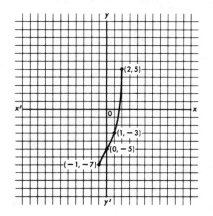

The graph indicates that the curve crosses the x-axis between the values $x = 1$ and $x = 2$.

In order to help locate this crossing place more precisely we shall assume that the portion of the graph between A and B is a straight line. Let the distance between A and C be designated by the letter p. Then

$$CB_1 = 1 - p$$

It is clear that

$$\triangle A_1 CA \sim \triangle B_1 BC$$

$$\therefore \quad \frac{p}{1 - p} = \frac{3}{5}$$

$$5p = 3 - 3p$$

$$8p = 3$$

$$p = .375 \text{ or } .4 \text{ to the nearest tenth}$$

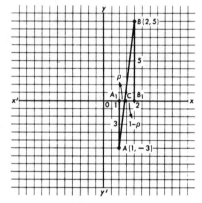

This indicates that the graph crosses the x-axis at approximately the point where $x = 1.4$.   Finding $f(1.4)$ by synthetic division.

$$
\begin{array}{rrrr|r}
1 + 0 & + 1 & - 5 & & \underline{1.4} \\
+ 1.4 & + 1.96 & + 4.144 & & \\
\hline
1 + 1.4 & + 2.96 & \underline{-\quad .856} & &
\end{array}
$$

Finding $f(1.5)$ by synthetic division.

$$
\begin{array}{rrrr|r}
1 + 0 & + 1 & - 5 & & \underline{1.5} \\
+ 1.5 & + 2.25 & + 4.875 & & \\
\hline
1 + 1.5 & + 3.25 & \underline{-\quad .125} & &
\end{array}
$$

Finding $f(1.6)$ by synthetic division

$$
\begin{array}{rrrr|r}
1 + 0 & + 1 & - 5 & & \underline{1.6} \\
+ 1.6 & + 2.56 & + 5.696 & & \\
\hline
1 + 1.6 & + 3.56 & \underline{+\quad .696} & &
\end{array}
$$

The sign of the value of $f(x)$ changes between $x = 1.5$ and $x = 1.6$.

Therefore, the root lies between 1.5 and 1.6. Note that the approximation 1.4 was slightly inaccurate. This is true because the assumption that the graph is a straight line between the points $(1, -3)$ and $(2, 5)$ is not accurate.

The inaccuracy of the assumption is corrected by locating the root precisely between 1.5 and 1.6. The same method is now used to find the next decimal place. We shall again assume that the portion of the graph between $D(1.5, -.125)$ and $E(1.6, .696)$ is a straight line. Let the distance between D and F be designated by the letter q. Then $FE_1 = .1 - q$. Since the triangles are similar

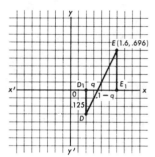

$$\frac{q}{.1 - q} = \frac{.125}{.696}$$
$$.696q = .0125 - .125q$$
$$.821q = .0125$$
$$q = \frac{.0125}{.821} = .02, \text{ to the nearest hundredth.}$$

Finding $f(1.52)$ by synthetic division

$$
\begin{array}{rrrr|l}
1 + 0 & + 1 & - 5 & & \underline{1.52} \\
+ 1.52 & + 2.3104 & + 5.031808 & & \\
\hline
1 + 1.52 & + 3.3104 & + \quad .031808 & & \\
\end{array}
$$

Finding $f(1.51)$ by synthetic division

$$
\begin{array}{rrrr|l}
1 + 0 & + 1 & - 5 & & \underline{1.51} \\
+ 1.51 & + 2.3104 & + 4.952951 & & \\
\hline
1 + 1.51 & + 3.2801 & - \quad .047049 & & \\
\end{array}
$$

The sign of the value of $f(x)$ changes between $x = 1.51$ and $x = 1.52$. This indicates that the graph crosses the x-axis between $x = 1.51$ and $x = 1.52$. Note that the approximation 1.52 was slightly too large. This slight inaccuracy is attributable to the as-

sumption that the graph is a straight
line between the points $(1.5, -.125)$
and $(1.6, .696)$.

In order to find the next decimal
place we shall again assume that the
portion of the graph between $G(1.51,$
$-.047049)$ and $H(1.52, .031808)$ is
a straight line. Let the distance between G_1 and J be designated by the
letter r. Then $JH = .01 - r$. Since
the triangles are similar,

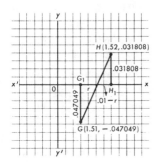

$$\frac{r}{.01 - r} = \frac{.047049}{.031808}$$

$$.00047049 - .047049r = .031808r$$

$$.078857r = .00047049$$

$$r\frac{.00047049}{.078857} = .006 \text{ to the nearest hundredth.}$$

Finding $f(1.516)$ by synthetic division

$$
\begin{array}{rrrr|r}
1 & +\ 0 & +\ 1 & -\ 5 & \underline{1.516} \\
  & +\ 1.516 & +\ 2.298256 & +\ 5.000156096 & \\
\hline
1 & +\ 1.516 & +\ 3.298256 & \!\!|+\ \ .000156096 &
\end{array}
$$

Finding $f(1.515)$ by synthetic division

$$
\begin{array}{rrrr|r}
1 & +\ 0 & +\ 1 & -\ 5 & \underline{1.515} \\
  & +\ 1.515 & +\ 2.295225 & +\ 4.992265875 & \\
\hline
1 & +\ 1.515 & +\ 3.295225 & \!\!|-\ \ .007734125 &
\end{array}
$$

The sign of the value of $f(x)$ changes between $x = 1.515$ and
$x = 1.516$.

This indicates that the graph crosses the x-axis between $x = 1.515$
and $x = 1.516$.

∴   The real root, correct to the nearest hundredth, is 1.52.

EXERCISE 15

Find, correct to the nearest hundredth, the real roots of the following equations, using the method of linear interpolation.

1. $x^3 + 5x - 9 = 0$ 3. $x^3 - 9x - 5 = 0$
2. $x^3 - 3x - 12 = 0$ 4. $x^3 - x^2 + x - 10 = 0$

Decreasing Each of the Roots of an Equation by a Constant

The method of decreasing each root of an equation by a constant is illustrated in the following example.

Example: Find the equation whose roots are less by 2 than the roots of the equation $x^3 - 3x^2 + 5x - 9 = 0$.

▶ To perform this transformation the method of synthetic division is used.

$$
\begin{array}{rrrr|r}
1 & -\ 3 & +\ 5 & -\ 9 & \underline{\ 2} \\
  & +\ 2 & -\ 2 & +\ 6 & \\
\hline
1 & -\ 1 & +\ 3 & \underline{-\ 3} = R_1 & \\
  & 1 & -\ 1 & +\ 3 & \\
  &   & +\ 2 & +\ 2 & \\
\hline
  & 1 & +\ 1 & \underline{+\ 5} = R_2 & \\
  &   & +\ 2 & & \\
\hline
  &   & 1 & \underline{+\ 3} = R_3 & \\
\end{array}
$$

The new equation is formed by using the number left after the successive divisions are performed (in this case 1) and the remainders (R_3, R_2, and R_1 in that order) as the coefficients of the new equation. The new equation is $y^3 + 3y^2 + 5y - 3 = 0$. Each root of the new equation is 2 less than the corresponding root of the original equation.

Example: Write the equation whose roots are less by 3 than the roots of the equation $x^3 - 10x^2 + 27x - 26 = 0$.

$$
\begin{array}{rrrr|r}
1 & -\ 10 & +\ 27 & -\ 26 & \underline{\ 3} \\
  & +\ 3 & -\ 21 & +\ 18 & \\
\hline
1 & -\ 7 & +\ 6 & \underline{-\ 8} & \\
  & +\ 3 & -\ 12 & & \\
\hline
1 & -\ 4 & \underline{-\ 6} & & \\
  & +\ 3 & & & \\
\hline
1 & \underline{-\ 1} & & & \\
\end{array}
$$

The new equation is $y^3 - y^2 - 6y - 8 = 0$.

Example: Write the equation whose roots are greater by 1 than the roots of the equation $x^4 - 3x^3 + 2x^2 - 5x + 7 = 0$.

▶ Diminishing the roots by -1 is equivalent to increasing the roots by 1.

$$
\begin{array}{rrrrrr}
1 & - 3 & + 2 & - 5 & + 7 & \underline{\mid -1} \\
 & - 1 & + 4 & - 6 & + 11 & \\
\hline
1 & - 4 & + 6 & - 11 & \underline{\mid + 18} \\
 & - 1 & + 5 & - 11 & \\
\hline
1 & - 5 & + 11 & \underline{\mid - 22} \\
 & - 1 & + 6 & \\
\hline
1 & - 6 & \underline{\mid + 17} \\
 & - 1 & \\
\hline
1 & \underline{\mid - 7} \\
\end{array}
$$

The new equation is $y^4 - 7y^3 + 17y^2 - 22y + 18 = 0$.

Example: Transform the equation $x^4 - 8x^3 + 14x^2 - 12x + 18 = 0$ into an equation in which the third degree term is missing.

▶ The sum of the roots of this equation is 8 and there are 4 roots. Thus, if each root is reduced by 2 the sum of the roots will become 0 and the third degree term will vanish.

$$
\begin{array}{rrrrrr}
1 & - 8 & + 14 & - 12 & + 18 & \underline{\mid 2} \\
 & + 2 & - 12 & + 4 & - 16 & \\
\hline
1 & - 6 & + 2 & - 8 & \underline{\mid + 2} \\
 & + 2 & - 8 & - 12 & \\
\hline
1 & - 4 & - 6 & \underline{\mid - 20} \\
 & + 2 & - 4 & \\
\hline
1 & - 2 & \underline{\mid - 10} \\
 & + 2 & \\
\hline
1 & \underline{\mid \ \ 0} \\
\end{array}
$$

The new equation is $y^4 - 10y^2 - 20y + 2 = 0$.

EXERCISE 16

In each case form equations whose roots are equal to those of the given equation diminished by the numbers in parentheses.

  1. $4x^3 - 20x^2 + 17x + 10 = 0$ (3)
  2. $2x^4 - 5x^3 + 4x^2 - 3x + 6 = 0$ (2)
  3. $x^4 - 30x^2 + 49x + 15 = 0$ (5)

    4. $3x^4 - 2x^3 + 7x^2 - 9 = 0$ (1)

    5. $4x^3 - 2x^2 - 6x - 7 = 0$ (1.5)

In each case form equations whose roots are equal to those of the given equation increased by the number in the parentheses.

    6. $2x^4 + 7x^3 + 8x^2 - 5x + 4 = 0$ (2)

    7. $3x^3 - 6x^2 + 5x + 8 = 0$ (1)

    8. $2x^4 + 8x^3 + 7x - 5 = 0$ (3)

In each case form equations as indicated:

    9. $x^4 - 4x^3 + 2x^2 - 3x - 7 = 0$

                                     (third degree term missing)

    10. $x^3 - 9x^2 + 7x - 5 = 0$

                               (second degree term missing)

    11. $x^3 + 6x^2 - 9x - 2 = 0$

                               (second degree term missing)

    12. $2x^4 + 8x^3 - 5x^2 + 4x - 8 = 0$

                                 (third degree term missing)

Horner's Method for Irrational Roots

Horner's method for finding irrational roots is illustrated by the following examples.

Example: Find the positive root of $x^3 + x^2 - 4x - 15 = 0$ correct to three decimal places.

▶ By Descartes' Rule of Signs this equation has 1 positive root.
To locate the positive root we note that $f(x)$ changes from negative to positive between $x = 2$ and $x = 3$.

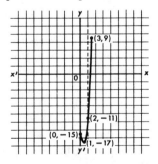

x	0	1	2	3
$f(x)$	-15	-17	-11	9

∴ The positive root is located between 2 and 3 as shown on the graph.

Now transform the equation into a new equation whose roots are less by 2 than the roots of the original equation.

$$1 + 1 - 4 - 15 \quad \underline{|\,2}$$
$$ + 2 + 6 + 4$$
$$\overline{1 + 3 + 2 \,\underline{|- 11}}$$
$$ + 2 + 10$$
$$\overline{1 + 5 \,\underline{|+ 12}}$$
$$ + 2$$
$$\overline{1 + 7}$$

The transformed equation is $y^3 + 7y^2 + 12y - 11 = 0$. Since the roots of the y-equation are each 2 less than the roots of the original equation, the root which was located between 2 and 3 in the original equation lies between 0 and 1 in the y-equation. The effect of the transformation is to move the original y-axis two units to the right as shown by the broken line on the graph.

In the y-equation we find, by trial, that $f(.6) = -1.064$ and $f(.7) = +1.173$. Thus, the y-equation has a root between .6 and .7 and the original equation has a root between 2.6 and 2.7.

Now transform the y-equation into a new equation whose roots are less by .6 than the roots of the y-equation.

$$1 + 7 + 12 - 11 \qquad \underline{|\,.6}$$
$$ + .6 + 4.56 + 9.936$$
$$\overline{1 + 7.6 + 16.56 \,\underline{|- 1.064}}$$
$$ + .6 + 4.92$$
$$\overline{1 + 8.2 \,\underline{|+ 21.48}}$$
$$ + .6$$
$$\overline{1 \,\underline{|+ 8.8}}$$

The transformed equation is $z^3 + 8.8z^2 + 21.48z - 1.064 = 0$. In order to estimate the root to the nearest hundredth solve the linear part of the z-equation.

$$21.48z - 1.064 = 0$$
$$z = .04$$
$$f(.04) = -.190656 \quad f(.05) = .032125$$

Thus, the z-equation has a root between .04 and .05 and the original equation has a root between 2.64 and 2.65. Now transform the z-equation into a new equation whose roots are less by .04 than the roots of the z-equation:

```
1 +  8.8  +  21.48   −  1.064        |.04
   +  .04  +   .3536  +  .873344
1 +  8.84 +  21.8336 |−  .190656
   +  .04  +   .3552
1 +  8.88 |+  22.1888
   +  .04
1 |+  8.92
```

The transformed equation is

$$w^3 + 8.2w^2 + 22.1888w - .190656 = 0$$

In order to estimate the root to the nearest thousandth solve the linear part of the w-equation.

$$22.1888w - .190656 = 0$$
$$w = .008$$
$$f(.008) = -.12574208 \quad f(.009) = +.009766449$$

Thus, the w-equation has a root between .008 and .009 and the original equation has a root between 2.648 and 2.649. Now transform the w-equation into a new equation whose roots are less by .008 than the roots of the w-equation.

```
1 +  8.92  +  22.1888     −  .190656        |.008
   +  .008  +    .071424  +  .178081792
1 +  8.928 +  22.260224 |−  .012574208
   +  .008  +    .071488
1 +  8.936 |+  22.331712
   +  .008
1 |+  8.944
```

The transformed equation is

$$v^3 + 8.944v^2 + 22.331712v - .012574208 = 0$$

In order to estimate the root to the nearest ten thousandth solve the linear part of the v-equation.

$$22.331712v - .012574208 = 0$$
$$v = .0005$$
$$f(.0005) = -.012406116875$$

∴ The root of the original equation is 2.6485+ or $x = 2.649$, correct to 3 decimal places, since the negative value of $f(.0005)$ indicates that the graph cuts the x-axis above 2.6485.

Example: Find the negative root of the equation $x^3 - 5x^2 - x + 7 = 0$ to three decimal places.

▶ $$f(x) = x^3 - 5x^2 - x + 7$$

Then $f(-x) = -x^3 - 5x^2 + x + 7$, or $x^3 + 5x^2 - x - 7 = 0$. The positive roots of $f(-x) = 0$ are the negative roots of $f(x) = 0$. By Descartes' Rule of Signs, the equation $x^3 + 5x^2 - x - 7 = 0$ has one variation, and one positive root. This positive root is the required negative root of the original equation. To locate the positive root we note that $g(x) = x^3 + 5x - x - 7$ changes from negative to positive between $x = 1$ and $x = 2$.

x	0	1	2
$g(x)$	-7	-2	$+19$

Transform the equation into a new equation whose roots are less by 1 than the roots of the original equation.

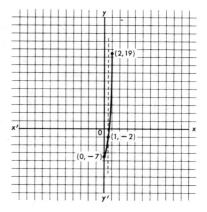

```
1 + 5 -   1 - 7 |1
    + 1 +   6 + 5
  1 + 6 +   5 |- 2
    + 1 +   7
  1 + 7 |+ 12
    + 1
  1 |+ 8
```

The transformed equation is $y^3 + 8y^2 + 12y - 2 = 0$. The root which was between 1 and 2 in $g(x) = 0$ is between 0 and 1 in the y-equation.

In the y-equation we find, by trial, that $f(.1) = -.819$ and $f(.2) = +.728$. Thus, the y-equation has a root between .1 and .2 and the original equation has a root between 1.1 and 1.2. Now transform the y-equation into a new equation whose roots are less by .1 than the roots of the y-equation.

$$\begin{array}{rrrrr}
1 & + \ 8 & + \ 12 & - \ 2 & \quad \underline{\lfloor .1}\\
& + \ .1 & + \ .81 & + \ 1.281 &\\
\hline
1 & + \ 8.1 & + \ 12.81 & \underline{\lfloor - \quad .819} &\\
& + \ .1 & + \ .82 &&\\
\hline
1 & + \ 8.2 & \underline{\lfloor + \ 13.63} &&\\
& + \ .1 &&&\\
\hline
1 & \underline{\lfloor + \ 8.3} &&&
\end{array}$$

The transformed equation is $z^3 + 8.3z^2 + 13.63z - .819 = 0$. In order to estimate the root to the nearest hundredth solve the linear part of the equation:

$$13.63z - .819 = 0$$
$$z = .06$$
$$f(.05) = -.116725 \quad f(.06) = +.028896$$

Thus the z-equation has a root between .05 and .06 and the original equation has a root between 1.15 and 1.16. Now, transform the z-equation into a new equation whose roots are less by .05 than the roots of the z-equation.

$$\begin{array}{rrrrr}
1 & + \ 8.3 & + \ 13.63 & - \ .819 & \quad \underline{\lfloor .05}\\
& + \ .05 & + \ .4175 & + \ .702375 &\\
\hline
1 & + \ 8.35 & + \ 14.0475 & \underline{\lfloor - .116125} &\\
& + \ .05 & + \ .4200 &&\\
\hline
1 & + \ 8.40 & \underline{\lfloor + \ 14.4675} &&\\
& + \ .05 &&&\\
\hline
1 & \underline{\lfloor + \ 8.45} &&&
\end{array}$$

The transformed equation is
$$w^3 + 8.45w^2 + 14.4675w - .116725 = 0.$$
In order to estimate the root to the nearest thousandth solve the linear part of the w-equation:

$$14.4675w - .116725 = 0$$
$$w = .008$$
$$f(.008) = -.000443688 \quad f(.009) = .004167679$$

Thus, the w-equation has a root between .008 and .009 and the original equation has a root between 1.158 and 1.159. Now, transform the w-equation into a new equation whose roots are less by .008 than the roots of the w-equation.

$$1 \; + \; 8.45 \;\; + \;\; 14.4657 \;\;\; - \; .116725 \qquad \underline{|\,.008}$$
$$\;\;\; + \;\; .008 \; + \;\;\;\; .067664 \; + \; .116266912$$
$$\overline{1 \; + \; 8.458 \; + \; 14.533364 \; \underline{|- \; .000458088}}$$
$$\;\;\; + \;\; .008 \; + \;\;\;\; .067728$$
$$\overline{1 \; + \; 8.466 \; \underline{|+ \; 14.601092}}$$
$$\;\;\; + \;\; .008$$
$$\overline{1 \; \underline{|+ \; 8.474}}$$

The transformed equation is

$$v^3 + 8.474v^2 + 14.601092v - .000458088 = 0.$$

In order to estimate the root to the nearest ten thousandth solve the linear part of the v-equation.

$$14.601092v - .000458088 = 0.$$
$$v = .00003$$

The root of $g(x) = 0$ is 1.1580 or 1.158, correct to 3 decimal places.

∴ The root of $f(x) = 0$ is -1.158.

Example: Find $\sqrt[3]{19}$ correct to the nearest hundreth.

▶ Let $x = \sqrt[3]{19}$

Then $x^3 = 19$ and $f(x) = x^3 - 19 = 0$. Descartes' Rule of Signs indicates that $f(x)$ has one positive root and no negative roots.

$$f(0) = -19, f(1) = -18, f(2) = -11, f(3) = 8$$

∴ $f(x)$ has a root between 2 and 3.

Decrease the roots of $f(x) = 0$ by 2.

$$1 \; + \; 0 \; + \;\; 0 \; - \; 19 \qquad \underline{|\,2}$$
$$\;\;\; + \; 2 \; + \;\; 4 \; + \;\; 8$$
$$\overline{1 \; + \; 2 \; + \;\; 4 \; \underline{|- \; 11}}$$
$$\;\;\; + \; 2 \; + \;\; 8$$
$$\overline{1 \; + \; 4 \; \underline{|+ \; 12}}$$
$$\;\;\; + \; 2$$
$$\overline{1 \; \underline{|+ \; 6}}$$

The transformed equation is $y^3 + 6y^2 + 12y - 11 = 0$. In the y-equation $f(.6) = -1.424$ and $f(.7) = +.683$. Thus, the y-equation has a root between .6 and .7 and the original equation has a root between 2.6 and 2.7. Now, transform the y-equation into a new equation whose roots are less by .6 than the roots of the y-equation.

$$
\begin{array}{rrrrl}
1 + 6 & + 12 & - 11 & & \underline{\;.6\;} \\
+ .6 & + 3.96 & + 9.576 & & \\
\hline
1 + 6.6 & + 15.96 & \underline{\;- 1.424} & & \\
+ .6 & + 4.32 & & & \\
\hline
1 + 7.2 & \underline{\;+ 20.28} & & & \\
+ .6 & & & & \\
\hline
1 \underline{\;+ 7.8} & & & &
\end{array}
$$

The transformed equation is $z^3 + 7.8z^2 + 20.28z - 1.424 = 0$. In order to estimate the root to the nearest hundreth solve the linear part of the z-equation.

$$20.28z - 1.424 = 0$$
$$z = .07$$
$$f(.06) = -.178904 \quad f(.07) = +.38123$$

Thus, the z-equation has a root between .06 and .07 and the original equation has a root between 2.66 and 2.67. Now, transform the z-equation into a new equation whose roots are less by .06 than the roots of the z-equation.

$$
\begin{array}{rrrrl}
1 + 7.8 & + 20.28 & - 1.424 & & \underline{\;.06\;} \\
+ .06 & + .4716 & + 1.245096 & & \\
\hline
1 + 7.86 & + 20.7516 & \underline{\;- .178904} & & \\
+ .06 & + .4752 & & & \\
\hline
1 + 7.92 & \underline{\;+ 21.2268} & & & \\
+ .06 & & & & \\
\hline
1 + \underline{\;7.98\;} & & & &
\end{array}
$$

The transformed equation is
$$w^3 + 7.98w^2 + 21.2268w - .178904 = 0.$$
In order to obtain the root correct to the nearest hundreth we need only determine whether the digit in the thousandth place is 5 or greater, or less than 5. $f(.005) = -.072570357$. The root of the original equation is 2.665, or 2.67, correct to the nearest hundreth.
$$\therefore \quad \sqrt[3]{19} = 2.67.$$

EXERCISE 17

Find by Horner's method correct to three decimal places the real roots of each of the following equations:

1. $x^3 + x^2 - 4x - 2 = 0$ (the positive root only)
2. $x^3 + x^2 - 2x - 11 = 0$ (the positive root only)
3. $x^3 - 6x - 12 = 0$ (the positive root only)
4. $x^3 - 3x^2 + 5x - 9 = 0$ (positive root)
5. $x^3 - 2x + 6 = 0$ (negative root)
6. $x^3 - x^2 + 7 = 0$ (negative root)

By Horner's method extract each of the following roots correct to two decimal places:

7. $\sqrt{7}$
8. $\sqrt[3]{39}$
9. $\sqrt[4]{51}$

ANSWERS

EXERCISE 1

1. a. -4
   b. -46
2. $4 - 3a$
3. $a^2 - 5a + 4$
4. a. $-4x^3 + 5x - 7$
   b. -6
   c. 15

EXERCISE 2

1. 40
2. -12
3. -32
4. 1
5. -9

EXERCISE 3

5. -4 6. 3

EXERCISE 4

1. a. quotient $3x^2 + 4x + 1$, remainder -3.
   b. quotient $2x^3 - x^2 + 2x - 2$, remainder $+4$.
   c. quotient $2x^2 + 4x + 6$, remainder -5.
   d. quotient $6x^3 + 3x^2 - 3x + 9$, remainder -2.
   e. quotient $x^4 + 3x^3 + 7x^2 + 21x + 63$, remainder $+182$.
2. a. 61 4. a. Yes
   b. -74 b. Yes
3. a. 15 c. No
   b. 68 d. Yes

EXERCISE 5

1. 3, 4

2. $\frac{1}{2}$, 2

3. $-1 \pm \sqrt{2}\, i$

4. $\pm \frac{1}{3}\sqrt{15}$

5. 3, $\frac{1}{2}$

6. $\dfrac{-3 \pm \sqrt{5}}{2}$

7. $\dfrac{3 \pm \sqrt{11}\, i}{2}$

8. $\frac{1}{4}$, $\frac{1}{2}$

EXERCISE 6

1. a. $x^3 - 2x^2 - 5x + 6 = 0$
  b. $x^3 - x^2 - 22x + 40 = 0$
  c. $x^3 + 7x^2 + 15x + 9 = 0$
  d. $x^4 + x^3 - 22x^2 - 40x = 0$

2. a. $x^4 - 6x^3 + 13x^2 - 14x + 6 = 0$
  b. $x^4 - 3x^3 - 12x - 16 = 0$
  c. $x^5 - 6x^4 + 28x^2 - 9x - 30 = 0$
  d. $x^5 + 2x^4 - 4x^3 - 4x^2 - 5x - 6 = 0$
  e. $12x^4 - 47x^3 + 55x^2 + 9x - 5 = 0$
  f. $24x^4 + 14x^3 - 51x^2 - 28x + 6 = 0$

3. a. $x^4 - 18x^3 + 120x^2 - 354x + 391 = 0$
  b. $x^4 - 6x^3 + 5x^2 - 6x + 4 = 0$
  c. $x^4 - 12x^3 + 58x^2 - 156x + 169 = 0$
  d. $x^4 - 12x^3 + 38x^2 + 52x - 319 = 0$

4. a. $2x^3 - 3x^2 - 23x + 12 = 0$
  b. $6x^3 - 31x^2 + 3x + 10 = 0$
  c. $48x^4 - 100x^3 + 60x^2 - 5x - 3 = 0$
  d. $72x^4 - 162x^3 + 7x^2 + 53x + 10 = 0$

EXERCISE 7

1. a. 5
  b. 7
  c. 9

2. 2

3. -15

4. 24

5. 1, $\frac{1}{2}$, $\frac{3}{2}$

6. 2, -2

7. 9

8. a. $\frac{2}{3}$
  b. $\frac{7}{3}$

9. a. -2
  b. 5
  c. 12
  d. -6
  e. $\frac{5}{12}$
  f. $-\frac{1}{6}$

10. $\pm\frac{3}{2}, -4$

11. a. -14

    b. -15

    c. -7

12. a. $x^3 - 7x^2 + 7x + 15 = 0$

    b. $x^4 + 4x^3 - 7x^2 - 22x + 24 = 0$

    c. $4x^3 - 12x^2 - 19x + 12 = 0$

EXERCISE 8

1. $y^3 - 6y^2 - 9y + 27 = 0$

2. $2y^3 + y^2 - y + 1 = 0$

3. $y^4 - 4y^3 - 4y^2 + 64 = 0$

4. $3y^3 - 20y^2 + 300y - 5{,}000 = 0$

5. $3y^3 - 6y^2 - y + 2 = 0$

6. $6y^5 + 2y^3 - y - 1 = 0$

7. $2y^3 - 7y^2 + 3y + 5 = 0$

8. $3y^4 + 2y^3 - 7y^2 - 2y + 4 = 0$

EXERCISE 9

1. $2, -2, -3$

2. $4, -3, -5$

3. $2, \dfrac{5 \pm \sqrt{33}}{2}$

4. $-2, 2 \pm \sqrt{2}i$

5. $2, 1 \pm \sqrt{3}$

6. $-1, 2, -1 \pm \sqrt{2}i$

7. $1, 1, -2, -2$

8. $-1, 2, 2 \pm i$

9. $-2, 3, \dfrac{1 \pm \sqrt{7}i}{2}$

10. $2, 2, 1 \pm 2i$

EXERCISE 10

1. $-\dfrac{3}{2}, 4, -2$

2. $\dfrac{1}{2}, 1 \pm \sqrt{2}i$

3. $1, -\dfrac{5}{3}, -\dfrac{3}{2}$

4. $\dfrac{1}{3}, -2, \dfrac{-1 \pm \sqrt{3}i}{2}$

5. $\dfrac{1}{2}, -2, \dfrac{-1 \pm \sqrt{3}i}{2}$

6. $\dfrac{1}{2}, \dfrac{2}{3}, -\dfrac{3}{4}$

7. $\dfrac{1}{2}, \dfrac{1}{2}, \dfrac{-1 \pm \sqrt{11}i}{2}$

8. $\dfrac{1}{2}, -3, -1 \pm \sqrt{3}i$

9. $-1, \dfrac{2}{3}, \dfrac{-1 \pm \sqrt{3}i}{2}$

10. $-\dfrac{1}{2}, 3, \dfrac{-1 \pm \sqrt{3}i}{2}$

11. $\frac{1}{2}, \frac{1}{3}, \pm i$

12. $3, 3, -1, \frac{1}{3}$

13. $-\frac{1}{4}, -\frac{3}{2}, 1 \pm \sqrt{3}$

14. $-\frac{2}{3}, \frac{2}{3}, 1 \pm \sqrt{2}$

15. $1, 1, \pm\frac{3}{2}$

EXERCISE 11

1. $2, -2$
2. $3, -2$
3. $3, -1$
4. $5, -2$
5. $4, -2$
6. $3, -2$

EXERCISE 12

1.

Pos.	Neg.	Complex
1	2	0
1	0	2

2.

Pos.	Neg.	Complex
0	1	4

3. 6 complex

4.

Pos.	Neg.	Complex
2	1	0
0	1	2

5.

Pos.	Neg.	Complex
1	1	4

6.

Pos.	Neg.	Complex
4	0	0
2	0	2
0	0	4

7.

Pos.	Neg.	Complex
3	2	0
1	2	2
1	0	4
3	0	2

8.

Pos.	Neg.	Complex
2	0	2
0	0	4

9.

Pos.	Neg.	Complex
4	0	2
2	0	4
0	0	6

10.

Pos.	Neg.	Complex
1	0	4
1	2	2

EXERCISE 13

1. 1.7

2. 2.2

3. $- .7, -1.8, 2.5$

4. 2.3

5. 4.5

6. $1.6, -.3, -4.3$

EXERCISE 14

1. Between 3 and 4

2. Between 1 and 2

3. Between 0 and 1, between 0 and -1, between -4 and -5

4. Between 0 and -1, between 2 and 3, between 3 and 4

5. Between 1 and 2, between -1 and -2

6. Between -1 and 0, between -1 and -2, between 3 and 4

EXERCISE 15

1. 1.32

2. 2.10

3. 3.25

4. 2.37

EXERCISE 16

1. $4y^3 + 16y^2 + 5y - 11 = 0$
2. $2y^4 + 11y^3 + 22y^2 + 17y + 8 = 0$
3. $y^4 + 20y^3 + 120y^2 + 249y + 135 = 0$
4. $3y^4 + 10y^3 + 19y^2 + 20y - 1 = 0$
5. $4y^3 + 16y^2 + 15y - 7 = 0$
6. $2y^4 - 9y^3 + 14y^2 - 17y + 22 = 0$
7. $3y^3 - 15y^2 + 26y - 6 = 0$
8. $2y^4 - 16y^3 + 36y^2 + 7y - 80 = 0$
9. $y^4 - 4y^2 - 7y - 11 = 0$
10. $y^3 - 20y - 38 = 0$
11. $y^3 - 21y + 32 = 0$
12. $2y^4 - 17y^2 + 30y - 23 = 0$

EXERCISE 17

1. 1.814
2. 2.195
3. 3.131
4. 2.456
5. -2.180
6. -1.631
7. 2.65
8. 3.39
9. 2.67

CHAPTER 16

Logarithms

Introduction

The development of science, navigation, and astronomy necessitated complicated and arduous calculations. The discovery of logarithms, based upon the theory of exponents, simplified these calculations immeasurably. In the following sections, the theory of logarithms is developed. In order to clarify the theory, its basis in the theory of exponents is presented.

Using Exponents in Computation

Table of Powers of 2

$2^1 = 2$	$2^5 = 32$	$2^9 = 512$	$2^{13} = 8{,}192$
$2^2 = 4$	$2^6 = 64$	$2^{10} = 1{,}024$	$2^{14} = 16{,}384$
$2^3 = 8$	$2^7 = 128$	$2^{11} = 2{,}048$	$2^{15} = 32{,}768$
$2^4 = 16$	$2^8 = 256$	$2^{12} = 4{,}096$	$2^{16} = 65{,}536$

The following computations may be performed mentally by using the above table.

Multiplication

Problem: Multiply 64 by 256

▶ $64 = 2^6, 256 = 2^8$
$64 \times 256 = 2^6 \times 2^8 = 2^{14}$
$2^{14} = 16,384$

Note that the only computation performed was the addition of the exponents 6 and 8.

Division

Problem: Divide 32,768 by 512
▶ $32,768 = 2^{15}, 512 = 2^9$
$32,768 \div 512 = 2^{15} \div 2^9 = 2^6$
$2^6 = 64$

Raising to a Power

Problem: Find the value of $(16)^3$
▶ $16 = 2^4$
$(16)^3 = (2^4)^3 = 2^{12}$
$2^{12} = 4,096$

Finding a Root

Problem: Find the value of $\sqrt[3]{32,768}$

$\sqrt[3]{32,768} = (32,768)^{1/3}$
$32,768 = 2^{15}$
$(32,768)^{1/3} = 2^5$
$2^5 = 32$

In the above computations each number was written as a power of 2 before the indicated operations were performed. By using exponents:

a. *Multiplication* was performed by *adding* exponents.
b. *Division* was performed by *subtracting* exponents.
c. *Raising to a power* was performed by *multiplying* exponents.
d. *Finding a root* was performed by *dividing* exponents.

These principles form the foundation for the study of logarithms.

EXERCISE 1

Use the table of powers of 2 to perform the indicated operations. These exercises can be worked out mentally.

1. $32 \times 1,024$ **5.** $(32)^3$
2. 128×512 **6.** $(16)^4$
3. $8,192 \div 256$ **7.** $\sqrt[4]{65,536}$
4. $16,384 \div 128$ **8.** $\sqrt[3]{4,096}$

It is clear that only certain numbers may be written as integral powers of the base 2. For example, the number 12 lies somewhere between 2^3 and 2^4.

In actual practice, the base that is most convenient for computation by means of exponents is the number 10. The reason for this will appear later. In the following sections the theory and practice of computation by the use of exponents will be developed.

DEFINITION: The logarithm of a number N is the exponent a indicating the power to which the base b must be raised to yield the number N. The base a must be a positive number.

Example: Write $2^5 = 32$ in logarithmic form.

▶   The number $N = 32$
    The base $b = 2$
    The exponent $a = 5$
    $\therefore \quad \log_2 32 = 5$

Example: Write $3^4 = 81$ in logarithmic form.

▶   The number $N = 81$
    The base $b = 3$
    The exponent $a = 4$
    $\therefore \quad \log_3 81 = 4$

Example: Write $5^{-3} = \dfrac{1}{125}$ in logarithmic form

▶   The number $N = \dfrac{1}{125}$

    The base $b = 5$
    The exponent $a = -3$
    $\therefore \quad \log_5 \dfrac{1}{125} = -3$

Example: Write $8^{-2/3} = \frac{1}{4}$ in logarithmic form.

▶ The number $N = \frac{1}{4}$
The base $b = 8$
The exponent $a = -\frac{2}{3}$
∴ $\log_8 \frac{1}{4} = -\frac{2}{3}$

Example: Write $\log_6 216 = 3$ in exponential form.

▶ The number $N = 216$
The base $b = 6$
The exponent $a = 3$
∴ $216 = 6^3$ or $6^3 = 216$

Example: Write $\log_2 \frac{1}{8} = -3$ in exponential form.

▶ The number $N = \frac{1}{8}$
The base $b = 2$
The exponent $a = -3$
∴ $\frac{1}{8} = 2^{-3}$ or $2^{-3} = \frac{1}{8}$

Example: Write $\log_9 \frac{1}{27} = -\frac{3}{2}$ in exponential form.

▶ The number $N = \frac{1}{27}$
The base $b = 9$
The exponent $a = -\frac{3}{2}$
∴ $\frac{1}{27} = 9^{-3/2}$ or $9^{-3/2} = \frac{1}{27}$

EXERCISE 2

Write each of the following in logarithmic form:

1. $2^7 = 128$
2. $3^5 = 243$
3. $5^2 = 25$
4. $4^{-2} = \frac{1}{16}$
5. $7^0 = 1$
6. $10^2 = 100$
7. $2^{-5} = \frac{1}{32}$
8. $9^{3/2} = 27$
9. $8^{-4/3} = \frac{1}{16}$

Write each of the following in exponential form:

10. $\text{Log}_2 16 = 4$
11. $\text{Log}_7 49 = 2$
12. $\text{Log}_4 64 = 3$
13. $\text{Log}_{10} .1 = -1$
14. $\text{Log}_5 125 = 3$
15. $\text{Log}_4 2 = \frac{1}{2}$
16. $\text{Log}_8 16 = \frac{4}{3}$
17. $\text{Log}_9 \frac{1}{3} = -\frac{1}{2}$
18. $\text{Log}_{\sqrt{3}} 3 = 2$

Example: Find the value of $\log_2 64$.

▶   Let $\log_2 64 = x$

Then $2^x = 64$

$2^x = 2^6$

∴ $x = 6$

Example: Find the value of $\log_{81} 27$.

▶   Let $\log_{81} 27 = x$

Then $81^x = 27$

Since $81 = 3^4$, $(3^4)^x = 27$

$3^{4x} = 3^3$

$4x = 3$ or $x = \frac{3}{4}$

Example: Find the value of $\log_4 \frac{1}{2}$.

▶   Let $\log_4 \frac{1}{2} = x$

Then $4^x = \frac{1}{2}$

$(2^2)^x = \frac{1}{2}$, $2^{2x} = \frac{1}{2}$

$2^{2x} = 2^{-1}$

$2x = -1$ or $x = -\frac{1}{2}$

Example: Find the value of x if $\log_4 x = \frac{5}{2}$.

▶ $\text{Log}_4 x = \frac{5}{2}$

$x = 4^{5/2}$

$x = 32$

Example: Find the value of x if $\log_{\sqrt{5}} 25 = x$.

▶ $\text{Log}_{\sqrt{5}} 25 = x$

$(\sqrt{5})^x = 25$ or $(5^{1/2})^x = 25$

$5^{x/2} = 5^2$

∴ $\frac{x}{2} = 2$ or $x = 4$

EXERCISE 3

Find the value of each of the following:

1. $\text{Log}_2 8$	5. $\text{Log}_6 1$	9. $\text{Log}_{16} 8$
2. $\text{Log}_3 27$	6. $\text{Log}_8 32$	10. $\text{Log}_2 \frac{1}{8}$
3. $\text{Log}_{10} 100$	7. $\text{Log}_{10} .01$	11. $\text{Log}_{25} \frac{1}{5}$
4. $\text{Log}_4 256$	8. $\text{Log}_9 3$	12. $\text{Log}_{16} \frac{1}{8}$

Find the value of x in each of the following:

13. $\text{Log}_3 x = 2$	**17.** $\text{Log}_4 64 = x$	**21.** $\text{Log}_3 \sqrt{27} = x$
14. $\text{Log}_7 x = 1$	**18.** $\text{Log}_{12} 1 = x$	**22.** $\text{Log}_{1/2} x = 3$
15. $\text{Log}_x 32 = 5$	**19.** $\text{Log}_3 x = \frac{1}{2}$	**23.** $\text{Log}_x 36 = 2$
16. $\text{Log}_x 81 = 4$	**20.** $\text{Log}_x 4 = \frac{1}{3}$	**24.** $\text{Log}_9 27 = x$

Properties of Logarithms

Because a logarithm is an exponent, the properties of logarithms derived below are based upon the laws of exponents. It will be recalled that:
$$a^m \cdot a^n = a^{m+n}, \ a^m \div a^n = a^{m-n}, \ (a^m)^n = a^{mn}$$

Theorem 1: The logarithm of the product of two numbers is equal to the sum of the logarithm of the numbers, or
$$\text{Log}_a MN = \log_a M + \log_a N$$

Proof: Let $x = \log_a M$, and $y = \log_a N$
Then $M = a^x$ and $N = a^y$
$$MN = a^{x+y} \quad \text{or} \quad a^{x+y} = MN$$
If this is written as a logarithmic equation we have
$$\log_a MN = x + y \quad \text{or} \quad \log_a MN = \log_a M + \log_a N$$

This theorem may be extended to include the product of three or more numbers, i.e. $\log_a MNP = \log_a M + \log_a N + \log_a P$.

Application: $\text{Log}_a (78)(43) = \log_a 78 + \log_a 43$
$\text{Log}_a 21 = \log_a (7)(3) = \log_a 7 + \log_a 3$
$\text{Log}_a 125 = \log_a (5)(5)(5) = \log_a (5) + \log_a (5)$
$\qquad + \log_a 5 = 3 \log_a 5$

Theorem 2: The logarithm of the quotient of two numbers is equal to the logarithm of the dividend minus the logarithm of the divisor, or:
$$\text{Log} \, a \, \frac{M}{N} = \log_a M - \log_a N$$

Proof: Let $x = \log_a M$, and $y = \log_a N$
Then $M = a^x$ and $N = a^y$
$$\frac{M}{N} = \frac{a^x}{a^y} = a^{x-y} \quad \text{or} \quad a^{x-y} = \frac{M}{N}$$

If this is written as a logarithmic equation we have

$$\log a \frac{M}{N} = x - y \quad \text{or} \quad \log a \frac{M}{N} = \log_a M - \log_a N$$

Application: $\quad \log a \dfrac{57}{29} = \log_a 57 - \log_a 29$

$$\log a \frac{37}{(46)(23)} = \log_a 37 - (\log_a 46 + \log_a 23)$$

Theorem 3: The logarithm of the kth power of a number is equal to the logarithm of the number multiplied by k, or

$$\log_a M^k = k \log_a M$$

Proof: $\quad$ Let $x = \log_a M$

$\quad\quad$ Then $M = a^x$

Raising both sides of this equation to the kth power, we obtain

$$M^k = (a^x)^k = a^{kx}$$

If this is written in logarithmic form, we have

$$\log_a M^k = kx \quad \text{or} \quad \log_a M^k = k \log_a M$$

Application: $\quad \log_a 5^6 = 6 \log_a 5$

The exponent k may be a fraction. In this case, we have $\log_a 3^{1/4} = \frac{1}{4} \log_a 3$. This enables us to use the theorem to express the logarithm of the root of a number.

Example: Express $\log_a 35$ in terms of logarithms of prime numbers.

▶ $\quad\quad \log_a 35 = \log_a 7 \times 5 = \log_a 7 + \log_a 5$

Example: Express $\log_a 24$ in terms of logarithms of prime numbers.

▶ $\quad \log_a 24 = \log_a (3 \times 2 \times 2 \times 2) = \log_a 3 + \log_a 2 + \log_a 2 + \log_a 2 = \log_a 3 + 3 \log_a 2.$

Example: Express $\log_a \sqrt{98}$ in terms of logarithms of prime numbers.

▶ $\quad \log_a \sqrt{98} = \log_a 98^{1/2} = \frac{1}{2} \log_a 98 = \frac{1}{2}(\log_a 2 \cdot 7 \cdot 7)$

$$= \frac{1}{2}(\log_a 2 + \log_a 7 + \log_a 7) = \frac{1}{2}(\log_a 2 + 2 \log_a 7)$$

Example: Express $\log_a \dfrac{\sqrt[3]{14}}{\sqrt[4]{15}}$ in terms of logarithms of prime numbers.

▶ $\log_a \dfrac{\sqrt[3]{14}}{\sqrt[4]{15}} = \log_a \sqrt[3]{14} - \log_a \sqrt[4]{15}$

$= \log_a (14)^{1/3} - \log_a (15)^{1/4} = \log_a (7 \cdot 2)^{1/3} - \log_a (3 \cdot 5)^{1/4}$

$= \tfrac{1}{3}(\log_a 7 + \log_a 2) - \tfrac{1}{4}(\log_a 3 + \log_a 5)$

EXERCISE 4

Express the logarithms of the following in terms of the logarithms of prime numbers:

1. $\log_a 14$

2. $\log_a 33$

3. $\log_a 20$

4. $\log_a 54$

5. $\log_a \dfrac{7}{19}$

6. $\log_a \dfrac{8}{9}$

7. $\log_a \dfrac{64}{25}$

8. $\log_a \sqrt{45}$

9. $\log_a \sqrt[3]{250}$

10. $\log_a \dfrac{\sqrt[3]{17}}{\sqrt[4]{23}}$

11. $\log_a \dfrac{\sqrt[5]{19}}{\sqrt[3]{31}}$

12. $\log_a \dfrac{(29)^2 \cdot 8}{\sqrt[4]{43}}$

In the following examples it is given that $\log_{10} 2 = .3010$, $\log_{10} 3 = .4771$, and $\log_{10} 7 = .8451$.

Example: Find the value of $\log_{10} 6$

▶ $\log_{10} 6 = \log_{10}(3)(2) = \log_{10} 3 + \log_{10} 2$

∴ $\log_{10} 6 = .4771 + .3010 = .7781$

Example: Find the value of $\log_{10} 21$

▶ $\log_{10} 21 = \log_{10}(7)(3) = \log_{10} 7 + \log_{10} 3$

∴ $\log_{10} 21 = .8451 + .4771 = 1.3222$

Example: Find the value of $\log_{10} 5$

▶ $\log_{10}5 = \log_{10}\dfrac{10}{2} = \log_{10}10 - \log_{10}2$

   ∴ $\log_{10}5 = 1 - .3010 = .6990$

Example: Find the value of $\log_{10}\sqrt[3]{18}$

▶ $\log_{10}\sqrt[3]{18} = \log_{10}(18)^{1/3} = \tfrac{1}{3}\log_{10}18$
   $= \tfrac{1}{3}\log_{10}(3)(3)(2) = \tfrac{1}{3}(\log_{10}3 + \log_{10}3 + \log_{10}2)$
   $= \tfrac{1}{3}(2\log_{10}3 + \log_{10}2) = \tfrac{1}{3}[2(.4771) + .3010]$
   $= \tfrac{1}{3}[1.2552] = .4184$

Example: Find the value of $\log_{10}\tfrac{9}{16}$

▶ $\log_{10}\tfrac{9}{16} = \log_{10}9 - \log_{10}16 = \log_{10}3^2 - \log_{10}2^4$
   $= 2\log_{10}3 - 4\log_{10}2 = 2(.4771) - 4(.3010)$
   $= .9542 - 1.2040$
   $= -.2498$

EXERCISE 5

It is given that $\log_{10}2 = .3010$, $\log_{10}3 = .4771$, $\log_{10}7 = .8451$. Find the values of the following:

1. $\log_{10}14$
2. $\log_{10}42$
3. $\log_{10}12$
4. $\log_{10}35$
5. $\log_{10}36$

6. $\log_{10}\sqrt[3]{21}$
7. $\log_{10}\sqrt{50}$
8. $\log_{10}\tfrac{7}{27}$
9. $\log_{10}90$
10. $\log_{10}\sqrt[3]{\tfrac{49}{4}}$

Logarithmic Equations

An equation which involves logarithmic functions is called a logarithmic equation. In solving logarithmic equations, we usually try to combine the given logarithmic expressions into a single logarithm by using the properties described above. The solution then readily follows.

Example: Solve the equation $\log_{10}(2x+3) = 1$

▶ If we write $\log_{10}(2x+3) = 1$ in exponential form, we obtain the equation $10^1 = 2x + 3$, since $2x + 3 = 10$, $2x = 7$ or $x = \tfrac{7}{2}$

Example: Solve the equation $\log_{10}x + \log_{10}2 - \log_{10}7 = 1$

▶ $\log_{10}x + \log_{10}2 - \log_{10}7 = 1$

Combining the terms of the left member of the equation $\log_{10}\dfrac{2x}{7} = 1$

∴ $10^1 = \dfrac{2x}{7}$, $2x = 70$, or $x = 35$

Example: Solve the equation $\log_{10}(x+1) - \log_{10}(x-3) = 2$

▶ $\log_{10}(x+1) - \log_{10}(x-3) = 2$

Combining the terms of the left member of the equation

$\log_{10}\dfrac{x+1}{x-3} = 2.$

∴ $10^2 = \dfrac{x+1}{x-3}$ or $100 = \dfrac{x+1}{x-3}$

$100x - 300 = x + 1$, $99x = 301$, or $x = \frac{301}{99}$

EXERCISE 6

Solve the following logarithmic equations:
1. $\log_{10}(3x + 5) = 1$
2. $\log_2(2x - 7) = 4$
3. $\log_{10}x + \log_{10}5 - \log_{10}3 = 2$
4. $\log_{10}(2x + 1) - \log_{10}(3x - 4) = 1$
5. $\log_2(3x + 7) - \log_2(2x - 5) = 5$
6. $\log_2 3 - \log_2 x + \log_2 5 = 3$

Finding the Logarithm of a Number

Any positive number except 1 can be used as the base of a system of logarithms. For computational purposes, it is convenient to use 10 as the base. Logarithms to the base 10 are called *common* logarithms.

When the base is omitted in writing a logarithm, the base 10 is understood. For example, log 79 means $\log_{10}79$. Since

$10^0 = 1$, $\log 1 = 0$

$10^1 = 10$, $\log 10 = 1$

$10^2 = 100$, $\log 100 = 2$

$10^3 = 1,000, \log 1,000 = 3$

$10^{-1} = .1, \log .1 = -1$

$10^{-2} = .01, \log .01 = -2$

Obviously, the logarithms of numbers which are integral powers of 10 may be written at sight. However, let us consider log 486. is clear that log 486 lies somewhere between 10^2 and 10^3, or $\log 486 = 10^{2 + a\,decimal}$. The whole number part of the logarithm is called the *characteristic*. The decimal part of the logarithm is called the *mantissa*. The characteristic for log 486 was determined at sight. The mantissa for log 486 is found by reference to the table on page 682. In using the table, we note that the first two digits of the number are located in the vertical column marked N and the third digit of the number is located in the horizontal row along the top of the table. Thus, the *mantissa* for log 486 is .6866 and log 486 = 2.6866.

The characteristic of the logarithm of a number can be determined at sight but the following rules will be helpful.

1. If a number is greater than 1, the characteristic is positive and is one less than the number of digits to the left of the decimal point.

Examples:

A. log 5843. has the characteristic 3, since the number 5843 has 4 digits to the left of the decimal point.

B. log 28.7 has the characteristic 1 since the number 28.7 has 2 digits to the left of the decimal point.

C. log 4.305 has the characteristic 0 since the number 4.305 has one digit to the left of the decimal point.

2. If a number is less than 1, the characteristic is negative and is numerically one more than the number of zeros between the decimal point and the first digit.

Examples:

A. log .037 has the characteristic -2 since the number .037 has one zero between the decimal point and the first digit.

Note that .037 lies between .01 and .1. Therefore, the logarithm of .037 lies between -2 and -1. A mantissa is always a positive quantity. Thus, the logarithm of .037 is written as $8.5682 - 10$

which represents $-2 + .5682$. The form $8.5682 - 10$ is used because it is more convenient in computation.

B. log .000786 has the characteristic -4 (usually written 6. $- 10$) since the number .000786 has three zeros between the decimal point and the first digit.

C. log .00503 has the characteristic -3 (usually written 7. $- 10$) since the number .00503 has two zeros between the decimal point and the first digit.

EXERCISE 7

State the characteristic of the following numbers:

1. 258	**5.** .046	**9.** .00005
2. 7.49	**6.** 69,000	**10.** 6.05
3. .843	**7.** .00345	**11.** 70.02
4. 23	**8.** 5,841.7	**12.** .0102

The logarithms of the numbers .00258, .0258, .258, 2.58, 25.8, 258, 2,580, 25,800. etc., have the same mantissas. They differ only in their characteristics. For example:

log .00258 $= 7.4116 - 10$
log .0258 $= 8.4116 - 10$
log 2,580 $= 3.4116$

This is true because any one of these numbers may be obtained from another by multiplying the latter by an integral power of 10.

Interpolation

The table is so constructed that the mantissas for logarithms of numbers containing more than three digits cannot be found directly. The method of finding the mantissas for logarithms of numbers of 4 or more digits is illustrated by the following examples:

Example: Find the logarithm of 3847.

▶ Since the number has 4 digits to the left of the decimal point the characteristic is 3. To find the mantissa, the following tabular arrangement may be used.

Number	Mantissa
10 ⎰ 7 ⎱ →3850 / →3847 / →3840	5855← / ? / 5843← ⎱ x ⎰ 12

The mantissas for 3840 and 3850 are found under 384 and 385 respectively in the table.

The assumption is made that the mantissa for 3847 is .7 of the numerical difference between the mantissas for 3840 and 3850. This yields the proportion

$$\frac{7}{10} = \frac{x}{12}$$
$$10x = 84$$
$$x = 8.4, \text{ or 8 to the nearest integer.}$$

When 8 is added to the lower mantissa, the result is 5851.

∴   log 3847 = 3.5851

Example: Find log .02763

►   Characteristic is 8. − 10.

To find mantissa:

Number	Mantissa
10 ⎰ 3 ⎱ →2770 / →2763 / →2760	4425← / ? / 4409← ⎱ x ⎰ 16

$$\frac{3}{10} = \frac{x}{16}$$
$$10x = 48$$
$$x = 4.8, \text{ or 5 to the nearest integer.}$$

The mantissa of 2763 = .4414.

∴   log .02763 = 8.4414 − 10.

EXERCISE 8

Find the logarithms of the following numbers:

1. 5267
2. 62.14
3. .8427
4. .003886
5. 9.053
6. 146.5
7. .6529
8. .02152
9. 9.563
10. 59.11

Finding the Number Which Corresponds to a Given Logarithm

In computation it is frequently necessary to find the number which corresponds to a given logarithm. The number corresponding to a given logarithm is called the *anti-logarithm*. The method of finding the anti-logarithm is illustrated in the following examples.

Example: Find the number whose logarithm is 2.8340.

▶ The mantissa .8340 does not appear in the table. It lies between the mantissas .8338 and .8344. The following arrangement is suggested.

Number		Mantissa
10 x ┌→6830 ┌→ ? └→6820		.8344← .8340← ┐2 .8338← ┘ 6

The number lies between 6820 and 6830. The following proportion will help to locate it more precisely:

$$\frac{x}{10} = \frac{2}{6}$$
$$6x = 20$$
$$x = 3 \text{ to the nearest integer}$$

Therefore, the number is 682.3 (The given characteristic 2 indicates that there are 3 digits to the left of the decimal point).

Example: Find the number whose logarithm is 9.5981 − 10.

▶ Locate the mantissa .5979 between .5977 and .5988.

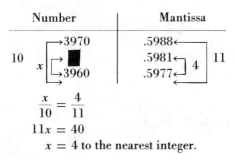

Number		Mantissa
10 x ┌→3970 ┌→ ■ └→3960		.5988← .5981← ┐ 4 .5977← ┘ 11

$$\frac{x}{10} = \frac{4}{11}$$
$$11x = 40$$
$$x = 4 \text{ to the nearest integer.}$$

The number is 0.3964. The given characteristic -1 indicates that there are no zeros between the decimal point and the first digit to the right.

EXERCISE **9**

Find the numbers whose logarithms are given below:

1. 2.8078	**6.** 9.6143 $-$ 10
2. 1.7021	**7.** 2.2163
3. 8.9408 $-$ 10	**8.** 8.4751 $-$ 10
4. 0.7410	**9.** 3.0673
5. 3.5368	**10.** 0.3606

Computation with Logarithms

Methods of computation with logarithms are illustrated by the following examples.

Example: Multiply 46.73 by .917.

▶ $N = 46.73 \times .917$
$\log N = \log 46.73 + \log .917$

A. Number | **Mantissa**

$$\frac{3}{10} = \frac{x}{9}$$
$$10x = 27, x = 3$$

$$\log 46.73 = 1.6696 \quad \text{(A)}$$
$$\log .917 = \underline{9.9624 - 10}$$
$$\log N = 11.6320 - 10$$
$$\log N = 1.6320$$
$$N = 42.85 \quad \text{(B)}$$

B. Number | **Mantissa**

$$\frac{x}{10} = \frac{6}{11}$$
$$11x = 60, x = 5$$

Example: Divide 8.64 by 63.27.

▶ $N = 8.64 \div 63.27$

$\log N = \log 8.64 - \log 63.27$

$\log 8.64 = 10.9365 - 10$

$\log 63.27 = 1.8012$ (A)

  $\log N = 9.1353 - 10$

    $N = .1366$ (B)

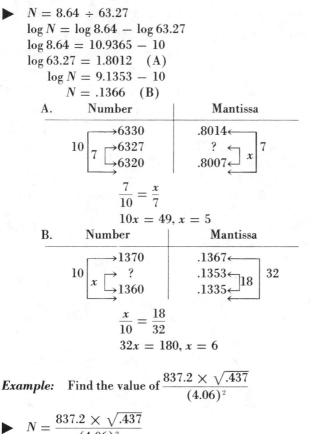

A.

Number	Mantissa

    →6330   .8014←

10 | 7 →6327   ? ← | 7

    →6320   .8007← x

$$\frac{7}{10} = \frac{x}{7}$$

$$10x = 49, \; x = 5$$

B.

Number	Mantissa

    →1370   .1367←

10 | x → ?   .1353← 18 | 32

    →1360   .1335←

$$\frac{x}{10} = \frac{18}{32}$$

$$32x = 180, \; x = 6$$

Example: Find the value of $\dfrac{837.2 \times \sqrt{.437}}{(4.06)^2}$

▶ $N = \dfrac{837.2 \times \sqrt{.437}}{(4.06)^2}$

$\log N = \log 837.2 + \tfrac{1}{2} \log .437 - 2 \log 4.06$

$\log 837.2 = 2.9228$

$\tfrac{1}{2} \log .437 = \dfrac{9.8203 - 10}{12.7431 - 10}$ (A)

$2 \log 4.06 = \dfrac{1.2170}{}$ (B)

  $\log N = 11.5261 - 10$

  $\log N = 1.5261$

     $N = 33.58$

(A) $\log .437 = 9.6405 - 10$

     $\log .437 = 19.6405 - 20$

     $\tfrac{1}{2} \log .437 = 9.8203 - 10$

(B) $\log 4.06 = 0.6085$

     $2 \log 4.06 = 1.2170$

Example: Find the value of $\dfrac{\sqrt[3]{-.8437} \times (.3014)^2}{.06042}$

▶ $\sqrt[3]{-.8437}$ is a real number. However, a negative number has no logarithm. Therefore, $\sqrt[3]{-.8473}$ will be treated as $\sqrt[3]{.8437}$ for purposes of logarithic computation. The adjustment in sign will be made in the final result.

$$N = \frac{\sqrt[3]{.8437} \times (.3014)^2}{.06042}$$

Log $N = \frac{1}{3}$ log .8437 $+ 2$ log .3014 $-$ log .6042

$\frac{1}{3}$ log .8437 $=$	9.9754 $-$ 10	(A)	(A) log .8437 $=$	9.9262 $-$ 10
2 log .3014 $=$	8.9584 $-$ 10	(B)	log .8437 $=$	29.9262 $-$ 30
	18.9338 $-$ 20		$\frac{1}{3}$ log .8437 $=$ 3| 29.9262 $-$ 30	
log .06042 $=$	8.7812 $-$ 10			9.9754 $-$ 10
	10.1526 $-$ 10			
N $=$	1.421			

The sign of the result is negative.   (B) log .3014 $=$ 9.4792 $-$ 10
∴ The final result is -1.421.   2
                                         2 log .3014 $=$ 18.9584 $-$ 20
                                                    $=$ 8.9584 $-$ 10

Example: If $y = Ae^{kt}$, find, correct to the nearest hundredth, the value of y if $A = .46$, $e = 2.718$, $k = 60$, and $t = .04$.

▶
$$y = Ae^{kt}$$
$$y = (.46)(2.718)^{(60)(.04)}$$
$$y = (.46)(2.718)^{2.4}$$
$$\log y = \log .46 + 2.4 \log 2.718$$
$$\log .46 = 9.6628 - 10$$
$$2.4 \log 2.718 = 1.0423$$
$$\log y = 10.7051 - 10$$
$$\log y = 0.7051$$
$$y = 5.07$$

Example: Find the value of $\dfrac{\log .23}{\log 4.09}$

▶
$$n = \frac{\log .23}{\log 4.09}$$

This is not the logarithm of a quotient but the division of one logarithm by another.

$$n = \frac{\log .23}{\log 4.09} = \frac{9.3617 - 10}{0.6117}$$

The logarithm 9.1367 − 10 is really −1 + the positive mantissa .3617. Therefore, 9.3167 − 10 is equal to −.6833.

$$n = \frac{-.6833}{.6117}$$

By division, $n = -1.12$ to the nearest hundredth.

Since the evaluation of n involves a lengthy long division the computation may be performed by logarithms, as follows. The fact that the numbers to be divided are themselves logarithms does not affect the computation.

$$n = -\frac{.6833}{.6117}$$
$$\log n = \log .6833 - \log. .6117$$
$$\log .6833 = 9.8346 - 10$$
$$\log .6117 = \underline{9.7866 - 10}$$
$$\log n = .0480$$
$$n = -1.116 \text{ or } -1.12$$

Cologarithms

The cologarithm of a number is the logarithm of its reciprocal.

$$\text{colog } N = \log \tfrac{1}{N} = \log 1 - \log N = 0 - \log N = -\log N$$

Thus, the colog of N may be found by subtracting log N from 0 or from 10. −10.

Cologarithms are used in computations involving division. In such cases, instead of subtracting one logarithm from another, one logarithm is added to the cologarithm of the other.

Example: Find colog 37.8.

▶
$$\text{colog } 37.8 = \log \frac{1}{37.8} = \log 1 - \log 37.8$$
$$= 0 - \log 37.8$$
$$0 = 10.0000 - 10$$
$$\log 37.8 = \underline{1.5775}$$
$$\text{colog } 37.8 = \overline{8.4225 - 10}$$

Example: Use cologarithms to find the value of

$$\frac{\sqrt{.7638}}{\sqrt[5]{9.08}(.472)^3}$$

▶ $N = \dfrac{\sqrt{.7638}}{\sqrt[5]{9.08}(.472)^3}$

$\log N = \frac{1}{2}\log.7638 + \frac{1}{5}\text{colog }9.08 + 3\text{ colog }.472$

$\frac{1}{2}\log.7638 =$ $9.9415 - 10$	(A) colog $9.08 = \log 1 - \log 9.08$
$\frac{1}{5}\text{colog }9.08 =$ $9.8084 - 10$ (A)	$= 0 - \log 9.08$
$3\text{ colog }.472 =$ $.9783$ (B)	$10.0000 - 10$

$$
\begin{array}{ll}
& 20.7282 - 20 \\
& 0.7282 \\
N = & 5.348
\end{array}
$$

(A)   colog $9.08 = \log 1 - \log 9.08$
$= 0 - \log 9.08$

$$
\begin{array}{r}
10.0000 - 10 \\
.9581 \\
\hline
5\ \vert\ 49.0419 - 50 \\
\hline
9.8084 - 10
\end{array}
$$

(B)   colog $.472 = 0 - \log.472$

$$
\begin{array}{r}
10.0000 - 10 \\
9.6739 - 10 \\
\hline
.3261
\end{array}
$$

$3\text{ colog }.472 = \ .9783$

EXERCISE 10

Perform the following computations:

1. $53.76 \times .853$

2. $6.52 \times .8514 \times 32.5$

3. $\dfrac{965.2}{35.7}$

4. $\dfrac{42.83 \times .786}{.5639}$

5. $\dfrac{.3874 \times 9852}{25.67}$

6. $\dfrac{6.423 \times 1053}{829 \times 2.057}$

7. $\dfrac{35.2 \times \sqrt{982}}{45.2}$

8. $\dfrac{405.6\ \sqrt[3]{385.3}}{(29)^2}$

9. $\dfrac{74.84\sqrt{.8526}}{.9043}$

10. $\dfrac{57.87\ \sqrt[3]{.6054}}{.6296}$

11. $\dfrac{256.3(.305)^3}{\sqrt[3]{.8754}}$

12. $\dfrac{6035}{\sqrt{.285} \times (4.062)^2}$

13. $\dfrac{\sqrt[3]{-.4378} \times (50.6)^2}{.915}$

14. $\dfrac{(2.906)^4 \times \sqrt[3]{-3.567}}{2.053}$

15. $\dfrac{\log 5.82}{\log .97}$

16. $\dfrac{\log .53}{\log 8.72}$

17. $\dfrac{\sqrt[3]{3.579} \times (6.92)^2}{.00856}$

18. Given the formula $t = a\sqrt{\dfrac{kp}{s}}$. Find, correct to the nearest hundredth, the value of t if $a = 37$, $k = .17$, $p = 4.9$, and $s = 8,000$.

19. Given the formula $Q = Pe^{-nr}$. Find Q to the nearest hundredth, if $P = 760$, $e = 2.718$, $n = 5$, and $r = .14$.

20. Given the formula $N = \dfrac{1}{2L}\sqrt{\dfrac{Mg}{m}}$. Find N if $L = 76.5$, $M = 5,470$ $g = 980$, and $m = .0045$.

Exponential Equations

An exponential equation is an equation which has the unknown or unknowns as exponents.

Example: Solve the equation $3^{x+2} = 81^x$

▶ $3^{x+2} = 81^x$

Replacing 81 by 3^4, we have $3^{x+2} = (3^4)^x = 3^{4x}$

$x + 2 = 4x$, or $x = \frac{2}{3}$

Example: Solve the equation $4^{2x} = 8^{2x+3}$

▶ $4^{2x} = 8^{2x+3}$

Writing 4 and 8 as powers of 2, we have $(2^2)^{2x} = (2^3)^{2x+3}$

$2^{4x} = 2^{6x+9}$

$4x = 6x + 9$, or $x = -\frac{9}{2}$

Example: Solve the equation $7^x = 31$

▶ $7^x = 31$

The bases may be made identical by writing the numbers 7 and 31 as powers of 10, as follows:

$(10^{0.8451})^x = 10^{1.4914}$

$.8451\ x = 1.4914$

$x = \dfrac{1.4914}{.8451} = 1.76$

Note: When we write the numbers as powers of base 10, we are writing their logarithms. Thus, an equation like the one above may be solved by taking the logarithms of both sides of the equation. The latter method is used in the following example.

Example: Solve the equation $13^{2x+2} = 5^x$

▶ $13^{2x+2} = 5^x$
Taking the logarithms of both sides of the equation, we have
$(2x+2) \log 13 = x \log 5$
$(2x+2)(1.1139) = x(.6990)$
$2.2278x + 2.2278 = .6990x$
$1.5288x = -2.2278$
$x = -1.46$

EXERCISE 11

Solve the following exponential equations:

1. $2^x = 32$
2. $3^x = \frac{1}{81}$
3. $5^{x+1} = 125$
4. $8^{2x} = 16^{3x-5}$
5. $9^{2x-2} = 3^{8x}$
6. $4^{-x} = 2^{5-x}$
7. $5^x = 73$
8. $3^{x+2} = 157$
9. $(2.8)^{2x-3} = 69$
10. $(.73)^{x-1} = (5.1)^{1-2x}$
11. $6^{2x+1} = (4.3)^{x-1}$
12. $2^{x+1} \cdot 3^x = 3{,}815$

Change of Base

As indicated earlier in the chapter, bases other than 10 may be used in work with logarithms. Specifically, the irrational number $e = 2.718 \ldots$ has important applications as a base. The formula developed below may be used to change from one base to another.

Let $\log_a N = x$
Then $a^x = N$
Taking logarithms of both sides of the equation,
$x \log a = \log N$
$x = \dfrac{\log_b N}{\log_b a}$ or $\log_a N = \dfrac{\log_b N}{\log_b a}$

Example: Find the value of $\log_e 3$ to the nearest tenth. (Let $e = 2.718$)

▶ $\log_a N = \dfrac{\log_b N}{\log_b a}$

     $\log_e 3 = \dfrac{\log_{10} 3}{\log_{10} e} = \dfrac{.4771}{.4343} = 1.1$ to the nearest tenth

Example: Find the value of $\log_4 11$ to the nearest tenth.

▶ $\log_a N = \dfrac{\log_b N}{\log_b a}$; $\log_4 11 = \dfrac{\log_{10} 11}{\log_{10} 4} = \dfrac{1.0414}{.6021} = 1.7$

EXERCISE **12**

Find the values of the following: (Let $e = 2.718$.)

1. $\log_e 9$ **3.** $\log_e 100$ **5.** $\log_4 19$
2. $\log_5 12$ **4.** $\log_{12} 4$ **6.** $\log_e 14.6$

Graphs of Logarithmic Functions

Let us consider the equation $y = \log_2 x$. In order to plot the graph of this equation, we note that $y = \log_2 x$ is equivalent to $x = 2^y$. The following table of values may now be worked out.

x	$\frac{1}{16}$	$\frac{1}{8}$	$\frac{1}{4}$	$\frac{1}{2}$	1	2	4	8	16	32
y	-4	-3	-2	-1	0	1	2	3	4	5

When these points are plotted we obtain the graph below:

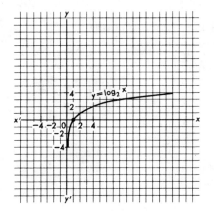

This curve is characteristic of the shape of all curves of the form $y = \log_a x$ where a is greater than 1. The following facts about this type of curve are noted.

a. x cannot be negative since a negative number does not have a real number for its logarithm.

b. The logarithm of a number greater than 1 is positive, the logarithm of 1 is zero, and the logarithm of a number less than 1 is negative.

c. As x approaches zero, $\log x$ decreases without limit.

d. As x increases greatly, $\log x$ increases without limit.

EXERCISE 13

Plot the following graphs:

1. $y = \log_3 x$
2. $y = \log_{10} x$ (Use log tables)
3. $y = \log_e x$ (Change to base 10)
4. $y = \log_5 x$

Exponential Functions

A function in which the variable appears as an exponent while the base is a constant is called an exponential function. We shall restrict ourselves to positive values of the constant.

Examples: 2^x, 3^{y+2}

Graphs of Exponential Functions

In order to plot the graph of the equation $y = 2^x$ we use the following table of values:

x	-2	-1	0	1	2	3	4
y	.25	.5	1	2	4	8	16

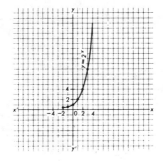

EXERCISE 14

Plot the graphs of the following exponential functions:

1. $y = 3^x$
2. $y = 5^x$
3. $y = (\frac{1}{2})^x$
4. $x = 2^y$

ANSWERS

EXERCISE 1

1. 32,768
2. 65,536
3. 32
4. 128
5. 32,768
6. 65,536
7. 16
8. 16

EXERCISE 2

1. $\log_2 128 = 7$
2. $\log_3 243 = 5$
3. $\log_5 25 = 2$
4. $\log_4 \frac{1}{16} = -2$
5. $\log_7 1 = 0$
6. $\log_{10} 100 = 2$
7. $\log_2 \frac{1}{32} = -5$
8. $\log_9 27 = \frac{3}{2}$
9. $\log_8 \frac{1}{16} = -\frac{4}{3}$
10. $2^4 = 16$
11. $7^2 = 49$
12. $4^3 = 64$
13. $10^{-1} = .1$
14. $5^3 = 125$
15. $4^{1/2} = 2$
16. $8^{4/3} = 16$
17. $9^{-1/2} = \frac{1}{3}$
18. $(\sqrt{3})^2 = 3$

EXERCISE 3

1. 3
2. 3
3. 2
4. 4
5. 0
6. $\frac{5}{3}$
7. -2
8. $\frac{1}{2}$
9. $\frac{3}{4}$
10. -3
11. $-\frac{1}{2}$
12. $-\frac{3}{4}$

13. 9
14. 7
15. 2
16. 3
17. 3
18. 0
19. $\sqrt{3}$
20. 64
21. $\frac{3}{2}$
22. $\frac{1}{8}$
23. 6
24. $\frac{3}{2}$

EXERCISE 4

1. $\log_a 7 + \log_a 2$
2. $\log_a 3 + \log_a 11$
3. $2 \log_a 2 + \log_a 5$
4. $3 \log_a 3 + \log_a 2$
5. $\log_a 7 - \log_a 19$
6. $3 \log_a 2 - 2 \log_a 3$
7. $6 \log_a 2 - 2 \log_a 5$
8. $\log_a 3 + \frac{1}{2} \log_a 5$
9. $\log_a 5 + \frac{1}{3} \log_a 2$
10. $\frac{1}{3} \log_a 17 - \frac{1}{4} \log_a 23$
11. $\frac{1}{15} \log_a 19 - \frac{1}{3} \log_a 31$
12. $2 \log_a 29 + 3 \log_a 2 - \frac{1}{4} \log_a 43$

EXERCISE 5

1. 1.1461
2. 1.6232
3. 1.0791
4. 1.5441
5. 1.5562

6. .4407
7. .8495
8. $-.5862$
9. 1.9542
10. .3627

EXERCISE 6

1. $\frac{5}{3}$
2. $\frac{23}{2}$
3. 60
4. $1\frac{13}{28}$
5. $\frac{167}{61}$
6. $\frac{15}{8}$

EXERCISE 7

1. 2
2. 0
3. -1
4. 1
5. -2
6. 4
7. -3
8. 3
9. -5
10. 0
11. 1
12. -2

EXERCISE 8

1. 3.7216
2. 1.7934
3. $9.9257 - 10$
4. $7.5895 - 10$
5. 0.9568
6. 2.1659
7. $9.8148 - 10$
8. $8.3328 - 10$
9. 0.9806
10. 1.7717

EXERCISE 9

1. 642.4
2. 50.36
3. .08726
4. 5.508
5. 3442.
6. .4115
7. 164.6
8. .02986
9. 1168
10. 2.294

EXERCISE 10

1. 45.86
2. 180.4
3. 27.03
4. 59.69
5. 148.7
6. 3.965
7. 24.41
8. 3.508
9. 76.42
10. 77.77

11. 7.602
12. 685.3
13. −2125
14. −53.07
15. −57.95

16. −.293
17. 8556.
18. .38
19. 377.6
20. 225.6

EXERCISE 11

1. 5
2. −4
3. 2
4. $3\frac{1}{3}$
5. −1
6. −5

7. 2.7
8. 2.6
9. 3.6
10. .45
11. −1.5
12. 4.22

EXERCISE 12

1. 2.2
2. 1.5
3. 4.6

4. .56
5. 2.1
6. 2.7

CHAPTER **17**

Mathematics of Finance

SIMPLE INTEREST

DEFINITION: *Interest* is the amount paid for the use of money.

DEFINITION: The *Principal* is the sum of money upon which the interest is paid.

DEFINITION: The *Rate of Interest* is the interest charged for the use of one unit of money for one unit of time.

DEFINITION: The *Amount* is the sum of the principal and the interest.

DEFINITION: *Simple interest* is interest computed upon the original principal.

Example: Mr. Kent borrows $600 for 1 year at 5% interest.   At the end of the year he repays the $600 and $30 interest, a total of $630.

In this case, $600 represents the Principal, $30 the Interest, 5% the Rate, and $630 the Amount.

The interest is computed by using the formula $I = PRT$, where I = interest, R = rate, and T = time in years.

Since the amount is the sum of the principal and the interest, the amount $A = P + PRT$ or $A = P(1 + RT)$.

Example: Mr. Logan borrows $800 from a friend for a period of 3 years at a rate of 4% simple interest. How much interest does Mr. Logan pay?

▶ To compute the interest, use the formula $I = PRT$.

Since $P = \$800$, $R = .04$, and $T = 3$, $I = \$800(.04)(3) = \$96$

Example: Mr. Brown borrows $450 for a period of 2 years and 4 months at the rate of $4\frac{1}{2}$% simple interest. What amount does Mr. Brown pay at the expiration of the loan period?

▶ To find the amount, use the formula $A = P(1 + RT)$. The rate of $4\frac{1}{2}$% is written as .045 and the time is $2\frac{4}{12}$ or $2\frac{1}{3}$ years.

$$A = 450\left[1 + (.045)\left(\tfrac{7}{3}\right)\right] = \$497.25.$$

Example: A man borrowed $350 and paid back $413 at the end of 4 years. What rate of simple interest did he pay?

▶ Use the formula $\quad A = P(1 + RT)$
$$413 = 350(1 + 4R)$$
$$413 = 350 + 1400R$$
$$R = \frac{63}{1400} = \frac{9}{200} = .045 \text{ or } 4\tfrac{1}{2}\%$$

Example: A man borrowed $1260 at $5\frac{1}{2}$% simple interest. At the expiration of the loan he paid back $1502.55. What was the period of the loan?

▶ Subtracting $1260 from $1502.55 we find that the interest on the loan was $242.55.

Using the formula $I = PRT$, $\$242.55 = \$1260(.055)T$
$$T = \frac{242.55}{(1260)(.055)} = 3\tfrac{1}{2} \text{ years.}$$

Example: A sum of money amounts to $1929.75 if invested for $2\frac{1}{2}$ years at $6\frac{1}{2}$% simple interest. What is the sum of money invested?

▶ In this problem it is required to find the principal. Using the formula $A = P(1 + RT)$, we have

$$\$1929.75 = P[1 + (.065)(2\tfrac{1}{2})]$$
$$P = \frac{1929.75}{[1 + (.065)(2.5)]} = \frac{1929.75}{1.1625} = \$1660$$

Example: A man borrows \$600 from a loan company for one year at 6% interest payable in advance. What rate of interest does he actually pay?

▶ Since the interest must be paid in advance the borrower receives $\$600 - (\$600)(.06) = \$600 - \$36 = \$564$.

Therefore, the borrower is paying \$36 interest for the use of \$564 for one year. The rate may be computed as follows.

$$A = P(1 + RT), \quad 600 = 564[1 + R(1)]$$
$$600 = 564 + 564R, \quad R = \frac{36}{564} = 6.4\%$$

Example: A man buys a house and has a mortgage debt on it of \$2,000. He arranges to pay \$100 each three months together with 6% simple interest on the unpaid balance.

a. How much interest does he pay during the five year period of the loan?

b. What is the total amount paid?

▶ After the first 3 months the man pays interest on \$2000 at 6% for the 3 month period.

$$I = PRT, \quad I = \$2000(.06)(\tfrac{1}{4}) = \$30.$$

After the second 3 months the man pays interest on \$1900 for the 3 month period.

$$I = PRT, \quad I = 1900(.06)(\tfrac{1}{4}) = \$28.50.$$

For each succeeding payment the principal is reduced by \$100 and the interest reduced by \$1.50. Thus, the interest payment for the third 3 month period is \$27.00.

The interest payments form an Arithmetic Progression with a first term of \$30.00, a common difference of $-\$1.50$, and 20 payment periods.

$$S = \frac{n}{2}[2a + d(n - 1)]$$

$$S = \frac{20}{2}[60 - 1.5(20 - 1)] = 10[60 - 28.5] = \$315$$

b. The total amount paid is $2,000 + $315 = $2,315.

EXERCISE 1

1. Find the interest on $950 at 5% simple interest for 6 years.

2. Find the interest on $640 at $3\frac{1}{2}$% simple interest for 2 years and 6 months.

3. Find the interest on $1600 at $4\frac{1}{4}$% simple interest for 3 years and 3 months.

4. A man borrowed $848 at 6% simple interest for 9 months. What amount did the man pay at the expiration of the loan period?

5. A man borrowed $450 and paid back $540 at the end of 4 years. What rate of simple interest did he pay?

6. A man borrowed $720 and paid back $858.60 at the end of $3\frac{1}{2}$ years. What rate of simple interest did he pay?

7. A man borrowed $980 at 5% simple interest. He paid back $1323. What was the period of the loan?

8. A man invested $780 at $6\frac{1}{2}$% simple interest. At the end of the investment period he received $1,008.15. What was the period of his investment?

9. A man invests a sum of money for $2\frac{1}{2}$ years at the rate of 5%. At the end of the investment period this sum amounted to $1,687.50. What was the original sum invested?

10. After investing a sum of money at $4\frac{1}{2}$% for 5 years and 3 months a man received a total of $2967. How much was his original investment?

11. A man borrows $800 from a bank for one year at 5% payable in advance. What rate of interest does he actually pay?

12. A man borrows $480 for one year at $5\frac{1}{2}$% payable in advance. What rate of interest does he actually pay?

13. A man pays off a loan of $4,000 in 5 years. He pays $200 quarterly together with 6% on the unpaid balance.

a. How much interest does the man pay during the five year period of the loan?

b. What is the total amount paid?

14. A man buys a house and has a mortgage debt on it of $4,480 payable in 7 years. He arranges to pay $160 each 3 months together with 5% simple interest on the unpaid balance.

a. How much interest does he pay during the seven year period of the loan?

b. What is the total amount paid?

COMPOUND INTEREST

When interest is added to the principal at stated times then the interest is said to be compounded. For example, let us suppose that $100 is invested at 6% compounded semiannually. In this case, the interest is computed at the end of 6 months. Then the interest of $3 is added to the original capital. Now, the interest for the second half-year is computed on a capital of $103. The interest for the second half-year is $3.09. The interval between successive computations of interest is called the *interest period* or *conversion period*. Interest rates are ordinarily given as year rates. For example, a savings bank may pay a rate of 3% compounded semiannually. Actually, this rate corresponds to an annual rate slightly higher than 3%. In this case, the 3% rate is called the *nominal rate* and the slightly higher rate the *effective rate*.

The compound interest may be computed by using the formula $S = P(1 + i)^n$ where $S =$ the compound amount

$\qquad P =$ the principal

$\qquad i =$ the rate of interest per conversion period

$\qquad n =$ the number of conversion periods.

Example: If $840 is placed in a bank at 2% compounded quarterly it will amount to $948.60 in six years.

In this case, $948.60 is the compound amount (S), $840 is the principal (P), the rate of interest per conversion period is $2\% \div 4$ or $\frac{1}{2}\%(i)$, the number of conversions periods is $6 \times 4 = 24(n)$.

PRESENT VALUE

The present value of a sum of money due at a future date is the amount which, put at interest at a given rate, will amount to the specified sum at the stated date.

Example: A man must pay a debt of $1,000 due in 8 years. If he can obtain a rate of 4% compounded semiannually how much should he invest in order that the principal invested will amount to $1,000 in 8 years? The principal needed is computed to be $728.50.

In this case, the *present value* is $728.50 i.e. the principal that, invested at 4%, compounded semiannually will amount to $1,000 in 8 yrs.

Example: A man invests $1,800 at 4% compounded semi-annually for a period of 7 years.   Find the amount and the interest.

▶   Use the formula $S = P(1 + i)^n$

In this problem $P = \$1800, i = 2\%, n = 14$.
$$S = \$1800(1.02)^{14}$$

S may be computed by logarithms, as follows

$$\log S = \log 1800 + 14 \log 1.02$$
$$\log 1800 = 3.2553$$
$$14 \log 1.02 = \underline{.1204}$$
$$\log S = 3.3757$$
$$S = \$2375.26$$

S may be computed by using table 2.
$$S = \$1800(1.02)^{14}$$

Under the 2% conversion rate column read 1.3195 alongside $n = 14$. Now, multiply 1800×1.3195. The result is $2375.10.

The slight difference in the results is due to the fact that logarithms and tables are approximate in the fourth decimal place.

Thus, the amount is $2,375.10 and the interest is $575.10.

Example: How long will it take $1200 to amount to $1500 if invested at 5% compounded semiannually?

▶   Use the formula $S = P(1 + i)^n$

In this problem $S = \$1500$, $P = \$1200$, $i = 2\frac{1}{2}\%$

$$\log S = \log P + n \log (1 + i)$$
$$\log 1500 = \log 1200 + n \log 1.025$$
$$n = \frac{\log 1500 - \log 1200}{\log 1.025}$$
$$n = \frac{3.1761 - 3.0792}{.0107} = \frac{.0969}{.0107}$$
$$n = 9 \text{ approximately}$$

n represents the number of conversion periods. Therefore, it would take about $4\frac{1}{2}$ years.

Example: A man wishes to provide \$6500 for his son's college education to be available 8 years from the present date. How much must he invest now at 5% compounded semiannually so that he will have the required sum in 8 years?

▶ In this problem we wish to find the *present value* (*P*) which will amount (*A*) to \$6500 in 8 years at 5% compounded semi-annually.

Use the formula $S = P(1 + i)^n$

$$\text{or } P = \frac{S}{(1 + i)^n} \qquad \text{or } P = S(1 + i)^{-n}$$

By logarithms

$$\log P = \log S - n\log (1 + i)$$
$$\log P = \log 6500 - 16 \log (1 + .025)$$
$$\log 6500 = 3.8129$$
$$16 \log 1.025 = \underline{.1712}$$
$$\log P = 3.6417$$
$$P = \$4,382$$

By use of table 3.

$$P = S(1 + i)^{-n}$$
$$P = 6500 \,(.67362)$$
$$P = \$4,378.53$$

The slight difference in results is due to the fact that the logarithm table and the present value table are approximate.

Example: A boy is left a legacy of $10,000 to be paid on his
twenty-first birthday, seven years from now. What is the present
value of the legacy if money is worth 3% compounded semiannually?

▶ Use the formula $S = P (1 + i)^n$

$$\text{or } P = S (1 + i)^{-n}$$
$$P = 10,000 (1 + .015)^{-n}$$

By use of table 3.

$$P = 10,000 (.81185) = 8,118.5$$
$$P = \$8,118.50$$

Example: A note for $7500 is payable in eight years. If money
is worth 6% compounded semiannually, what is the present value of
the note? What is the discount?

▶ Use the formula $S = P (1 + i)^n$

$$P = S (1 + i)^{-n}, P = 7,500 (1 + .03)^{-16}$$

Use table 3.

$$P = 7,500 (.62317)$$
$$P = \$4,673.78$$

The discount is $7,500 − $4,673.78 = $2,826.22

EXERCISE 2

1. A man invests $2,300 at 6% compounded semiannually for 6
years. Find the amount and the interest.

2. A man invests $4,324 at 3% compounded semiannually for 9
years. Find the amount and interest.

3. A man invests $1,865 at 5% compounded quarterly for 5 years.
Find the amount and the interest.

4. How long will it take $1,000 to amount to $1,400 if invested at
4% compounded semiannually?

5. How long will it take $3,250 to amount to $6,020 if invested at
5% compounded semiannually?

6. A man wishes to have $10,000 in cash when he retires in 12
years. How much must he invest now at 4% compounded semiannually so that he will have the required sum in 12 years?

7. A father wishes to provide a fund of $7,500 for his son when the son reaches the age of 18. If the son is now 4 years old, what sum must the father invest at 5% compounded semiannually?

8. A man wishes to have a sum of $12,000 at the age of 55. If he is now 39 years of age how much must he invest at 6% compounded semiannually?

9. A note for $9,600 is payable in six years. If money is worth 4% compounded semiannually, what is the present value of the note? What is the discount?

10. A note for $6,350 is payable in nine years. If money is worth 5% compounded semiannually, what is the present value of the note? What is the discount?

Effective Rate

Interest rates are usually stated as annual rates although conversion periods may be semiannual or quarterly. If the conversion period is not a year then the annual interest rate is called a *nominal rate*. The *effective rate* is the actual annual rate.

Example: Find the effective rate corresponding to a 6% nominal rate, compounded quarterly.

▶ If $1 is invested at 6% nominal, compounded quarterly, it amounts to $1.0614. (Use table 2.) The effective rate is therefore 6.14%.

ANNUITIES

An *annuity* is a series of equal payments, made at the ends of equal intervals of time, usually a year.

There are two basic problems involving annuities.

1. What will be the accumulated sum of the compounded amounts of all the payments to the end of the term, i.e., what single payment made at the end of the term would be equivalent to the compound sum of all the payments?

This single payment is called the *amount* of the annuity.

2. What is the *present value* of the annuity, i.e., what single payment made at the beginning of the term would be equivalent to the sum of all the payments?

Formula for the Amount of an Annuity

Let us suppose that a series of n payments are made, each payment being equivalent to R. Each of these payments will accumulate interest from the date it is made until the end of the term.

The first payment will therefore amount to $R(1 + i)^{n-1}$

The second payment will amount to $R(1 + i)^{n-2}$

The final payment will amount to R.

If the accumulated amount be $S_{n]i}$, then

$$S_{n]}i = R + R(1 + i) + R(1 + i)^2 \ldots R(1 + i)^{n-1}$$

This is a geometric progression and the sum is

$$S_{n]\,i} = R\frac{(1 + i)^n - 1}{i}$$

The expression $\dfrac{(1 + i)^n - 1}{i}$ is designated as $s_{n]\,i}$ in table 4.

Formula for the Present Value of an Annuity

Let us consider a series of n payments, each of amount a, the first payment being due a year from the present. How may we compute the total present value?

The individual present values are $\dfrac{a}{1 + i}, \dfrac{a}{(1 + i)^2}, \dfrac{a}{(1 + i)^3}, \ldots$

$\ldots \dfrac{a}{(1 + i)^n}$

If the total present value is denoted by $a_{n]}$, then

$$a_{n]i} = \frac{a}{1 + i} + \frac{a}{(1 + i)^2} + \frac{a}{(1 + i)^3} + \ldots + \frac{a}{(1 + i)^n}$$

This is a geometric progression with ratio $\dfrac{1}{(1 + i)}$. Using the sum formula we have

$$A_{n]i} = \frac{a}{1 + i} \cdot \frac{1 - \dfrac{1}{(1 + i)^n}}{1 - \dfrac{1}{1 + i}} = a \cdot \frac{1 - (1 + i)^{-n}}{i}$$

The expression $\dfrac{1 - (1 + i)^{-n}}{i}$ is designated as $a_{n\,i}$ in the table on page 688.

Example: A man deposited \$300 every six months in a bank that pays 3% interest, compounded semiannually. What is the value of the man's fund after his twelfth payment?

▶ We wish to find $S_{\overline{n}|i}$ where $n = 12$ and $i = 1\frac{1}{2}\%$. This may be written as $S_{\overline{12}|.015}$. The value of $S_{\overline{12}|1\,1/2}$ is most easily determined by using table 4. In this table locate $n = 12$ on the vertical column and $i = 1\frac{1}{2}\%$ on the horizontal row. The number corresponding to both these values is 13.0412.

If we multiply 13.0412 by R (the value of each payment) we have $(13.0413) \times (300)$ or \$3,912.36.

The amount of the annuity is \$3,912.36.

Example: A man bought a house for \$8,000 plus \$500 a year for 12 years.

a. What was the amount of the man's payments if money was worth 5%?

b. What was the equivalent cash price?

▶ a. We wish to find $S_{\overline{n}|i}$ where $n = 12$ and $i = 5\%$, i.e. we wish to find $S_{\overline{12}|.05}$. In table 4 the number corresponding to $n = 12$ and $i = 5\%$ is 15.9171.

In this case, $R = \$500$. Therefore, $(\$500)(15.9171) = \$7,958.55$. The total amount of the man's payments was $\$8,000 + \$7,958.55 = \$15,958.55$.

b. In order to find the equivalent cash price, we must find the present value of the 12 payments. In table 5 for $a_{\overline{n}|i}$ the number corresponding to $n = 12$ and $i = 5\%$. Therefore $a_{\overline{12}|.05} = 8.8633$.

In this case $a = 500$. Therefore $(\$500)(8.8633) = \$4431.65$. The equivalent cash price is $\$8,000 + \$4,431.65 = \$12,431.65$.

Example: A bank holds a note that calls for 8 annual payments of \$600 each, the first payment being due one year from the present date. The man who gave the note wishes to exchange it for a note for a single payment due in 5 years. What should be the amount of the new note if money is worth 4%?

▶ In order to provide a fair exchange we will first find the present value of the 8 annual payments. Then we will find the amount of this present value payable as a lump sum 5 years hence.

$a_{8\rceil.04} = 6.7327$

$$\$600 \times 6.7327 = \$4,039.62$$
$$A = P(1 + R)^n, \quad A = P(1 + .04)^5, \text{ using table 2}$$
$$A = 4,039.62(1.2167) = \$4,915.01.$$

Example: A clerical job costs a company \$8,500 a year for labor. The same job can be performed by an automatic machine with no labor cost. The automatic machine can be installed for \$44,625.00, payable immediately. In how many years will the machine pay for itself, if money is worth 5% compounded annually?

▶ The present value of the \$8,500 per year outlay can be expressed as $a_{n\rceil.05}$. This must be equal to \$44,625.00.

Let $a_{n\rceil.05} = x$

Then $8,500x = \$44,625$

$x = 5.2494$

If we examine the table in the 5% column we can see that n is approximately equal to 6.

The machine will pay for itself in about 6 years.

Example: A certain piece of heavy machinery costs \$16,000 and must be replaced in 9 years. How much should be set aside each year, starting a year from the present, to provide for the replacement, if money is worth 4%, compounded annually?

▶ In this case, \$16,000 is the amount of an annuity for a period of 9 years at 4% interest.

Using the formula, $S_{n\rceil i} = R\dfrac{(1 + i)^n - 1}{i} = Rs_{n\;i}$.

$$R = \frac{S_{n\rceil i}}{s_{n\rceil i}} = \frac{\$16,000}{s_{9\rceil.04}} = \frac{\$16,000}{10.5828} = \$1,511.89.$$

It is necessary to set aside \$1,511.89 each year.

Example: A township has \$50,000 in bonds due in 15 years. A sinking fund is created to provide for satisfaction of the debt. How much must be deposited in the fund at the end of each year if money is worth 5%?

▶ In this case, $50,000 is the amount of an annuity for a period of 15 years at 5% interest.

Using the formula, $S_{\overline{n}|i} = R\dfrac{(1 + i)^n - 1}{i} = Rs_{\overline{n}|i}$

$$R = \frac{S_{\overline{n}|i}}{s_{\overline{n}|i}} = \frac{\$50,000}{s_{\overline{15}|.05}} = \frac{\$50,000}{21.5786} = \$2317.11$$

$2317.11 must be deposited in the sinking fund annually.

Example: A man purchases a house and takes a mortgage on it for $10,000 to be amortized (paid off) in 12 years by equal annual payments. If the interest rate is 5% compounded annually, what is the value of each of the equal annual payments?

▶ In this case, $10,000 is the present value of an annuity for a period of 12 years at 5% interest. Using the formula

$$A_{\overline{n}|i} = a \cdot \frac{1 - (1 + i)^{-n}}{i} = a \cdot a_{\overline{n}|i}$$

$$a = \frac{A_{\overline{n}|i}}{a_{\overline{n}|i}} = \frac{\$10,000}{a_{\overline{12}|.05}} = \frac{\$10,000}{8.8633} = \$1,128.25.$$

The annual equal payments are $1,128.25.

Example: A man took out a life insurance policy for which he agreed to pay $150 at the beginning of each year for 20 years. What was the present value of his agreement if money was worth 5%?

▶ In solving this problem, the first payment is part of the present value since it is paid at the start of the agreement. The problem then is to find the present value of the annuity for 19 payments and then add to the result the first payment of $150.

$A_{\overline{n}|i} = A_{\overline{19}|.05} = 12.0853$

$(12.0853)(\$150) = \$1,812.80$

$\$1,812.80 + \$150 = \$1,962.50.$

Example: A man leaves his wife a legacy of $15,000. She wishes to receive equal payments at the end of each 6 months for the next 8 years. What will be the value of each payment if money is worth 4%?

▶ $A_{\overline{n}|i} = a \cdot \dfrac{1 - (1 + i)^{-n}}{i} = a \cdot a_{\overline{n}|i}$

$a = \dfrac{A_{\overline{n}|i}}{a_{\overline{n}|i}} = \dfrac{\$15,000}{a_{\overline{16}|.02}} = \dfrac{\$15,000}{13.5777} = \$1,104.71$

Each payment will be \$1,104.71.

EXERCISE 3

1. A man invested \$800 every 6 months at 4% interest, compounded semiannually. What was his investment worth after 9 years?

2. A man established a savings fund for his son at the son's fourth birthday. If the man paid \$400 into the fund every 6 months and received interest at 3% compounded semiannually, what was the fund worth at the son's 18th birthday?

3. A firm bought a building for \$12,000 plus \$750 a year for 15 years. What was the amount of the firm's payments if money was worth 5%?

4. A company buys a machine for \$2,500 cash and 8 semiannual payments of \$600 each. What is the equivalent cash price if money is worth 4%?

5. Find the effective rate equivalent to a 4% nominal rate compounded semiannually.

6. A bank holds a note that calls for 12 annual payments of \$1,500 each, the first payment being due one year from the present date. The man who gave the note wishes to exchange it for a note for a single payment being due in 3 years. What should be the amount of the new note, if money is worth 5%?

7. Find the effective rate equivalent to a 5% nominal rate compounded quarterly.

8. A man deposited \$5,000 in a savings bank paying 3%, compounded semiannually. In another fund paying 3% compounded annually, he deposited \$520 at the end of each year. Which fund contained the greater amount at the end of 10 years? By how much?

9. A man plans to replace his car at the end of 7 years. How much should he deposit in the bank at the end of each year at 3% compounded annually so that he will have \$4,500 at the end of 7 years?

10. A house is sold for $5,000 cash and $800 payable semiannually for 9 years. What is the equivalent cash price of the house if money is worth 5% compounded semiannually?

11. A business nets a profit of $8,600 annually. What is the present cash value of the business if it is agreed that the profits are to pay for the business in 8 years and money is worth 6% compounded annually?

12. By installing a modern machine a firm can save $7,200 a year for labor. The modern machine can be installed for $76,200 payable immediately. In how many years will the machine pay for itself if money is worth 4% compounded annually?

13. A piece of machinery costs a firm $24,000 and must be replaced at the end of 8 years. How much should be set aside each year, starting a year from the present, to provide for the replacement, if money is worth 5% compounded annually?

14. A business house has a bank debt of $35,000 due in 10 years. A special fund is created to provide for payment of the bank debt. How much must be deposited in the fund at the end of each year if money is worth 3%, compounded annually?

15. A man purchases a house and takes a mortgage on it for $12,000 to be amortized in 15 years by equal annual payments. If the interest rate is 4% compounded annually, what is the value of each of the equal annual payments?

16. A man took out a life insurance policy for which he agreed to pay $180 at the beginning of each year for 25 years. What was the present value of his agreement if money was worth 4%?

17. A man leaves his wife a legacy of $18,000. She wishes to receive equal payments at the end of each 6 months for the next 12 years. What will be the value of each payment if money is worth 5%?

18. Approximately how long will it take to accumulate $8,200 by depositing $600 at the end of each half year into a savings fund which pays 5%, compounded semiannually?

ANSWERS

EXERCISE 1

1. $285
2. $56
3. $221
4. $886.16
5. 5%
6. $5\frac{1}{2}\%$
7. 7 years
8. $4\frac{1}{2}$ years
9. $1,500

10. $2,400
11. 5.3%, correct to the nearest tenth of a per cent
12. 5.8%, correct to the nearest tenth of a per cent
13. a. $630
    b. $4,630
14. a. $812
    b. $5,292

EXERCISE 2

1. $3,275.71 (logs), $3,279.34 (table)
2. $5,661.11 (logs), $5,652.77 (table)
3. $2,391.63 (logs)
4. $8\frac{1}{2}$ years
5. $12\frac{1}{2}$ years
6. $6,217.14 (logs), $6,217.20 (table)
7. $3,762.73 (logs), $3,706.60 (table)
8. $4,673.33 (logs), $4,660.08 (table)
9. $7,569.50 (present value), $2,028.50 (discount)
10. $4,071.43 (present value), $2,278.57 (discount)

EXERCISE 3

1. $17,129.84
2. $13,792.60
3. $28,183.95
4. $6,895.30
5. 4.04%
6. $15,390.23
7. 5.09%
8. The first fund by $773.27
9. $587.28

10. $16,482.72
11. $53,404.28
12. 14 years
13. $2,513.33
14. $3,053.06
15. $1,079.29
16. $2,924.46
17. $1,006.43
18. 6 years

CHAPTER 18

Permutations and Combinations

PERMUTATIONS

Fundamental Principle

If an act can be performed in *m* ways and if, after this first act has been performed, a second act can be performed in *n* ways then the number of ways in which both acts can be performed, in the order given is *m* × *n* ways.

Example: A man has 3 sport jackets and 4 pairs of slacks. In how many different ways can he combine a sport jacket with a pair of slacks?

▶   The jacket can be chosen in 3 ways (*m* ways)
The slacks can be chosen in 4 ways (*n* ways)
Therefore, the two can be chosen in
$$3 \times 4 = 12 \text{ ways } (m \times n \text{ ways})$$

Example: A baseball manager has 6 pitchers and 3 catchers on his team. In how many ways may he select a pitcher and a catcher.

▶   The pitcher can be chosen in 6 ways (*m* ways)
The catcher can be chosen in 3 ways (*n* ways)

Therefore, the pitcher and catcher can be chosen in
$$6 \times 3 = 18 \text{ ways } (m \times n \text{ ways})$$

Example: In a restaurant, a meal consists of appetizer, entree, and desert. A menu has 5 appetizers, 8 entrees, and 4 deserts. In how many ways can a meal be selected from this menu?

▶ The appetizer can be selected in 5 ways (m ways).
The entree can be selected in 8 ways (n ways)
The dessert can be selected in 4 ways (p ways)
Therefore, a meal can be selected in
$$5 \times 8 \times 4 = 160 \text{ ways } (m \times n \times p \text{ ways})$$

Example: A dinner dance is attended by 9 couples. In how many ways may these couples dance if husband and wife are not to dance together?

▶ 1: The total number of ways which couples may dance is 9 $\times 9 = 81$ ways since the man may be selected in 9 ways and the woman may be selected in 9 ways. Included in this number are the 9 cases which a husband and wife dance together. Therefore
$$81 - 9 = 72 \text{ ways}$$

▶ 2: The following arrangement may be used to arrive at the same conclusion.

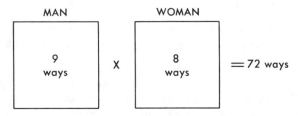

Once the man has been chosen, he has the choice of only 8 partners if he is not to dance with his wife.

Example: In a club of 15 boys and 10 girls, a boy is to be chosen as president, a girl as vice-president, and either a boy or a girl as secretary-treasurer. In how many ways may this be done?

▶ The president may be selected 15 ways and the vice-president in 10 ways. After this has been done, the secretary-treasurer may be selected from the twenty-three remaining members.

Therefore, the selection may be made in $15 \times 10 \times 23 = 3,450$ ways.

PRESIDENT		VICE-PRES.		SEC. TREAS.	
15 ways	X	10 ways	X	23 ways	= 3,450

Using the *nPr* Formula

A set of *n* distinct objects may be arranged in some order in a straight line in a number of ways. Any one of these arrangements is called a *permutation* of these objects. For example, we can arrange a cent, a nickel, and a dime in a straight line in the following 6 ways. Any one of these 6 arrangements is called a permutation.

Also, if we have *n* distinct objects we may wish to arrange *r* of these objects in some order in a straight line. Any one of such arrangements is called a *permutation* of the *n* distinct objects taken *r* a

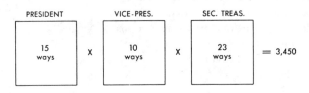

time.   For example, we can arrange two of the coins mentioned
above in a straight line in the following six ways:

In this case, *n* represents the number
3 (objects), and *r* represents the number
2 (2 objects taken at a time.)

In the above example, we saw that the
first coin was chosen in 3 ways and that
after the first coin was chosen it was possible to choose the second coin in 2 ways.
The number of arrangements was therefore $3 \times 2 = 6$ ways.

Example: A firm has 4 job openings.
In how many ways may these jobs be filled
from a list of 9 applicants?

▶   The first job can be filled by any
one of the 9 applicants.   After this job has
been filled the second job can be filled by
any one of the 8 remaining applicants. The
third job can be filled by any one of the 7
applicants remaining after the first two jobs have been filled.   The
last job can be filled by any one of the 6 unplaced applicants.   Thus,
the four jobs can be filled in $9 \times 8 \times 7 \times 6 = 3{,}024$ ways.

In the above examples, the number of ways in which a set of objects can be arranged in order or a number of positions filled in order
are called *permutations*.   In general, if we wish to find the number
of permutations that can be obtained from *n* objects taken *r* at a time
the result would be $n(n - 1)(n - 2) \ldots$ to *r* factors.

Since the second factor is $(n - 1)$, the third factor $(n - 2)$ etc.,
the rth factor is $n - (r - 1)$ or $(n - r + 1)$.

These results may be summarized as follows

$$_nP_r = n(n - 1)(n - 2)(n - 3) \ldots\ldots\ldots (n - r + 1)$$

Example: How many numbers of 3 digits can be written using
the digits 2, 3, 4, 5, 6, 7, 8 if no digits are repeated?

▶ In this problem, $n = 7$ and $r = 3$.
Therefore $_nP_r = {}_7P_3 = 7 \cdot 6 \cdot 5 = 210$ numbers

Example: There are 12 candidates for first, second, and third prizes in a mathematics contest. In how many ways may these prizes be awarded, if no candidate may receive more than one prize?

▶ In this problem, $N = 12$ and $r = 3$
Therefore $_nP_r = {}_{12}P_3 = 12 \cdot 11 \cdot 10 = 1{,}320$ ways

Example: If 8 different signal flags are available, how many signals can be displayed using 4 flags in a row?

▶ 1: In this problem, $n = 8$ and $r = 4$.
Therefore $_nP_r = {}_8P_4 = 8 \cdot 7 \cdot 6 \cdot 5 = 1{,}680$ signals

▶ 2: Problems such as this may also be solved by using the following arrangement

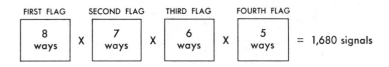

Example: How many odd numbers of 3 digits can be written using the digits 5, 6, 7, 8, 9 if no digits are repeated?

▶ Since the numbers are to be odd the digit on the right must be 5, 7, or 9. After this digit has been selected, the digit on the left can be chosen in 4 ways, and by similar reasoning the digit in the middle can be chosen in 3 ways.
The following diagram shows the solution:

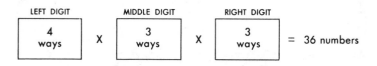

Note: In a problem where there is a restriction on any of the choices, the choice with the restriction should be taken into account first.

Example: How many even numbers less than 1,000 can be formed with the digits 1, 2, 3, 4, 5, 6, 7 if no digit may be repeated?

▶ First, let us consider three-digit numbers. The number of three-digit numbers can be found as follows:

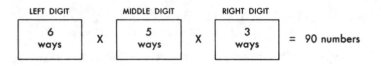

Note: The right digit must be an even number. Therefore, it must be 2, 4, or 6. Since the restriction is on the right digit, this digit must be taken into account first. Second, let us consider two-digit numbers

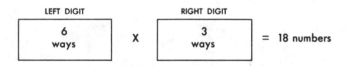

Third, we note that the number of one-digit even numbers = 3 numbers. Total number of even numbers = 111 numbers.

Example: In how many ways may the letters of the word *DOMAIN* be arranged if vowels, and consonants are to alternate?

▶ If the first, third, and fifth places are to be occupied by vowels, the number of arrangements can be shown as follows:

The first place may be filled by any of the three vowels, the third place by any of the remaining vowels, and the fifth place by the one vowel left. The consonants may be selected similarly.

Another possibility is that the first, third and fifth places will be occupied by consonants. The number of arrangements in this case is also

| 3 | X | 3 | X | 2 | X | 2 | X | 1 | X | 1 | = 36 ways |

The total number of arrangements is $36 + 36 = 72$ ways.

Example: A football coach has 15 linemen. In how many ways may the coach arrange the line if only two of the men can play center.

(There are 7 men in the line in football.)

▶ The solution is given by the diagram below

| 14 | X | 13 | X | 12 | X | 2 | X | 11 | X | 10 | X | 9 | = 4,324,320 ways |

The center position can be filled in two ways. There are no restrictions on the other positions.

FACTORIAL NOTATION

Example: In how many ways may 7 books be arranged on a shelf?

▶ In this problem $n = 7$ and $r = 7$

Therefore $_nP_r = {_7P_7} = 7 \cdot 6 \cdot 5 \cdot 4 \cdot 3 \cdot 2 \cdot 1 = 5,040$ ways.

In this case, $n = r$ and the formula $_nP_r$ may be written as $_nP_n$. $_nP_n$ is usually written $n!$ and is read "factorial n."

DEFINITION: In general, the symbol $n!$ is defined for positive integral values of n to be the product of all the integers from 1 to n inclusive.

Examples: $5! = 5 \cdot 4 \cdot 3 \cdot 2 \cdot 1 = 120, 4! = 4 \cdot 3 \cdot 2 \cdot 1 = 24$

Example: Find the value of $\dfrac{10! \times 4!}{6!}$

▶ $\dfrac{10! \times 4!}{6!} \quad \dfrac{10\cdot 9\cdot 8\cdot 7\cdot 6\cdot 5\cdot 4\cdot 3\cdot 2\cdot 1 \times 4\cdot 3\cdot 2\cdot 1}{6\cdot 5\cdot 4\cdot 3\cdot 2\cdot 1}$

Example: In how many ways may 6 students be arranged in a single file in a gymnasium?

▶ In this problem $n = 6$ and $r = 6$
Therefore, $_nP_r = {_nP_n} = {_6P_6} = 6! = 720$ ways.

Permutations of a Set of Objects Whose Elements Are Not All Different

Example: In how many ways can the letters in the word *ERROR* be arranged?

▶ In this case, $n = 5$ and $r = 5$. Therefore, the result would be 5! if the R's could be distinguished from one another. Since the three R's can be arranged in 3! ways we must divide 5! by 3! to obtain the final result 20. In general, the number of permutations of n objects of which u are of one kind, v of another kind, w of a third kind, etc. is $\dfrac{n!}{u!\, v!\, w!}$.

Example: In how many ways may 3 history books, 5 French books, and 4 science books be arranged on a shelf?

▶ There are 12 books in all. These can be arranged in 12! ways. Since 3 are of one kind, 5 of a second kind, and 4 of a third kind we must divide 12! by 3! 5! and 4!

$$\frac{12!}{3!\,5!\,4!} = 27,720 \text{ ways.}$$

Circular Permutations

In arranging objects in a circle one object must be set in place first in order to have a starting point. Therefore, in arranging n objects in a circle we have $(n - 1)!$ arrangements.

Example: In how many ways may 8 people be seated around a dinner table?

▶ In this case, $n = 8$. Since we are finding the number of circular permutations the result is $(n - 1)!$ or $7! = 5,040$ ways.

Example: In how many ways may 5 men and 5 women be seated at a round table so that the men and women alternate?

▶ First, let the women be seated. This may be done in 4! ways. Next, the men may be seated in 5! ways. Therefore, the total number of ways in 4! 5! = 2,880.

Additional Examples in Permutations

Example: If $_nP_2 = 56$, find n

▶ $_nP_2 = n(n - 1)$
$n^2 - n = 56$
$n^2 - n - 56 = 0$
$(n - 8)(n + 7) = 0$
$n = 8, \quad n = -7$ reject.

Example: If $_nP_4 = 5{_n}P_3$, find n

▶ $_nP_4 = n(n - 1)(n - 2)(n - 3)$
$_nP_3 = n(n - 1)(n - 2)$
Therefore, $n - 3 = 5$ or $n = 8$.

Example: In how many ways may 5 different coins be inserted in three toy banks?

▶ The first coin may be inserted in 3 banks, the second coin may be inserted in 3 banks, etc. Thus the five acts may be performed in $3 \cdot 3 \cdot 3 \cdot 3 \cdot 3 = 3^5 = 243$ ways.

Example: How many arrangements can be made from the letters of the word STREAM, if the vowels are not to be separated?

▶ Since the vowels are not to be separated let us regard them as one letter for the time being. Then the result would be 5!. However, the vowels EA may be arranged as EA or as AE. Therefore, the result is $2 \times 5! = 240$ ways.

Example: A scholar has 7 different books on a shelf.   In how many ways can he arrange these books if two of them are not to stand together?

▶   The total number of ways of arranging these books on the shelf is 7!

The number of arrangements in which the two books in question stand together is 2 × 6!

Therefore, the number of arrangements in which the two books in question do not stand together is 7! − 2 × 6! = 3,600 ways.

Example: (a) How many arrangements can be made from the letters of the word TURNPIKE?

(b) How many of these arrangements begin with T?

(c) How many of these arrangements begin with a vowel and end with a consonant?

(d) How many of these arrangements have the vowels in the last 3 places?

▶   (a) Since the letters are all different the number of arrangements is 8! = 40,320 ways.

(b) The T is fixed and the other 7 letters may be arrangement at will.   Therefore, the number of arrangements beginning with T is 7! = 5,040 ways.

(c) The solution is shown in the diagram below.   The first and last box are filled in first.   = 10,800 ways.

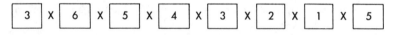

(d) The solution is shown in the diagram below.   The last three boxes are filled in first.   = 720 ways.

Example: A boat's crew consists of 8 men, of whom 2 can row only on the port side, and 3 on the starboard side.   In how many ways can the crew be arranged?

▶ The two men who can row only on the port side can be placed in 4 ways and 3 ways respectively. Similarly, the three men who can row only on the starboard side can be placed in 4 ways, 3 ways and 2 ways respectively. Thus, the men who are restricted may be placed in $4 \times 3 \times 4 \times 3 \times 2 = 288$ ways.

The remaining men may be placed in 3! or 6 ways. Therefore, the number of ways of arranging the crew is $288 \times 6 = 1,728$ ways.

Example: a. How many different numbers of 5 digits may be made from the digits 1, 2, 3, 4, 5?

b. How many of these numbers will be even? odd?

c. How many of these numbers will be greater than 30,000?

d. How many of these numbers will fall between 10,000 and 30,000?

e. How many of these numbers start with 54? In this problem, digits may not be repeated.

▶ a. The number of different numbers is 5! = 120 numbers.

b. The number of even numbers is shown by the diagram below:

EVEN NUMBERS

| 4 | X | 3 | X | 2 | X | 1 | X | 2 | = 48 numbers |

The first box on the right is filled in first.

The number of odd numbers may be obtained by subtracting 48 from 120. There are 72 odd numbers.

The number of odd numbers may also be found by using the diagram below:

ODD NUMBERS

| 4 | X | 3 | X | 2 | X | 1 | X | 3 | = 72 numbers |

The first box on the right is filled in first.

c. The numbers greater than 30,000 must have 3, 4, or 5 as the first digit on the left. The number of such numbers is shown by the diagram below.

| 3 | X | 4 | X | 3 | X | 2 | X | 1 | = 72 numbers |

The first box on the left is filled in first.

d. The number between 10,000 and 30,000 must have 1 or 2 as the first digit on the left. The number of such numbers is shown by the diagram below:

$$\boxed{2} \times \boxed{4} \times \boxed{3} \times \boxed{2} \times \boxed{1} = 72 \text{ numbers}$$

e. The numbers starting with 54 have 5 and 4 as the first two digits on the left. The number of such numbers is shown by the diagram below:

$$\boxed{1} \times \boxed{1} \times \boxed{3} \times \boxed{2} \times \boxed{1} = 6$$

Example: a. How many arrangements can be made from the letters of the word MATHEMATICS?

b. How many of these arrangements begin with an S?

c. How many of these arrangements begin with an M and end with an M?

▶ a. Since there are 11 letters in the word with two M's, two A's, and two T's the number of arrangements is

$$\frac{11!}{2!\,2!\,2!} \text{ or } 4,989,600.$$

b. In this case, we have one letter fixed. Therefore, the number of arrangements is

$$\frac{10!}{2!\,2!\,2!} \text{ or } 453,600.$$

c. In this case, we have two identical letters fixed. Therefore, the number of arrangements is

$$\frac{9!}{2!\,2!} \text{ or } 90,720.$$

Example: A girl has 8 beads of different colors. She wishes to make a necklace of these beads by stringing them together. How many different necklaces can she make?

▶ The beads are to be arranged in circular fashion. Therefore, there would normally be 7! arrangements. However, only half of these arrangements are distinct since the other half of these can be obtained simply by turning the necklace over. Therefore, the result is $\frac{1}{2}(7!)$ or 2,520 arrangements.

Example: A boat seats 6 people on each side. Five people enter the boat, two of whom must sit on the left side. In how many ways may these people be seated?

▶ The first person who must sit on the left side has a choice of 6 places. The second person who must sit on the left side has a choice of 5 places. The others are unrestricted and may sit in the remaining 10, 9, and 8 places respectively. Therefore, the total numbers of ways is $6 \times 5 \times 10 \times 9 \times 8 = 21,600$ ways.

Example: A party of 5 men and 5 women enter a living room which contains a sofa seating three, two armchairs, and five other chairs. In how many ways may these people be seated if two of the women must sit on the sofa, and one of the men must occupy an armchair?

▶ The first woman has a choice of 3 places on the sofa, the second woman has a choice of 2 places on the sofa and the man has a choice of 2 armchairs. The other members may be seated in the remaining places without restriction. Therefore, the total number of ways is $3 \cdot 2 \cdot 2 \cdot 7 \cdot 6 \cdot 5 \cdot 4 \cdot 3 \cdot 2 \cdot 1 = 60,480$ ways.

Example: An instructor writes an examination paper containing ten questions. He wishes to place three of the simpler questions as the first three questions and two of the difficult questions at the end of the examination. In how many ways may he arrange the examination paper?

▶ The diagram below shows the solution: = 1,440 ways.

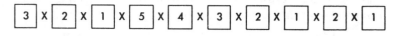

$$\boxed{3} \times \boxed{2} \times \boxed{1} \times \boxed{5} \times \boxed{4} \times \boxed{3} \times \boxed{2} \times \boxed{1} \times \boxed{2} \times \boxed{1}$$

Note that the easy questions and the difficult questions are accounted for first.

EXERCISE 1

Solve the following problems:

1. Find the value of n if $4 \cdot {_nP_4} = 12 \cdot {_{n-1}P_3}$

2. Find the value of n if ${_nP_5} = 10 \cdot {_nP_4}$

3. A girl has 8 shirts and 5 skirts. How many different outfits can she wear?

4. A large office building has 8 exits. In how many ways may a man enter the building and leave by a different exit?

5. At a college, there are 6 instructors who teach freshman English, 4 instructors who teach freshman history, and 3 instructors who teach freshman science. If a freshman takes English, history, and science, how many different sets of instructors may he have?

6. In how many ways can 3 different prizes be awarded among 8 people if no person is to receive more than one prize?

7. In how many ways can the letters of the word CALIPER be arranged?

8. Using the digits 3, 4, 5, 6, 7, 8, 9 how many four-digit numbers can be written if no digits are repeated?

9. A bridge match has 16 contestants. In how many ways may three prizes be awarded if no player is to receive more than one prize?

10. Five travelers reach a city which has 3 good hotels. In how many ways may these travelers select a hotel?

11. Using the digits 1, 2, 3, 4, 5, 6 how many three-digit numbers can be written if no digits may be repeated?

12. A square dance was attended by 8 couples. In how many ways may these couples dance if no husband is to dance with his wife?

13. How many odd number of 3 digits can be written using the digits 4, 5, 6, 7, 8 if no digits are repeated?

14. At a club election there were 3 candidates for president, 5 candidates for vice-president and 6 candidates for secretary-treasurer. In how many ways may these officers be chosen?

15. License plates are designated by a letter followed by four digits. How many such designations can be made? (Digits may be repeated and 0 is considered a digit)

16. A room contains 6 chairs. In how many ways may 4 people be seated in these chairs?

17. In how many ways may 8 books be arranged on a shelf?

18. How many even numbers less than 200 may be formed from the digits 1, 2, 3, 4, 5, 6, if no digit may be repeated?

19. In how many ways may the letters of the word PRECIOUS be arranged if vowels and consonants are to alternate?

20. A basketball coach has a squad of 12 players. In how many ways may the coach arrange his lineup if only 3 players can play center and the others can each play any of the other 4 positions?

21. In how many ways may the letters of the word PARALLEL be arranged?

22. In how many different ways may 4 green flags and 3 red flags be arranged one above another?

23. In how many ways may 6 people be seated around a dinner table?

24. In how many ways may 5 mathematics books and 4 science books be arranged on a shelf?

25. In how many ways may 6 men and 6 women be seated at a round table so that the men and the women alternate?

26. In how many ways may 4 different articles of clothing be placed in 3 closets?

27. In how many ways may 6 boys be arranged in a line if 2 of the boys must be kept together?

28. How many arrangements can be made from the letters of the word CHEMIST if each arrangement is to start with a consonant and end with a vowel?

29. a. How many different numbers of 4 digits may be made from the digits 3, 4, 5, 6, 7? (Digits may not be repeated.)

b. How many of these numbers will be even? odd?

c. How many of these numbers will be greater than 5,000?

d. How many of these numbers will be less than 6,000?

e. How many of these numbers will start with an even number and end in an odd number?

30. a. How many arrangements can be made from the letters of the word MUSICAL?

b. How many of these arrangements begin with M?

c. How many of these arrangements begin with a consonent and end with a vowel?

d. How many of these arrangements begin with the letters MUS?

31. A girl has 7 beads of different colors. She wishes to make a bracelet of these beads by stringing them together. How many different bracelets can she make?

32. In how many ways may 2 dimes, 5 nickels, and 1 penny be arranged in a line?

33. A bus seats 6 passengers on each side. In how many ways can 12 passengers be seated if two of them must sit on the right side and three must sit on the left side?

34. a. How many arrangements can be made from the letters of the word CURRICULUM?

b. How many of these arrangements begin with the letter M?

c. How many of these arrangements begin with the letters CC?

d. How many of these arrangements begin with an R and end with an R?

COMBINATIONS

In permutation problems we are interested in discovering the number of *arrangements* of a number of objects. In combination problems we are interested in discovering the number of ways in which a number of objects may be *selected* without regard to their arrangement. For example, if we wish to find the number of ways in which 9 boys may be seated on 4 chairs we have a problem in permutations. If, however, we wish to find the number of ways in which we can select committees of 4 from a group of 9 boys (without regard to arrangement) we have a problem in combinations.

Identify each of the following as a problem in permutations or as a problem in combinations.

1. A basketball squad is composed of 12 players. How many teams of five players each may be chosen from this squad if each player is capable of playing all positions.

2. Five points lie on the circumference of a circle. How many inscribed triangles can be drawn having these points as vertices?

3. There are 5 panel speakers at a discussion meeting. In how many ways may the chairman call upon the five speakers?

4. A supermarket advertises 8 meat specials. In how many ways can a housewife select 5 of these specials?

5. In an art gallery, 6 pictures are to be arranged in a horizontal line. In how many ways may this be done?

Answers—1. Combination—arrangement is not involved. 2. Combination—arrangement is not involved since $\triangle$ ABC is the same as $\triangle$ BAC or $\triangle$ CAB. 3. Permutation—the order in which the speakers are called upon is involved. 4. Combination—choice, but not order, is involved. 5. Permutation—arrangement is involved.

Formula for Combinations

Combinations involve the number of selections while permutations involve the number of arrangements. Let us consider the following examples.

Example: In how many ways may 7 people be seated on 3 chairs?

▶ $_7P_3 = 7 \cdot 6 \cdot 5 = 210$ ways

Example: In how many ways may a committee of 3 be selected to form a group of 7 people?

▶ This problem is similar to example 19 except that we are not interested in the arrangement of the people chosen.

Since each committee can be arranged in 3! ways the solution to the problem is $\dfrac{_7P_3}{3!} = 35$ ways. Therefore,

$_7C_3 = \dfrac{_7P_3}{3!}$ (where C represents the number of combinations).

In general, $_nC_r = \dfrac{_nP_r}{r!} = \dfrac{n(n-1)(n-2)\ldots\ldots(n-r+1)}{r!}$

Example: From a group of 15 soldiers 3 are to be chosen for special duty. In how many ways may this choice to be made?

▶ $_{15}C_3 = \dfrac{15 \cdot 14 \cdot 13}{3!} = \dfrac{15 \cdot 14 \cdot 13}{1 \cdot 2 \cdot 3} = 455$ ways

Example: A student must read 4 books from a list of 12. In how many ways may he select these books?

▶ $_{12}C_4 = \dfrac{12 \cdot 11 \cdot 10 \cdot 9}{4!} = 495$ ways

Example: A firm received 15 applications for 10 openings in its sales department. In how many ways may the firm fill the openings?

▶ $_{15}C_{10} = \dfrac{15 \cdot 14 \cdot 13 \cdot 12 \cdot 11 \cdot 10 \cdot 9 \cdot 8 \cdot 7 \cdot 6}{10 \cdot 9 \cdot 8 \cdot 7 \cdot 6 \cdot 5 \cdot 4 \cdot 3 \cdot 2 \cdot 1}$

$= 3{,}003$ ways

In the solution of this problem it is clear that there is overlapping in the numerator and denominator (i.e., from 10 down to 6). This will be true when r is greater than $\frac{1}{2}$ of n. In order to simplify computation we make use of the following theorem $_nC_r = {_nC_{n-r}}$. In the above case we see that $_{15}C_{10} = {_{15}C_5}$.

Example: Find the value of $_{24}C_{20}$.

▶ In this case, $n = 24$ and $r = 20$. Therefore, $n - r = 4$.

Since $_nC_r = {_nC_{n-r}}$, $_{24}C_{20} = {_{24}C_4} = \dfrac{24 \cdot 23 \cdot 22 \cdot 21}{1 \cdot 2 \cdot 3 \cdot 4} = 10{,}626$

Example: The Brown Taxi Co. has a fleet of 30 cabs. During the early evening it has 25 cabs in service. In how many ways can the 25 cabs be selected?

▶ In this case, $n = 30$, $r = 25$, and $n - r = 5$.
$_{30}C_{25} = {_{30}C_5} = \dfrac{30 \cdot 29 \cdot 28 \cdot 27 \cdot 26}{1 \cdot 2 \cdot 3 \cdot 4 \cdot 5} = 142{,}506$ ways.

Total Number of Combinations of *n* Different Objects

Example: A man has 5 close friends. In how many ways may he invite 1 or more of them to accompany him on a hunting trip?

▶ He may invite 1 friend in 5 ways. He may invite 2 friends in $_5C_2$ ways, or 10 ways. He may invite 3 friends in $_5C_3$ ways, or 10 ways. He may invite 4 friends in $_5C_4$ ways, or 5 ways. He may invite all 5 friends in $_5C_5$ ways, or 1 way. The total is 31 ways.
The same result may be obtained by using the formula
$2^n - 1 = 2^5 - 1 = 32 - 1 = 31$ ways

In general, the total number of combinations of *n* things taking any number at a time from 1 to *n* is $2^n - 1$ ways.

Example: A woman has in her purse a dollar, a half-dollar, a quarter, a dime, a nickel, and a penny. In how many ways may she draw a sum of money from this purse?

▶ It is required to find the total number of combinations of 6 things taking from 1 to 6 things at a time. Therefore, the result is $2^6 - 1 = 64 - 1 = 63$ ways.

Example: From a group of 7 Republicans and 5 Democrats how many different committees can be formed consisting of 3 Republicans and 2 Democrats?

▶ The Republicans members may be chosen in $_7C_3$ or 35 ways. The Democrats may be chosen in $_5C_2$ or 10 ways. Using the fundamental principal, the Republicans may be chosen in 35 ways (*m* ways) and the Democrats may be chosen in 10 ways (*n* ways). Therefore, both choices may be made in 35×10 (*m* × *n*) ways $= 350$ ways.

Example: Out of 6 consonants and 3 vowels, how many words can be made each containing 3 consonants and 2 vowels.

▶ The number of ways in which 3 consonants and 2 vowels can be chosen is $_6C_3 \times _3C_{22} = 20 \times 3 = 60$ ways. Each of these 60 selections can now be arranged in $_5P_5$ ways. Therefore, the number of words that can be formed is $60 \times _5P_5 = 60 \times 120 = 7,200$ words.

Example: How many different juries, each of 12 people, can be selected from a panel of 15 people?

▶   In this case, $n = 15$ and $r = 12$, $_nC_n = {_{15}C_{12}}$.
Since $_nC_r = {_nC_{n-r}}$, $_{15}C_{12} = {_{15}C_3} = 455$

Example: If $_nC_2 = 45$, find n

▶ $_nC_2 = \dfrac{n(n-1)}{2} = 45$

$n^2 - n = 90, \quad n^2 - n - 90 = 0$

$(n - 10)(n + 9) = 0$

$n = 10, \quad n = -9$ reject.

Example: How many different straight lines are determined by 10 points if no three of the points lie in the same straight line?

▶   Since two points determine a straight line, the number of lines determined is $_{10}C_2 = 45$ lines.

Example: The executive board of a certain society consists of 8 members. How many committees of 4 can be chosen from this board if a certain member of the board is to serve on every committee?

▶   Since one member must always be included we have a choice of 3 members out of 7, or $_7C_3$.

$$_7C_3 = \frac{7 \cdot 6 \cdot 5}{1 \cdot 2 \cdot 3} = 35 \text{ choices}$$

Example: A basketball squad consists of 10 men. In how many ways can a team of 5 be selected to form the squad if two of the men can only play center?

▶   For the positions except center we have a choice of 4 men out of 8, or $_8C_4$

$$_8C_4 = \frac{8 \cdot 7 \cdot 6 \cdot 5}{1 \cdot 2 \cdot 3 \cdot 4} = 70$$

Since we can associate 2 different centers with each of the 70 groups of 4 men, the total number of ways is $2 \times 70 = 140$ ways.

Example: A polygon has n sides. How many diagonals does the polygon have?

▶ Since the polygon has n sides it also has n vertices. Therefore, the total number of straight lines joining these points is $_nC_2 = \dfrac{n(n-1)}{2}$. Of this total, n lines represent the sides of the polygon. Therefore, the number of diagonals is $\dfrac{n(n-1)}{2} - n = \dfrac{n^2 - 3n}{2}$.

Example: There are 18 men attending a director's meeting. If each man shakes the hand of every other man, how many handshakes are there?

▶ Each handshake represents a combination of two men. Therefore, the total number of handshakes is $_{18}C_2 = \dfrac{18 \cdot 17}{1 \cdot 2} = 153$ handshakes.

Example: At a succeeding meeting there were 105 handshakes with each man shaking the hand of every other man. How many men attended the meeting?

▶ $_nC_2 = 105, \quad \dfrac{n(n-1)}{2} = 105$

$n^2 - n = 210, \qquad (n-15)(n+14) = 0$

$n = 15, \qquad n = -14 \text{ reject}$

15 men attended the meeting.

Example: If $_nC_3 = {_nC_{12}}$ find $_{18}C_n$.

▶ Since $_nC_r = {_nC_{n-r}}, r = 3$ and $n - r = 12$.

Therefore $n - 3 = 12$, and $n = 15$.

$$_{18}C_n = {_{18}C_{15}} = \dfrac{18 \cdot 17 \cdot 16}{1 \cdot 2 \cdot 3} = 816$$

Example: If $_{2n}C_5 = {_{2n}P_4}$ find the value of n.

▶ $_{2n}C_5 = \dfrac{2n(2n-1) \cdot (2n-2) \cdot (2n-3) \cdot (2n-4)}{1 \cdot 2 \cdot 3 \cdot 4 \cdot 5}$

$_{2n}P_4 = 2n(2n-1)(2n-2)(2n-3)(2n-4)$

$\dfrac{2n(2n-1)(2n-2)(2n-3)(2n-4)}{1 \cdot 2 \cdot 3 \cdot 4 \cdot 5}$

$= 2n(2n-1)(2n-2)(2n-3)$

$$2n - 4 = 120$$
$$2n = 124$$
$$n = 62$$

Example: A girl has 9 dresses and 3 coats. In how many ways may she select 6 dresses and 2 coats for a trip?

▶ She may select the dresses in $_9C_6$ ways = 84 ways.
She may select the coats in $_3C_2$ ways = 3 ways.
Both selections may be made in $84 \times 3 = 252$ ways.

Example: There are 8 men in a work crew. In how many ways may 3 men be assigned to one task and 5 men be assigned to another task?

▶ The three men may be selected in $_8C_3$ ways = 56 ways. After this is done the remaining 5 men are assigned in $_5C_5$ ways = 1 way $= 56 \times 1 = 56$ ways.

Example: Out of a squad of 12 baseball players, in how many ways may a team be selected? Every change of position by a player represents a new team.

▶ Nine players may be selected in $_{12}C_9$ ways = 220 ways. Each selection of 9 players may be arranged as a team in 9! ways. Therefore, the total number of teams is $220 \times 9! = 79,833,600$ ways.

Example: In how many ways may 12 books be divided equally among 4 students?

▶ The selection for the first student may be made in $_{12}C_3$ ways. After this is done, the selection for the second student may be made in $_9C_3$ ways. The selection for the third student may be made in $_6C_3$ ways and the selection for the fourth student may be made in $_3C_3$ ways.

$$_{12}C_3 \times {}_9C_3 \times {}_6C_3 \times {}_3C_3 = 369,600 \text{ ways.}$$

Example: A pack of playing cards contains 52 cards.
a. How many selections of 3 cards may be made from a pack?
b. Of these selections, how many will contain one king?

  c. Of these selections, how many will contain one king and one queen?

  d. Of these selections, how many will contain all spades?

  e. Of these selections, how many will contain all red cards?

▶

  a. $_{52}C_3 = 22,100$ ways.

  b. A king may be selected in $_4C_1$ ways. The other 2 cards may be selected in $_{48}C_2$ ways since the kings are excluded in this selection. $_4C_1 \times _{48}C_2 = 4,512$ ways.

  c. One king and one queen may each be selected in $_4C_1$ ways. This leaves 44 cards for the third selection.
$_4C_1 \times _4C_1 \times _{44}C_1 = 704$ ways.

  d. Since there are 13 spades in a deck, the selection may be made in $_{13}C_3$ ways $= 286$ ways.

  e. Since half the cards in the deck are red cards, the selection may be made in $_{26}C_3$ ways $= 2,600$ ways.

  Example: A Congressional committee consists of 7 Democrats and 5 Republicans.

  a. How many sub-committees of 4 members may be chosen from the larger committee?

  b. How many of these committees will consist of 3 Democrats and 1 Republican?

  c. How many of these committees will consist of 2 Democrats and 2 Republicans?

  d. How many of these committees will consist of 1 Democrat and 3 Republicans?

  e. How many of these committees will contain no Republicans?

  f. How manyof these committees will contain no Democrats?

▶

  a. Sub-committees may be selected in $_{12}C_4 = 495$ ways.

  b. Sub-committees consisting of 3 Democrats and 1 Republican may be selected in $_7C_3 \times _5C_1$ ways $= 175$ ways.

  c. Sub-committees consisting of 2 Democrats and 2 Republicans may be selected in $_7C_2 \times _5C_2$ ways $= 210$ ways.

  d. Sub-committees consisting of 1 Democrat and 3 Republicans may be selected by $_7C_1 \times _5C_3$ ways $= 70$ ways.

e. If a sub-committee contains no Republicans it will contain 4 Democrats. Sub-committees consisting of 4 Democrats may be selected in $_7C_4$ ways = 35 ways.

f. Committees of 4 Republicans may be selected in $_5C_4 = 5$ ways.

Example: Out of 7 consonants and 5 vowels, how many five letter words can be formed, each word containing at least 2 vowels?

▶ The selection requirements may be met by having 2 consonants and 3 vowels or 3 consonants and 2 vowels. After the selections are made each selection may be arranged in 5! ways.

$$_7C_2 \times _3C_3 \times 5! = \quad 2,520$$
$$_7C_3 \times _3C_2 \times 5! = 12,600$$
$$\text{Total} = 15,120$$

Example: In how many ways may 6 travelers select hotels so that 3 select one hotel, 2 select a second hotel, and 1 selects a third hotel?

▶ A group of 3 travelers may select a hotel in $_6C_3$ ways. This leaves 3 travelers from whom to select 2 for the second hotel. After five travelers have been placed in hotels, one traveler is left for the third hotel.

$$_6C_3 \times _3C_2 \times _1C_1 = 60 \text{ ways.}$$

Example: From 10 sophomores and 12 freshmen a committee of 6 was chosen so that the number of sophomores should not exceed the number of freshmen. In how many ways could this be done?

▶ The requirement may be met as follows:

3 sophomores and 3 freshmen	$_{10}C_3 \times _{12}C_3 =$	26,400
2 sophomores and 4 freshmen	$_{10}C_2 \times _{12}C_4 =$	22,275
1 sophomore and 5 freshmen	$_{10}C_1 \times _{12}C_5 =$	7,920
no sophomores and 6 freshmen	$_{12}C_6 =$	924
	Total =	57,519

Example: In a bowl there are 10 tickets of which 3 are prizes and 7 are blanks.

a. In how many ways can a person draw 3 tickets from the bowl?

b. How many of these drawings will contain 3 prizes?

c. How many of these drawings will contain 2 prizes and a blank?

d. How many of these drawings will contain 1 prize and 2 blanks?

e. How many of these drawings will contain 3 blanks?

▶ a. Three tickets may be drawn in $_{10}C_3 = 120$ ways.

b. Since there are only three prize winning tickets they may be selected in $_3C_3 = 1$ way.

c. Two prizes and 1 blank may be drawn in $_3C_2 \times _7C_1$ ways $= 21$ ways.

d. One prize and 2 blanks may be drawn in $_3C_1 \times _7C_2$ ways $= 63$ ways.

e. Three blanks may be selected in $_7C_3$ ways $= 35$ ways.

The reader will note that the 4 possible ways of drawing 3 tickets add up to the total number of ways of selecting 3 tickets i.e., 120 ways.

Example: A student club has a membership of 24 including the officers. The officers include a president, a vice-president, and a secretary.

a. In how many ways may a committee of 3 be chosen?

b. How many committees of 3 can be chosen if the president must be a member of the committee?

c. How many committees of 3 can be chosen if the president and the secretary must be members of each committee?

d. How many committees of 3 can be chosen if one of the officers may be on a committee?

e. How many committees of 5 can be chosen if all 3 of the officers are on each committee?

▶

a. $_{24}C_3 = 2,024$ ways.

b. Since the president must be on each committee, the number of committees is $_1C_1 \times _{23}C_2 = 253$ committees.

c. $_1C_1 \times _1C_1 \times _{22}C_1 = 22$ committees.

d. $_{21}C_3 = 1,330$ committees.

e. $_{21}C_2 = 210$ committees.

Example: A board of directors consists of 10 members. In how many ways may a committee of the board of directors be selected?

▶ Since the number of members of a committee is not specified we may consider that a committee may consist of 1 to 10 members. The number of committees is

$$_{10}C_1 + _{10}C_2 + _{10}C_3 + _{10}C_4 + _{10}C_5 + _{10}C_6 + _{10}C_7 + _{10}C_8 + _{10}C_9 + _{10}C_{10} = 1,023$$

The same result may be obtained by using the formula $2^n - 1$. In this case, $2^n - 1 = 2^{10} - 1 = 1,023$.

EXERCISE 2

Solve the following problems:

1. An examination paper contains 10 questions. In how many ways may a student select 8 of these questions?

2. From a group of 6 people, in how many ways may 4 be selected for a bridge game?

3. How many triangles can be formed from 12 points no 3 of which are in a straight line?

4. In how many ways can a baseball nine be selected from a group of 15 players if every player can play every position?

5. How many diagonals can be drawn in a decagon (a polygon of 10 sides)?

6. A college building has 28 classrooms. During the 9 o'clock hour, 25 of these classrooms are in use. In how many ways may these 25 rooms be selected?

7. A woman has 6 close friends. In how many ways may she invite 1 or more of them to a luncheon?

8. From a group of 8 seniors and 6 juniors how many different committees can be formed consisting of 4 seniors and 3 juniors?

9. Out of 7 consonants and 5 vowels, how many words can be made each containing 4 consonants and 2 vowels?

10. A faculty committee consists of 9 professors. How many sub-committees of 4 can be chosen from this board if a certain professor on the committee must be on each sub-committee?

11. In how many ways may 15 Christmas gifts be divided equally among 3 children?

12. A pack of playing cards contains 52 cards. In how many ways may 4 cards be selected if one and only one card is a king and one and only one card is a queen?

13. A Senatorial committee consists of 5 Democrats and 4 Republicans. How many sub-committees of 3 Democrats and 2 Republicans may be selected from the larger committee if the Democratic chairman of the larger committee must be on each sub-committee?

14. A student club has a membership of 20, consisting of 12 seniors and 8 juniors. In how many ways may a slate of 3 officers, president, vice-president, and secretary be selected if the president must be a senior?

15. At a convention meeting each guest shook the hand of every other guest. If there were a total of 105 handshakes, how many guests attended the meeting?

16. A baseball league contains 8 teams. Each team plays each of the other teams 12 times. How many games are played?

17. A box contains 18 cards of which two are prize cards and the rest are blanks. In how many ways may a person draw 3 cards such that 1 is a prize and the other 2 are blanks?

18. Find the value of n if $_{(n+2)}C_5 = {_n}C_3$.

19. A class contains 6 seniors, 12 juniors, and 3 sophomores

a. In how many ways may a committee of 5 be chosen from the members of the class?

b. In how many ways may a committee of 5 be chosen if the committee is to be composed of 2 seniors, 2 juniors and 1 sophomore?

c. In how many ways may a committee of 5 be chosen if the committee is to contain 1 and only 1 senior?

d. In how many ways may a committee of 5 be chosen if the committee is to contain no sophomores?

e. In how many ways may a committee of 3 be chosen if the committee is to contain at least 1 senior?

20. A wallet contains a $1 bill, a $2 bill, a $5 bill, a $10 bill, a $20 bill, and a $50 bill. How many different sums of money can be formed from the bills in the wallet?

21. At a bridge party attended by 8 people, two tables of 4 players each are selected. In how many ways may this be done if the order in which the players are seated at tables is not considered?

22. A store employs 12 workers. Of these 9 are assigned to sales work and the other 3 to other tasks. In how many ways may this be done?

23. From a detachment of 5 officers and 12 men, in how many ways may 2 officers and 3 men be selected and then arranged in a line?

24. An examination paper consists of 3 parts. The first part contains 7 questions, the second part contains 5 questions, and the third part contains 3 questions. The student is instructed to omit 2 questions from each group, thus answering a total of 9 questions. In how many ways may a student make his selection?

25. How many different planes are determined by 18 points in space if no four of the points are in the same plane?

ANSWERS

EXERCISE 1

1. 3	**23.** 120
2. 14	**24.** 126
3. 40	**25.** 86,400
4. 56	**26.** 81
5. 72	**27.** 240
6. 336	**28.** 1,200
7. 5,040	**29.** a. 120
8. 840	b. 48, 72
9. 3,360	c. 72
10. 243	d. 72
11. 120	e. 36
12. 56	**30.** a. 5,040
13. 24	b. 720
14. 90	c. 1,440
15. 260,000	d. 24
16. 360	**31.** 360
17. 40,320	**32.** 168
18. 30	**33.** 18,144,000
19. 1,152	**34.** a. 151,200
20. 23,760	b. 15,120
21. 3,360	c. 3,360
22. 35	d. 3,360

EXERCISE 2

1. 45
2. 15
3. 220
4. 5,005
5. 35
6. 3,276
7. 63
8. 1,400
9. 252,000
10. 56
11. 756,756
12. 15,136
13. 36
14. 2,052
15. 15

16. 336
17. 240
18. 3
19. a. 20,349
    b. 2,970
    c. 8,190
    d. 8,568
    e. 875
20. 63
21. 70
22. 220
23. 264,000
24. 630
25. 816

CHAPTER 19

Probability

DEFINITION: If an event can happen in *s* ways and fail to happen if *f* ways, and if all these ways $(s + f)$ are assumed to be equally likely, then the probability (P) that the event will happen is

$$P = \frac{s}{s + f} \quad \frac{\text{(successful ways)}}{\text{(total ways)}}$$

Example: If a letter is selected at random from the word LIBRARY, what is the probability that the letter will be an R?

▶ The number of ways in which an R will be selected is 2 (successful ways). The total number of ways in which a letter may be selected is 7. Therefore, the probability of selecting an R is

$$P = \frac{s}{s + f} \quad \frac{\text{(successful ways)}}{\text{(total ways)}} = \frac{2}{7}$$

Example: A box contains 4 red balls and 7 black balls. If 1 ball is drawn from the box what is the probability that the ball will be a black one?

▶ The number of ways in which a black ball may be drawn is 7 (successful ways). The total number of ways in which a ball may be drawn is 11. Therefore, the probability of selecting a black ball is:

$$P = \frac{s \text{ (successful ways)}}{s + f \text{ (total ways)}} = \frac{7}{11}$$

Example: A committee of 3 is to be chosen by lot from a group of 9 men, one of whom is Mr. Smith. What is the probability that Mr. Smith will be among those chosen?

▶ The number of ways in which a committee of 3 can be chosen from 9 men is $_9C_3$, i.e. the total number of ways is 84. The number of ways in which a committee of 3 containing Mr. Smith can be chosen is $_1C_1 \cdot {_8C_2} = 28$. Therefore, the probability of selecting a committee containing Mr. Smith is

$$P = \frac{s}{s + f} = \frac{28}{84} = \frac{1}{3}$$

DEFINITION: If the number of successful ways is equivalent to the total number of ways then the probability, $P = 1$. Thus, a probability of 1 represents certainty. Also, if an event is certain to fail then its probability, $P = 0$.

Example: A box contains 3 white balls and 5 red balls. If a ball is drawn at random, what is the probability that the ball drawn is either white or red?

▶ The number of ways in which a white or red ball can be drawn is 8. The total number of ways in which a ball can be drawn is 8. Therefore, the probability

$$P = \frac{\text{successful}}{\text{total}} = \frac{8}{8} = 1$$

DEFINITION: If the probability that an event will happen is $\frac{a}{b}$ then the probability that this event will not happen is $1 - \frac{a}{b}$.

Example: A lottery offers first and second prizes. If 25 tickets are sold and *A* buys 3 of them what is the probability that *A* will not win a prize?

▶ *A* may draw 2 prizes and 1 blank in $_2C_2 \cdot {}_{23}C_1 = 23$ ways. *A* may draw 1 prize and 2 blanks in $_2C_1 \cdot {}_{23}C_2 = 506$ ways. Hence, the total number of ways in which *A* may draw a prize is $23 + 506 = 529$ ways. The total number of ways of drawing 3 tickets out of 25 is $_{25}C_3 = 2300$. Therefore, the probability of *A* winning at least one prize is

$$P = \frac{\text{successful}}{\text{total}} = \frac{529}{2300}$$

Since the probability of *A* winning a prize is $\dfrac{529}{2300}$, the probability of *A* not winning a prize is

$$1 - \frac{529}{2300} = \frac{1771}{2300} = \frac{77}{100}$$

Odds

DEFINITION: If an event can happen in *s* ways and fail to happen in *f* ways then, if $s > f$ the odds are *s* to *f* in favor of the event happening; and if $f > s$ then the odds are *f* to *s* against the event happening.

Example: A man has 9 white shirts and 5 colored shirts in his closet. If he selects a shirt at random what are the odds in favor of his selection of a white shirt?

▶ The odds in this case are *s* to *f*. Since $s = 9$ and $f = 5$, the odds in favor of the event are $9 : 5$.

Example: A deck has 52 cards. If a card is selected at random what are the odds against selecting a king?

▶ The odds in this case are *f* to *s*. In a deck of 52 cards, there are 4 kings and 48 cards that are not kings. Since $f = 48$ and $s = 4$, the odds against the selection of a king are $48 : 4$ or $12 : 1$.

INDEPENDENT EVENTS

DEFINITION: Two or more events are said to be *independent* if the occurrence of one event has *no* effect upon the occurrence or non-occurrence of the other.

For example, suppose a bag contains 9 balls numbered 1 to 9. If one ball is drawn at random the probability that it is numbered 5 is $\frac{1}{9}$. Now if this ball is replaced and another ball is drawn at random the probability that it is again numbered 5 is $\frac{1}{9}$. The probability of drawing a 5 on the second draw is in no way dependent upon the results of the first draw.

We shall state the following theorem without proof.

THEOREM: If two events are independent and the probability of occurrence of the first event is p and the probability of occurrence of the second event is q, then the probability that both will happen in the order stated is pq.

Example: A coin is tossed five times in succession. What is the probability that five consecutive tails turn up?

▶ The probability of obtaining a tail on the first toss is $\frac{1}{2}$. The second toss is independent of the first and the probability of obtaining a tail again is $\frac{1}{2}$. Thus, the probabilities of the five independent events are $\frac{1}{2}, \frac{1}{2}, \frac{1}{2}, \frac{1}{2}$, and $\frac{1}{2}$. Therefore, the probability of obtaining 5 tails is $\frac{1}{2} \cdot \frac{1}{2} \cdot \frac{1}{2} \cdot \frac{1}{2} \cdot \frac{1}{2} = \frac{1}{32}$.

Example: On an examination, a student answers a series of 3 multiple-choice questions each of which has 4 choices. What is the probability that he answers all 3 questions correctly, if he does not know the correct answer to any of the questions?

▶ The probability of answering each question correctly is $\frac{1}{4}$. Since these are independent events the probability of obtaining 3 correct answers is $\left(\frac{1}{4}\right)^3 = \frac{1}{64}$.

Example: One bag contains 3 white balls and 5 red balls. Another bag contains 4 white balls and 7 red balls. If one ball is drawn from each bag, what is the probability that both balls will be white?

▶ The probability of drawing a white ball from the first bag is $\frac{3}{8}$. The probability of drawing a white ball from the second bag is $\frac{4}{11}$. The probability of drawing two white balls is $\frac{3}{8} \times \frac{4}{11} = \frac{3}{22}$.

DEPENDENT EVENTS

DEFINITION: Two or more events are said to be *dependent* if the occurrence of one event has *an* effect upon the occurrence or non-occurrence of the other.

For example, suppose a box contains 4 white balls and 7 red balls. If 3 balls are drawn in succession without replacing any ball after it has been drawn the probability of drawing 3 white balls is affected by the fact that each successive drawing changes the probability of drawing a white ball. We shall use the following theorem to solve this problem.

THEOREM: If the probability of occurrence of one event is p and the probability of the occurrence of a second event is q then the probability that both events will happen in the order stated is pq.

Example: A box contains 4 white balls and 7 red balls. If 3 balls are drawn in succession without replacing any ball after it has been drawn, what is the probability of drawing 3 white balls?

▶ The probability of drawing a white ball on the first draw is $\frac{4}{11}$. There are now 10 balls left in the box of which 3 are white. Therefore, the probability of drawing a white ball on the second draw is $\frac{3}{10}$. In the same manner, the probability of drawing a white ball on the third draw is $\frac{2}{9}$. Therefore, the probability of drawing 3 white balls in succession is the product of the 3 individual probabilities
$\frac{4}{11} \times \frac{3}{10} \times \frac{2}{9} = \frac{4}{165}$.

MUTUALLY EXCLUSIVE EVENTS

DEFINITION: Two events are said to be mutually exclusive if the occurrence of one of them precludes the occurrence of the other.

For example, if two dice are thrown and the sum of the dice is 5 then it is impossible that the sum of the dice be any other number, such as 7 or 9.

The following theorem is used in solving problems involving mutually exclusive events.

THEOREM: The probability that either one of two mutually exclusive events will occur is the *sum* of the probabilities of the separate events.

Example: If two dice are thrown what is the probability that the sum of the dice will be either 5 or 8?

▶ Each one of the dice has 6 faces. Therefore, the total number of ways in which the two dice may turn up is $6 \times 6 = 36$ ways. The two dice will yield a sum of 5 in the following ways

$$4+1, \ 1+4, \ 3+2, \text{ and } 2+3, \text{ or 4 ways.}$$

Therefore, the probability of obtaining a sum of 5 is $\frac{4}{36}$. In like manner, the two dice will yield a sum of 8 in the following ways

$$2+6, \ 6+2, \ 3+5, \ 5+3, \text{ and } 4+4 \text{ or 5 ways.}$$

Therefore, the probability of obtaining a sum of 8 is $\frac{5}{36}$. The probability of obtaining either a 5 or an 8 is the sum of the individual probabilities i.e., $\frac{4}{36} + \frac{5}{36} = \frac{9}{36} = \frac{1}{4}$.

Example: From a group of 6 men and 8 women a committee of 3 is chosen. What is the probability that the committee will consist of all men or all women?

▶ The total number of ways in which a committee of 3 can be chosen from a group of 14 people is $_{14}C_3 = 364$. The number of ways in which a committee of 3 men can be chosen is $_6C_3 = 20$. Therefore, the probability of selecting a committee composed of men only is $\frac{20}{364}$. The number of ways in which a committee of 3 women can be chosen is $_8C_3 = 56$. Therefore, the probability of obtaining a committee composed of women only is $\frac{56}{364}$. The probability of obtaining a committee composed entirely of men or entirely of women is the sum of the individual probabilities i.e.,
$\frac{20}{364} + \frac{56}{364} = \frac{76}{364} = \frac{19}{91}$.

EXPECTATION

DEFINITION: If the probability of acquiring an amount of money (A) is p, then the expectation is the product of p and A ($E = p \cdot A$).

For example, if the probability of winning a \$10 bet is $\frac{1}{4}$ then the expectation is $\frac{1}{4} \times 10 = \$2.50$.

Example: A television set worth \$160 is raffled off.   If 75 tickets are sold and A buys 2 of them what is his expectation?

▶   The probability of A winning is $\frac{2}{75}$.   The amount he may win is \$160.   Therefore, the expectation $(E) = \frac{2}{75} \times 160 = \$4.27$.

Example: First prize in a golf tournament is \$2,000.   The odds against a certain player winning the tournament are 35 to 1.   What is his expectation?

▶   Since the odds are 35 to 1 against the player, his probability of success is $\frac{1}{36}$.   Therefore, his expectation $(E) = \frac{1}{36} \times \$2000 = \$55.56$.

REPEATED TRIALS

The following theorems are used to solve problems involving repeated trials.

THEOREM: If p is the probability that an event will happen in a single trial and q is the probability that this event will fail in this trial then $_nC_r p^r q^{n-r}$ is the probability that this event will happen exactly r times in n trials.

Example: If a die is thrown 5 times in succession, what is the probability that 4 will be thrown exactly twice?

▶   In order that this event take place, 4 must turn up on two throws and fail to turn up on three throws.   On a specific throw the probability of obtaining a 4 is $\frac{1}{6}$ and the probability of failing to obtain a 4 is $\frac{5}{6}$.   If we isolate the set of throws which yield 4 exactly twice the probability that this will happen in any one member of the set is $\left(\frac{1}{6}\right)^2\left(\frac{5}{6}\right)^3$.   However, there are $_5C_2$ different ways in which the particular throws for obtaining two 4's may be designated. Therefore, the probability is $_5C_2 \left(\frac{1}{6}\right)^2 \left(\frac{5}{6}\right)^3 = \frac{625}{3888}$.

▶ Using the formula $_nC_r p^r q^{n-r}$ we note that $n = 5$ and $r = 2$. Since the probability that a 4 will be thrown in a single trial is $\frac{1}{6}$, $p = \frac{1}{6}$. If the probability of throwing a 4 is $\frac{1}{6}$ then the probability of failing to throw a 4 is $\frac{5}{6}$, or $q = \frac{5}{6}$.

$$_nC_r p^r q^{n-r} = {}_5C_2 \left(\tfrac{1}{6}\right)^2 \left(\tfrac{5}{6}\right)^3 = \tfrac{625}{3888}$$

THEOREM: The probability that an event will occur at least r times in n trials is $p^n + {}_nC_1 p^{n-1}q + {}_nC_2 p^{n-2}q^2 + \cdots\cdots + {}_nC_r p^r q^{n-r}$.

This expression is the sum of the first $n - r + 1$ terms of the expansion of the binomial $(p+q)^n$.

Example: If a die is thrown 5 times in succession, what is the probability that 4 will be thrown at least 3 times?

▶ Using the method explained previously, we shall find the probabilities of throwing 4 exactly three times, exactly four times, and exactly five times. The sum of these probabilities will yield the probability of obtaining 4 at least 3 times.

The probability of obtaining 4 exactly 3 times is $_5C_3 \left(\frac{1}{6}\right)^3 \left(\frac{5}{6}\right)^2 = \frac{250}{7776}$

The probability of obtaining 4 exactly 4 times is $_5C_4 \left(\frac{1}{6}\right)^4 \left(\frac{5}{6}\right)^1 = \frac{25}{7776}$

The probability of obtaining 4 exactly 5 times is $_5C_5 \left(\frac{1}{6}\right)^5 = \frac{1}{7776}$

The sum of these probabilities is $\frac{276}{7776} = \frac{23}{648}$

▶ Using the theorem, and the expansion of $(p+q)^n$. In this case $p = \frac{1}{6}$, $q = \frac{5}{6}$, and $n = 5$. We wish to find the sum of the first 3 terms in the expansion, since the first term gives the probability of obtaining five 4's, the second term gives the probability of obtaining four 4's, and the third term gives the probability of obtaining three 4's.

$$\left(\frac{1}{6} + \frac{5}{6}\right)^5 = \left(\frac{1}{6}\right)^5 + {}_5C_1 \left(\frac{1}{6}\right)^4 \left(\frac{5}{6}\right) + {}_5C_2 \left(\frac{1}{6}\right)^3 \left(\frac{5}{6}\right)^2$$

$$= \frac{1}{7,776} + \frac{25}{7,776} + \frac{250}{7,776} = \frac{276}{7,776} = \frac{23}{648}$$

Additional Problems in Probability

Example: From a pack of 52 cards two cards are drawn at random. What is the probability that one is a king and one is a queen?

▶ The number of ways in which 2 cards can be drawn is $_{52}C_2$ = 1326. The number of ways in which a king and a queen can be drawn is $_4C_1 \cdot _4C_1 = 16$. Therefore, the probability $p = \frac{16}{1326}$ = $\frac{8}{663}$.

Example: Ten persons take their place at a round table. What are the odds against two particular persons sitting together?

▶ The number of ways in which ten persons may sit at a round table is 9! If two particular persons sit together then the number of ways in which they may be arranged at a round table is $2 \cdot 8!$ Therefore, the number of ways in which two particular people do not sit together is $9! - 2 \cdot 8!$ The odds against two particular persons sitting together are

$$f{:}s = \frac{9! - 2 \cdot 8!}{9! - (9! - 2 \cdot 8!)} = \frac{9! - 2 \cdot 8!}{2 \cdot 8!.} = \frac{282,240}{80,640} = \frac{7}{2} \text{ or } 7{:}2$$

Example: From a pack of 52 cards, 5 cards are drawn. What is the probability that the 5 cards will be honors of the same suit? (Honor cards are 10, J, Q, K, A).

▶ The total number of ways of drawing 5 cards from 52 is $_{52}C_5$ = 2,598,960. The number of ways in which honors of the same suit may be drawn is 4, one in each suit. Therefore, the probability $p = \dfrac{4}{2,598,960} = \dfrac{1}{648,160}$

Example: A bag contains 7 white balls, 5 black balls, and 4 red balls.

a. If three balls are drawn at random what is the probability that all the balls are white?

b. If five balls are drawn at random what is the probability that 2 balls are white and 3 balls are red?

c. If six balls are drawn at random what is the probability that 2 balls of each color are drawn?

d. If 4 balls are drawn at random what is the probability that none is black?

► a. $p = \dfrac{{}_7C_3}{{}_{16}C_3} = \dfrac{35}{560} = \dfrac{1}{16}$

b. $p = \dfrac{{}_7C_2 \cdot {}_4C_3}{{}_{16}C_5} = \dfrac{84}{4368} = \dfrac{1}{52}$

c. $p = \dfrac{{}_7C_2 \cdot {}_5C_2 \cdot {}_4C_2}{{}_{16}C_6} = \dfrac{1260}{8008} = \dfrac{45}{286}$

d. The total number of choices of 4 is ${}_{16}C_4$. If black balls are excluded then the number of choices of 4 is ${}_{11}C_4$.

$$p = \dfrac{{}_{11}C_4}{{}_{16}C_4} = \dfrac{330}{1820} = \dfrac{33}{182}$$

Example: One bag contains 6 blue balls and 4 red balls. Another bag contains 5 blue balls and 3 red balls. In drawing one ball from each bag what are the probabilities of the following events?

a. 2 blue balls   b. 2 red balls   c. 1 blue ball and 1 red ball

► a. $p = \dfrac{6}{10} \cdot \dfrac{5}{8} = \dfrac{3}{8}$

b. $p = \dfrac{4}{10} \cdot \dfrac{3}{8} = \dfrac{3}{20}$

c. One blue ball and one red ball may be drawn in 2 ways i.e., one blue ball may be drawn from the first bag and one red ball from the second bag or one red ball from the first bag and one blue ball from the second bag.

The probability of drawing one blue ball from the first bag and one red ball from the second bag is $p = \frac{6}{10} \cdot \frac{3}{8} = \frac{9}{40}$.

The probability of drawing one red ball from the first bag and one blue ball from the second bag is $p = \frac{4}{10} \cdot \frac{5}{8} = \frac{1}{4}$.

The sum of these probabilities is $\frac{9}{40} + \frac{1}{4} = \frac{19}{40}$.

Example: If two dice are thrown, what is the probability of getting either 7 or 11?

► If two dice are thrown there are $6 \cdot 6$ or 36 ways in which the dice may turn up.

A sum of 7 may be obtained as follows, $1 + 6, 6 + 1, 5 + 2,$ $2 + 5, 4 + 3, 3 + 4,$ or 6 ways. Therefore, the probability of obtaining a 7 is $\frac{6}{36}$.

A sum of 11 may be obtained as follows, $6 + 5, 5 + 6,$ or 2 ways. Therefore, the probability of obtaining an 11 is $\frac{2}{36}$.

The probability of obtaining a 7 or 11 is the sum of these two probabilities or

$$\frac{6 + 2}{36} = \frac{2}{9}$$

Example: Eight books are placed on a shelf at random. What is the probability that two particular books will be together?

▶   The total number of ways in which 8 books may be arranged on a shelf is 8! The number of ways in which two particular books of the 8 will be together is $2 \cdot 7!$ The probability that two particular books will be together is

$$\frac{2 \cdot 7!}{8!} = \frac{1}{4}$$

Example: From a group of 5 Democrats and 3 Republicans, a committee of 3 is chosen. What is the probability that the committee will contain more Republicans than Democrats?

▶   The probability of a committee of 2 Republicans and 1 Democrat is

$$\frac{{}_3C_2 \cdot {}_5C_1}{{}_8C_3} = \frac{15}{56}$$

The probability of a committee of 3 Republicans is

$$\frac{{}_3C_3}{{}_8C_3} = \frac{1}{56}$$

Therefore, the probability of a committee of more Republicans than Democrats is

$$\frac{15}{56} + \frac{1}{56} = \frac{16}{56} = \frac{2}{7}$$

Example: The 26 letters of the alphabet are placed on individual cardboard squares. If 5 of the squares are placed at random in a horizontal row, what is the probability that the name FRANK is spelled out?

▶ The number of ways of arranging 5 letters out of 26 is $_{26}P_5 \doteq 7,893,600$. Since only one of these ways spells out FRANK, the probability

$$p = \frac{1}{7,893,600}$$

Example: A man bets $50 that he will throw a 6 or 9 on a single throw with two dice. What is his expectation?

▶ Two dice can turn up in $6 \cdot 6 = 36$ ways. The ways in which a 6 may be obtained are $5 + 1, 1 + 5, 4 + 2, 2 + 4$, and $3 + 3$. The ways in which a 9 may be obtained are $6 + 3, 3 + 6, 5 + 4$, and $4 + 5$. Thus, a 6 or 9 may be obtained in 9 ways. Therefore, the probability of obtaining a 6 or 9 is $\frac{9}{36} = \frac{1}{4}$. The expectation is $\frac{1}{4} \times 50 = \$12.50$.

Example: From a bag containing 3 white balls and 2 red balls a ball is drawn and replaced in the bag. This procedure is followed 6 times. What is the probability that on exactly 4 of the drawings the ball drawn is white?

▶ We use the formula for repeated trials $p = {}_nC_r p^r q^{n-r}$. In this formula $n = 6, r = 4, p = \frac{3}{5}, q = \frac{2}{5}$.

$$p = {}_6C_4 \left(\frac{3}{5}\right)^4 \left(\frac{2}{5}\right)^2 = \frac{972}{3125}$$

Example: What is the probability of obtaining 9 at least 3 times on 4 throws of two dice?

▶ We shall use the theorem involving the expansion $(p + q)^n$. The probability of obtaining a 9 with one throw of two dice is $\frac{4}{36}$ or $\frac{1}{9}$. This is true because, of the 36 ways in which two dice may turn up, 9 may be obtained as $6 + 3, 3 + 6, 5 + 4$, and $4 + 5$.

In this case, $p = \frac{1}{9}$, $q = \frac{8}{9}$, and $n = 4$. The first term in the expansion will yield the probability of obtaining 9 four times and the second term in the expansion will yield the probability of obtaining 9 three times. Thus, the desired probability will be the sum of the first two terms.

$$\left(\frac{1}{9} + \frac{8}{9}\right)^4 = \left(\frac{1}{9}\right)^4 + 4C_1\left(\frac{1}{9}\right)^3\left(\frac{8}{9}\right)$$

$$= \frac{1}{6561} + \frac{32}{6561} = \frac{33}{656} = \frac{11}{2187}$$

Example: From a pack of 52 cards, 4 cards are drawn at random.

   a. What is the probability that two are aces and two are kings?

   b. What is the probability that all the cards are diamonds?

   c. What is the probability that no card is a spade?

   d. What is the probability that there is one card of each suit?

▶   a. The number of selections is $_{52}C_4 = 1{,}082{,}900$. The number of ways of selecting 2 aces and 2 kings is $_4C_2 \cdot {}_4C_2 = 36$. Therefore, the probability

$$p = \frac{36}{1{,}082{,}900} = \frac{9}{270{,}725}$$

b. The number of ways of selecting 4 diamonds is $_{13}C_4 = 715$. Therefore, the probability of obtaining 4 diamonds is

$$\frac{715}{1{,}082{,}900} = \frac{143}{216{,}580}$$

c. The number of ways to select cards from the 39 that are not spades, $_{39}C_4 = 82{,}251$. Therefore, the probability of selecting 4 cards that are not spades is

$$\frac{82{,}251}{1{,}082{,}900}$$

d. The number of ways of selecting one card from each suit is $_{13}C_1 \cdot {}_{13}C_1 \cdot {}_{13}C_1 \cdot {}_{13}C_1 = 28{,}561$. Therefore, the probability of selecting one card from each suit is

$$p = \frac{28{,}561}{1{,}082{,}900}$$

Example: If a coin is tossed 6 times what is the probability of obtaining exactly 4 heads and 2 tails?

▶ Using the formula for repeated trials $_nC_rp^rq^{n-r}$, $n = 6$, $r = 4$, p and q are each $\frac{1}{2}$ since there is equal likelihood of obtaining a head and a tail on each toss.

$$_6C_4 \left(\frac{1}{2}\right)^4 \left(\frac{1}{2}\right)^2 = \frac{15}{64}$$

Example: If a coin is tossed 5 times what is the probability of obtaining at least 2 heads?

▶ Using the expansion of $(p + q)^n$,

$$(p + q)^5 = p^5 + {_5C_1}p^4q + {_5C_2}p^3q^2 + {_5C_3}p^2q^3$$

In this expansion, p^5 gives the probability of obtaining 5 heads, $5C_1p^4q$ gives the probability of obtaining 4 heads, etc.

$$\left(\frac{1}{2}\right)^5 + 5\left(\frac{1}{2}\right)^4\left(\frac{1}{2}\right) + 10\left(\frac{1}{2}\right)^3\left(\frac{1}{2}\right)^2 + 10\left(\frac{1}{2}\right)^2\left(\frac{1}{2}\right)^3$$

$$= \frac{1}{32} + \frac{5}{32} + \frac{10}{32} + \frac{10}{32} = \frac{26}{32} = \frac{13}{16}$$

Example: In a golf match, the probability of A's winning is $\frac{1}{4}$, the probability of B's winning is $\frac{1}{8}$, and the probability of C's winning is $\frac{1}{6}$. What is the probability that no one of these three will win?

▶ The probability of A's losing is $1 - \frac{1}{4} = \frac{3}{4}$, of B's losing is $1 - \frac{1}{8} = \frac{7}{8}$, of C's losing is $1 - \frac{1}{6} = \frac{5}{6}$. The probability of A, B, and C losing is $\frac{3}{4} \cdot \frac{7}{8} \cdot \frac{5}{6} = \frac{35}{64}$.

Example: A bag contains 8 balls one of which is marked with a star. A ball is to be removed by 3 persons in turn, each person replacing the ball after it is drawn. The first person removing the ball marked with a star is to receive a prize of \$64.

a. What is the expectation of each person?
b. What is the probability that the prize will not be won?

▶ a. The probability of the first person winning the prize is $\frac{1}{8}$. His expectation is $\frac{1}{8} \times 64 = \$8$. The probability of the first person not winning the prize is $1 - \frac{1}{8} = \frac{7}{8}$. Therefore, the probability of the second person winning the prize is $\frac{7}{8} \cdot \frac{1}{8} = \frac{7}{64}$. His expectation is $\frac{7}{64} \cdot 64 = \$7$. The probability of neither the first person nor the second person winning is $1 - (\frac{1}{8} + \frac{7}{64}) = 1 - \frac{15}{64} = \frac{49}{64}$. Therefore, the probability of the third person winning the prize $\frac{49}{64} \times \frac{1}{8}$ $= \frac{49}{512}$. His expectation is $\frac{49}{512} \times 64 = \$6.13$.

b. The respective probabilities of winning for the three men are $\frac{1}{8}, \frac{7}{64}$, and $\frac{49}{512}$. Therefore, the probability of the first person winning, or the second person winning, or the third person winning is $\frac{1}{8} + \frac{7}{64} + \frac{49}{512}$. Therefore, the probability of no person winning is $1 - \frac{169}{512} = \frac{343}{512}$.

Example: The probability of a student receiving an A in one course is $\frac{1}{2}$, the probability of his receiving an A in a second course is $\frac{1}{3}$, and the probability of his receiving an A in a third course is $\frac{1}{4}$.

a. What is the probability of his receiving an A in just one course?

b. What is the probability of his receiving an A in at least one course?

▶ a. The probability of his receiving an A in just the first course is $(\frac{1}{2})(\frac{2}{3})(\frac{3}{4}) = \frac{1}{4}$ since he must not receive an A in the other two courses. The probability of his receiving an A in just the second course is $(\frac{1}{2})(\frac{1}{3})(\frac{3}{4}) = \frac{1}{8}$. The probability of his receiving an A in just the third course is $(\frac{1}{2})(\frac{2}{3})(\frac{1}{4}) = \frac{1}{12}$. These events are mutually exclusive and the probability that he will receive one A is $\frac{1}{4} + \frac{1}{8} + \frac{1}{12} = \frac{11}{24}$.

b. The probability that he will receive at least one A is obtained by subtracting from 1 the probability that he will receive no A's. The probability of receiving no A's is $(\frac{1}{2})(\frac{2}{3})(\frac{3}{4}) = \frac{1}{4}$. Therefore, the probability that he will receive one A is $1 - \frac{1}{4} = \frac{3}{4}$.

Example: Three dice are thrown once.
a. What is the probability of obtaining 3 aces?
b. What is the probability of obtaining at least one ace?
c. What is the probability that the sum of the three dice will be 6?

▶ a. Three dice may turn up in $6 \cdot 6 \cdot 6 = 216$ ways. Three aces may turn up in only 1 way. Therefore, the probability of obtaining 3 aces is $\frac{1}{216}$.

b. The probability of not obtaining an ace in one throw is $1 - \frac{1}{6} = \frac{5}{6}$. The probability that no ace will turn up is $(\frac{5}{6})(\frac{5}{6})(\frac{5}{6}) = \frac{125}{216}$. Therefore, the probability of obtaining at least one ace is $1 - \frac{125}{216} = \frac{91}{216}$. This result may also be obtained by adding the first 3 terms of the expansion $(\frac{1}{6} + \frac{5}{6})^3$.

c. The total number of ways that three dice may turn up is 216. A total of 6 may be obtained in the following ways, $2 + 2 + 2$, $1 + 3 + 2$, $1 + 2 + 3$, $2 + 1 + 3$, $2 + 3 + 1$, $3 + 1 + 2$, $3 + 2 + 1$, $4 + 1 + 1$, $1 + 4 + 1$, $1 + 1 + 4$, i.e., 10 ways. Therefore, the probability of obtaining the sum of 6 is $\frac{10}{216} = \frac{5}{108}$.

Example: At the start of the baseball season the odds are 7 to 5 in favor of team A, 4 to 1 against team B, and 5 to 1 against team C.

a. What are the probabilities of each of these teams winning?

b. What is the probability that one of these 3 teams will win?

▶ a. The probability of A's winning is $\dfrac{7}{7 + 5} = \dfrac{7}{12}$.

The probability of B's winning is $\dfrac{1}{1 + 4} = \dfrac{1}{5}$.

The probability of C's winning is $\dfrac{1}{1 + 5} = \dfrac{1}{6}$.

b. The probability of the 3 teams losing is $1 - \frac{7}{12}, 1 - \frac{1}{5}, 1 - \frac{1}{6}$ or $\frac{5}{12}$, $\frac{4}{5}$, and $\frac{5}{6}$. The probability that all 3 teams will lose is $(\frac{5}{12} \times \frac{4}{5} \times \frac{5}{6}) = \frac{5}{18}$. Therefore, the probability that one of the teams will win is $1 - \frac{5}{18} = \frac{13}{18}$.

Example: Two golfers play 4 matches. The odds in favor of A winning any one match from B are 3 to 2.

a. Find the probability that A wins 3 of the matches.

b. Find the probability that B wins at least 2 of the matches.

▶ a. The probability of A winning any one match is
$$\frac{3}{3+2} = \frac{3}{5}$$
The probability of A losing any one match is $\frac{2}{5}$. The probability that A wins 3 of the matches is given by the third term in the expansion $\left(\frac{3}{5} + \frac{2}{5}\right)^4 = {}_4C_2\left(\frac{3}{5}\right)^2\left(\frac{2}{5}\right)^2 = \frac{216}{625}$.

b. The probability that B wins any one match is $\frac{2}{5}$. The probability of B losing any one match is $\frac{3}{5}$. The probability that B wins at least two matches is the sum of the first two terms of the expansion $\left(\frac{2}{5} + \frac{3}{5}\right)^4$.
$$\left(\frac{2}{5}\right)^4 + {}_4C_1\left(\frac{2}{5}\right)^3\left(\frac{3}{5}\right) = \frac{16}{625} + \frac{96}{625} = \frac{112}{625}$$

Example: The probability that a certain student will be elected president of his class is $\frac{4}{9}$. And the probability that his friend will be elected secretary of his class is $\frac{2}{3}$.

a. What is the probability that both will be elected?

b. What is the probability that both will be defeated?

c. What is the probability that the first student will be elected and the second student defeated?

▶ a. The probability that both will be elected is $\left(\frac{4}{9}\right)\left(\frac{2}{3}\right) = \frac{8}{27}$.

b. The probability that both will be defeated is $\left(1 - \frac{4}{9}\right)\left(1 - \frac{2}{3}\right) = \frac{5}{27}$.

c. The probability that the first student will be elected and the second student defeated is $\left(\frac{4}{9}\right)\left(1 - \frac{2}{3}\right) = \frac{4}{27}$.

EXERCISE 1

Solve the following problems:

1. From a group of 4 men and 3 women a committee of 3 is chosen. What is the probability that the committee will consist of 3 men?

2. From a bag containing 2 black and 3 white balls, 2 balls are drawn at random. What is the probability of drawing one ball of each color?

3. One letter is to be taken at random from each of the words *factor* and *father*. What is the probability that the same letter will be taken from each?

4. A committee of 3 is to be chosen by lot from a group of 10 men. What is the probability that a certain individual in the group will be among those chosen?

5. In a lottery 20 tickets are sold and two prizes are offered. Mr. Stone buys 3 tickets. What are the odds against Mr. Stone winning a prize?

6. A coin is tossed 4 times in succession. What is the probability
    a. that 4 heads will turn up?
    b. that 2 heads and 2 tails will turn up?
    c. that 1 head and 3 tails will turn up?

7. One bag contains 4 white balls and 5 black balls. Another bag contains 3 white balls and 8 black balls. If one ball is drawn from each bag, what is the probability that both balls will be black?

8. A box contains 5 white balls and 7 black balls. If 3 balls are drawn in succession without replacing any ball after it has been drawn, what is the probability of drawing 3 white balls?

9. If two dice are thrown what is the probability that the sum of the dice will be either 4 or 10?

10. From a shelf containing 5 history books and 7 science books 3 books are chosen at random. What is the probability that the 3 books chosen are all history books or all science books?

11. The odds in favor of a player winning a tennis match are 5 to 3. If he bets $20 on the outcome, what is his expectation?

12. Eight men stand in a line. What is the probability that two of the men A and B will not stand together?

13. From a pack of 52 cards, 4 cards are drawn. What is the probability that these cards will be 3 aces and a king?

14. A bag contains 5 white balls, 6 black balls, and 3 red balls.
    a. If two balls are drawn at random what is the probability that they are black?
    b. If four balls are drawn at random what is the probability that two balls are white and two balls are red?
    c. If 3 balls are drawn at random what is the probability that one ball of each color is drawn?
    d. If two balls are drawn what is the probability that no balls are white?

15. One bag contains 5 blue balls and 6 red balls. Another bag contains 4 blue balls and 7 red balls. In drawing one ball from each bag what are the probabilities of the following events?

    a. 2 blue balls

    b. 2 red balls

    c. 1 blue ball and 1 red ball

16. From a group of 4 Democrats and 5 Republicans a committee of 3 is chosen. What is the probability that the committee will contain more Democrats than Republicans?

17. From the 26 letters in the alphabet 5 letters are drawn at random. What is the probability that the selection will consist of 3 consonants and 2 vowels?

18. From a pack of 52 cards, 3 cards are drawn at random.

    a. What is the probability that 2 aces and a king are drawn?

    b. What is the probability that all the cards are spades?

    c. What is the probability that no card is a heart?

    d. What is the probability that the cards are JQK of the same suit?

19. If a coin is tossed 7 times what is the probability

    a. of obtaining exactly 5 heads and 2 tails?

    b. of obtaining at least 3 heads?

20. In a basketball league, the probability of team A winning first place is $\frac{1}{3}$, the probability of team B winning first place is $\frac{1}{4}$, and the probability of team C winning first place is $\frac{1}{6}$. What is the probability that no one of these three will win?

21. A box contains counters marked 1 to 10. A counter is to be removed by 3 persons in turn each person replacing the counter after it is drawn. The first person removing the counter marked 7 is to receive a prize of $50.

    a. What is the expectation of each person?

    b. What is the probability that the prize will not be won?

22. The probability of a college football team winning the first game on its schedule is $\frac{2}{3}$, the probability of the team winning the second game is $\frac{3}{4}$, and the probability of winning the third game is $\frac{5}{6}$.

    a. What is the probability of the team winning all three games?

    b. What is the probability of the team losing all three games?

    c. What is the probability of the team winning just one game?

    d. What is the probability of the team winning at least one game?

23. In an intercollegiate basketball league the odds are 5 to 3 in favor of team *A*, 3 to 1 against team *B*, and 5 to 1 against team *C*.

    a. What are the probabilities of each of these teams winning?

    b. What is the probability that at least one of these 3 teams will win?

24. The probability that *A* will be elected to an honor society is $\frac{7}{12}$ and the probability that *B* will be elected to this society is $\frac{5}{9}$.

    a. What is the probability that both *A* and *B* will be elected?

    b. What is the probability that both *A* and *B* will not be elected?

    c. What is the probability that *A* will be elected and *B* will not be elected?

    d. What is the probability that *A* will not be elected and that *B* will be elected?

25. If two dice are thrown what is the probability that the sum of the dice will be greater than 9?

ANSWERS

EXERCISE 1

1. $\frac{4}{35}$

2. $\frac{3}{5}$

3. $\frac{1}{9}$

4. $\frac{3}{10}$

5. 68 to 27

6. a. $\frac{1}{16}$
   b. $\frac{3}{8}$
   c. $\frac{1}{4}$

7. $\frac{40}{99}$

8. $\frac{1}{22}$

9. $\frac{1}{6}$

10. $\frac{9}{44}$

11. $12.50

12. $\frac{3}{4}$

13. $\dfrac{16}{270,725}$

14. a. $\frac{15}{91}$
   b. $\frac{30}{1001}$
   c. $\frac{45}{182}$
   d. $\frac{36}{91}$

15. a. $\frac{20}{121}$
   b. $\frac{42}{121}$
   c. $\frac{59}{121}$

16. $\frac{17}{42}$

17. $\frac{665}{3289}$

18. a. $\frac{6}{5525}$

    b. $\frac{11}{850}$

    c. $\frac{703}{1700}$

    d. $\frac{1}{5525}$

19. a. $\frac{21}{128}$

    b. $\frac{99}{128}$

20. $\frac{5}{12}$

21. a. $5.00, $4.50, $4.09\frac{1}{2}$

    b. $\frac{729}{1000}$

22. a. $\frac{5}{12}$

    b. $\frac{1}{72}$

    c. $\frac{5}{36}$

    d. $\frac{7\frac{1}{2}}{72}$

23. a. $\frac{5}{8}, \frac{1}{4}, \frac{1}{6}$

    b. $\frac{49}{64}$

24. a. $\frac{35}{108}$

    b. $\frac{5}{27}$

    c. $\frac{7}{27}$

    d. $\frac{25}{108}$

25. $\frac{1}{6}$

CHAPTER **20**

Mathematical Induction and the Binomial Theorem

MATHEMATICAL INDUCTION

Mathematical induction is a method of proof. It is useful in proving theorems in algebra which state relations that are true for all positive integers. The method consists of two steps.

1. Prove that the proposition in question is true for the case $n = 1$. This is usually done by direct substitution.

2. Prove that if the proposition is true for some value of n, for example $n = k$, then it must also be true for $n = k + 1$.

Example: Prove by mathematical induction that the following theorem is true for all positive integral values of n.

$$2 + 4 + 6 + \ldots + 2n = n(n+1)$$

Proof: Step 1—The theorem is true for $n = 1$, since $2 = 1(1+1)$

Step 2—Prove that if the theorem is true for $n = k$, where k represents some positive integer then it must also be true for $n = k + 1$. That is, that if

$$2 + 4 + 6 + \ldots + 2k = k(k+1), \text{ then}$$
$$2 + 4 + 6 + \ldots + 2(k+1) = (k+1)(k+2)$$

To prove this we note that

$$2 + 4 + 6 + \ldots + 2(k+1) = k(k+1) + 2(k+1)$$
$$= k^2 + 3k + 2$$
$$= (k+1)(k+2)$$

Hence, if the theorem is true for the case $n = k$, then it is true for the case $n = k + 1$. We have verified the fact that the original theorem is true for $n = 1$. Therefore, it is true for the next greater integer, $n = 2$. Since it is true for $n = 2$, it is true for the next greater integer, $n = 3$, etc. Therefore, the theorem is true for every positive integral value of n.

Example: Prove by mathematical induction that the following formula is true for all positive integral values of n.

$$1^2 + 2^2 + 3^2 + \ldots + n^2 = \frac{n(n + 1)(2n + 1)}{6}$$

► Step 1—The formula is true for $n = 1$ since

$$1^2 = \frac{1(1 + 1)(2 + 1)}{6}, \; 1 = \frac{1(2)(3)}{6}, \; 1 = 1$$

Step 2—Prove that if the formula is true for $n = k$, where k represents some positive integer then it must also be true for $n = k + 1$. That is, that if

$$1^2 + 2^2 + 3^2 + \ldots + k^2 = \frac{k(k + 1)(2k + 1)}{6}, \text{ then}$$

$$1^2 + 2^2 + 3^2 + \ldots + k^2 + (k + 1)^2 = \frac{k(k + 1)(2k + 1)}{6}$$
$$+ (k + 1)^2$$

$$= \frac{k(k + 1)(2k + 1) + 6(k + 1)^2}{6}$$

= Remove the common factor $(k + 1)$,

$$\frac{(k + 1)[k(2k + 1) + 6(k + 1)]}{6}$$

$$= \frac{(k + 1)(2k^2 + 7k + 1)}{6}$$

$$= \frac{(k + 1)(k + 2)(2k + 3)}{6}$$

Note that the right member $\dfrac{(k+1)(k+2)(2k+3)}{6}$ is equivalent to the original right member when $(k+1)$ is substituted for k.

Hence, if the formula is true for the case $n = k$, then it is true for the case $n = k + 1$. We have verified the fact that the original formula is true for $n = 1$. Therefore, it is true for the next greater integer, $n = 2$. Since it is true for $n = 2$, it is true for the next greater integer $n = 3$, etc. Therefore, the formula is true for every positive integral value of n.

Example: Prove by mathematical induction that the following formula is true for all positive integral values of n.

$$\frac{1}{1 \cdot 2} + \frac{1}{2 \cdot 3} + \frac{1}{3 \cdot 4} + \cdots + \frac{1}{n(n+1)} = \frac{n}{n+1}$$

▶ Step 1—The formula is true for $n = 1$ since

$$\frac{1}{1 \cdot 2} = \frac{1}{1+1}, \; \frac{1}{2} = \frac{1}{2}$$

Step 2—Prove that if the formula is true for $n = k$, where k represents some positive integer then it must also be true for $n = k + 1$. That is, that if

$$\frac{1}{1 \cdot 2} + \frac{1}{2 \cdot 3} + \frac{1}{3 \cdot 4} + \cdots + \frac{1}{k(k+1)} = \frac{k}{k+1}, \text{ then}$$

$$\frac{1}{1 \cdot 2} + \frac{1}{2 \cdot 3} + \frac{1}{3 \cdot 4} + \cdots + \frac{1}{k(k+1)} + \frac{1}{(k+1)(k+2)}$$

$$= \frac{k}{k+1} + \frac{1}{(k+1)(k+2)}$$

$$= \frac{k(k+2)+1}{(k+1)(k+2)} = \frac{k^2+2k+1}{(k+1)(k+2)}$$

$$= \frac{(\cancel{k+1})(k+1)}{(\cancel{k+1})(k+2)} = \frac{k+1}{k+2}$$

Note that the right member $\dfrac{k+1}{k+2}$ is equivalent to the original right member $\dfrac{k}{k+1}$ when $k+1$ is substituted for k.

Hence, if the formula is true for the case $n = k$, then it is true for the case $n = k + 1$. We have verified the fact that the original formula is true for $n = 1$. Therefore, it is true for the next greater integer, $n = 2$. Since it is true for $n = 2$, it is true for the next greater integer $n = 3$, etc. Therefore, the formula is true for every positive integral value of n.

Example: Prove by mathematical induction that the following formula is true for all positive integral values of n.

$$1 \cdot 3 + 2 \cdot 4 + 3 \cdot 5 + \ldots + n(n + 2) = \frac{1}{6} n(n + 1)(2n + 7)$$

▶ Step 1—The formula is true for $n = 1$, since

$$1 \cdot 3 = \frac{1}{6}(1)(1 + 1)(2 \cdot 1 + 7), \quad 3 = \frac{1}{6}(1)(2)(9), \quad 3 = \frac{1}{6} \times 18.$$

Step 2—Prove that if the formula is true for $n = k$, where k represents some positive integer then it must also be true for $n = k + 1$. That is, that if

$$1 \cdot 3 + 2 \cdot 4 + 3 \cdot 5 + \ldots + k(k + 2)$$
$$= \frac{1}{6}(k + 1)(k + 2)(2k + 9) \text{ then}$$
$$1 \cdot 3 + 2 \cdot 4 + 3 \cdot 5 + \ldots + k(k + 2) + (k + 1)(k + 3)$$
$$= \frac{1}{6}(k + 1)(k + 2)(2k + 7) + (k + 1)(k + 3)$$
$$= \frac{k(k + 1)(2k + 7)}{6} + \frac{6(k + 1)(k + 3)}{6}$$

Removing the factor
$$\frac{1}{6}(k + 1) = \frac{1}{6}(k + 1)[k(2k + 7) + 6(k + 3)].$$

$$= \frac{1}{6}(k + 1)[2k^2 + 7k + 6k + 18]$$

$$= \frac{1}{6}(k + 1)[2k^2 + 13k + 18]$$

$$= \frac{1}{6}(k + 1)(k + 2)(2k + 9)$$

Note that the right hand member $\frac{1}{6}(k + 1)(k + 2)(2k + 9)$ is equivalent to the original right hand member $\frac{1}{6}k(k + 1)(2k + 7)$ when $(k + 1)$ is substituted for k.

Hence, we have completed Step 2 in the proof. We have shown that, if the formula is true for the case $n = k$, then it is also true for the case $n = k + 1$.

Example: Prove by mathematical induction that $(a^n - b^n)$ is divisible by $(a - b)$ for all positive integral values of n.

▶ Step 1—$(a^n - b^n)$ is divisible by $(a - b)$ when $n = 1$, since $(a - b)$ is divisible by $a - b$.

Step 2—Prove that if the theorem is true for $n = k$, where k represents some positive integer then it must also be true for $n = k + 1$. That is, that if
$$(a^k - b^k) \text{ is divisible by } (a - b)$$
then
$$(a^{k+1} - b^{k+1}) \text{ is divisible by } a - b.$$
The following identity can easily be verified $a^{k+1} - b^{k+1}$
$$= a(a^k - b^k) + ab^k - b^{k+1} = a(a^k - b^k) + b^k(a - b)$$
The right hand member of this identity is divisible by $(a - b)$ since both $(a^k - b^k)$ and $(a - b)$ are divisible by $(a - b)$. Therefore, the left hand member of this identity i.e., $(a^{k+1} - b^{k+1})$ is also divisible by $(a - b)$.

Hence, we have completed Step 2 in the proof. We have shown that if the divisibility is true for the case $n = k$, then it is also true for the case $n = k + 1$.

EXERCISE 1

Prove each of the following by mathematical induction. In each case, n represents a positive integer.

1. $1 + 2 + 3 + \ldots + n = \dfrac{n}{2}(n + 1)$

2. $1 + 3 + 5 + \ldots + (2n - 1) = n^2$

3. $4 + 8 + 12 + \ldots + 4n = 2n(n + 1)$

4. $1 + 2 + 4 + 8 + \ldots + 2^{n-1} = 2^n - 1$

5. $1^2 + 3^2 + 5^2 + \ldots + (2n - 1)^2 = \dfrac{n}{3}(2n + 1)(2n - 1)$

6. $1 \cdot 2 + 2 \cdot 3 + 3 \cdot 4 + \ldots + n(n+1) = \dfrac{n}{3}(n+1)(n+2)$

7. $2 + 5 + 10 + \ldots + (n^2 + 1) = \dfrac{n}{6}(2n^2 + 3n + 7)$

8. $1^3 + 2^3 + 3^3 + \ldots + n^3 = \dfrac{n^2(n+1)^2}{4}$

9. $\dfrac{1}{1 \cdot 3} + \dfrac{1}{3 \cdot 5} + \dfrac{1}{5 \cdot 7} + \ldots + \dfrac{1}{(2n-1)(2n+1)} = \dfrac{n}{2n+1}$

10. Prove that $(a^{2n} - b^{2n})$ is divisible by $(a+b)$

11. Prove that $(a^{2n-1} + b^{2n-1})$ is divisible by $(a+b)$

12. Prove that $a + (a+d) + (a+2d) + \ldots [a+(n-1)d]$
$= \dfrac{n}{2}[2a+(n-1)d]$

BINOMIAL THEOREM FOR POSITIVE INTEGRAL EXPONENTS

The Binomial Theorem is a method by means of which a binomial may be raised to any positive integral power. By actual multiplication we note that

$$(a + b)^1 = a + b$$
$$(a + b)^2 = a^2 + 2ab + b^2$$
$$(a + b)^3 = a^3 + 3a^2b + 3ab^2 + b^3$$
$$(a + b)^4 = a^4 + 4a^3b + 6a^2b^2 + 4ab^3 + b^4$$
$$(a + b)^5 = a^5 + 5a^4b + 10a^3b^2 + 10a^2b^3 + 5ab^4 + b^5$$

From these results we may surmise that the following conclusions about $(a+b)^n$ are true.

1. The first term is a^n, the last term is b^n, and there are $(n+1)$ terms in the expansion.

2. The coefficient of the second term is n.

3. The exponents of a decrease by one from the first term on. The exponents of b increase by one from the second term on.

4. The coefficient of any term is of the form $_nC_r$ where n represents the power to which the binomial is being raised and $r = 0, 1, 2, 3, \ldots, n-1, n$.

In general, the binomial expansion may be written

$$(a + b)^n = a^n + {}_nC_1a^{n-1}b + {}_nC_2a^{n-2}b^2 + {}_nC_3a^{n-3}b^3 + \cdots\cdots$$
$$+ {}_nC_{n-1}b^{n-1} + b^n$$

where the coefficients ${}_nC_1 = n$, ${}_nC_2 = \dfrac{n(n-1)}{1 \cdot 2}$,

${}_nC_3 = \dfrac{n(n-1)(n-2)}{1 \cdot 2 \cdot 3}$ etc. It will be recalled that ${}_nC_r = {}_nC_{n-r}$

${}_nC_{n-r}$, e.g., ${}_8C_6 = {}_8C_2$.

The coefficients of the binomial expansion may be arranged in triangular array as follows

$(a + b)^1$	$\rightarrow$					1		1				
$(a + b)^2$	$\rightarrow$				1		2		1			
$(a + b)^3$	$\rightarrow$			1		3		3		1		
$(a + b)^4$	$\rightarrow$		1		4		6		4		1	
$(a + b)^5$	$\rightarrow$	1		5		10		10		5		1
$(a + b)^6$	$\rightarrow$	1		6	15		20		15	6		1
$(a + b)^7$	$\rightarrow$	1	7	21		35		35		21	7	1

This array is known as Pascal's Triangle. Each number, except those at the ends, is the sum of the two nearest numbers in the line above. When the exponent n is not high Pascal's Triangle may be conveniently used to find the coefficients of the terms of the expansion.

Proof of the Binomial Theorem for Positive Integral Exponents by Mathematical Induction

Proof: Step 1—The theorem is true for $n = 1$, since $(a + b)^1 = a + b$

Step 2—Prove that if the theorem is true for $n = k$, where k represents some positive integer then it must also be true for $n = k + 1$

If $(a + b)^k = a^k + {}_kC_1a^{k-1}b + {}_kC_2a^{k-2}b^2 + \cdots\cdots\cdots$
$+ {}_kC_{r-1}a^{k-r+1}b^{r-1} + {}_kCra^{k-r}b^r + \cdots + b^k$, then $(a + b)^{k+1} = a^{k+1}$
$+ {}_{k+1}C_1a^kb + {}_{k+1}C_2a^{k-1}b^2 + \cdots + {}_{k+1}C_ra^{k-r+1}b^2 + \cdots + b^{k+1}$.

To prove this we multiply both sides of the expansion of $(a + b)^k$ by $(a + b^-)$.

The left member becomes $(a + b)^{k+1}$. For the right member we consider a representative term in the product, for example, the term involving b^r. This term is the sum of two terms, the first obtained as the product of a and the term involving b^k, and the second obtained as the product of b and the term involving b^{k-1}. These two products are $_kC_r\, a^{k-r+1}b^r$ and $_kC_{r-1}\, a^{k-r+1}b^r$. Their sum is

$$a^{k-r+1}b^r[_kC_r + {}_kC_{r-1}].$$

But $_kC_r + {}_kC_{r-1} = {}_{k+1}C_r$

Therefore, the representative term in the expansion $(a + b)^{k+1}$
$= {}_{k+1}C_r a^{k-r+1}b^r$.

Hence, we have completed Step 2 in the proof. We have shown that, if the formula is true for the case $n = k$, then it is also true for the case $n = k + 1$.

Example:　Expand by the binomial theorem $(a + 2b)^5$.

▶　$(a + 2b)^5 = (a)^5 + {}_5C_1(a)^4(2b) + {}_5C_2(a)^3(2b)^2$
$+ {}_5C_3(a)^2(2b)^3 + {}_5C_4(a)(2b)^4 + (2b)^5$
$= a^5 + 5a^4(2b) + \dfrac{5 \cdot 4}{1 \cdot 2}a^2(2b)^2 + \dfrac{5 \cdot 4}{1 \cdot 2}a^2(2b)^3 + 5a(2b)^4$
$+ (2b)^5$
$= a^5 + 5a^4(2b) + 10a^3(4b^2) + 10a^2(8b^3) + 5a(16b^4) + 32b^5$
$= a^5 + 10a^4b + 40a^3b^2 + 80a^2b^3 + 80ab^4 + 32b^5$

Example:　Expand by the binomial theorem $(3x - y)^6$.

▶　$(3x - y)^6 = [3x + (-y)]^6 = (3x)^6 + {}_6C_1(3x)^5(-y)$
$+ {}_6C_2(3x)^4(-y)^2 + {}_6C_3(3x)^3(-y)^3 + {}_6C_4(3x)^2(-y)^4$
$+ {}_6C_5(3x)(-y)^5 + (-y)^6$
$= (3x)^6 + 6(3x)^5(-y) + 15(3x)^4(-y)^2 + 20(3x)^3(-y)^3$
$+ 15(3x)^2(-y)^4 + 6(3x)(-y)^5 + (-y)^6$
$= 729x^6 - 1{,}458x^5y + 1{,}215x^4y^2 - 540x^3y^3 + 135x^2y^4 - 18xy^5$
$+ y^6$

Example: Expand by the binomial theorem $\left(2x - \dfrac{1}{y}\right)^4$.

▶ $\left(2x - \dfrac{1}{y}\right)^4 = \left[2x + \left(-\dfrac{1}{y}\right)\right]^4 = (2x)^4 + {}_4C_1(2x)^3$

$\left(-\dfrac{1}{y}\right) + {}_4C_2(2x)^2 \left(-\dfrac{1}{y}\right)^2 + {}_4C_3(2x)\left(-\dfrac{1}{y}\right)^3 + \left(-\dfrac{1}{y}\right)^4$

$= 16x^4 - \dfrac{32x^3}{y} + \dfrac{24x^2}{y^2} - \dfrac{8x}{y^3} + \dfrac{1}{y^4}$

Example: Expand by the binomial theorem $(\sqrt{x} - 1)^7$.

▶ $\sqrt{x} - 1)^7 = [x^{1/2} + (-1)]^7 = (x^{1/2})^7 + {}_7C_1(x^{1/2})^6(-1)$
$+ {}_7C_2(x^{1/2})^5(-1)^2 + {}_7C_3(x^{1/2})^4(-1)^3 + {}_7C_4(x^{1/2})^3(-1)^4$
$+ {}_7C_5(x^{1/2})^2(-1)^5 + {}_7C_6(x^{1/2})(-1)^6 + (-1)^7$
$= (x^{1/2})^7 + 7(x^{1/2})^6(-1) + 21(x^{1/2})^5(-1)^2 + 35(x^{1/2})^4(-1)^3$
$+ 35(x^{1/2})^3(-1)^4 + 21(x^{1/2})^2(-1)^5 + 7(x^{1/2})(-1)^6 + (-1)^7$
$= x^{7/2} - 7x^3 + 21x^{5/2} - 35x^2 + 35x^{3/2} - 21x + 7x^{1/2} - 1$

Example: Expand by the binomial theorem $\left(2 + \dfrac{x^2}{2}\right)^5$.

▶ $\left(2 + \dfrac{x^2}{2}\right)^5 = (2)^5 + {}_5C_1(2)^4 \left(\dfrac{x^2}{2}\right) + {}_5C_2(2)^3 \left(\dfrac{x^2}{2}\right)^2$

$+ {}_5C_3(2)^2 \left(\dfrac{x^2}{2}\right)^3 + {}_5C_4(2)\left(\dfrac{x^2}{2}\right)^4 + \left(\dfrac{x^2}{2}\right)^5$

$= 32 + 40x^2 + 20x^4 + 5x^6 + \dfrac{5}{8}x^8 + \dfrac{x^{10}}{32}$

Example: Expand by the binomial theorem $\left(a^2 - \dfrac{2}{\sqrt{b}}\right)^6$.

▶ Write $\left(a^2 - \dfrac{2}{\sqrt{b}}\right)$ as $[a^2 + (-2b^{-1/2})]^6$

$(a^2 - 2b^{-1/2})^6 = (a^2)^6 + {}_6C_1(a^2)^5(-2b^{-1/2}) + {}_6C_2(a^2)^4(-2b^{-1/2})^2$
$+ {}_6C_3(a^2)^3(-2b^{-1/2})^3 + {}_6C_4(a^2)^2(-2b^{-1/2})^4 + {}_6C_5(a^2)(-2b^{-1/2})^5$
$+ (-2b^{-1/2})^6$
$= a^{12} - 12a^{10}b^{-1/2} + 60a^8b^{-1} - 160a^6b^{-3/2} + 240a^4b^{-2}$
$- 192a^2b^{-5/2} + 64b^{-3}$

Example: Expand by the binomial theorem $\left(\dfrac{x}{a} - \dfrac{a}{x}\right)^7$.

$$\blacktriangleright \quad \left(\frac{x}{a} - \frac{a}{x}\right)^7 = \left[\frac{x}{a} + \left(-\frac{x}{a}\right)\right]^7 = \left(\frac{x}{a}\right)^7 + {}_7C_1\left(\frac{x}{a}\right)^6$$

$$\left(-\frac{a}{x}\right) + {}_7C_2\left(\frac{x}{a}\right)^5\left(-\frac{a}{x}\right)^2 + {}_7C_3\left(\frac{x}{a}\right)^4\left(-\frac{a}{x}\right)^3$$

$$+ {}_7C_4\left(\frac{x}{a}\right)^3\left(-\frac{a}{x}\right)^4 + {}_7C_5\left(\frac{x}{a}\right)^2\left(-\frac{a}{x}\right)^5$$

$$+ {}_7C_6\left(\frac{x}{a}\right)\left(-\frac{a}{x}\right)^6 + \left(-\frac{a}{x}\right)^7$$

$$= \frac{x^7}{a^7} - \frac{7x^5}{a^5} + \frac{21x^3}{a^3} - \frac{35x}{a} + \frac{35a}{x} - \frac{21a^3}{x^3} + \frac{7a^5}{x^5} - \frac{a^7}{x^7}$$

Example: Expand by the binomial theorem $(x^{-2} + y^{2/3})^5$.

$$\blacktriangleright \quad (x^{-2} + y^{2/3})^5 = (x^{-2})^5 + {}_5C_1(x^{-2})^4(y^{2/3}) + {}_5C_2(x^{-2})^3(y^{2/3})^2$$
$$+ {}_5C_3(x^{-2})^2(y^{2/3})^3 + {}_5C_4(x^{-2})(y^{2/3})^4 + (y^{2/3})^5$$
$$= x^{-10} + 5x^{-8}y^{2/3} + 10x^{-6}y^{4/3} + 10x^{-4}y^2 + 5x^{-2}y^{8/3} + y^{10/3}$$

Example: Expand by the binomial theorem $(x + y - z)^3$.

$\blacktriangleright \quad x + y - z = [(x+y) + (-z)]$. Thus, we treat $(x+y)$ as one term and $(-z)$ as the other term.

$$[(x + y) + (-z)]^3 = (x + y)^3 + {}_3C_1(x + y)^2(-z)$$
$$+ {}_3C_2(x + y)(-z)^2 + (-z)^3$$
$$= x^3 + 3x^2y + 3xy^2 + y^3 - 3x^2z - 6xyz - 3y^2z + 3xz^2$$
$$+ 3yz^2 - z^3$$

Example: Expand by the binomial theorem $(y^2 + y + 1)^4$.

$\blacktriangleright \quad (y^2 + y + 1)^4 = [y^2 + (y + 1)]^4$. We treat y^2 as the first term and $(y + 1)$ as the second term. $(y^2 + y +)^4 =$

$$= [y^2 + (y + 1)]^4 = (y^2)^4 + {}_4C_1(y^2)^3(y + 1)$$
$$+ {}_4C_2(y^2)^2(y + 1)^2 + {}_4C_3(y^2)(y + 1)^3 + (y + 1)^4$$
$$= y^8 + 4y^7 + 4y^6 + 6y^6 + 12y^5 + 6y^4 + 4y^5 + 12y^4 + 12y^3$$
$$+ 4y^2 + y^4 + 4y^3 + 6y^2 + 4y + 1$$
$$= y^8 + 4y^7 + 10y^6 + 16y^5 + 19y^4 + 16y^3 + 10y^2 + 4y + 1$$

EXERCISE 2

Expand the following by the binomial theorem:

1. $(a + b)^7$

2. $(x - 3b)^5$

3. $(a^2 + b^3)^6$

4. $(x^2 - 2a)^4$

5. $(1 - \sqrt{y})^6$

6. $\left(3 - \dfrac{y^2}{3}\right)^4$

7. $\left(x^2 - \dfrac{1}{y}\right)^5$

8. $\left(a^2 - \dfrac{1}{\sqrt{b}}\right)^6$

9. $\left(\dfrac{3x}{y} - \dfrac{y}{x}\right)^4$

10. $(y^{-3} + a^{3/2})^5$

11. $(a^{1/2} + 3b^{-2})^4$

12. $\left(x^{-2} + \dfrac{2x}{a}\right)^5$

13. $(a + b + c)^3$

14. $(x - y + z)^3$

15. $(x^2 + x + 1)^3$

Finding a Specific Term in a Binomial Expansion

Sometimes, it is required to find a specified term in a binomial expansion. For example, it may be required to find the 4th term in the expansion $(a - 2b)^{15}$. In finding this result we must concern ourselves with four details:

  a. the sign of the required term

  b. the numerical coefficient of the required term

  c. the exponent of the first term

  d. the exponent of the second term

a. *The Sign of the Required Term*

The sign of the required term will be controlled by the expression that must be simplified. In the case of the 4th term of $(a - 2b)^{15}$, $(-2b)$ will be raised to the third power, yielding a negative sign.

b. *The Numerical Coefficient of the Required Term*

In our previous expansions we have noted that the numerical coefficient of the second term is $_nC_1$, the numerical coefficient of the third term is $_nC_2$, the numerical coefficient of the fourth term is $_nC_3$, etc. In this case, the numerical coefficient of the fourth term is $_{15}C_3$. In general, the coefficient of the $(r + 1)$th term is $_nC_r$. Note that we have written the general term as the $(r + 1)$th term.

c. *The Exponent of the First Term*

In the expansion $(a - 2b)^{15}$, the exponent of the first term, a, is 15 and is reduced by 1 for each succeeding term. The fourth term will therefore have the exponent 12 for a. In general, the exponent of the first term of the expansion for the $(r + 1)$th term is $n - r$. Up to this point, we note that the $(r + 1)$th term is $_nC_r a^{n-r}$. We must now find a designation for the exponent of the second term of the expansion.

d. *The Exponent of the Second Term*

The exponent of the second term, $2b$, in the expansion is 1 and is increased by 1 for each succeeding term. In this case, the fourth term will have the exponent 3. In general, the exponent of the second term of the expansion for the $(r + 1)$th term is r.

Thus, the $(r + 1)$th term is given by the expression $_nC_r a^{n-r} b^r$.

To find the fourth term of $(a - 2b)^{15}$, we note that $r + 1 = 4$, or $r = 3$. Therefore, in this case, $_nC_r a^{n-r} b^r = {}_{15}C_3(a)^{15-3}(-2b)^3 = -3{,}640a^{12}b^3$.

Note: In the expression, $_{15}C_3(a)^{12}(-2b)^3$, the number after C is identical with the exponent of the second term. Also, the sum of the exponents is n, or 15.

Example: Write the 9th term of the expansion $(x^2 - y)^{12}$.

▶   In this case, $n = 12$, and $r + 1 = 9$. The $(r + 1)$th or 9th term is given by the expression $_nC_r a^{n-r} b^r$. Since $r + 1 = 9$, $r = 8$. $_nC_r a^{n-r} b^r = {}_{12}C_8(x^2)^{12-8}(-y)^8 = {}_{12}C_4(x^2)^4(-y)^8 = 495x^8y^8$.

Example: Write the 16th term of the expansion $\left(x + \dfrac{y}{x}\right)^{18}$.

▶   In this case, $n = 18$, and $r + 1 = 16$. The $(r + 1)$th term, or 16th term, is given by the expression $_nC_r a^{n-r} b^r$. Since $r + 1 = 16$, $r = 15$.

$$_nC_r a^{n-r} b^r = {}_{18}C_{15}(x)^{18-15}\left(\frac{y}{x}\right)^{15} = \frac{816y^{15}}{x^{12}}$$

Example: Write the fourth term of the expansion $(y^{1/2} + y^{1/3})^{10}$.

▶ In this case, $n = 10$, and $r + 1 = 4$. The $(r + 1)$th, or 4th term, is given by the expression $nCra^{n-r}b^r$.

Since $r + 1 = 4$, $r = 3$.

$$_nC_ra^{n-r}b^r = {}_{10}C_3(y^{1/2})^{10-3}(y^{1/3})^3 = 120y^{9/2}$$

Example: Write the 6th term of the expansion $(2a^{-3} - \sqrt{b})^8$.

▶ In this case, $n = 8$, and $r + 1 = 6$.

Since $r + 1 = 6$, $r = 5$.

$$_nC_ra^{n-r}b^r = {}_8C_5(2a^{-3})^{8-5}(-b^{1/2})^5 = -448a^{-9}b^{5/2}$$

Example: Write the middle term of the expansion
$$\left(3x - \frac{y^2}{2}\right)^6.$$

▶ Since the expansion has 7 terms, the middle term is the 4th. In this case, $n = 6$, and $r + 1 = 4$.

$$_nC_ra^{n-r}b^r = {}_6C_3(3x)^{6-3}\left(\frac{-y^2}{2}\right)^3 = \frac{-135x^3y^6}{2}$$

Example: Write the term which is independent of y in the expansion $\left(y^2 + \dfrac{1}{y}\right)^9$.

▶ The required or $(r + 1)$th term is given by the expression $_nC_ra^{n-r}b^r$. Since $n = 9$, this expression becomes $_9C_ra^{9-r}b^r$.

Investigating powers of y we have

$$(y^2)^{9-r}\left(\frac{1}{y}\right)^r \quad \text{or} \quad \frac{y^{18-2r}}{y^r} = y^{18-3r}$$

Since the exponent of y must be 0, $18 - 3r = 0$. Therefore, $r = 6$ and we must find the 7th term. The desired term

$$= {}_9C_6(y^2)^{9-3}\left(\frac{1}{y}\right)^6 = 84$$

Example: Write the term containing x^6 in the expansion

$$\left(x - \frac{a}{x^2} \right)^{15}$$

▶ The required or $(r + 1)$th term may be written as

$$_{15}C_r(x)^{15-r} \left(-\frac{a}{x^2} \right)^r$$

Investigating powers of x, we have

$$x^{15-r} \cdot \frac{-a^r}{x^{2r}} = \frac{x^{15-}}{x^{2r}} = x^{15-3r}$$

Since the exponent of x must be 6, $15 - 3r = 6$. Therefore, $r = 3$ and we must find the 4th term. The desired term

$$= {}_{15}C_3(x)^{15-3} \left(-\frac{a}{x^2} \right)^3 = -455x^6a^3.$$

Example: Use the binomial theorem to compute to two decimal places the value of $(1.01)^8$.

▶ $(1.01)^8 = (1 + .01)^8 = (1)^8 + {}_8C_1(1)^7(.01)^1$
$$+ {}_8C_2(1)^6(.01)^2$$
$= 1 + 8(.01) + 28(.0001) = 1 + .08 + .0028 = 1.0828 = 1.08$
to two decimal places. It is unnecessary to take more than 3 terms because the additional terms will not affect the hundredths place.

EXERCISE 3

Write the indicated term in the following expansions:

1. $(x + y)^7$—4th term
2. $(2x - y)^8$—6th term
3. $(a - 3b)^{12}$—3rd term
4. $(x^2 - y)^{10}$—7th term
5. $(a - 2b^2)^{14}$—5th term
6. $\left(a - \dfrac{1}{b} \right)^9$—6th term
7. $(\sqrt{y} - 2)^{12}$—5th term

8. $(x^{2/3} + 1)^{15}$—12th term
9. $\left(3 - \dfrac{y^2}{z} \right)^8$—3rd term
10. $\left(x^2 - \dfrac{3}{\sqrt{y}} \right)^7$—4th term
11. $\left(\dfrac{a}{b} - \dfrac{b}{a} \right)^{16}$—12th term
12. $(a^{-3} + b^{1/2})^{14}$—10th term

13. $(x^{-2} + 2y^{1/3})^8$—3rd term

14. $(2b^{-1} - c^2)^{12}$—9th term

Compute, to two decimal places, by use of the binomial theorem:

15. $(1.02)^9$

16. $(.99)^6$

Combinations of *n* Different Objects Taken Any Number at a Time

For positive integral exponents the binomial formula may be written as

$$(a + b)^n = {}_nC_0a^n + {}_nC_1a^{n-1}b + {}_nC_2a^{n-2}b^2 + \cdots + {}_nC_ra^{n-r}b^r$$
$$+ \cdots + {}_nC_nb^n$$

In this identity if $a = b = 1$, then

$$(1 + 1)^n = 2^n = {}_nC_0 + {}_nC_1 + {}_nC_2 + \cdots + {}_nC_n = 2^n$$

i.e., ${}_nC_0 + {}_nC_1 + {}_nC_2 + \cdots + {}_nC_n = 2^n$.

Since ${}_nC_0 = 1$, we have

$${}_nC_1 + {}_nC_2 + {}_nC_3 + \cdots + {}_nC_n = 2^n - 1$$

Thus, the total number of combinations C of n different objects taken $1, 2, 3, \cdots n$ at a time is $C = 2^n - 1$.

THE BINOMIAL SERIES

The binomial formula may be used to expand $(a + b)^n$ where n is fractional or negative. In these cases, we obtain an infinite series and it is desirable to be able to write the terms at the beginning of the series. When n is fractional or negative $(a + b)^n$ may be expanded if, and only if, the absolute value of b is less than the absolute value of a.

Example: Write the first four terms of the expansion $(a + b)^{\frac{1}{2}}$.

$$\blacktriangleright \quad (a + b)^{1/2} = (a)^{1/2} + \frac{1}{2}(a)^{1/2}b + \frac{(\frac{1}{2})(-\frac{1}{2})}{1 \cdot 2}(a)^{-3/2}b^2$$

$$+ \frac{(\frac{1}{2})(-\frac{1}{2})(-\frac{3}{2})}{1 \cdot 2 \cdot 3}(a)^{-5/2}b^3$$

$$= a^{1/2} + \frac{1}{2}a^{-1/2}b - \frac{1}{8}a^{-3/2}b^2 + \frac{1}{16}a^{-5/2}b^3$$

Example: Write the first 4 terms of the expansion $(1 - x)^{1/3}$.

▶ $(1 - x)^{1/3} = (1)^{1/3} + \dfrac{1}{3}(1)^{-2/3}(-x)$

$+ \dfrac{(\frac{1}{3})(-\frac{2}{3})}{1 \cdot 2}(1)^{-5/3}(-x)^2 + \dfrac{(\frac{1}{3})(-\frac{2}{3})(-\frac{5}{3})}{1 \cdot 2 \cdot 3}(1)^{-7/3}(-x)^3$

$= 1 - \dfrac{x}{3} - \dfrac{1}{9}x^2 - \dfrac{5}{81}x^3$

Example: Write the first 4 terms of the expansion $(a + b)^{-1}$.

▶ $(a + b)^{-1} = (a)^{-1} + (-1)(a)^{-2}(b)$

$+ \dfrac{(-1)(-2)}{1 \cdot 2}(a)^{-3}(b)^2 + \dfrac{(-1)(-2)(-3)}{1 \cdot 2 \cdot 3}(a)^{-4}(b)^3$

$= a^{-1} - a^{-2}b + a^{-3}b^2 - a^{-4}b^3$

Example: Write the first 4 terms of the expansion $(x - y)^{-2}$.

▶ $(x - y)^{-2} = (x)^{-2} + (-2)(x)^{-3}(-y)$

$+ \dfrac{(-2)(-3)}{1 \cdot 2}(x)^{-4}(-y)^2 + \dfrac{(-2)(-3)(-4)}{1 \cdot 2 \cdot 3}(x)^{-5}(-y)^3$

$= x^{-2} + 2x^{-3}y + 3x^{-4}y^2 + 4x^{-5}y^3$

Example: Write the first 4 terms of the expansion $(4 + y)^{-\frac{1}{2}}$.

▶ $(4 + y)^{-1/2} = (4)^{-1/2} + \left(-\dfrac{1}{2}\right)(4)^{-3/2}(y)$

$+ \dfrac{(-\frac{1}{2})(-\frac{3}{2})}{1 \cdot 2}(4)^{-5/2}(y)^2 + \dfrac{(-\frac{1}{2})(-\frac{3}{2})(-\frac{5}{2})}{1 \cdot 2 \cdot 3}(4)^{-7/2}(y)^3$

$= \dfrac{1}{4^{1/2}} - \dfrac{y}{2 \cdot 4^{3/2}} + \dfrac{3y^2}{8 \cdot 4^{5/2}} - \dfrac{5y^3}{16 \cdot 4^{7/2}}.$

$= \dfrac{1}{2} - \dfrac{y}{16} + \dfrac{3y^2}{256} - \dfrac{5y^3}{2{,}048}$

Example: Calculate $\sqrt{50}$ to 4 significant figures.

▶ Write $\sqrt{50}$ as $(7^2 + 1)^{1/2}$. $\sqrt{50}$ may not be written as $(1 + 7^2)^{1/2}$ since $(a + b)^n$ may be expanded for fractional or negative values of n if, and only if, the absolute value of a is greater than the absolute value of b.

$$\sqrt{50} = (7^2 + 1)^{1/2} = (7^2)^{1/2} + \tfrac{1}{2}(7^2)^{-1/2}(1)$$

$$+ \frac{(\tfrac{1}{2})(-\tfrac{1}{2})}{1 \cdot 2}(7^2)^{-3/2}(1)^2 = 7 + \tfrac{1}{2}(7)^{-1} - \tfrac{1}{8}(7)^{-3}$$

$$= 7 + \frac{1}{14} - \frac{1}{2,744} = 7 + .0714 - .0004 = 7.071$$

Example: Calculate $\sqrt[3]{63}$ to 4 significant figures.

▶ Write $\sqrt[3]{63}$ as $(8^2 - 1)^{1/3}$.

$$(8^2 - 1)^{1/3} = (8^2)^{1/3} + (\tfrac{1}{3})(8^2)^{-2/3}(-1)$$

$$+ \frac{(\tfrac{1}{3})(-\tfrac{2}{3})}{1 \cdot 2}(8^2)^{-5/3}(-1)^2$$

$$= 8^{2/3} - \frac{1}{3 \cdot 8^{4/3}} - \frac{1}{9 \cdot 8^{10/3}}$$

$$= 4 - \frac{1}{3 \cdot 16} - \frac{1}{9 \cdot 1024} = 4 - \frac{1}{48} - \frac{1}{9,218}$$

$$= 4 - .02083 - .00011 = 3.979$$

EXERCISE 4

Write the first 4 terms of the following expansions:

1. $(x + y)^{1/3}$
2. $(1 - a)^{1/2}$
3. $(a^2 + b)^{-1}$
4. $(1 - y)^{-2}$
5. $(y - \sqrt{x})^{-1/2}$
6. $(a + \sqrt{b})^{1/2}$
7. $(8 + y)^{-1/3}$
8. $\left(y^{2/3} + \dfrac{1}{y} \right)^{-3}$

9. $\left(a^2 + \dfrac{1}{a} \right)^{-2}$
10. $(9 - x)^{1/2}$
11. $\left(x^3 - \dfrac{1}{x} \right)^{1/3}$
12. $\left(4 - \dfrac{x}{2} \right)^{1/2}$

Calculate to 4 significant figures:

13. $\sqrt{37}$
14. $\sqrt{17}$
15. $\sqrt[3]{7}$
16. $\sqrt[3]{48}$

ANSWERS

EXERCISE 2

1. $(a + b)^7 = a^7 + 7a^6b + 21a^5b^2 + 35a^4b^3 + 35a^3b^4 + 21a^2b^5 + 7ab^6 + b^7$

2. $(x - 3b)^5 = x^5 - 15x^4b + 90x^3b^2 - 270x^2b^3 + 405xb^4 - 243b^5$

3. $(a^2 + b^3)^6 = a^{12} + 6a^{10}b^3 + 15a^8b^6 + 20a^6b^9 + 15a^4b^{12} + 6a^2b^{15} + b^{18}$

4. $(x^2 - 2a)^4 = x^8 - 8x^6a + 24x^4a^2 - 32x^2a^3 + 16a^4$

5. $(1 - \sqrt{y})^6 = 1 - 6y^{1/2} + 15y - 20y^{3/2} + 15y^2 - 6y^{5/2} + y^3$

6. $\left(3 - \dfrac{y^2}{3}\right)^4 = 81 - 36y^2 + 6y^4 - \dfrac{4}{9}y^6 + \dfrac{y^8}{81}$

7. $\left(x^2 - \dfrac{1}{y}\right)^5 = x^{10} - \dfrac{5x^8}{y} + \dfrac{10x^6}{y^2} - \dfrac{10x^4}{y^3} + \dfrac{5x^2}{y^4} - \dfrac{1}{y^5}$

8. $\left(a^2 - \dfrac{1}{\sqrt{b}}\right)^6 = a^{12} - 6a^{10}b^{-1/2} + 15a^8b^{-1} - 20a^6b^{-3/2} + 15a^4b^{-2} - 6a^2b^{-5/2} + b^{-3}$

9. $\left(\dfrac{3x}{y} - \dfrac{y}{x}\right)^4 = \dfrac{81x^4}{y^4} - \dfrac{108x^2}{y^2} + 54 - \dfrac{12y^2}{x^2} + \dfrac{y^4}{x^4}$

10. $(y^{-3} + a^{3/2})^5 = y^{-15} + 5y^{-12}a^{3/2} + 10y^{-9}a^3 + 10y^{-6}a^{9/2} + 5y^{-3}a^6 + a^{15/2}$

11. $(a^{1/2} + 3b^{-2})^4 = a^2 + 12a^{3/2}b^{-2} + 54ab^{-4} + 108a^{1/2}b^{-6} + 81b^{-8}$

12. $\left(x^{-2} + \dfrac{2x}{a}\right)^5 = x^{-10} + \dfrac{10x^{-7}}{a} + \dfrac{40x^{-4}}{a^2} + \dfrac{80x^{-1}}{a^3} + \dfrac{80x^2}{a^4} + \dfrac{32x^5}{a^5}$

13. $(a + b + c)^3 = a^3 + 3a^2b + 3ab^2 + b^3 + 3a^2c + 6abc + 3b^2c + 3ac^2 + 3bc^2 + c^3$

14. $(x - y + z)^3 = x^3 - 3x^2y + 3xy^2 - y^3 + 3x^2z - 6xyz + 3y^2z + 3xz^2 - 3yz^2 + z^3$

15. $(x^2 + x + 1)^2 = x^6 + 3x^5 + 6x^4 + 7x^3 + 6x^2 + 3x + 1$

EXERCISE 3

1. $35x^4y^3$
2. $-448x^3y^5$
3. $594a^{10}b^2$
4. $210x^8y^6$
5. $16,016a^{10}b^8$
6. $\dfrac{-126a^4}{b^5}$
7. $7,920y^4$
8. $1,365x^{8/3}$
9. $\dfrac{20,412y^4}{z^2}$

10. $\dfrac{-945x^8}{y^{3/2}}$
11. $\dfrac{-4,368b^6}{a^6}$
12. $2,002a^{-15}b^{9/2}$
13. $112x^{-12}y^{2/3}$
14. $7,920b^{-4}c^{16}$
15. 1.19
16. $.94$

EXERCISE 4

1. $x^{1/3} + \frac{1}{3}x^{-2/3}y - \frac{1}{9}x^{-5/3}y^2 + \frac{5}{81}x^{-8/3}y^3$
2. $1 - \dfrac{a}{2} - \dfrac{a^2}{8} - \dfrac{a^3}{16}$
3. $a^{-2} - a^{-4}b + a^{-6}b^2 - a^{-8}b^3$
4. $1 + 2y + 3y^2 + 4y^3$
5. $y^{-1/2} + \frac{1}{2}y^{-3/2}x^{1/2} + \frac{3}{8}y^{-5/2}x + \frac{5}{16}y^{-7/2}x^{3/2}$
6. $a^{1/2} + \frac{1}{2}a^{-1/2}b^{1/2} - \frac{1}{8}a^{-3/2}b + \frac{1}{16}a^{-5/2}b^{3/2}$
7. $\dfrac{1}{2} - \dfrac{y}{48} + \dfrac{y^2}{576} - \dfrac{7y^3}{41,472}$
8. $y^{-2} - 3y^{-11/3} + 6y^{-16/3} - 10y^{-7}$
9. $a^{-4} - 2a^{-7} + 3a^{-10} - 4a^{-13}$
10. $3 - \dfrac{x}{6} - \dfrac{x^2}{216} - \dfrac{x^3}{3,888}$
11. $x - \dfrac{1}{3x^3} - \dfrac{1}{9x^7} - \dfrac{5}{81x^{11}}$
12. $2 - \dfrac{x}{8} - \dfrac{x^2}{256} - \dfrac{x^3}{4,096}$
13. 6.083
14. 4.123
15. 1.913
16. 3.639

CHAPTER 21

Determinants

DETERMINANTS OF THE SECOND ORDER

The symbol $\begin{vmatrix} a_1 & b_1 \\ a_2 & b_2 \end{vmatrix}$

consisting of a square arrangement (i.e., two rows and two columns) is called a determinant of the second order.

The value of the determinant

$$\overset{+}{\underset{}{}} \quad \overset{-}{\underset{}{}}$$

$$\begin{vmatrix} a_1 & b_1 \\ a_2 & b_2 \end{vmatrix} \quad \text{is } a_1 b_2 - a_2 b_1$$

i.e., the difference of the diagonal products indicated by the arrows.

Example: Find the value of the determinant $\begin{vmatrix} 6 & 4 \\ 2 & -7 \end{vmatrix}$

▶ $\begin{vmatrix} 6 & 4 \\ 2 & -7 \end{vmatrix} = 6(-7) - 2(4) = -42 - 8 = -50$

Example: Find the value of the determinant $\begin{vmatrix} 5\frac{1}{2} & -2\frac{3}{4} \\ 0 & 6 \end{vmatrix}$

542

▶ $\begin{vmatrix} 5\frac{1}{2} & -2\frac{3}{4} \\ 0 & 6 \end{vmatrix} = 5\frac{1}{2}(6) - (0)(-2\frac{3}{4}) = 33 + 0 = 33$

Example: Find the value of the determinant $\begin{vmatrix} x & x-5 \\ x-3 & 2x+3 \end{vmatrix}$

▶ $\begin{vmatrix} x & x-5 \\ x-3 & 2x+3 \end{vmatrix} \begin{aligned} &= x(2x+3) - (x-3)(x-5) \\ &= 2x^2 + 3x - x^2 + 8x - 15 \\ &= x^2 + 11x - 15 \end{aligned}$

Using Determinants of the Second Order in Solving Sets of Linear Equations

A set of two linear simultaneous equations may be represented as:

$$a_1x + b_1y = c_1$$
$$a_2x + b_2y = c_2$$

If these equations are solved by algebraic methods, we obtain

$$x = \frac{c_1b_2 - c_2b_1}{a_1b_2 - a_2b_1}, \text{ and } y = \frac{a_1c_2 - a_2c_1}{a_1b_2 - a_2b_1}$$

These solutions may be written in determinant form, as follows:

$$x = \frac{\begin{vmatrix} c_1 & b_1 \\ c_2 & b_2 \end{vmatrix}}{\begin{vmatrix} a_1 & b_1 \\ a_2 & b_2 \end{vmatrix}} \qquad\qquad y = \frac{\begin{vmatrix} a_1 & c_1 \\ a_2 & c_2 \end{vmatrix}}{\begin{vmatrix} a_1 & b_1 \\ a_2 & b_2 \end{vmatrix}}$$

Note that the denominators are equivalent and are composed of the coefficients of x and y.

The numerator of the value of x is formed from the denominator by substituting c_1 and c_2 for the coefficients of x in the equations (i.e. a_1 and a_2).

Likewise, the numerator of the value of y is formed from the denominator by substituting c_1 and c_2 for the coefficients of y in the equations (i.e. b_1 and b_2).

Example: Solve the equations $2x + 3y = -5$
$x - 2y = 8$

▶

$$x = \dfrac{\begin{vmatrix} c_1 & b_1 \\[6pt] c_2 & b_2 \end{vmatrix}}{\begin{vmatrix} a_1 & b_1 \\[6pt] a_1 & b_2 \end{vmatrix}} = \dfrac{\begin{vmatrix} -5 & 3 \\[6pt] 8 & -2 \end{vmatrix}}{\begin{vmatrix} 2 & 3 \\[6pt] 1 & -2 \end{vmatrix}} = \dfrac{10 - 24}{-4 - 3} = \dfrac{-14}{-7} = 2$$

$$y = \dfrac{\begin{vmatrix} a_1 & c_1 \\[6pt] a_2 & c_2 \end{vmatrix}}{\begin{vmatrix} a_1 & b_1 \\[6pt] a_2 & b_2 \end{vmatrix}} = \dfrac{\begin{vmatrix} 2 & -5 \\[6pt] 1 & 8 \end{vmatrix}}{-7} = \dfrac{16 + 5}{-7} = \dfrac{21}{-7} = -3$$

Note: Since the denominators are identical it is necessary to compute this value only once.

Example: Solve $9x = 4y - 11$
$5y = 8 - 6x$

▶   First, arrange the equations in standard form

$$9x - 4y = -11$$
$$6x + 5y = 8$$

$$x = \dfrac{\begin{vmatrix} c_1 & b_1 \\[6pt] c_2 & b_2 \end{vmatrix}}{\begin{vmatrix} a_1 & b_1 \\[6pt] a_2 & b_2 \end{vmatrix}} = \dfrac{\begin{vmatrix} -11 & -4 \\[6pt] 8 & +5 \end{vmatrix}}{\begin{vmatrix} 9 & -4 \\[6pt] 6 & +5 \end{vmatrix}} = \dfrac{-55 + 32}{45 + 24} = \dfrac{-23}{69} = \dfrac{-1}{3}$$

$$y = \dfrac{\begin{vmatrix} a_1 & c_1 \\[6pt] a_2 & c_2 \end{vmatrix}}{\begin{vmatrix} a_1 & b_1 \\[6pt] a_2 & b_2 \end{vmatrix}} = \dfrac{\begin{vmatrix} 9 & -11 \\[6pt] 6 & 8 \end{vmatrix}}{69} = \dfrac{72 + 66}{69} = \dfrac{138}{69} = 2$$

Example: Solve $\dfrac{2x - 5}{4} + \dfrac{y}{5} = -2$

$$\dfrac{4x}{3} - \dfrac{2y - 10}{4} = 5\tfrac{2}{3}$$

▶ If these equations are cleared of fractions and written in standard form, we obtain

$$10x + 4y = -15$$
$$16x - 6y = 38$$

$$x = \dfrac{\begin{vmatrix} -15 & 4 \\ 38 & -6 \end{vmatrix}}{\begin{vmatrix} 10 & 4 \\ 16 & -6 \end{vmatrix}} = \dfrac{90 - 152}{-60 - 64} = \dfrac{-62}{-124} = \dfrac{1}{2}$$

$$y = \dfrac{\begin{vmatrix} 10 & -15 \\ 16 & 38 \end{vmatrix}}{\begin{vmatrix} 10 & 4 \\ 16 & -6 \end{vmatrix}} = \dfrac{380 + 240}{-124} = \dfrac{620}{-124} = -5$$

Example: Solve $\dfrac{3}{x} + \dfrac{8}{y} = -11$

$$\dfrac{4}{x} - \dfrac{9}{y} = -9\tfrac{3}{4}$$

▶

$$\dfrac{3}{x} + \dfrac{8}{y} = -11$$

$$\dfrac{4}{x} - \dfrac{9}{y} = \dfrac{-39}{4}$$

These equations may be solved for $\dfrac{1}{x}$ and $\dfrac{1}{y}$ as follows:

$$\dfrac{1}{x} = \dfrac{\begin{vmatrix} -11 & 8 \\ \dfrac{-39}{4} & -9 \end{vmatrix}}{\begin{vmatrix} 3 & 8 \\ 4 & -9 \end{vmatrix}} = \dfrac{99 + 78}{-27 - 32} = \dfrac{177}{-59} = -3$$

Since $\dfrac{1}{x} = -3$, $-3x = 1$, and $x = -\dfrac{1}{3}$

$$\frac{1}{y} = \frac{\begin{vmatrix} 3 & -11 \\ 4 & \dfrac{-39}{4} \end{vmatrix}}{\begin{vmatrix} 3 & 8 \\ 4 & -9 \end{vmatrix}} = \frac{\dfrac{-117}{4} + 44}{-27 - 32} = \frac{\dfrac{59}{4}}{-59} = -\frac{1}{4}$$

Since $\dfrac{1}{y} = \dfrac{-1}{4}$, $y = -4$

Example: Solve the equations $px + qy = t$
$$mx + sy = v$$

$$x = \frac{\begin{vmatrix} t & q \\ v & s \end{vmatrix}}{\begin{vmatrix} p & q \\ m & s \end{vmatrix}} = \frac{ts - qv}{ps - qm}$$

$$y = \frac{\begin{vmatrix} p & t \\ m & v \end{vmatrix}}{\begin{vmatrix} p & q \\ m & s \end{vmatrix}} = \frac{pv - tm}{ps - qm}$$

EXERCISE 1

Find the values of the following determinants:

1. $\begin{vmatrix} 3 & -6 \\ 5 & 2 \end{vmatrix}$

2. $\begin{vmatrix} \frac{1}{2} & -\frac{1}{3} \\ 6 & 4 \end{vmatrix}$

3. $\begin{vmatrix} a + 2 & 2a + 1 \\ a - 3 & a - 5 \end{vmatrix}$

Solve the following equations by determinants:

4. $2x - y = 8$
   $x + 2y = 9$
5. $2x + 7y = -5$
   $5x - y = 6$
6. $3x + 2y = 8$
   $x - 4y = 5$

7. $9x + 8y = 3$
$3x - 4y = -4$

8. $2x + 10 = -3y$
$4y - 5x = 2$

9. $x - y = \dfrac{4x + 5y}{40}$

$2y - \tfrac{1}{2} = \dfrac{y - 2x}{3}$

10. $x - 5 = \dfrac{y - 2}{7}$

$4y - 3 = \dfrac{x + 10}{3}$

11. $\dfrac{2}{x} + \dfrac{6}{y} = 5$

$\dfrac{7}{x} - \dfrac{4}{y} = -20$

12. $ax + by = c$
$cx - dy = a$

DETERMINANTS OF THE THIRD ORDER

The symbol $\begin{vmatrix} a_1 & b_1 & c_1 \\ a_2 & b_2 & c_2 \\ a_3 & b_3 & c_3 \end{vmatrix}$

consisting of a square arrangement of elements (i.e. three rows and three columns) is called a determinant of the third order. This symbol represents the number $a_1b_2c_3 + b_1c_2a_3 + c_1a_2b_3 - a_3b_2c_1 - b_3c_2a_1 - c_3a_2b_1$ which is called the value of the determinant.

Finding the Value of a Third Order Determinant

A convenient method for finding the value of a third order determinant is shown in the diagram at the right. Note that the first two columns are rewritten to the right of the original determinant.

$$\begin{array}{ccc|cc} a_1 & b_1 & c_1 & a_1 & b_1 \\ a_2 & b_2 & c_2 & a_2 & b_2 \\ a_3 & b_3 & c_3 & a_3 & b_3 \end{array}$$

Then there will be three positive diagonal products ($a_1b_2c_3$, $b_1c_2a_3$, and $c_1a_2b_3$). From the sum of these three positive diagonal products is subtracted the three diagonal products ($a_3b_2c_1$, $b_3c_2a_1$, and $c_3a_2b_1$). The result is $(a_1b_2c_3 + b_1c_2a_3 + c_1a_2b_3) - (a_3b_2c_1 + b_3c_2a_1 + c_3a_2b_1)$.

Example: Find the value of the determinant

$$\begin{vmatrix} 3 & 1 & 6 \\ 4 & -1 & 8 \\ 2 & 5 & 7 \end{vmatrix}$$

▶

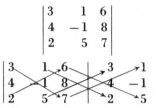

The three positive diagonal products are

$$(3)(-1)(7) + (1)(8)(2) + (6)(4)(5) = +115$$

The three negative diagonal products are

$$-(2)(-1)(6) - (5)(8)(3) - (7)(4)(1) = -136$$

The value of the determinant $= +115 - 136 = -21$.

Example: Find the value of the determinant

$$\begin{vmatrix} x & 2x & 3 \\ x+1 & 2x-1 & 2 \\ x-1 & 2x+1 & 1 \end{vmatrix}$$

▶

$$\begin{vmatrix} x & 2x & 3 \\ x+1 & 2x-1 & 2 \\ x-1 & 2x+1 & 1 \end{vmatrix} \quad \begin{matrix} x & 2x \\ x+1 & 2x-1 \\ x-1 & 2x+1 \end{matrix}$$

The positive factors are

$$x(2x-1)(1) + (2x)(2)(x-1) + 3(x+1)(2x+1)$$

The negative factors are

$$-(x-1)(2x-1)(3) - (2x+1)(2)(x) - (1)(x+1)(2x)$$

When the results are multiplied out and combined the result is $9x$.

Example: Solve the equation

$$\begin{vmatrix} 3 & 5 & -2 \\ 1 & 2 & -1 \\ x & -3 & 4x \end{vmatrix} = 0$$

▶   In solving the equation, the determinant is expanded and the result set equal to zero. The equation thus formed can readily be solved.

$$\begin{vmatrix} 3 & 5 & -2 \\ 1 & 2 & -1 \\ x & -3 & 4x \end{vmatrix} \begin{matrix} 3 & 5 \\ 1 & 2 \\ x & -3 \end{matrix} = \begin{aligned} &(3)(2)(4x) + (5)(-1)(x) \\ &+ (-2)(1)(3) - (x)(2)(-2) \\ &- (-3)(-1)(3) - (4x)(1)(5) \end{aligned}$$

$$24x - 5x + 6 + 4x - 9 - 20x = 0$$
$$3x - 3 = 0$$
$$x = 1$$

EXERCISE 2

Evaluate the following determinants:

1. $\begin{vmatrix} 5 & -1 & 6 \\ 2 & 3 & -2 \\ 1 & 4 & 5 \end{vmatrix}$

2. $\begin{vmatrix} 3 & -2 & 5 \\ -1 & 4 & -6 \\ 0 & 7 & -3 \end{vmatrix}$

3. $\begin{vmatrix} -8 & 3 & 2 \\ 6 & -5 & 1 \\ 4 & 3 & 0 \end{vmatrix}$

4. $\begin{vmatrix} 7 & 0 & -3 \\ -5 & 1 & 2 \\ -4 & 6 & -8 \end{vmatrix}$

5. $\begin{vmatrix} a-1 & a & -1 \\ a & a+1 & 1 \\ a+2 & a-1 & 2 \end{vmatrix}$

6. $\begin{vmatrix} x+2 & 1 & x \\ x & 3 & x+1 \\ x-1 & -2 & x-2 \end{vmatrix}$

Solve the following equations:

7. $\begin{vmatrix} y & -2 & 4 \\ 1 & 3 & -1 \\ 3y & 5 & 5 \end{vmatrix} = 0$

8. $\begin{vmatrix} 3 & 2y & 3y \\ 1 & -2 & -1 \\ 7 & 4 & 5 \end{vmatrix} = 12$

Using Determinants of the Third Order in Solving Sets of Linear Equations

A set of three linear simultaneous equations may be represented as follows:

$$a_1x + b_1y + c_1z = d_1$$
$$a_2x + b_2y + c_2z = d_2$$
$$a_3x + b_3y + c_3z = d_3$$

If these equations are solved by algebraic methods, we obtain

$$x = \frac{d_1b_2c_3 + b_1c_2d_3 + c_1d_2b_3 - d_3b_2c_1 - b_3c_2d_1 - c_3d_2b_1}{a_1b_2c_3 + b_1c_2a_3 + c_1a_2b_3 - a_3b_2c_1 - b_3c_2a_1 - c_3a_2b_1}$$

This solution may be written in determinant form as follows:

$$x = \frac{\begin{vmatrix} d_1 & b_1 & c_1 \\ d_2 & b_2 & c_2 \\ d_3 & b_3 & c_3 \end{vmatrix}}{\begin{vmatrix} a_1 & b_1 & c_1 \\ a_2 & b_2 & c_2 \\ a_3 & b_3 & c_3 \end{vmatrix}}$$

In similar manner, solutions for y and z in determinant form are written as follows:

$$y = \frac{\begin{vmatrix} a_1 & d_1 & c_1 \\ a_2 & d_2 & c_2 \\ a_3 & d_3 & c_3 \end{vmatrix}}{\begin{vmatrix} a_1 & b_1 & c_1 \\ a_2 & b_2 & c_2 \\ a_3 & b_3 & c_3 \end{vmatrix}} \qquad z = \frac{\begin{vmatrix} a_1 & b_1 & d_1 \\ a_2 & b_2 & d_2 \\ a_3 & b_3 & d_3 \end{vmatrix}}{\begin{vmatrix} a_1 & b_1 & c_1 \\ a_2 & b_2 & c_2 \\ a_3 & b_3 & c_3 \end{vmatrix}}$$

Note that the denominators are equivalent. The determinant of the denominators is composed of the coefficients of the variables x, y, and z in that order.

The numerator of the value of x is formed from the denominator by substituting d_1, d_2, and d_3 for the coefficients of x (i.e. a_1, a_2, and a_3). The numerators of the values of y and z are formed similarly.

This method of solving sets of linear equations by determinants is known as *Cramer's Rule*.

Example: Solve the equations

$$\begin{aligned} 2x + 3y - z &= -9 \\ 3x - 2y + z &= 16 \\ x + 4y + 3z &= 2 \end{aligned}$$

▶

$$\begin{aligned} 2x + 3y - z &= -9 \\ 3x - 2y + z &= 16 \\ x + 4y + 3x &= 2 \end{aligned}$$

$$x = -\frac{\begin{vmatrix} -9 & 3 & -1 \\ 16 & -2 & +1 \\ 2 & 4 & +3 \end{vmatrix}}{\begin{vmatrix} 2 & 3 & -1 \\ 3 & -2 & +1 \\ 1 & 4 & +3 \end{vmatrix}}$$

$$= \frac{[54 + 6 - 64] - [4 - 36 + 144]}{[-12 + 3 - 12] - [2 + 8 + 27]} = \frac{(-4) - (112)}{(-21) - (37)}$$

$$= \frac{-116}{-58} = +2$$

$$y = \frac{\begin{vmatrix} 2 & -9 & -1 \\ 3 & 16 & +1 \\ 1 & 2 & +3 \end{vmatrix}}{\text{Denominator is the same} = -58} = \frac{[96 - 9 - 6] - [-16 + 4 - 81]}{-58}$$

$$= \frac{(81) - (-93)}{-58} = \frac{174}{-58} = -3$$

$$z = \frac{\begin{vmatrix} 2 & 3 & -9 \\ 3 & -2 & 16 \\ 1 & 4 & -2 \end{vmatrix}}{\text{Denominator is the same} = -58}$$

$$= \frac{[-8 + 48 - 108] - [18 + 128 + 18]}{-58} = \frac{(-68) - (164)}{-58}$$

$$= \frac{-232}{-58} = 4$$

Example: Solve the equations

$$2x + y + 3z = -2$$
$$5x + 2y = 5$$
$$2y + 3z = -13$$

▶ The method of solution is essentially the same as that of example 1. Zeros are used as coefficients of missing terms.

$$
x = \frac{\begin{vmatrix} -2 & 1 & 3 \\ 5 & 2 & 0 \\ -13 & 2 & 3 \end{vmatrix} \begin{matrix} -2 & 1 \\ 5 & 2 \\ -13 & 2 \end{matrix}}{\begin{vmatrix} 2 & 1 & 3 \\ 5 & 2 & 0 \\ 0 & 2 & 3 \end{vmatrix} \begin{matrix} 2 & 1 \\ 5 & 2 \\ 0 & 2 \end{matrix}}
$$

$$
= \frac{[-12 + 0 + 30] - [-78 + 0 + 15]}{[12 + 0 + 30] - [0 + \quad 0 + 15]} = \frac{(18) - (-63)}{(42) - (15)}
$$

$$
= \frac{81}{27} = 3
$$

$$
y = \frac{\begin{vmatrix} 2 & -2 & 3 \\ 5 & 5 & 0 \\ 0 & -13 & 3 \\ \text{Denominator} \\ = 27 \end{vmatrix} \begin{matrix} 2 & -2 \\ 5 & 5 \\ 0 & -13 \end{matrix}}{} = \frac{[30 + 0 - 195] - [0 + 0 - 30]}{27}
$$

$$
= \frac{(-165) - (-30)}{27} = \frac{-135}{27} = -5
$$

$$
z = \frac{\begin{vmatrix} 2 & 1 & -2 \\ 5 & 2 & 5 \\ 0 & 2 & -13 \\ \text{Denominator} \\ = 27 \end{vmatrix} \begin{matrix} 2 & 1 \\ 5 & 2 \\ 0 & 2 \end{matrix}}{} = \frac{[-52 + 0 - 20] - [0 + 20 - 65]}{27}
$$

$$
= \frac{(-72) - (-45)}{27} = \frac{-27}{27} = -1
$$

EXERCISE 3

Solve the following systems of equations by determinants:

1. $x + y + z = 1$
    $2x + 3y + z = 4$
    $4x + 9y + z = 16$

2. $3x - 2y + z = 12$
    $x + 3y - 2z = -9$
    $2x - 4y - 3z = -4$

3. $2x + y - 2z = 1$ **4.** $4x + 3y = 1$
$x + y - z = 0$ $6y - z = -6$
$3x - 4z = 1$ $2x - 9y = 4$

DETERMINANTS OF HIGHER ORDER

DEFINITION: An *inversion* of the positive integers occurs whenever a greater number precedes a smaller number. For example, in the order 3142 there are 3 inversions because 3 precedes 1 and 2, and 4 precedes 2.

DEFINITION: The *principal diagonal* of a determinant is the diagonal from the upper left-hand to the lower right-hand corner of the square arrangement.

DEFINITION: A *determinant of order n* is a square arrangement of elements containing n rows and n columns, written as follows:

$$\begin{vmatrix} a_1 & b_1 & c_1 & \ldots\ldots.q_1 \\ a_2 & b_2 & c_2 & \ldots\ldots.q_2 \\ \cdot & \cdot & \cdot & \ldots\ldots \\ \cdot & \cdot & \cdot & \ldots\ldots \\ \cdot & \cdot & \cdot & \ldots\ldots \\ a_n & b_n & c_n & q_n \end{vmatrix}$$

It is assumed that the letter q is the nth letter of the alphabet.

The above symbol represents the algebraic sum of all products which can be formed

a. by taking one and only one element from each row and from each column as factors, and

b. by affixing to each term a positive or a negative sign according as the number of inversions of the subscripts of the term is even or odd, when the letters have the same order as they have in the principal diagonal.

In practice, determinants of orders above 3 are usually expanded by minors as explained below.

DEFINITION: The *minor* of any element of a determinant is the determinant obtained by erasing the row and column which contains the element. The minor of an element of a determinant is a determinant of the next lower order.

Example: The minor of the element -2 in the determinant

$$\begin{vmatrix} 3 & -2 & 6 \\ 1 & 5 & -3 \\ 4 & 2 & 1 \end{vmatrix} \quad \text{is the determinant} \quad \begin{vmatrix} 1 & -3 \\ 4 & 1 \end{vmatrix}$$

DEFINITION: The *co-factor* of an element of a determinant is the minor of the element preceded by a $+$ or $-$ sign.

The sign of the co-factor of an element is determined as follows. *The number of the row* and the *number of the column* of the element are added.   If this sum is an even number then the minor preceded by a $+$ sign is the co-factor.   If this sum is an odd number then the minor preceded by a $-$ sign is the co-factor.

Example: Write the co-factor of 2 in the determinant

$$\begin{vmatrix} 3 & 1 & 9 \\ -4 & 5 & 8 \\ 2 & 6 & -3 \end{vmatrix}$$

▶   2 is in column 1 and row 3

$$1 + 3 = 4, \text{ an even number}$$

Therefore, the co-factor of 2 is $\quad + \begin{vmatrix} 1 & 9 \\ 5 & 8 \end{vmatrix}$

Example: Write the co-factor of 8 in the above determinant.

▶   8 is in column 3 and row 2

$$3 + 2 = 5, \text{ an odd number}$$

Therefore, the co-factor of 8 is $\quad - \begin{vmatrix} 3 & 1 \\ 2 & 6 \end{vmatrix}$

Expanding by Co-Factors

THEOREM: A determinant of *any* order is equal to the sum of the products obtained by multiplying the elements of any row or column by their respective co-factors.

Example: Evaluate the determinant

$$\begin{vmatrix} 3 & 4 & 2 & -1 \\ 1 & 1 & 3 & 5 \\ 2 & 5 & 4 & 3 \\ -1 & 3 & -2 & 6 \end{vmatrix}$$

▶ According to the theorem we select any row or column and write the products of the elements of the row or column and their co-factors. Let us take the fourth row. The products of the elements and their co-factors are

$$-(-1)\begin{vmatrix} 4 & 2 & -1 \\ 1 & 3 & 5 \\ 5 & 4 & 3 \end{vmatrix} + 3\begin{vmatrix} 3 & 2 & -1 \\ 1 & 3 & 5 \\ 2 & 4 & 3 \end{vmatrix} -(-2)\begin{vmatrix} 3 & 4 & -1 \\ 1 & 1 & 5 \\ 2 & 5 & 3 \end{vmatrix}$$

$$+6\begin{vmatrix} 3 & 4 & 2 \\ 1 & 1 & 3 \\ 2 & 5 & 4 \end{vmatrix}$$

The combined values of these determinants will give us the value of the original determinant.

$$+ 1(11) + 3(-17) + 2(-41) + 6(-19) = -236$$

Example: Evaluate the determinant

$$\begin{vmatrix} 2 & 1 & -1 & -4 \\ 3 & 0 & 2 & 5 \\ -1 & -5 & 3 & -3 \\ 4 & -2 & 1 & 6 \end{vmatrix}$$

▶ In selecting a row or column for expansion by co-factors it is an advantage to choose a row or column containing one or more zeros. This will make it simpler to evaluate the resulting determinants. For example, if we select the second column in the above determinant the second of the resulting determinants will vanish, as shown below.

$$-(+1)\begin{vmatrix} 3 & 2 & 5 \\ -1 & 3 & -3 \\ 4 & 1 & 6 \end{vmatrix} +0\begin{vmatrix} 2 & -1 & -4 \\ -1 & 3 & -3 \\ 4 & 1 & 6 \end{vmatrix}$$

$$-(-5)\begin{vmatrix} 2 & -1 & -4 \\ 3 & 2 & 5 \\ 4 & 1 & 6 \end{vmatrix} + (-2)\begin{vmatrix} 2 & -1 & -4 \\ 3 & 2 & 5 \\ -1 & 3 & -3 \end{vmatrix}$$

$-(54 - 24 - 5 - 60 + 9 + 12) + 0 + 5(24 - 20 - 12 + 32$
$- 10 + 18) - 2(-12 + 5 - 36 - 8 - 30 - 9)$
$= +14 + 0 + 160 + 180 = 354$

EXERCISE 4

Expand the following determinants

1. $\begin{vmatrix} 2 & 6 & 3 & 0 \\ 1 & 0 & -2 & 4 \\ -3 & 2 & 1 & 1 \\ 5 & 1 & 4 & 3 \end{vmatrix}$

4. $\begin{vmatrix} 6 & 3 & 1 & -3 \\ -1 & 0 & -5 & -2 \\ 4 & 5 & 2 & 7 \\ 2 & -4 & 3 & 1 \end{vmatrix}$

2. $\begin{vmatrix} 4 & -2 & 1 & -5 \\ 2 & 5 & 0 & 3 \\ -1 & 3 & 2 & -1 \\ 3 & 1 & 4 & 2 \end{vmatrix}$

5. $\begin{vmatrix} -4 & 8 & -1 & 2 \\ 2 & 0 & 3 & 1 \\ -3 & 1 & -5 & -5 \\ 5 & 2 & 4 & 6 \end{vmatrix}$

3. $\begin{vmatrix} 1 & 6 & 2 & 1 \\ 3 & -4 & 3 & 5 \\ 7 & 1 & -1 & 2 \\ -2 & 5 & 4 & -3 \end{vmatrix}$

PROPERTIES OF DETERMINANTS

The following theorems describe properties that are true for determinants of order *n*. Third order determinants are used for illustrative purposes as a matter of convenience.

THEOREM 1:   The value of a determinant is not changed when the rows are written as columns and the columns are written as rows.

Example: Consider the determinant

$$\begin{vmatrix} 4 & 1 & -1 \\ 3 & 6 & 0 \\ 5 & 2 & 7 \end{vmatrix}$$

If we write the rows as columns and the columns as rows we obtain the determinant

$$\begin{vmatrix} 4 & 3 & 5 \\ 1 & 6 & 2 \\ -1 & 0 & 7 \end{vmatrix}$$

The above theorem states that these two determinants are equal in value. This may be verified by the reader by expanding each determinant. Because of this theorem any new theorems which are true for the columns of a determinant will also be true for the rows of the determinant.

THEOREM 2: The sign of a determinant is changed by interchanging any two columns or rows of the determinant.

Example: Consider the determinant

$$\begin{vmatrix} 3 & 4 & 6 \\ -1 & 5 & 1 \\ 2 & -3 & -2 \end{vmatrix}$$

If we interchange the second and third columns we obtain the determinant

$$\begin{vmatrix} 3 & 6 & 4 \\ -1 & 1 & 5 \\ 2 & -2 & -3 \end{vmatrix}$$

The theorem states that the above determinants are equal in value except for the sign, or that

$$\begin{vmatrix} 3 & 4 & 6 \\ -1 & 5 & 1 \\ 2 & -3 & -2 \end{vmatrix} = - \begin{vmatrix} 3 & 6 & 4 \\ -1 & 1 & 5 \\ 2 & -2 & -3 \end{vmatrix}$$

This result may be verified by evaluating each determinant.

THEOREM 3: If two columns or rows of a determinant are identical then the value of the determinant is zero.

Example: Consider the determinant

$$\begin{vmatrix} 7 & 3 & 3 \\ 1 & -2 & -2 \\ 4 & 5 & 5 \end{vmatrix}$$

According to the theorem this determinant is equal to zero because two of the columns (the second and third) are identical.

THEOREM 4:   If all the elements of a column or row of a determinant are multiplied by a factor, then the determinant is multiplied by that factor.

Example: Consider the determinant

$$\begin{vmatrix} 5 & 2 & -3 \\ -1 & -4 & 6 \\ 3 & 7 & 4 \end{vmatrix}$$

Let us multiply the second row by 3.   The determinant is now

$$\begin{vmatrix} 5 & 2 & -3 \\ -3 & -12 & 18 \\ 3 & 7 & 4 \end{vmatrix}$$

The theorem states that this new determinant is equal in value to three times the value of the original determinant.   The multiplying factor may be a fraction as well as a whole number.

THEOREM 5:   If a determinant has a row or column consisting completely of zeros, then the value of the determinant is zero.

Example: The value of the determinant

$$\begin{vmatrix} 1 & 0 & -2 \\ 5 & 0 & 4 \\ 8 & 0 & 3 \end{vmatrix}$$

is zero since the second column contains three zeros.

THEOREM 6:   If each element of a row or column is multiplied by the same number, and the products be added to or subtracted from the corresponding elements of another row or column, then the value of the determinant is not changed.

Example: Consider the determinant

$$\begin{vmatrix} 1 & 2 & 4 \\ 3 & -1 & -2 \\ 6 & 5 & 3 \end{vmatrix}$$

Let us multiply the elements of the second column by 2 and add the results to the first column, as follows

$$\begin{vmatrix} 1 + 4 & 2 & 4 \\ 3 - 2 & -1 & -2 \\ 6 + 10 & 5 & 3 \end{vmatrix}$$

According to the theorem, the value of the new determinant is equal to the value of the original determinant. If each determinant is expanded the common value is -49.

The student should note that the column or row whose elements are multiplied is left intact. It is the row or column to which the multiplied row or column is added that is changed. This theorem is most useful in expanding determinants. By its use we can obtain one or more zeros in a row or column thus simplifying expansion by co-factors.

Applications of the Theorems

In the following examples, the theorems will be applied in the expansion of determinants.

Example: Find the value of the determinant

$$\begin{vmatrix} 35 & 7 & 1 \\ 25 & 5 & 3 \\ 30 & 6 & -4 \end{vmatrix}$$

▶ We may rewrite this determinant as follows:

$$\begin{vmatrix} 35 & 7 & 1 \\ 25 & 5 & 3 \\ 30 & 6 & -4 \end{vmatrix} = \begin{vmatrix} 7 \cdot 5 & 7 & 1 \\ 5 \cdot 5 & 5 & 3 \\ 6 \cdot 5 & 6 & -4 \end{vmatrix} = 5 \begin{vmatrix} 7 & 7 & 1 \\ 5 & 5 & 3 \\ 6 & 6 & -4 \end{vmatrix}$$

Example: Find the value of the determinant:

▶ This determinant could be expanded by minors. However, the work of expansion can be simplified greatly by obtaining zeros in the row or column used for expansion. The follow-

$$\begin{vmatrix} 4 & 1 & 5 & 3 \\ -2 & -3 & 1 & 2 \\ 1 & 2 & -3 & -1 \\ 3 & -1 & -2 & 4 \end{vmatrix}$$

ing steps will indicate how this is done by using Theorem 6.

a. We shall use the first column as the operating column to obtain three zeros in the third row, since the number 1 is a convenient number to work with.

Multiply the elements of the first column by -2 and add the results to the corresponding elements of the second column, as follows:

$$
\begin{vmatrix}
4 & -8+1 & 5 & 3 \\
-2 & 4-3 & 1 & 2 \\
1 & -2+2 & -3 & -1 \\
3 & -6-1 & -2 & 4
\end{vmatrix}
=
\begin{vmatrix}
4 & -7 & 5 & 3 \\
-2 & 1 & 1 & 2 \\
1 & 0 & -3 & -1 \\
3 & -7 & -2 & 4
\end{vmatrix}
$$

b. Add the elements of the first column to the corresponding elements of the fourth column, as follows:

$$
\begin{vmatrix}
4 & -7 & 5 & 4+3 \\
-2 & 1 & 1 & -2+2 \\
1 & 0 & -3 & 1+(-1) \\
3 & -7 & -2 & 3+4
\end{vmatrix}
=
\begin{vmatrix}
4 & -7 & 5 & 7 \\
-2 & 1 & 1 & 0 \\
1 & 0 & -3 & 0 \\
3 & 7 & -2 & 7
\end{vmatrix}
$$

c. In order to obtain a zero in the element representing the third row and third column, we multiply the elements in the first column by 3 and add the results to the corresponding elements in the third column, as follows:

$$
\begin{vmatrix}
4 & -7 & 12+5 & 7 \\
-2 & 1 & -6+1 & 0 \\
1 & 0 & 3-3 & 0 \\
3 & -7 & 9-2 & 7
\end{vmatrix}
=
\begin{vmatrix}
4 & -7 & 17 & 7 \\
-2 & 1 & -5 & 0 \\
1 & 0 & 0 & 0 \\
3 & -7 & 7 & 7
\end{vmatrix}
$$

d. Now we may conveniently expand the resulting determinant in terms of the elements of the third row, as follows:

$$
1\begin{vmatrix}
-7 & 17 & 7 \\
1 & -5 & 0 \\
-7 & 7 & 7
\end{vmatrix}
-0\begin{vmatrix}
4 & 17 & 7 \\
-2 & -5 & 0 \\
3 & 7 & 7
\end{vmatrix}
+0\begin{vmatrix}
4 & -7 & 7 \\
-2 & 1 & 0 \\
3 & -7 & 7
\end{vmatrix}
-0\begin{vmatrix}
4 & -7 & 17 \\
-2 & 1 & -5 \\
3 & -7 & 7
\end{vmatrix}
$$

Since the second, third, and fourth determinants are equal to zero it is necessary to evaluate only the first determinant.

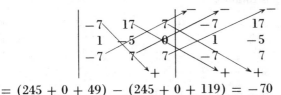

$$= (245 + 0 + 49) - (245 + 0 + 119) = -70$$

Example: Evaluate the following determinant

$$\begin{vmatrix} 6 & 3 & 9 & 12 \\ -8 & -3 & 4 & -5 \\ 12 & 1 & 2 & -1 \\ -4 & -2 & 1 & 3 \end{vmatrix}$$

$$\begin{vmatrix} 6 & 3 & 9 & 12 \\ -8 & -3 & 4 & -5 \\ 12 & 1 & 2 & -1 \\ -4 & -2 & 1 & 3 \end{vmatrix}$$

a. Remove the factor 3 from the first row

$$3 \begin{vmatrix} 2 & 1 & 3 & 4 \\ -8 & -3 & 4 & -5 \\ 12 & 1 & 2 & -1 \\ -4 & -2 & 1 & 3 \end{vmatrix}$$

b. Remove the factor 2 from the first column

$$3 \cdot 2 \begin{vmatrix} 1 & 1 & 3 & 4 \\ -4 & -3 & 4 & -5 \\ 6 & 1 & 2 & -1 \\ -2 & -2 & 1 & 3 \end{vmatrix}$$

c. We may obtain three zeros in the first row by using the first column as the pivotal column. First, we will multiply the first column by -1 and add the result to column 2. Then, we will multiply the first column by -3 and add the result to column 3. Finally, in order to obtain a zero in the fourth row we will multiply the first column by -4 and add the result to column 4. This may be done in one step, as follows:

$$6\begin{vmatrix} 1 & -1+1 & -3+3 & -4+4 \\ -4 & +4-3 & +12+4 & +16-5 \\ 6 & -6+1 & -18+2 & -24-1 \\ -2 & +2-2 & +6+1 & +8+3 \end{vmatrix} = 6\begin{vmatrix} 1 & 0 & 0 & 0 \\ -4 & 1 & 16 & 11 \\ 6 & -5 & -16 & -25 \\ -2 & 0 & 7 & 11 \end{vmatrix}$$

d. Now, we shall expand the resulting determinant in terms of elements of the first row

$$+6\begin{vmatrix} 1 & 16 & 11 \\ -5 & -16 & -25 \\ 0 & 7 & 11 \end{vmatrix}$$

In order to avoid multiplication with large numbers we shall simplify this third order determinant by adding the first row to the second row

$$6\begin{vmatrix} 1 & 16 & 11 \\ -4 & 0 & -14 \\ 0 & 7 & 11 \end{vmatrix} = 6(494) = +2{,}964$$

Example: Evaluate the determinant

$$\begin{vmatrix} 3 & 6 & 4 & 1 & -3 \\ 1 & 2 & 3 & 0 & 2 \\ 0 & 3 & 1 & 0 & 1 \\ 4 & 1 & 1 & 0 & -1 \\ 1 & -1 & 5 & 0 & 2 \end{vmatrix}$$

▶   a. If we expand in terms of the elements of the fourth column, we obtain the determinant

$$-1\begin{vmatrix} 1 & 2 & 3 & 2 \\ 0 & 3 & 1 & 1 \\ 4 & 1 & 1 & -1 \\ 1 & -1 & 5 & 2 \end{vmatrix}$$

b. In this fourth order determinant we shall obtain three zeros in the second row. Then we shall expand the resulting determinant in terms of the elements of the second row. First, multiply the elements of the third column by -3 and add the results to the corresponding elements of the second column.

$$-1\begin{vmatrix} 1 & -7 & 3 & 2 \\ 0 & 0 & 1 & 1 \\ 4 & -2 & 1 & -1 \\ 1 & -16 & 5 & 2 \end{vmatrix}$$

c. Second, multiply the elements of the third column by -1 and add the results to the corresponding elements of the fourth column

$$-1\begin{vmatrix} 1 & -7 & 3 & -1 \\ 0 & 0 & 1 & 0 \\ 4 & -2 & 1 & -2 \\ 1 & -16 & 5 & -3 \end{vmatrix}$$

d. Expand the above determinant in terms of the elements of the second row

$$(-1)(-1)\begin{vmatrix} 1 & -7 & -1 \\ 4 & -2 & -2 \\ 1 & -16 & -3 \end{vmatrix} = +1(-34) = -34$$

Example: Evaluate the determinant

$$\begin{vmatrix} -9 & -2 & 2 & 5 \\ 12 & 20 & -16 & 28 \\ 3 & 5 & -1 & -4 \\ 15 & 4 & 3 & -1 \end{vmatrix}$$

▶ a. The factor 3 may be removed from the first column and the factor 4 may be removed from the second row. The resulting determinant is

$$12\begin{vmatrix} -3 & -2 & 2 & 5 \\ 1 & 5 & -4 & 7 \\ 1 & 5 & -1 & -4 \\ 5 & 4 & 3 & -1 \end{vmatrix}$$

b. Multiply the elements of the first column by -5 and add the result to the corresponding elements of the second column. Then, add the elements of the first column to the corresponding elements of the third column.   Then multiply the elements of the first column by 4 and add the results to the corresponding elements of the fourth column.   This may be accomplished in one step, as follows:

$$12 \begin{vmatrix} -3 & 15-2 & -3+2 & -12+5 \\ 1 & -5+5 & 1-4 & 4+7 \\ 1 & -5+5 & 1-1 & 4-4 \\ 5 & -25+4 & 5+3 & 20-1 \end{vmatrix} = 12 \begin{vmatrix} -3 & 13 & -1 & -7 \\ 1 & 0 & -3 & 11 \\ 1 & 0 & 0 & 0 \\ 5 & -21 & 8 & 19 \end{vmatrix}$$

c. Now, expand the determinant in terms of the elements of the third row

$$(12)(1) \begin{vmatrix} 13 & -1 & -7 \\ 0 & -3 & 11 \\ -21 & 8 & 19 \end{vmatrix}$$

In order to simplify computation multiply the elements of the first row by 2 and add the results to the corresponding elements of the third row

$$12 \begin{vmatrix} 13 & -1 & -7 \\ 0 & -3 & 11 \\ 5 & 6 & 5 \end{vmatrix}$$

Now, multiply the elements of the first column by -1 and add the results to the corresponding elements of the third column

$$12 \begin{vmatrix} 13 & -1 & -20 \\ 0 & -3 & 11 \\ 5 & 6 & 0 \end{vmatrix} = 12(-1,213) = -14,556$$

EXERCISE 5

Evaluate each of the following determinants:

1. $\begin{vmatrix} 3 & -5 & 4 & 7 \\ -2 & 6 & -1 & -2 \\ 4 & 3 & 5 & -3 \\ 1 & 0 & 2 & -4 \end{vmatrix}$
2. $\begin{vmatrix} 2 & -3 & -4 & -7 \\ -1 & -2 & 3 & 4 \\ 4 & 1 & 2 & 5 \\ 5 & 6 & 8 & 1 \end{vmatrix}$

3.
$$
\begin{vmatrix}
0 & 0 & 1 & 0 & 0 \\
3 & -2 & -4 & 1 & 2 \\
1 & 5 & -3 & 4 & 1 \\
4 & -3 & 2 & 7 & 5 \\
2 & -6 & 5 & -3 & 4
\end{vmatrix}
$$

5.
$$
\begin{vmatrix}
3 & 5 & 2 & 1 & 4 \\
-1 & 2 & -3 & 5 & -1 \\
0 & 0 & 2 & 0 & 0 \\
6 & -4 & 3 & -1 & -3 \\
-2 & 5 & 2 & 3 & 4
\end{vmatrix}
$$

4.
$$
\begin{vmatrix}
6 & 4 & -8 & 10 \\
9 & 1 & -2 & 3 \\
-15 & 0 & -4 & 5 \\
+12 & 3 & -1 & 4
\end{vmatrix}
$$

APPLICATIONS OF DETERMINANTS TO SYSTEMS OF LINEAR EQUATIONS

In previous sections it has been shown that determinants may be used to solve systems of linear equations in two unknowns. If two linear equations in two unknowns be plotted on a graph the result will be two straight lines. The two lines thus plotted may intersect in one point, they may be parallel, or they may coincide. If the lines intersect then the coordinates of the point of intersection determine the pair of values which satisfy the two equations *simultaneously*. If the lines are parallel, they have no point in common and there is no solution; the equations are said to be *inconsistent*. If the lines coincide there are an infinite number of solutions and the equations are said to be *dependent*.

It will be recalled that for the system of equations

$$
a_1x + b_1y = c_1
$$
$$
a_2x + b_2y = c_2
$$

the solutions in determinant form are

$$
x = \frac{\begin{vmatrix} c_1 & b_1 \\ c_2 & b_2 \end{vmatrix}}{\begin{vmatrix} a_1 & b_1 \\ a_2 & b_2 \end{vmatrix}}
\qquad
y = \frac{\begin{vmatrix} a_1 & c_1 \\ a_2 & c_2 \end{vmatrix}}{\begin{vmatrix} a_1 & b_1 \\ a_2 & b_2 \end{vmatrix}}
$$

If the determinant $\begin{vmatrix} a_1 & b_1 \\ a_2 & b_2 \end{vmatrix} \neq 0$

then the equations have a unique solution. If the determinant

$$\begin{vmatrix} a_1 & b_1 \\ a_2 & b_2 \end{vmatrix} = 0$$

then the equations are either *inconsistent* or *dependent*.

For example, the equations

$$2x + 3y = 5$$
$$3x - 2y = -9$$

have a unique solution because the determinant

$$\begin{vmatrix} 3 & -4 \\ 6 & -8 \end{vmatrix} \neq 0$$

However, the equations $3x - 4y = 8$
$$6x - 8y = 9$$

have no solution because the determinant

$$\begin{vmatrix} 3 & -4 \\ 6 & -8 \end{vmatrix} = 0$$

A similar analysis holds true for three equations in three unknowns.

$$a_1x + b_1y + c_1z = d_1$$
$$a_2x + b_2y + c_2z = d_2$$
$$a_3x + b_3y + c_3z = d_3$$

If the determinant formed from the coefficients of the unknowns

$$\begin{vmatrix} a_1 & b_1 & c_1 \\ a_2 & b_2 & c_2 \\ a_3 & b_3 & c_3 \end{vmatrix} \neq 0$$

then the three equations have a unique common solution.

If this determinant is equal to zero then the three equations are either *inconsistent* or *dependent*. If the equations are inconsistent they have no solution. If they are dependent they have an infinite number of solutions.

Example: Does the following system of equations have a common solution?

$$x - y + 2z = -6$$
$$2x + y + z = 3$$
$$x - y + 3z = -9$$

▶ The determinant formed from the coefficients of the unknown quantities is

$$\begin{vmatrix} 1 & -1 & 2 \\ 2 & 1 & 1 \\ 1 & -1 & 3 \end{vmatrix} = 3$$

If this determinant is evaluated we obtain the value 3.

Since this determinant $\neq 0$ the above system of equations has a common solution.

Example: Does the following system of equations have a common solution?

$$2x - y + z = 3$$
$$x + 3y - 3z = 1$$
$$3x + 2y - 2z = 4$$

▶ The determinant formed from the coefficients of the unknowns is

$$\begin{vmatrix} 2 & -1 & 1 \\ 1 & 3 & -3 \\ 3 & 2 & -2 \end{vmatrix} = 0$$

Since the value of the determinant is zero, the above equations are either inconsistent or dependent.

EXERCISE **6**

Test each of the following systems of equations to determine whether they have unique solutions or are inconsistent or dependent.

1. $x + y + 2z = 5$
  $2x + 3y - 5z = -4$
  $3x - y + z = 2$

3. $x - y - z = 2$
  $2x - 2y - 2z = 5$
  $x + 2y + 3z = 4$

2. $5x - 3y - 2z = 4$
  $x + y - z = 3$
  $2x - 6y + z = -5$

4. $2x + 3y + z = 1$
  $x - 2y + 4z = 17$
  $3x - y - 5z = 6$

Systems of Homogeneous Linear Equations

Let us consider the set of equations

$$a_1 x + b_1 y + c_1 z = 0$$
$$a_2 x + b_2 y + c_2 z = 0$$
$$a_3 x + b_3 y + c_3 z = 0$$

These equations are called linear homogeneous equations; their constant terms are zero. Obviously, $x = 0$, $y = 0$, and $z = 0$ is a solution of this system of equations. This solution is referred to as the *trivial solution.*

If the determinant of the coefficients is different from zero, the trivial solution is the only one. If the determinant of the coefficients is equal to zero, the system has an infinite set of solutions.

Example: Test the following system of equations to determine whether it has only one solution (the trivial solution) or whether it has an infinite number of solutions.

$$x + y + z = 0$$
$$3x - y + z = 0$$
$$2x + 3y - 2z = 0$$

▶ The determinant of the coefficients is

$$\begin{vmatrix} 1 & 1 & 1 \\ 3 & -1 & 1 \\ 2 & 3 & -2 \end{vmatrix} = 18$$

Since the determinant of the coefficients $\neq 0$ the trivial solution $x = 0, y = 0, z = 0$ is the only one.

Example: Test the following system of equations to determine whether it has only one solution or whether it has an inflnite set of solutions.

$$2x - y + z = 0$$
$$x + 3y - 2z = 0$$
$$9x - y + 2z = 0$$

▶ The determinant

$$\begin{vmatrix} 2 & -1 & 1 \\ 1 & 3 & -2 \\ 9 & 1 & 2 \end{vmatrix} = 0$$

Since the determinant of the coefficients $= 0$ the system has an infinite set of solutions. The method of solution of this set of equations is worked out below.

Method of Solution of Sets of Homogeneous Linear Equations

If homogeneous linear equations have non-trivial solutions the method described below may be used to find these solutions. The above equations will be solved

$$2x - y + z = 0$$
$$x + 3y - 2z = 0$$
$$9x - y + 2z = 0$$

Solve for x and y in terms of z in the first two equations

$$2x - y = -z$$
$$x + 3y = 2z$$

Express the solutions for x and y as second order determinants

$$x = \frac{\begin{vmatrix} -z & -1 \\ 2z & 3 \end{vmatrix}}{\begin{vmatrix} 2 & -1 \\ 1 & 3 \end{vmatrix}} = \frac{-3z + 2z}{+1} = \frac{-z}{7}$$

$$y = \frac{\begin{vmatrix} 2 & -z \\ 1 & 2z \end{vmatrix}}{7} = \frac{4z + z}{7} = \frac{5z}{7}$$

Since $x = \dfrac{-z}{7}, \dfrac{x}{-1} = \dfrac{z}{7}$

Since $y = \dfrac{5z}{7}, \dfrac{y}{5} = \dfrac{z}{7}$

Therefore, $\dfrac{x}{-1} = \dfrac{y}{5} = \dfrac{z}{7} = K$

And, $x = -K, y = 5K, z = 7K$ is the general solution of the equations.

If we assign values to K we will obtain specific solutions. For example, if $K = 1, x = -1, y = 5$, and $z = 7$ is one such solution. If $K = 2, x = -2, y = 10$, and $z = 14$ is another solution. Since K may be assigned an infinite number of values, there are an infinite number of solutions.

EXERCISE 7

Test the following systems of equations to determine whether they have one solution or whether they have an infinite set of solutions. Solve the system if it has an infinite set of solutions.

1. $3x + 2y - z = 0$
   $x - 3y - 4z = 0$
   $2x + y - 3z = 0$

3. $x - y + z = 0$
   $2x + y + z = 0$
   $2x + 7y - z = 0$

2. $x - y - z = 0$
   $2x - 3y + z = 0$
   $3x - 2y - 6z = 0$

4. $2x - 3y + 4z = 0$
   $5x + 2y + z = 0$
   $3x - 4y + 2z = 0$

Systems of Equations Containing Fewer Variables than Equations

Consider the equations

$$3x + 4y = 6$$
$$x - 3y = -11$$
$$5x + 2y = -4$$

When plotted, each of these equations represents a straight line. In general, three straight lines will not intersect in the same point. This is another way of saying that three equations in two unknowns will not ordinarily have a common solution. However, under the conditions stated in the following theorem, three equations in two unknowns may have a common solution.

THEOREM: In order that n equations in $(n - 1)$ variables have a common solution it is necessary that the *eliminant* be equal to zero. The *eliminant* is the determinant formed from the coefficients of the variables and the constant terms.

Example: Write the eliminant of the set of equations and find its value. What conclusion can be drawn from this value?

$$3x + 4y = 6$$
$$x - 3y = -11$$
$$5x + 2y = -4$$

▶ The eliminant of the system is

$$\begin{vmatrix} 3 & 4 & 6 \\ 1 & -3 & -11 \\ 5 & 2 & -4 \end{vmatrix}$$

When this determinant is expanded its value is found to be zero. According to the theorem, the necessary condition that this set of equations has a solution has been met.

In finding the possible common solution the following method is used.

Solve any two of the above equations simultaneously, e.g.

$$3x + 4y = 6$$
$$x - 3y = -11$$

The values obtained are $x = -2$ and $y = 3$.

These values will satisfy the third equation $5x + 2y = -4$.

Therefore, the common solution of the three equations is $x = -2$, $y = 3$.

Example: Write the eliminant of the set of equations and find its value.   What conclusion can be drawn from this value?

$$x - y = 3$$
$$2x - 2y = 5$$
$$x - y = -2$$

▶   The eliminant of this system is

$$\begin{vmatrix} 1 & -1 & 3 \\ 2 & -2 & 5 \\ 1 & -1 & -2 \end{vmatrix}$$

When this determinant is expanded its value is found to be zero. According to the theorem, the necessary condition that this set of conditions has a solution has been met.

However, no two of these equations can be solved simultaneously. Therefore, this system of equations has no solution even though the eliminant is equal to zero.   This example indicates that the condition that the *eliminant* be equal to zero is a necessary condition but it is not sufficient.

Example: Write the eliminant of the system of equations and find its value.   What conclusion can be drawn from this value?

$$x + 2y = 3$$
$$3x - y = 4$$
$$2x + 5y = 1$$

▶   The eliminant of this system is

$$\begin{vmatrix} 1 & 2 & 3 \\ 3 & -1 & 4 \\ 2 & 5 & 1 \end{vmatrix}$$

When this determinant is expanded its value is found to be 40. Since the value of the eliminant is $\neq 0$, the necessary condition that this system of equations has a common solution has not been met.

EXERCISE 8

Examine each system of equations for the existence of a solution. Find the solutions where they exist.

1. $2x + 3y = -4$
   $x - 2y = 5$
   $3x - y = 5$

2. $x + y = -1$
   $3x + 2y = 1$
   $2x - y = 3$

3. $2x - y = -3$
   $3x + 2y = 4$
   $3x - 5y = -13$

4. $x - 2y = 5$
   $3x - 6y = 4$
   $2x - 4y = 7$

ANSWERS

EXERCISE 1

1. 36
2. 4
3. $-a^2 + 2a - 7$
4. $x = 5, y = 2$
5. $x = 1, y = -1$
6. $x = 3, y = -\frac{1}{2}$

7. $x = -\frac{1}{3}, y = \frac{3}{4}$
8. $x = -2, y = -2$
9. $x = \frac{1}{4}, y = \frac{1}{5}$
10. $x = 5, y = 2$
11. $x = -\frac{1}{2}, y = \frac{2}{3}$
12. $x = \dfrac{cd + ab}{ad + bc}, y = \dfrac{c^2 - a^2}{ad + bc}$

EXERCISE 2

1. 157
2. 61
3. 112
4. -62

5. $8a - 1$
6. $11x - 9$
7. $y = 3$
8. $y = 1$

EXERCISE 3

1. $x = -3, y = 3, z = 1$
2. $x = 2, y = -1, z = 4$
3. $x = 3, y = -1, z = 2$
4. $x = \frac{1}{2}, y = -\frac{1}{3}, z = 4$

EXERCISE 4

1. 517
2. 581
3. -978

4. 1,195
5. 349

EXERCISE 5

1. 342
2. 1,344
3. 320

4. 420
5. 448

EXERCISE 6

1. unique solution
2. inconsistent or dependent

3. inconsistent or dependent
4. unique solution

EXERCISE 7

1. trivial solution

2. $x = 4k, y = 3k, z = k$

3. $x = \dfrac{-2k}{3}, y = \dfrac{k}{3}, z = k$

4. trivial solution

EXERCISE 8

1. $x = 1, y = -2$
2. no common solution

3. $x = -\frac{2}{7}, y = \frac{17}{7}$
4. no common solution

CHAPTER **22**

Partial Fractions

The process of combining two or more algebraic fractions of the type $\dfrac{4}{x+1} - \dfrac{3}{x+2}$ into the single fraction $\dfrac{x+5}{(x+1)(x+2)}$ is a familiar one. However, it is sometimes necessary to perform the reverse process i.e., to take a fraction such as $\dfrac{12x+11}{x^2+x-6}$ and break it up into the fractions $\dfrac{7}{x-2} + \dfrac{5}{x+3}$. In this case, the fractions $\dfrac{7}{x-2}$ and $\dfrac{5}{x+3}$ are called partial fractions.

The work on partial fractions to follow will be based upon the following theorems. In these theorems we will deal with fractions of the form $\dfrac{P}{Q}$ where P and Q are relatively prime polynomials, and P is of lower degree than Q. Two polynomials are relatively prime if they have no common factors.

1. If a linear factor of the form $(ax+b)$ occurs as a factor of the denominator, Q, then there corresponds to $(ax+b)$ a partial fraction of the form $\dfrac{A}{ax+b}$, where A is a constant unequal to zero.

Example: $\dfrac{x + 2}{(x + 5)(3x - 1)} = \dfrac{A_1}{x + 5} + \dfrac{A_2}{3x - 1}$

2. If a linear factor of the form $(ax + b)$ occurs as a factor of the denominator, Q, to the nth degree, then there corresponds to $(ax + b), (ax + b)^2, (ax + b)^3, \ldots, (ax + b)^n$ the partial fractions

$$\frac{A_1}{(ax + b)}, \frac{A_2}{(ax + b)^2}, \frac{A_3}{(ax + b)^3}, \ldots \frac{A_n}{(ax + b)^n}$$

where $A_1, A_2, A_3, \ldots A_n$, are constants with $A_n \neq 0$.

Example: $\dfrac{x + 2}{(3x - 1)^2} = \dfrac{A_1}{(3x - 1)} + \dfrac{A_2}{(3x - 1)^2}$

3. If a quadratic factor of the form $ax^2 + bx + c$ occurs as a factor of the denominator, Q, then there corresponds to $ax^2 + bx + c$ a partial fraction of the form $\dfrac{A_1 x + B_1}{ax^2 + bx + c}$, where A_1 and B_1 are constants, and not both A_1 and B_1 are equal to zero. It is assumed that $ax^2 + bx + c$ is not factorable.

Example: $\dfrac{x + 2}{x(x^2 + 3x + 1)} = \dfrac{A_1 x + B_1}{x^2 + 3x + 1} + \dfrac{C}{x}$

4. If a quadratic factor of the form $ax^2 + bx + c$ occurs as a factor of the denominator, Q, to the nth degree, then there corresponds to $(ax^2 + bx + c), (ax^2 + bx + c)^2, (ax^2 + bx + c)^3, \ldots, (ax^2 + bx + c)^n$ the partial fractions

$$\frac{A_1 x + B_1}{ax^2 + bx + c}, \frac{A_2 x + B_2}{(ax^2 + bx + c)^2}, \frac{A_3 x + B_3}{(ax^2 + bx + c)^3}, \ldots,$$

$$\frac{A_n x + B_n}{(ax^2 + bx + c)^n},$$

where the A's and B's are constants, and either A or B is not equal to zero.

Example: Resolve into partial fractions

$$\frac{5x + 1}{(x + 2)(x - 1)}$$

▶
Method 1:

$$\text{Let } \frac{5x + 1}{(x + 2)(x - 1)} = \frac{A}{x + 2} + \frac{B}{x - 1}$$

Combining fractions,

$$\frac{5x + 1}{(x + 2)(x - 1)} = \frac{A(x - 1) + B(x + 2)}{(x + 2)(x - 1)}$$

Since the denominators of these equivalent fractions are equal the numerators must be equal, or

$$5x + 1 = A(x - 1) + B(x + 2)$$
$$5x + 1 = Ax - A + Bx + 2b = Ax + Bx - A + 2b$$
$$= (A + B)x - A + 2B$$

If $5x + 1 = (A + B)x - A + 2B$ then equating coefficients of x and constant terms, we have

$$A + B = 5 \quad \text{and} \quad -A + 2B = 1$$

Solving these equations simultaneously $A = 3, B = 2$.
Therefore,

$$\frac{5x + 1}{(x + 2)(x - 1)} = \frac{3}{x + 2} + \frac{2}{x - 1}$$

Method 2:

$$\text{Let } \frac{5x + 1}{(x + 2)(x - 1)} = \frac{A}{x + 2} + \frac{B}{x - 1}$$

$$\text{Then } \frac{5x + 1}{(x + 2)(x - 1)} = \frac{A(x - 1) + B(x + 2)}{(x + 2)(x - 1)}$$

$5x + 1 = A(x - 1) + B(x + 2)$ is an identity. Therefore, we may substitute any value for x and the equality will be maintained.

If we let $x = -2$, we have $-9 = A(-2 - 1) + B(-2 + 2)$, and $-9 = -3A$, and $A = 3$. If we let $x = 1$, we have $6 = A(1 - 1) + B(1 + 2)$, and $6 = 3B$, and $B = 2$.

Note that the values -2, and 1 were chosen because they yield results quickly since each of these values reduces part of the right member of the equation to zero.

Since $A = 3$, and $B = 2$,

$$\frac{5x + 1}{(x + 2)(x - 1)} = \frac{3}{x + 2} + \frac{2}{x - 1}$$

Example: Resolve into partial fractions

$$\frac{5x^2 + 19x - 18}{x(x + 3)(x - 2)}$$

Method 1:

Let $\dfrac{5x^2 + 19x - 18}{x(x + 3)(x + 2)} = \dfrac{A}{x} + \dfrac{B}{x + 3} + \dfrac{C}{x - 2}$

Combining fractions, $\dfrac{5x^2 + 19x - 18}{x(x + 3)(x - 2)}$

$$= \frac{A(x + 3)(x - 2) + Bx(x - 2) + Cx(x + 3)}{x(x + 3)(x - 2)}$$

Therefore,

$5x^2 + 19x - 18$
$= A(x + 3)(x - 2) + Bx(x - 2) + Cx(x + 3)$
$5x^2 + 19x - 18 = Ax^2 + Ax - 6A + Bx^2 - 2Bx + Cx^2 + 3Cx$
$5x^2 + 19x - 18$
$= (A + B + C)x^2 + (A - 2B + 3C)x + (-6A)$

Equating coefficients of like powers of x:

$$
\begin{aligned}
A + B + C &= 5 \\
A - 2B + 3C &= 19 \\
-6A \phantom{{}+ 2B + 3C} &= -18
\end{aligned}
$$

When these equations are solved simultaneously, $A = 3$, $B = -2$, $C = 4$.

Therefore, $\dfrac{5x^2 + 19x - 18}{x(x + 3)(x - 2)} = \dfrac{3}{x} - \dfrac{2}{x + 3} + \dfrac{4}{x - 2}$

Method 2:

Let $\dfrac{5x^2 + 19x - 18}{x(x + 3)(x - 2)} = \dfrac{A}{x} + \dfrac{B}{x + 3} + \dfrac{C}{x - 2}$

Then, $5x^2 + 19x - 18 = A(x + 3)(x - 2) + Bx(x - 2) + Cx(x + 3)$.

If we let $x = 0$, we have $-18 = -16A$, and $A = 3$.

If we let $x = -3$, we have $-30 = 15B$, and $B = -2$.

If we let $x = 2$, we have $40 = 10c$, and $C = 4$.

Example: Resolve into partial fractions

$$\frac{13x^2 + 12x - 4}{30x^3 + 5x^2 - 10x}$$

▶

Method 1: $\dfrac{13x^2 + 12x - 4}{30x^3 + 5x^2 - 10x} = \dfrac{13x^2 + 12x - 4}{5x(2x - 1)(3x + 2)}$

Let $\dfrac{13x^2 + 12x - 4}{5x(2x - 1)(3x + 2)} = \dfrac{A}{5x} + \dfrac{B}{2x - 1} + \dfrac{C}{3x + 2}$

Combining fractions,

$$\frac{13x^2 + 12x - 4}{5x(2x - 1)(3x + 2)}$$

$$= \frac{A(2x - 1)(3x + 2) + 5Bx(3x + 2) + 5Cx(2x - 1)}{5x(2x - 1)(3x + 2)}$$

$$\frac{13x^2 + 12x - 4}{5x(2x - 1)(3x + 2)}$$

$$= \frac{6Ax^2 + Ax - 2A + 15Bx^2 + 10Bx + 10Cx^2 - 5Cx}{5x(2x - 1)(3x + 2)}$$

Equating coefficients of like powers of x;

$$6A + 15B + 10C = 13$$
$$A + 10B - 5C = 12$$
$$-2A = -4$$

When these equations are solved simultaneously,
$$A = 2, \ B = \tfrac{3}{5}, \ C = -\tfrac{4}{5}$$

Therefore,

$$\frac{13x^2 + 12x - 4}{5x(2x - 1)(3x + 2)} = \frac{2}{5x} + \frac{\tfrac{3}{5}}{2x - 1} - \frac{\tfrac{4}{5}}{3x + 2} = \frac{2}{5x}$$
$$+ \frac{3}{5(2x - 1)} - \frac{4}{5(3x + 2)}$$

Method 2:

$$13x^2 + 12x - 4 = A(2x - 1)(3x + 2) + 5Bx(3x + 2)$$
$$+ 5Cx(2x - 1).$$

If we let $x = 0$, we have $-4 = -2A$, and $A = 2$.

If we let $x = -\dfrac{2}{3}$, we have $\dfrac{-56}{9} = \dfrac{70C}{9}$, and $C = \dfrac{-4}{5}$.

If we let $x = \dfrac{1}{2}$, we have $\dfrac{21}{4} = \dfrac{35B}{4}$, and $B = \dfrac{3}{5}$.

Therefore, $\dfrac{13x^2 + 12x - 4}{5x(2x - 1)(3x + 2)} = \dfrac{2}{5x} + \dfrac{3}{5(2x - 1)} - \dfrac{4}{5(3x + 2)}$

Example: Resolve into partial fractions

$$\frac{2x^2 + 16x + 29}{(x + 3)^2(x + 4)}$$

▶

Method 1:

Let $\dfrac{2x^2 + 16x + 29}{(x + 3)^2(x + 4)} = \dfrac{A}{x + 3} + \dfrac{B}{(x + 3)^2} + \dfrac{C}{x + 4}$

$$\frac{2x^2 + 16x + 29}{(x + 3)^2(x + 4)}$$
$$= \frac{A(x + 3)(x + 4) + B(x + 4) + C(x + 3)^2}{(x + 3)^2(x + 4)}$$
$$2x^2 + 16x + 29 = Ax^2 + 7Ax + 12A + Bx + 4B + Cx + 6Cx$$
$$+ 9C$$

Equating coefficients of like powers of x

$$A + C = 2$$
$$7A + B + 6C = 16$$
$$12A + 4B + 9C = 29$$

When these equations are solved simultaneously, $A = 5$, $B = -1$, $C = -3$.

Therefore, $\dfrac{2x^2 + 16x + 29}{(x + 3)^2(x + 4)} = \dfrac{5}{x + 3} - \dfrac{1}{(x + 3)^2} - \dfrac{3}{x + 4}$

Method 2:

Let $\dfrac{2x^2 + 16x + 29}{(x + 3)^2(x + 4)} = \dfrac{A}{x + 3} + \dfrac{B}{(x + 3)^2} + \dfrac{C}{x + 4}$

$2x^2 + 16x + 29 = A(x + 3)(x + 4) + B(x + 4) + C(x + 3)^2$

If we let $x = -3$, we have $-1 = B$, and $B = -1$.

If we let $x = -4$, we have $-3 = C$, and $C = -3$.

To find the value of A, let x have a convenient value such as 0.

If $x = 0$, we have $29 = 12A + 4B + 9C$.

If $B = -1$, and $C = -3$ in this equation $29 = 12A - 4 - 27$,

$12A = 60$, and $A = 5$.

Therefore, $\dfrac{2x^2 + 16x + 29}{(x + 3)^2(x + 4)} = \dfrac{5}{x + 3} - \dfrac{1}{(x + 3)^2} - \dfrac{3}{x + 4}$

Example: Resolve into partial fractions

$$\frac{-9x^3 + 13x^2 + 39x - 18}{x^2(x - 3)^2}$$

▶

Method 1:

Let $\dfrac{-9x^3 + 13x^2 + 39x - 18}{x^2(x - 3)^2} = \dfrac{A}{x} + \dfrac{B}{x^2} + \dfrac{C}{x - 3} + \dfrac{D}{(x - 3)^2}$

Combining fractions,

$\dfrac{-9x^3 + 13x^2 + 39x - 18}{x^2(x - 3)^2}$

$= \dfrac{Ax(x - 3)^2 + B(x - 3)^2 + Cx^2(x - 3) + Dx^2}{x^2(x - 3)^2}$

$-9x^3 + 13x^2 + 39x - 18$

$= Ax^3 - 6Ax^2 + 9Ax + Bx^2 - 6Bx + 9B + Cx^3 - 3Cx^2 + Dx^2$

Equating coefficients of like powers of x:

$$A + C = -9$$
$$-6A + B - 3C + D = 13$$
$$9A - 6B = 39$$
$$9B = -18$$

When these equations are solved simultaneously, $A = 3$, $B = -2$, $C = -12$, $D = -3$.
Therefore,

$$\frac{-9x^3 + 13x^2 + 39x - 18}{x^2(x - 3)^2} = \frac{3}{x} - \frac{2}{x^2} - \frac{12}{x - 3} - \frac{3}{(x - 3)^2}$$

Method 2:

Let $\dfrac{-9x^3 + 13x^2 + 39x - 18}{x^2(x - 3)^2} = \dfrac{A}{x} + \dfrac{B}{x^2} + \dfrac{C}{x - 3} + \dfrac{D}{(x - 3)^2}$

$-9x^3 + 13x^2 + 39x - 18$

$= Ax(x - 3)^2 + B(x - 3)^2 + Cx^2(x - 3) + Dx^2$

If we let $x = 0$, $-18 = 9B$, and $B = -2$
If we let $x = 3$, $-27 = 9D$, and $D = -3$
If we let $x = 1$, $+25 = 4A + 4B - 2C + D$

$$+25 = 4A - 8 - 2C - 3$$
$$+36 = 4A - 2C$$

If we let $x = 2$, $+40 = 2A + B - 4C + 4D$

$$+40 = 2A - 2 - 4C - 12$$
$$+54 = 2A - 4C$$

Solving for A and C, we have $A = 3$, $C = -12$.
Therefore,

$$\frac{-9x^3 + 13x^2 + 39x - 18}{x^2(x - 3)^2} = \frac{3}{x} - \frac{2}{x^2} - \frac{12}{x - 3} - \frac{3}{(x - 3)^2}$$

Example: Resolve into partial fractions

$$\frac{2x^2 - 14x + 8}{(x^2 + 3x - 2)(x - 3)}$$

▶

Let $\dfrac{2x^2 - 14x + 8}{(x^2 + 3x - 2)(x - 3)} = \dfrac{Ax + B}{x^2 + 3x - 2} + \dfrac{C}{x - 3}$

$$\frac{2x^2 - 14x + 8}{(x^2 + 3x - 2)(x - 3)} = \frac{(Ax + B)(x - 3) + C(x^2 + 3x - 2)}{(x^2 + 3x - 2)(x - 3)}$$

$2x^2 - 14x + 8$
$$= (Ax + B)(x - 3) + C(x^2 + 3x - 2)$$
$2x^2 - 14x + 8$
$$= Ax^2 + Bx - 3Ax - 3B + Cx^2 + 3Cx - 2C$$
$2x^2 - 14x + 8$
$$= (A + C)x^2 + (B - 3A + 3C)x + (-3B - 2C)$$

Equating coefficients of like powers of x:

$$A + C = 2$$
$$-3A + B + 3C = -14$$
$$-3B - 2C = 8$$

Solving these equations simultaneously we have $A = 3$, $B = -2$, $C = -1$.

Therefore, $\dfrac{2x^2 - 14x + 8}{(x^2 + 3x - 2)(x - 3)} = \dfrac{3x - 2}{x^2 + 3x - 2} - \dfrac{1}{x - 3}$

Example: Resolve into partial fractions

$$\frac{x^2}{(2x + 3)(x^2 + 9)}$$

▶

Let $\dfrac{x^2}{(2x + 3)(x^2 + 9)} = \dfrac{A}{2x + 3} + \dfrac{Bx + C}{x^2 + 9}$

$$x^2 = A(x^2 + 9) + (Bx + C)(2x + 3)$$
$$x^2 = Ax^2 + 9A + 2Bx^2 + 2Cx + 3Bx + 3C$$
$$x^2 = (A + 2B)x^2 + (2C + 3B)x + 9A + 3C$$

Equating coefficients of like powers of x:

$$A + 2B = 1$$
$$3B + 2C = 0$$
$$9A + 3C = 0$$

Solving these equations simultaneously we have $A = \frac{1}{5}$, $B = \frac{2}{5}$, $C = -\frac{3}{5}$.

Therefore, $\dfrac{x^2}{(2x + 3)(x^2 + 9)} = \dfrac{\frac{1}{5}}{2x + 3} + \dfrac{\frac{2}{5}x - \frac{3}{5}}{x^2 + 9} = \dfrac{1}{5(2x + 3)}$

$$+ \frac{2x - 3}{5(x^2 + 9)}$$

Example: Resolve into partial fractions

$$\frac{3x + 6}{(x - 1)(x^2 + 2)^2}$$

▶

Let $\dfrac{3x + 6}{(x - 1)(x^2 + 2)^2} = \dfrac{A}{x - 1} + \dfrac{Bx + C}{x^2 + 2} + \dfrac{Bx + E}{(x^2 + 2)^2}$

$$\frac{3x + 6}{(x - 1)(x^2 + 2)^2}$$
$$= \frac{A(x^2 + 2)^2 + (Bx + C)(x - 1)(x^2 + 2) + (Dx + E)(x - 1)}{(x - 1)(x^2 + 2)^2}$$

$3x + 6 = Ax^4 + 4Ax^2 + 4A + Bx^4 - Bx^3 + 2Bx^2 - 2Bx + Cx^3$
$\quad - Cx^2 + 2Cx - 2C + Dx^2 - Dx + Ex - E$

$3x + 6 = (A + B)x^4 + (-B + C)x^3 + (4A + 2B - C + D)x^2$
$\quad + (-2B + 2C - D + E)x + (4A - 2C - E)$

Equating coefficients of like powers of x, we have,

$$A + B = 0$$
$$-B + C = 0$$
$$4A + 2B - C + D = 0$$
$$-2B + 2C - D + E = 3$$
$$4A - 2C - E = 6$$

Solving these equations simultaneously we obtain, $A = 1$, $B = -1, C = -1, D = -3$, and $E = 0$.
Therefore,

$$\frac{3x + 6}{(x - 1)(x^2 + 2)^2} = \frac{1}{x - 1} - \frac{x + 1}{x^2 + 2} - \frac{3x}{(x^2 + 2)^2}$$

Example: Resolve into partial fractions

$$\frac{x^3 - 3x^2 - 4x + 5}{(x^2 + x - 1)(x^2 - 3)}$$

▶

Let $\dfrac{x^3 - 3x^2 - 4x + 5}{(x^2 + x + 1)(x^2 - 3)} = \dfrac{Ax + B}{(x^2 + x + 1)} + \dfrac{Cx + D}{x - 3}$

$\begin{aligned}
x^3 &- 3x^2 - 4x + 5 \\
&= (Ax + B)(x^2 - 3) + (Cx + D)(x^2 + x + 1) \\
&= Ax^3 + Bx^2 - 3Ax - 3B + Cx^3 + Cx^2 + Cx + Dx^2 \\
&\quad + Dx + D \\
&= (A + C)x^3 + (B + C + D)x^2 + (-3A + C + D)x \\
&\quad + (-3B + D)
\end{aligned}$

$$A + C = 1$$
$$B + C + D = -3$$
$$-3A + C + D = -4$$
$$-3B + D = 5$$

Solving these equations simultaneously we obtain, $A = 1$, $B = -2, C = 0, D = -1$.
Therefore,

$$\frac{x^3 - 3x^2 - 4x + 5}{(x^2 + x + 1)(x^2 - 3)} = \frac{x - 2}{x^2 + x + 1} - \frac{1}{x^2 - 3}$$

EXERCISE 1

Resolve into partial fractions:

1. $\dfrac{5x - 4}{(x + 1)(x - 2)}$

2. $\dfrac{3x - 26}{(x + 3)(x - 4)}$

3. $\dfrac{3x + 7}{2x^2 - 9x - 5}$

4. $\dfrac{2x - 1}{(x - 1)(x - 2)}$

5. $\dfrac{x - 1}{x^2 + 6x + 8}$

6. $\dfrac{5x - 11}{(2x - 3)(x + 2)}$

7. $\dfrac{3x + 20}{x^2 + 4x}$

8. $\dfrac{x - 4}{(x + 1)(x - 3)}$

9. $\dfrac{5x - 1}{x(x - 1)(x + 1)}$

10. $\dfrac{x^2 - 10x + 13}{(x - 1)(x^2 - 5x + 6)}$

11. $\dfrac{3x^2 + 21x - 84}{(x - 2)(x + 3)(x - 5)}$

12. $\dfrac{4x^2 - 11x + 12}{x^2(x - 4)}$

13. $\dfrac{9}{(x - 1)(x + 2)^2}$

14. $\dfrac{2x^2 - 11x + 5}{(x - 3)(x^2 + 2x - 5)}$

15. $\dfrac{4x^2 + 5x + 8}{(x^2 - 5)(x + 2)}$

16. $\dfrac{8x^2 - x - 4}{(2x^2 - x - 1)(x^2 + x + 1)}$

17. $\dfrac{13x}{(3x^2 - 2)(2x^2 + 3)}$

18. $\dfrac{5x^3 - 6x^2 - 15x + 5}{x^2(x^2 - 5)}$

ANSWERS

EXERCISE 1

1. $\dfrac{3}{x + 1} + \dfrac{2}{x - 2}$

2. $\dfrac{5}{x + 3} - \dfrac{2}{x - 4}$

3. $-\dfrac{1}{2x + 1} + \dfrac{2}{x - 5}$

4. $-\dfrac{1}{x - 1} + \dfrac{3}{x - 2}$

5. $\dfrac{\frac{5}{2}}{x + 4} - \dfrac{\frac{3}{2}}{x + 2} = \dfrac{5}{2(x + 4)} - \dfrac{3}{2(x + 2)}$

6. $-\dfrac{1}{2x - 3} + \dfrac{3}{x + 2}$

7. $\dfrac{5}{x} - \dfrac{2}{x + 4}$

8. $\dfrac{\frac{5}{4}}{x + 1} - \dfrac{\frac{1}{4}}{x - 3} = \dfrac{5}{4(x + 1)} - \dfrac{1}{4(x - 3)}$

9. $\dfrac{1}{x} + \dfrac{2}{x - 1} - \dfrac{3}{x + 1}$

10. $\dfrac{2}{x - 1} - \dfrac{4}{x - 3} + \dfrac{3}{x - 2}$

11. $\dfrac{2}{x - 2} - \dfrac{3}{x + 3} + \dfrac{4}{x - 5}$

12. $\dfrac{2}{x} - \dfrac{3}{x^2} + \dfrac{2}{x - 4}$

13. $\dfrac{1}{x - 1} - \dfrac{1}{x + 2} - \dfrac{3}{(x + 2)^2}$

14. $-\dfrac{1}{x - 3} + \dfrac{3x}{x^2 + 2x - 5}$

15. $\dfrac{18x - 31}{x^2 - 5} - \dfrac{14}{x + 2}$

16. $\dfrac{2x - 1}{2x^2 - x - 1} + \dfrac{3 - x}{x^2 + x + 1}$

17. $\dfrac{3x}{3x^2 - 2} - \dfrac{2x}{2x^2 + 3}$

18. $\dfrac{3}{x} - \dfrac{1}{x^2} + \dfrac{2x - 5}{x^2 - 5}$

CHAPTER **23**

Infinite Series

DEFINITION: A *sequence* is a set of numbers arranged in a definite order and formed according to a given law.

APPLICATION: 1, 3, 5, 7, 9 is a sequence. Each of the numbers 1, 3, 5, 7, and 9 is a term of the sequence. The law of formation of a sequence is usually given by an algebraic expression called the *general term* of the sequence. In this case, the general term is $(2n - 1)$. When $n = 1$, the first term 1 is obtained. When $n = 2$, the second term 3 is obtained, etc.

Example: Write the first five terms of the sequence having the general term $\dfrac{n}{2n - 1}$.

▶ For term 1, $n = 1$, $\dfrac{1}{2(1) - 1} = \dfrac{1}{2 - 1} = 1$

For term 2, $n = 2$, $\dfrac{2}{2(2) - 1} = \dfrac{2}{4 - 1} = \dfrac{2}{3}$

For term 3, $n = 3$, $\dfrac{3}{2(3) - 1} = \dfrac{3}{6 - 1} = \dfrac{3}{5}$

For term 4, $n = 4$, $\dfrac{4}{2(4) - 1} = \dfrac{4}{8 - 1} = \dfrac{4}{7}$

For term 5, $n = 5$, $\dfrac{5}{2(5) - 1} = \dfrac{5}{10 - 1} = \dfrac{5}{9}$

Example: Write the first five terms of the sequence having the general term $\dfrac{1}{2^n + 1}$.

▶ For term 1, $n = 1$, $\dfrac{1}{2^1 + 1} = \dfrac{1}{2 + 1} = \dfrac{1}{3}$

For term 2, $n = 2$, $\dfrac{1}{2^2 + 1} = \dfrac{1}{4 + 1} = \dfrac{1}{5}$

For term 3, $n = 3$, $\dfrac{1}{2^3 + 1} = \dfrac{1}{8 + 1} = \dfrac{1}{9}$

For term 4, $n = 4$, $\dfrac{1}{2^4 + 1} = \dfrac{1}{16 + 1} = \dfrac{1}{17}$

For term 5, $n = 5$, $\dfrac{1}{2^5 + 1} = \dfrac{1}{32 + 1} = \dfrac{1}{33}$

EXERCISE 1

Write the first 5 terms of the sequences having the indicated general terms.

1. $\dfrac{2}{3n + 1}$

2. $\dfrac{2n - 1}{n^2}$

3. $\dfrac{2^n}{n^2 + 3}$

4. $\dfrac{n + 1}{3^n - 1}$

5. $\dfrac{2^{n-1}}{n^3 + 5}$

6. $\dfrac{n^2 + 1}{2^{n+1}}$

WRITING A GENERAL TERM OF A SEQUENCE

It is sometimes necessary to write a general term of a given sequence. This is done by careful observation of the relationships existing among the terms of the sequence.

Example: Write a general (or n^{th}) term of the sequence $\frac{1}{3}$, $\frac{1}{5}$, $\frac{1}{7}$, $\frac{1}{9}$, ...

▶   A general term here, is a fraction whose numerator is 1. Observe that each term of the denominator may be obtained by multiplying the number representing the position of the term by 2 and adding 1. For example, the denominator of the first term is $2(1) + 1$, the denominator of the second term is $2(2) + 1$, the denominator of the third term is $2(3) + 1$. Therefore, the denominator of the nth term is $2n + 1$.

The nth term $= \dfrac{1}{2n + 1}$.

Example: Write the nth term of the sequence $\frac{1}{2}$, $\frac{4}{3}$, $\frac{9}{4}$, $\frac{16}{5}$, ...

▶   A general term here, is a fraction whose numerator is a perfect square; the square of the number representing the position of the term. Therefore, the numerator of the nth term is n^2. Each term of the denominator is 1 greater than the number representing the position of the term. Therefore, the denominator of the nth term is $n + 1$.

The nth term $= \dfrac{n^2}{n + 1}$.

Example: Write the nth term of the sequence $\frac{2}{1}$, $\frac{4}{3}$, $\frac{8}{5}$, $\frac{16}{7}$, ...

▶   A general term here, is a fraction whose numerator is a power of 2. The terms in the denominator are obtained by multiplying the number representing the position of the term and subtracting 1.

The nth term $= \dfrac{2^n}{2n - 1}$.

EXERCISE 2

Write an nth term for each of the following sequences.

1. 2, 4, 6, 8, ...

2. $\dfrac{1}{2}, \dfrac{1}{3}, \dfrac{1}{4}, \dfrac{1}{5}, \ldots$

3. $\dfrac{1}{2}, \dfrac{2}{3}, \dfrac{3}{4}, \dfrac{4}{5}, \ldots$

4. $\dfrac{1}{1 \cdot 2}, \dfrac{1}{2 \cdot 3}, \dfrac{1}{3 \cdot 4}, \dfrac{1}{4 \cdot 5}, \ldots$

5. $\dfrac{1}{4}, \dfrac{4}{5}, \dfrac{9}{6}, \dfrac{16}{7}, \ldots$ **7.** $\dfrac{2}{3}, \dfrac{4}{5}, \dfrac{8}{7}, \dfrac{16}{9}, \ldots$

6. $\dfrac{1}{1}, \dfrac{4}{3}, \dfrac{9}{5}, \dfrac{16}{7}, \ldots$ **8.** $\dfrac{3}{2}, \dfrac{9}{4}, \dfrac{27}{6}, \dfrac{81}{8}, \ldots$

DEFINITION: A *series* is the indicated sum of the numbers in a sequence.

If the numbers $u_1, u_2, u_3, u_4, \ldots$ represent a never-ending sequence of numbers, the expression $u_1 + u_2 + u_3 + u_4 + \ldots$ is called an *infinite series*. Each of the numbers $u_1, u_2, u_3, u_4, \ldots$ is called a term of the series.

Unless otherwise specified the term "series" will mean infinite series in this chapter. The following notation will be used.

$$S_1 = u_1$$
$$S_2 = u_1 + u_2$$
$$S_3 = u_1 + u_2 + u_3$$
$$S_n = u_1 + u_2 + u_3 + \cdots + u_n$$

The following notation is also frequently used.

$$\sum_{n=1}^{k} u_n = u_1 + u_2 + u_3 + u_4 + \cdots + u_k$$

The symbol Σ is a symbol of summation. In this case, $\sum_{n=1}^{k}$ represents the sum of the first k terms of the series *i.e.* $u_1 + u_2 + u_3 + \cdots + u_k$, as indicated above.

Example: For the following nth term write the first four terms of the corresponding series, $\dfrac{n}{n+2}$.

▶   When $n = 1$, the first term $= \dfrac{1}{1+2} = \dfrac{1}{3}$.

The series is $\dfrac{1}{3} + \dfrac{2}{4} + \dfrac{3}{5} + \dfrac{4}{6}$.

Example: For the following nth term write the first four terms of the corresponding series, $\dfrac{n}{2^n - 1}$.

▶ The series is $\dfrac{1}{1} + \dfrac{2}{3} + \dfrac{3}{7} + \dfrac{4}{15}$.

Example: For the following nth term write the first four terms of the corresponding series, $\dfrac{1}{n(n + 1)}$.

▶ The series is $\dfrac{1}{2} + \dfrac{1}{6} + \dfrac{1}{12} + \dfrac{1}{20}$.

EXERCISE 3

For each of the following nth terms write the first four terms of the corresponding series:

1. $\dfrac{n}{n + 5}$

2. $\dfrac{n}{3n - 1}$

3. $\dfrac{2^n}{n + 1}$

4. $\dfrac{n^2}{3n + 1}$

5. $\dfrac{1}{n(n + 2)}$

6. $\dfrac{n}{(n + 1)(n + 3)}$

An infinite series is not necessarily indicated by a finite number of terms. However, in the work below it will be assumed that a simple corresponding nth term can be derived from the few terms given.

Example: Write the nth term and the corresponding $(n + 1)$th term for the following series.

$$1 + 4 + 7 + 10 + \cdots\cdots$$

▶ The nth term $= 3n - 2$.
The $(n + 1)$th term $= 3(n + 1) - 2 = 3n + 3 - 2 = 3n + 1$.

Example: Write the nth term and the corresponding $(n + 1)$th term for the following series:

$$\dfrac{3}{2 \cdot 1} + \dfrac{4}{2 \cdot 3} + \dfrac{5}{3 \cdot 4} + \dfrac{6}{5 \cdot 4} + \cdots\cdots$$

▶ The nth term $= \dfrac{n+2}{n(n+1)}$.

The $(n+1)$th term $= \dfrac{(n+1)+2}{(n+1)(n+1+1)} = \dfrac{n+3}{(n+1)(n+2)}$.

Example: Write the nth term and the corresponding $(n+1)$th term for the following series:

$$\frac{x^2}{1!} - \frac{x^4}{2!} + \frac{x^6}{3!} - \frac{x^8}{4!}$$

▶ Since the signs alternate, the sign of the coefficient is $(-1)^{n-1}$. The nth term $= \dfrac{(-1)^{n-1}x^{2n}}{n!}$.

The $(n+1)$th term $= \dfrac{(-1)^{n+1-1}x^{2(n+1)}}{(n+1)!} = \dfrac{(-1)^n x^{2n+2}}{(n+1)!}$.

Example: Write the first four terms of the series represented by

$$\sum_{n=1}^{\infty} \frac{n^2}{n+2}.$$

▶ The first term is $\dfrac{(1)^2}{1+2} = \dfrac{1}{3}$

The second term is $\dfrac{(2)^2}{2+2} = \dfrac{4}{4}$

The series is $\dfrac{1}{3} + \dfrac{4}{4} + \dfrac{9}{5} + \dfrac{16}{6} + \cdots\cdots$

EXERCISE 4

For each of the following series write the nth term and the corresponding $(n+1)$th term.

1. $2 + 7 + 12 + 17 + \cdots$

2. $\dfrac{1}{5} + \dfrac{1}{10} + \dfrac{1}{15} + \dfrac{1}{20} + \cdots$

3. $\dfrac{1}{2\cdot 3} + \dfrac{1}{3\cdot 4} + \dfrac{1}{4\cdot 5} + \dfrac{1}{5\cdot 6} +$

4. $\dfrac{3}{1^2} + \dfrac{4}{2^2} + \dfrac{5}{3^2} + \dfrac{6}{4^2} + \cdots$

5. $\dfrac{x^3}{2!} + \dfrac{x^5}{3!} + \dfrac{x^7}{4!} + \dfrac{x^9}{5!} + \cdots\cdots$

6. $\dfrac{x}{1\cdot 2} + \dfrac{x^2}{2\cdot 3} + \dfrac{x^3}{3\cdot 4} + \dfrac{x^4}{4\cdot 5} + \cdots\cdots$

Write the first four terms of the series represented by each of the following:

7. $\displaystyle\sum_{n=1}^{\infty} (2 + 3n)$

8. $\displaystyle\sum_{n=1}^{\infty} \dfrac{(n+1)}{n!}$

9. $\displaystyle\sum_{n=1}^{\infty} \dfrac{n}{(2^n - 1)}$

LIMIT OF A SEQUENCE

Let us consider the sequence

$$1, \frac{1}{2}, \frac{1}{2^2}, \frac{1}{2^3}, \frac{1}{2^4}, \cdots\cdots, \frac{1}{2^n}.$$

As n increases without bound it is clear that the nth term has a value closer and closer to zero. In such a case, the sequence is said to approach zero as a limit.

If we inscribe a regular polygon of n sides in a circle and let n increase without bound, the sequence formed by the successive areas of the polygons will approach the area, πr^2, of the circle, as a limit.

DEFINITION: As n increases without bound u_n approaches the limit L (written $\lim_{n \to \infty} u_n = L$) provided that for any positive number ε, however small, there exists a number N such that $|u_n - L| < \varepsilon$ for all integers $n > N$.

Application of Definition

Let us consider the sequence whose general term is $\dfrac{n}{n+1}$. If we write the first five terms of this sequence we have $\frac{1}{2}, \frac{2}{3}, \frac{3}{4}, \frac{4}{5}, \frac{5}{6}$. Intuitively, we can see that, as we take more terms, we will get closer to 1, or that the limit of the sequence is 1. In order to prove that the limit of this sequence is 1 we would have to do so in terms of the definition or by means of the theorems on limits which will be given below.

Let us consider the sequence whose general term is $\dfrac{2n-1}{n}$. If we write the first five terms of this sequence we have $\frac{1}{1}, \frac{3}{2}, \frac{5}{3}, \frac{7}{4}, \frac{9}{5}$. A reasonable estimate of the limit of this sequence is 2. If we wish to prove this in terms of the definition we proceed as follows. We must show that, for a given positive number ε (no matter how small) there will be a positive number N such $|\, u_n - 2\,| < \varepsilon$ for all $n > N$. In this case,

$$\left|\frac{2n-1}{n} - 2\right| < \varepsilon \quad \text{or} \quad \left|\frac{2n-1-2n}{n}\right| < \varepsilon$$

$$\text{or} \quad \left|-\frac{1}{n}\right| < \varepsilon \quad \text{or} \quad \frac{1}{n} < \varepsilon.$$

This can be written as $n > \dfrac{1}{\varepsilon}$.

Now, if we take $\varepsilon = .1$ (i.e. if we wish to make the absolute value of the difference between u_n and the limit less than .1) we may use the inequality $n > \dfrac{1}{\varepsilon}$ to find the corresponding value of n. In this case, $n > \dfrac{1}{.1}$ or $n > 10$. For example, if $n = 11$ then

$$\left|\frac{2(11)-1}{11} - 2\right| = \left|\frac{21}{11} - 2\right| = |\, 1.909 - 2\,| = .091$$

Again, if we take $\varepsilon = .01$, $n > \dfrac{1}{.01}$, $n > 100$. It can easily be verified that the absolute value of the difference between any term of the sequence after the 100th and 2 is less than .01.

Let us consider the sequence whose general term is $\dfrac{2^n}{n}$. The first five terms are $\dfrac{2}{1}$, $\dfrac{4}{2}$, $\dfrac{8}{3}$, $\dfrac{16}{4}$, $\dfrac{32}{5}$. If we examine this sequence there is no apparent limit. If we write the fifteenth term, we obtain the result $\dfrac{2^{15}}{15}$ and it appears that the successive terms of the sequence increase without bound. In this case, the terms grow indefinitely large as n increases and $\lim\limits_{n \to \infty} = \infty$. The symbol ∞ is not to be regarded as a number.

Theorems on Limits

The following theorems on limits are stated without proof.
If $\lim\limits_{n \to \infty} u_n = u$ and $\lim\limits_{n \to \infty} v_n = v$ then,

1. $\lim\limits_{n \to \infty} (u_n \pm v_n) = \lim\limits_{n \to \infty} u_n \pm \lim\limits_{n \to \infty} v_n = u \pm v$.

2. $\lim\limits_{n \to \infty} (u_n \cdot v_n) = \lim\limits_{n \to \infty} u_n \cdot \lim\limits_{n \to \infty} v_n = uv$.

3. if c is a constant $\neq 0$, $\lim\limits_{n \to \infty} (cu_n) = c \lim\limits_{n \to \infty} u_n = cu$.

4. $\lim\limits_{n \to \infty} \dfrac{u_n}{v_n} = \dfrac{\lim\limits_{n \to \infty} u_n}{\lim\limits_{n \to \infty} v_n} = \dfrac{u}{v}$ (provided $v \neq 0$).

The following theorems are also useful in determining limits.

5. If c is a constant $\neq 0$, then $\lim\limits_{n \to \infty} \dfrac{c}{0}$ does not exist.

6. If c is a constant $\neq 0$, then $\lim\limits_{n \to \infty} cn = \infty$.

7. If c is a constant $\neq 0$, then $\lim\limits_{n \to \infty} \dfrac{n}{c} = \infty$.

8. If c is a constant $\neq 0$, then $\lim\limits_{n \to \infty} \dfrac{c}{n} = 0$.

9. If a is a positive number less than 1, then $\lim\limits_{n \to \infty} a^n = 0$.

10. If a is a positive number greater than 1, then $\lim\limits_{n \to \infty} a^n = \infty$.

Example: Evaluate $\displaystyle\lim_{n \to \infty} \frac{2n^2}{n^2 + 1}$.

▶ To find this limit we divide numerator and denominator of the fraction by the highest power of n. In this case, divide numerator and denominator by n^2.

$$\lim_{n \to \infty} \frac{2n^2}{n^2 + 1} = \lim_{n \to \infty} \frac{2}{1 + \dfrac{1}{n^2}} = \frac{\displaystyle\lim_{n \to \infty} 2}{\displaystyle\lim_{n \to \infty} \left(1 + \dfrac{1}{n^2}\right)}$$

$$= \frac{\displaystyle\lim_{n \to \infty} 2}{\displaystyle\lim_{n \to \infty} + 1 \ \lim_{n \to \infty}} \left(\frac{1}{n^2}\right) = \frac{2}{1 + 0} = 2$$

Example: Evaluate $\displaystyle\lim_{n \to \infty} \frac{2n + 1}{n^2}$.

▶ Divide numerator and denominator by n^2 to get

$$\lim_{n \to \infty} \frac{\dfrac{2}{n} + \dfrac{1}{n^2}}{1} = \frac{0 + 0}{1} = 0$$

Example: Evaluate $\displaystyle\lim_{n \to \infty} \frac{2n^2 + 5}{n^3}$

▶ Divide the numerator and denominator of the fraction by n^2 to get

$$\lim_{n \to \infty} \frac{2n^2 + 5}{n^3} = \lim_{n \to \infty} \frac{\dfrac{2}{n} + \dfrac{5}{n^3}}{1} = \frac{0 + 0}{1} = 0$$

Example: Evaluate $\displaystyle\lim_{n \to \infty} \frac{3n^2 + 7}{5n}$.

▶ Divide the numerator and denominator of the fraction by n^2

to get $\displaystyle\lim_{n \to \infty} \frac{3n^2 + 7}{5n} = \lim_{n \to \infty} \frac{3 + \dfrac{7}{n^2}}{\dfrac{5}{n}} = \frac{3 + 0}{0} = \frac{3}{0}$

In this case, the limit does not exist since the result $\frac{3}{0}$ is meaningless.

EXERCISE 5

Evaluate each of the following:

1. $\lim\limits_{n \to \infty} \dfrac{3n}{2n - 1}$

2. $\lim\limits_{n \to \infty} \dfrac{2n^2 - 4}{2n^2 + 5}$

3. $\lim\limits_{n \to \infty} \dfrac{n + 3}{2n^2 - 1}$

4. $\lim\limits_{n \to \infty} \dfrac{5n - 1}{n^2 + 2}$

5. $\lim\limits_{n \to \infty} \dfrac{2n^2 - 3}{n + 2}$

6. $\lim\limits_{n \to \infty} \dfrac{n^3 + 2n - 1}{n^3 - 2}$

7. $\lim\limits_{n \to \infty} \dfrac{3n^2 - 5}{2n^3 - 7}$

8. $\lim\limits_{n \to \infty} \dfrac{n^3 + 1}{3n + 5}$

CONVERGENCE AND DIVERGENCE OF SERIES

Let us consider the infinite series $u_1 + u_2 + u_3 + \cdots\cdots + u_n + \cdots\cdots$. If we let

$$S_1 = u_1$$
$$S_2 = u_1 + u_2$$
$$S_3 = u_1 + u_2 + u_3$$
$$S_n = u_1 + u_2 + u_3 + \cdots\cdots + u_n$$

then the numbers $S_1, S_2, S_3, \cdots\cdots S_n$ are called the partial sums of the infinite series. The following definition applies to the sequence composed of the partial sums of the infinite series.

DEFINITION: If the sequence of partial sums of an infinite series approaches a limit as the number of terms of the sequence increases without bound, then the series is said to be *convergent*. If a series is not *convergent* then it is *divergent*.

Application of Definition

1. Consider the series $2 + 1 + \frac{1}{2} + \frac{1}{4} + \frac{1}{8} + \frac{1}{16} + \cdots\cdots$

Then $S_1 = 2$

$$S_2 = 2 + 1 = 3$$
$$S_3 = 2 + 1 + \tfrac{1}{2} = 3\tfrac{1}{2}$$
$$S_4 = 2 + 1 + \tfrac{1}{2} + \tfrac{1}{4} = 3\tfrac{3}{4}$$
$$S_5 = 2 + 1 + \tfrac{1}{2} + \tfrac{1}{4} + \tfrac{1}{8} = 3\tfrac{7}{8}$$
$$S_6 = 2 + 1 + \tfrac{1}{2} + \tfrac{1}{4} + \tfrac{1}{8} + \tfrac{1}{16} = 3\tfrac{15}{16}$$

It appears that the sequence of partial sums approaches the limit 4 as the number of terms of the sequence increases without bound.

The convergency of this series may be proved as follows.

The sequence $2, 1, \frac{1}{2}, \frac{1}{4}, \cdots\cdots$ is a geometric progression whose first term is 2 and whose ratio is $\frac{1}{2}$. The sum of n terms of this sequence may be obtained by formula.

$$S_n = \frac{a - ar^n}{1 - r} = \frac{2 - 2(\frac{1}{2})^n}{1 - \frac{1}{2}} = \frac{2 - (2)(\frac{1}{2})^n}{\frac{1}{2}}$$

$$= 4 - 2^2 \left(\frac{1}{2}\right)^n = 4 - \frac{2^2}{2^n}$$

$$\lim_{n \to \infty} S_n = \lim_{n \to \infty} \left(4 - \frac{2^2}{2^n}\right) = \lim_{n \to \infty} (4) - \lim_{n \to \infty} \left(\frac{2^2}{2^n}\right)$$

$$= 4 - 0 = 4$$

The series is convergent since $\lim_{n \to \infty} S_n = 4$.

2. Consider the series $2 + 5 + 8 + 11 + 14 + 17 + \cdots\cdots$

Then
$$\begin{array}{ll} S_1 = 2 & S_4 = 26 \\ S_2 = 2 + 5 = 7 & S_5 = 40 \\ S_3 = 2 + 5 + 8 = 15 & S_6 = 57 \end{array}$$

Apparently, the sum of the first n terms of this infinite series does not approach a limit as n increases without bound.

The divergency of this series may be proven as follows.

The sequence $2, 5, 8, 11, 14, \ldots$ is an arithmetic progression. The sum of n terms of this sequence may be obtained by formula.

$$S_n = \frac{n}{2}[2a + (n - 1)d]$$

$$S_n = \frac{n}{2}[4 + (n - 1)3] = \frac{n}{2}[4 + 3n - 3] = \frac{n}{2}(3n + 1) = \frac{3n^2 + n}{2}$$

$$\lim_{n \to \infty} S_n = \lim_{n \to \infty} \left(\frac{3n^2 + n}{2}\right) = \infty$$

Since $\lim_{n \to \infty} S_n$ does not exist, the series is divergent.

3. Consider the series $1 - 1 + 1 - 1 + 1 - 1 + \cdots\cdots$

This series does not converge since the sum will be zero if an even number of terms are taken and the sum will be 1 if an odd number of terms are taken. This type of divergent series is called an *oscillating* series.

A Necessary Condition for Convergence

THEOREM: A necessary condition that an infinite series converge is that its nth term approach zero as a limit as n increases without bound.

This theorem states that an infinite series *cannot* converge unless its nth term approaches zero as n increases without bound. However, the fact that the nth term of an infinite series approaches zero as a limit as n increases without bound does *not* imply that the infinite series converges. In other words, this theorem states a necessary but not a sufficient condition.

Application of the Theorem

Consider the series $\frac{1}{2} + \frac{2}{5} + \frac{3}{8} + \frac{4}{11} + \cdots$. The nth term of this series is $\dfrac{n}{3n-1}$.

The limit of the nth term as n increases without bound is

$$\lim_{n \to \infty} \frac{n}{3n-1} = \lim_{n \to \infty} \frac{\dfrac{n}{n}}{\dfrac{3n}{n} - \dfrac{1}{n}} = \lim_{n \to \infty} \frac{1}{3 - \dfrac{1}{n}} = \frac{1}{3}.$$

Therefore, the above series is *divergent*. It cannot be convergent because it fails to meet the necessary condition of the above theorem.

Consider the series $1 + \frac{1}{2} + \frac{1}{3} + \frac{1}{4} + \frac{1}{5} + \cdots$. The nth term of this series is $\dfrac{1}{n}$. Now, $\lim_{n \to \infty} \dfrac{1}{n} = 0$. However, this series is divergent, as will be shown later.

EXERCISE 6

Show that each of the following series does not meet the necessary condition for convergence and is therefore divergent.

1. $\dfrac{3}{4} + \dfrac{4}{5} + \dfrac{5}{6} + \dfrac{6}{7} + \cdots$

2. $\dfrac{1}{4} + \dfrac{4}{7} + \dfrac{9}{12} + \dfrac{16}{19} + \cdots$

3. $\dfrac{2}{3} + \dfrac{5}{6} + \dfrac{10}{9} + \dfrac{17}{12} + \cdots$

4. $\dfrac{1}{3} + \dfrac{8}{5} + \dfrac{27}{7} + \dfrac{64}{9} + \cdots$

Comparison Tests for Convergence and Divergence of Series

Comparison tests are based upon the following two theorems which are stated without proof.

THEOREM 1: If, from some term on, each term of a given series of *positive* terms is equal to or less than the corresponding term of a known convergent series of positive terms, then the given series is convergent.

THEOREM 2: If, from a certain term on, each term of a series of *positive* terms is greater than or equal to the corresponding terms of a known divergent series of positive terms, then the given series is divergent.

In effect, these theorems enable us to show that a given series is convergent provided that a convergent series can be found whose terms are no less than the corresponding terms of the given series or to show that a given series is divergent, provided that a divergent series of positive terms can be found whose terms are no greater than the corresponding terms of the given series.

It does not follow that a given series is divergent because each term of the given series is greater than the corresponding term of a known convergent series. Also, it does not follow that a given series is convergent if each term of the given series is less than the corresponding term of a known divergent series.

In comparing series, a finite number of terms of either series may be dropped without affecting the conclusion regarding convergency or divergency.

The following series are especially useful for comparison.

Convergent Series

     1. $a + ar + ar^2 + \cdots + ar^{n-1} \cdots$ provided $\mid r \mid < 1$.

     2. $1 + \dfrac{1}{2^p} + \dfrac{1}{3^p} + \cdots + \dfrac{1}{n^p} + \cdots$ provided $p > 1$.

Divergent Series

     1. $a + ar + ar^2 + \cdots + ar^{n-1} \cdots$ provided $\mid r \mid \geqq 1$.

     2. $1 + \dfrac{1}{2^p} + \dfrac{1}{3^p} + \cdots + \dfrac{1}{n^p} + \cdots$ provided $p \leqq 1$.

If $p = 1$ in the series $1 + \dfrac{1}{2^p} + \dfrac{1}{3^p} + \cdots + \dfrac{1}{n^p} + \cdots$ we have

the *harmonic* series $1 + \dfrac{1}{2} + \dfrac{1}{3} + \cdots + \dfrac{1}{n} + \cdots$. This series

may be proved divergent as follows. Let us compare the following
two series:

a. $1 + \dfrac{1}{2} + \dfrac{1}{3} + \dfrac{1}{4} + \dfrac{1}{5} + \dfrac{1}{6} + \dfrac{1}{7} + \dfrac{1}{8} + \cdots$ (harmonic series)

b. $1 + \dfrac{1}{2} + \dfrac{1}{4} + \dfrac{1}{4} + \dfrac{1}{8} + \dfrac{1}{8} + \dfrac{1}{8} + \dfrac{1}{8} + \cdots$

Each term of series *a* is equal to or greater than the corresponding term of series *b*.

If we collect the terms of series *b* as bracketed we obtain
$1 + \dfrac{1}{2} + \dfrac{1}{2} + \dfrac{1}{2} + \cdots$. The sum of the terms of series *b* can be

represented as $1 + \dfrac{1}{2}n$ where *n* represents the number of $\dfrac{1}{2}$s collected.

Now, $1 + \dfrac{1}{2}n$ can be made as large as desired by taking a sufficiently

great number of terms of series *b*. Therefore, series *b* is divergent.

Since each term of series *a* is equal to or greater than the corresponding term of series *b* then series *a* is also divergent.

Example: Show that the series

$$\frac{1}{1 \cdot 2} + \frac{1}{2 \cdot 3} + \frac{1}{3 \cdot 4} + \cdots + \frac{1}{n(n + 1)} + \cdots$$

is convergent.

▶ Compare the series

$$\frac{1}{1 \cdot 2} + \frac{1}{2 \cdot 3} + \frac{1}{3 \cdot 4} + \cdots + \frac{1}{n(n + 1)} + \cdots$$

term by term with the convergent geometric series with $r = \dfrac{1}{2}$,

$\dfrac{1}{2} + \dfrac{1}{4} + \dfrac{1}{8} + \cdots + \dfrac{1}{2^n}$.

We note that each term of the given series is less than the corresponding term of the convergent geometric series. Therefore, the given series is convergent.

Example: Show that the series is

$$1 + \frac{1}{\sqrt[3]{2}} + \frac{1}{\sqrt[3]{3}} + \frac{1}{\sqrt[3]{4}} + \cdots + \frac{1}{\sqrt[3]{n}}$$

is divergent.

▶ Compare the series

$$1 + \frac{1}{\sqrt[3]{2}} + \frac{1}{\sqrt[3]{3}} + \frac{1}{\sqrt[3]{4}} + \cdots + \frac{1}{\sqrt[3]{n}}$$

with the p-series $1 + \frac{1}{2^p} + \frac{1}{3^p} + \frac{1}{4^p} + \cdots + \frac{1}{n^p} + \cdots$.

Since the given series is a p-series with $p < 1$ (in this case $p = \frac{1}{3}$) it is divergent.

Example: Show that the series

$$1 + \frac{2}{3} + \frac{3}{5} + \frac{4}{7} + \cdots \frac{n}{2n-1} + \cdots$$

is divergent.

▶ Compare the series

$$1 + \frac{2}{3} + \frac{3}{5} + \frac{4}{7} + \cdots + \frac{n}{2n-1} + \cdots$$

term by term with the harmonic series

$$1 + \frac{1}{2} + \frac{1}{3} + \frac{1}{4} + \cdots + \frac{1}{n} + \cdots$$

We note that each term of the given series after the first term is greater than the corresponding term of the harmonic series which is divergent. Therefore, the given series is divergent.

Example: Show that the series

$$\frac{1}{\sqrt{2}} + \frac{1}{\sqrt{4}} + \frac{1}{\sqrt{6}} + \frac{1}{\sqrt{8}} + \cdots + \frac{1}{\sqrt{2n}} + \cdots$$

is divergent.

 The series

$$\frac{1}{\sqrt{2}} + \frac{1}{\sqrt{4}} + \frac{1}{\sqrt{6}} + \frac{1}{\sqrt{8}} + \cdots + \frac{1}{\sqrt{2n}} + \cdots$$

may be written as

$$\frac{1}{\sqrt{2}} + \frac{1}{\sqrt{2} \cdot \sqrt{2}} + \frac{1}{\sqrt{2} \cdot \sqrt{4}} + \cdots \frac{1}{\sqrt{2} \cdot \sqrt{n}} + \cdots$$

This may be written as

$$\frac{1}{\sqrt{2}} \left(1 + \frac{1}{\sqrt{2}} + \frac{1}{\sqrt{3}} + \frac{1}{\sqrt{4}} + \cdots + \frac{1}{\sqrt{n}} + \cdots \right)$$

The series

$$1 + \frac{1}{\sqrt{2}} + \frac{1}{\sqrt{3}} + \frac{1}{\sqrt{4}} + \cdots + \frac{1}{\sqrt{n}} + \cdots$$

is divergent because it is a *p*-series with $p = \frac{1}{2}$. Therefore, the original series is divergent.

Example: Test the following series for convergence or divergence.

$$1 + \frac{1}{\log 2} + \frac{1}{\log 3} + \frac{1}{\log 4} + \cdots + \frac{1}{\log n} + \cdots$$

 Compare the series

$$1 + \frac{1}{\log 2} + \frac{1}{\log 3} + \frac{1}{\log 4} + \cdots + \frac{1}{\log n} + \cdots$$

with the harmonic series

$$1 + \frac{1}{2} + \frac{1}{3} + \frac{1}{4} + \cdots + \frac{1}{n} + \cdots$$

We note that each term of the given series is either equal to or greater than the corresponding term of the harmonic series which is known to be divergent. Therefore, the given series is divregent.

Example: Test the following series for convergence or divergence.

$$1 + \frac{1}{2\sqrt{2}} + \frac{1}{3\sqrt{3}} + \frac{1}{4\sqrt{4}} + \cdots + \frac{1}{n\sqrt{n}} + \cdots$$

▶ The series

$$1 + \frac{1}{2\sqrt{2}} + \frac{1}{3\sqrt{3}} + \frac{1}{4\sqrt{4}} + \cdots + \frac{1}{n\sqrt{n}} + \cdots$$

may be written

$$1 + \frac{1}{2^{3/2}} + \frac{1}{3^{3/2}} + \frac{1}{4^{3/2}} + \cdots + \frac{1}{n^{3/2}} + \cdots$$

The series is convergent because it is a *p*-series with $p > 1$.

Example: Examine the following series for convergence or divergence.

$$\frac{1+1}{1+1^2} + \frac{1+2}{1+2^2} + \frac{1+3}{1+3^2} + \frac{1+4}{1+4^2} + \cdots + \frac{1+n}{1+n^2} + \cdots$$

▶ Compare the given series with the harmonic series

$$\frac{1}{2} + \frac{1}{3} + \frac{1}{4} + \frac{1}{5} + \cdots + \frac{1}{n+1} + \cdots$$

We note that each term of the given series is equal to or greater than the corresponding term of the harmonic series. In fact, the *n*th term of the given series

$$\frac{1+n}{1+n^2} > \frac{1+n}{1+2n+n^2} \quad \text{or} \quad \frac{1+n}{(1+n)^2} \quad \text{or} \quad \frac{1}{1+n} \cdots$$

Since the harmonic series is divergent the given series is also divergent.

EXERCISE 7

Examine each of the following for convergence or divergence.

1. $1 + \dfrac{1}{2^2} + \dfrac{1}{3^2} + \dfrac{1}{4^2} + \dfrac{1}{5^2} + \cdots + \dfrac{1}{n^2} + \cdots$

2. $\dfrac{1}{2} + \dfrac{2}{3} + \dfrac{3}{4} + \dfrac{4}{5} + \cdots$

3. $1 + \dfrac{2}{3} + \dfrac{4}{9} + \dfrac{8}{27} + \cdots + \left(\dfrac{2}{3}\right)^{n-1} + \cdots$

4. $1 + \dfrac{1}{2 \cdot 3} + \dfrac{1}{3 \cdot 3^2} + \dfrac{1}{4 \cdot 3^3} + \dfrac{1}{5 \cdot 3^4} + \cdots$

5. $1 + \dfrac{1}{2\sqrt[3]{3}} + \dfrac{1}{3\sqrt[3]{3}} + \dfrac{1}{4\sqrt[3]{3}} + \dfrac{1}{5\sqrt[3]{3}} + \cdots$

6. $\dfrac{1}{2 + 3} + \dfrac{1}{2^2 + 3} + \dfrac{1}{2^3 + 3} + \dfrac{1}{2^4 + 3} + \cdots$

7. $\dfrac{1}{1 \cdot 2} + \dfrac{1}{3 \cdot 4} + \dfrac{1}{5 \cdot 6} + \dfrac{1}{7 \cdot 8} + \cdots$

8. $\dfrac{1}{2 - \sqrt{2}} + \dfrac{1}{3 - \sqrt{3}} + \dfrac{1}{4 - \sqrt{4}} + \dfrac{1}{5 - \sqrt{5}} + \cdots$

9. $\dfrac{1}{5} + \dfrac{2}{5} + \dfrac{2^2}{5} + \dfrac{2^3}{5} + \dfrac{2^4}{5} + \cdots$

10. $\dfrac{1}{\sqrt{10}} + \dfrac{1}{\sqrt{11}} + \dfrac{1}{\sqrt{12}} + \dfrac{1}{\sqrt{13}} + \dfrac{1}{\sqrt{14}} + \cdots$

11. $\dfrac{1}{2} + \dfrac{2}{5} + \dfrac{3}{10} + \dfrac{4}{17} + \cdots + \dfrac{n}{n^2 + 1} + \cdots$

12. $\dfrac{1}{2} + \dfrac{2}{17} + \dfrac{3}{82} + \cdots + \dfrac{n}{n^4 + 1} + \cdots$

13. $\dfrac{1}{2^2 - 1} + \dfrac{1}{3^2 - 1} + \dfrac{1}{4^2 - 1} + \cdots + \dfrac{1}{(n + 1)^2 - 1} + \cdots$

14. $1 + \dfrac{1}{\sqrt{2}} + \dfrac{1}{\sqrt{3}} + \cdots + \dfrac{1}{\sqrt{n}} + \cdots$

15. $\dfrac{1}{5 - \sqrt{1}} + \dfrac{1}{6 - \sqrt{2}} + \dfrac{1}{7 - \sqrt{3}} + \dfrac{1}{8 - \sqrt{4}} + \cdots$

Ratio Test for Convergence and Divergence

THEOREM: If, in the infinite series $u_1 + u_2 + u_3 + u_4 + \cdots + u_n$ with like or unlike signs $\lim\limits_{n \to \infty} \dfrac{u_n + 1}{u_n} = R$, then the series is convergent if $R < 1$, and is divergent if $R > 1$. If $R = 1$ the test fails.

Example: Use the ratio test to examine the following series for convergence or divergence.

$$\frac{1}{2} + \frac{2}{2^2} + \frac{3}{2^3} + \frac{4}{2^4} + \cdots$$

▶

$$u_n = \frac{n}{2^n}, \qquad u_{n+1} = \frac{n + 1}{2^{n+1}}$$

$$\frac{u_{n+1}}{u_n} = \frac{\dfrac{n + 1}{2^{n+1}}}{\dfrac{n}{2^n}} = \frac{n + 1}{2^{n+1}} \cdot \frac{2^n}{n} = \frac{n + 1}{2n}$$

$$\lim_{n \to \infty} \left| \frac{n + 1}{2n} \right| = \frac{1}{2} \qquad \text{Since } R < 1 \text{ the series is convergent.}$$

Example: Use the ratio test to examine the following series for convergence or divergence.

$$\frac{3^2}{1} + \frac{3^3}{2} + \frac{3^4}{3} + \frac{3^5}{4} + \cdots$$

▶

$$u_n = \frac{3^{n+1}}{n}, \qquad u_{n+1} = \frac{3^{n+2}}{n + 1}$$

$$\frac{u_{n+1}}{u_n} = \frac{\dfrac{3^{n+2}}{n + 1}}{\dfrac{3^{n+1}}{n}} = \frac{3^{n+2}}{n + 1} \cdot \frac{n}{3^{n+1}} = \frac{3n}{n + 1}$$

$$\lim_{n \to \infty} \left| \frac{3n}{n + 1} \right| = 3$$

Since $R > 1$ the series is divergent.

Example: Use the ratio test to examine the following series for convergence or divergence.

$$\frac{1}{3} + \frac{2^2}{3^2} + \frac{3^2}{3^3} + \frac{4^2}{3^4} + \cdots\cdot$$

▶ $$u_n = \frac{n^2}{3^n}, \qquad u_{n+1} = \frac{(n+1)^2}{3^{n+1}}$$

$$\frac{u_{n+1}}{u_n} = \frac{\dfrac{(n+1)^2}{3^{n+1}}}{\dfrac{n^2}{3^n}} = \frac{(n+1)^2}{3^{n+1}} \cdot \frac{3^n}{n^2} = \frac{(n+1)^2}{3n^2}$$

$$\lim_{n \to \infty} \left| \frac{n^2 + 2n + 1}{3n^2} \right| = \frac{1}{3}$$

Since $R < 1$ the series is convergent.

Example: Test the following series for convergence or divergence by the ratio test.

$$\frac{2}{1 \cdot 2} + \frac{2^2}{2 \cdot 3} + \frac{2^3}{3 \cdot 4} + \cdots\cdot + \frac{2^n}{n(n+1)} + \cdots\cdot$$

▶ $$u_n = \frac{2^n}{n(n+1)}, \qquad u_{n+1} = \frac{2^{n+1}}{(n+1)(n+2)}$$

$$\frac{u_{n+1}}{u_n} = \frac{\dfrac{2^{n+1}}{(n+1)(n+2)}}{\dfrac{2^n}{n(n+1)}} = \frac{2^{n+1}}{(n+1)(n+2)} \cdot \frac{n(n+1)}{2^n} = \frac{2n}{n+2}$$

$$\lim_{n \to \infty} \left| \frac{2n}{n+2} \right| = 2$$

Since $R > 1$ the series is divergent.

Example: Test the following series for convergence or divergence by the ratio test.

$$\frac{1^2}{2} + \frac{2^2}{3} + \frac{3^2}{4} + \frac{4^2}{5} + \cdots$$

▶ $\quad u_n = \frac{n^2}{n + 1}, \qquad u_{n+1} = \frac{(n + 1)^2}{n + 2}$

$$\frac{u_{n+1}}{u_n} = \frac{\dfrac{(n + 1)^2}{n + 2}}{\dfrac{n^2}{n + 1}} = \frac{(n + 1)^2}{n + 2} \cdot \frac{n + 1}{n^2} = \frac{(n + 1)^3}{n^2(n + 2)}$$

$$= \frac{n^3 + 3n^2 + 3n + 1}{n^3 + 2n^2}$$

$$\lim_{n \to \infty} \left| \frac{n^3 + 3n^2 + 3n + 1}{n^3 + 2n^2} \right| = 1$$

Since $R = 1$, the ratio test fails. We may try the comparison test. Let us compare the given series term by term with the harmonic series which is known to be divergent.

$$\frac{1^2}{2} + \frac{2^2}{3} + \frac{3^2}{4} + \frac{4^2}{5} + \cdots + \frac{n^2}{n + 1}$$

$$\frac{1}{2} + \frac{1}{3} + \frac{1}{4} + \frac{1}{5} + \cdots + \frac{1}{n}$$

We note that each term of the given series is greater than the corresponding term of the harmonic series.

Since the harmonic series is divergent, the given series is divergent.

EXERCISE 8

Apply the ratio test to each of the following series to determine its convergence or divergence. If the ratio test fails apply the comparison test.

1. $\dfrac{1}{2} + \dfrac{3}{2^2} + \dfrac{5}{2^3} + \dfrac{7}{2^4} + \cdots + \dfrac{2n - 1}{2^n} +$

2. $1 + \dfrac{2!}{2^2} + \dfrac{3!}{3^2} + \dfrac{4!}{4^2} + \cdots\cdots$

3. $\dfrac{10^2}{2} + \dfrac{10^3}{3} + \dfrac{10^4}{4} + \cdots\cdots + \dfrac{10^{n+1}}{n+1} + \cdots\cdots$

4. $1 + \dfrac{1}{2!} + \dfrac{1}{3!} + \dfrac{1}{4!} + \cdots\cdots$

5. $1 + \dfrac{1}{2^2} + \dfrac{1}{3^2} + \dfrac{1}{4^2} + \dfrac{1}{5^2} + \cdots\cdots$

6. $\dfrac{5}{1^2} + \dfrac{5^2}{2^2} + \dfrac{5^3}{3^2} + \dfrac{5^4}{4^2} + \cdots\cdots$

7. $1 + 2 + \dfrac{2^2}{2!} + \cdots\cdots + \dfrac{2^{n-1}}{(n-1)!} + \cdots\cdots$

8. $\dfrac{1 \cdot 3}{2} + \dfrac{3 \cdot 5}{2^2} + \dfrac{5 \cdot 7}{2^3} + \dfrac{7 \cdot 9}{2^4} + \cdots\cdots + \dfrac{(2n-1)(2n+1)}{2^n}$

$\qquad + \dfrac{(2n-1)(2n+1)}{2^n} + \cdots\cdots$

9. $\dfrac{7^2}{1} + \dfrac{7^3}{2} + \dfrac{7^4}{3} + \dfrac{7^5}{4} + \cdots\cdots$

10. $\dfrac{1}{1^2 + 5} + \dfrac{1}{2^2 + 5} + \dfrac{1}{3^2 + 5} + \dfrac{1}{4^2 + 5} + \cdots\cdots$

11. $\dfrac{2}{1 \cdot 2} + \dfrac{2^2}{2 \cdot 3} + \dfrac{2^3}{3 \cdot 4} + \dfrac{2^4}{4 \cdot 5}$

12. $\dfrac{2 \cdot 3}{1!} + \dfrac{3 \cdot 4}{2!} + \dfrac{4 \cdot 5}{3!} + \dfrac{5 \cdot 6}{4!} + \cdots\cdots$

ALTERNATING SERIES

DEFINITION: A series whose terms are alternately positive and negative is called an alternating series.

Example: $\dfrac{1}{2} - \dfrac{1}{4} + \dfrac{1}{6} - \dfrac{1}{8} + \dfrac{1}{10} + \cdots\cdots + (-1)^{n+1}\dfrac{1}{2n}$

THEOREM: An alternating series $u_1 - u_2 + u_3 - u_4 + \cdots\cdots$ is convergent if each term is numerically less than the preceding term and if the limit of the nth term as n becomes infinite is zero.

Example: Test the following series for convergence or divergence.

$$1 - \frac{1}{\sqrt{2}} + \frac{1}{\sqrt{3}} - \frac{1}{\sqrt{4}} + \frac{1}{\sqrt{5}} \text{ etc.}$$

▶ In order to show that this series is convergent we must show that

a. each term is numerically less than the preceding term. An examination of the given series shows that this requirement is met. $u_n = \frac{1}{\sqrt{n}}$ and $u_{n+1} = \frac{1}{\sqrt{n+1}}$.

b. $\lim\limits_{n \to \infty} u_n = 0$. In the given series, $u_n = \frac{1}{\sqrt{n}}$ and $\lim\limits_{n \to \infty} \frac{1}{\sqrt{n}} = 0$.

Since both requirements are met the series is convergent.

Example: Test the following series for convergence or divergence

$$\frac{3}{2} - \frac{4}{3} + \frac{5}{4} - \frac{6}{5} + \cdots$$

▶ In the given series each term is numerically less than the preceding term. $u_n = \frac{n+2}{n+1}$ and $\lim\limits_{n \to \infty} \frac{n+2}{n+1} = 1$.

Since the limit of the nth term as n becomes infinite is not zero the given series is divergent.

EXERCISE 9

Test the following series for convergence or divergence.

1. $1 - \frac{1}{3} + \frac{1}{5} - \frac{1}{7} + \cdots$

2. $\frac{3}{1} - \frac{5}{3} + \frac{7}{5} - \frac{9}{7} + \cdots$

3. $1 - \frac{1}{2} + \frac{1}{4} - \frac{1}{6} + \cdots$

4. $1 - \frac{2}{3!} + \frac{3}{5!} - \frac{4}{7!} + \cdots$

5. $1 - \frac{1}{2^3} + \frac{1}{3^3} - \frac{1}{4^3} + \cdots$

6. $\frac{3}{4} - \frac{5}{7} + \frac{7}{10} - \frac{9}{13} + \cdots$

ABSOLUTE AND CONDITIONAL CONVERGENCE

DEFINITION: A series is said to be *absolutely convergent* if the series formed from it by making all its terms positive is convergent.

A convergent series which is not absolutely convergent is called conditionally convergent.

Examples: The series

$$1 - \frac{1}{2^2} + \frac{1}{3^2} - \frac{1}{4^2} + \cdots \cdot$$

is absolutely convergent since the series formed by making all the terms positive

$$1 + \frac{1}{2^2} + \frac{1}{3^2} + \frac{1}{4^2} + \cdots \cdot$$

is convergent.

The series

$$1 - \frac{1}{2} + \frac{1}{3} - \frac{1}{4} + \cdots \cdot$$

is convergent if we apply the test for alternating series. However, the series formed by making all the terms positive

$$1 + \frac{1}{2} + \frac{1}{3} + \frac{1}{4} + \cdots \cdot$$

is divergent. In this case, the original series is said to be *conditionally* convergent.

Example: Test the following series for absolute convergence.

$$1 - \frac{2}{3} + \frac{3}{3^2} - \frac{4}{3^3} + \cdots \cdot$$

▶ The given series is

$$1 - \frac{2}{3} + \frac{3}{3^2} - \frac{4}{3^3} + \cdots \cdot$$

Write the series with all terms positive

$$1 + \frac{2}{3} + \frac{3}{3^2} + \frac{4}{3^3} + \cdots\cdot$$

Use the ratio test to test the latter series for convergence

$$u_n = \frac{n}{3^n}, \qquad u_{n+1} = \frac{n+1}{3^{n+1}}, \qquad \frac{u_{n+1}}{u_n} = \frac{\dfrac{n+1}{3^{n+1}}}{\dfrac{n}{3^n}}$$

$$= \frac{n+1}{3^{n+1}} \times \frac{3^n}{n} = \frac{n+1}{3^n}, \quad \lim_{n \to \infty} \frac{n+1}{3n} = \frac{1}{3}$$

Since this limit is less than 1, the series formed from the original is convergent and the original series is *absolutely* convergent.

Example: Test the following series for absolute convergence.

$$\frac{1}{\sqrt[4]{2}} - \frac{1}{\sqrt[4]{4}} + \frac{1}{\sqrt[4]{6}} - \frac{1}{\sqrt[4]{8}} + \cdots\cdot$$

▶ The given series is

$$\frac{1}{\sqrt[4]{2}} - \frac{1}{\sqrt[4]{4}} + \frac{1}{\sqrt[4]{6}} - \frac{1}{\sqrt[4]{8}} + \cdots\cdot$$

Write the series with all terms positive.

$$\frac{1}{\sqrt[4]{2}} + \frac{1}{\sqrt[4]{4}} + \frac{1}{\sqrt[4]{6}} + \frac{1}{\sqrt[4]{8}} + \cdots\cdot$$

The latter series is divergent because it is a p series with $p < 1$ (in this case $p = \frac{1}{4}$).

The given series is convergent since each term is less than the preceding term,

$$u_n = \frac{1}{\sqrt[4]{n}}, \quad u_{n+1} = \frac{1}{\sqrt[4]{n+1}} \qquad \text{also,} \qquad \lim_{n \to \infty} \frac{1}{\sqrt[4]{n}} = 0$$

Therefore, the original series is conditionally convergent.

Example: Test the following series for absolute convergence.

$$1 - \frac{2}{2!} + \frac{4}{3!} - \frac{8}{4!} + \frac{16}{5!} - \cdots$$

▶ The given series is

$$1 - \frac{2}{2!} + \frac{4}{3!} - \frac{8}{4!} + \frac{16}{5!} - \cdots$$

Write the series with all terms positive.

$$1 + \frac{2}{2!} + \frac{4}{3!} + \frac{8}{4!} + \frac{16}{5!} + \cdots$$

$$u_n = \frac{2^{n-1}}{n!}, \qquad u_{n+1} = \frac{2^n}{(n+1)!}$$

$$\frac{u_{n+1}}{u_n} = \frac{\dfrac{2^n}{(n+1)!}}{\dfrac{2^{n-1}}{n!}} = \frac{2^n}{(n+1)!} \cdot \frac{n!}{2^{n-1}} = \frac{2}{n+1}$$

$$\lim_{n \to \infty} \frac{2}{n+1} = 0$$

Since this limit is less than 1, the series formed from the original is convergent and the original series is *absolutely* convergent.

Example: Test the following series for absolute convergence.

$$\frac{1}{2} - \frac{1}{4} + \frac{1}{6} - \frac{1}{8} + \cdots$$

▶ The given series is

$$\frac{1}{2} - \frac{1}{4} + \frac{1}{6} - \frac{1}{8} + \cdots$$

Write the series with all terms positive.

$$\frac{1}{2} + \frac{1}{4} + \frac{1}{6} + \frac{1}{8} + \cdots$$

The latter series can be written as

$$\frac{1}{2}\left(1 + \frac{1}{2} + \frac{1}{3} + \frac{1}{4} + \cdots\right)$$

Since the series in parentheses is the harmonic series, the latter series is divergent.

In the given series

$$u_n = \frac{1}{2n}, \qquad u_{n+1} = \frac{1}{2(n+1)}$$

In this series $u_n > u_{n+1}$. Also, $\lim\limits_{n \to \infty} \frac{1}{2n} = 0$.

Therefore, the given series is conditionally convergent.

Example: Test the following series for absolute convergence.

$$\frac{1}{1 \cdot 2} - \frac{1}{2 \cdot 3} + \frac{1}{3 \cdot 4} - \frac{1}{4 \cdot 5} + \cdots$$

▶ The given series is

$$\frac{1}{1 \cdot 2} - \frac{1}{2 \cdot 3} + \frac{1}{3 \cdot 4} - \frac{1}{4 \cdot 5} + \cdots$$

Write the series with all terms positive.

$$\frac{1}{1 \cdot 2} + \frac{1}{2 \cdot 3} + \frac{1}{3 \cdot 4} + \frac{1}{4 \cdot 5} + \cdots$$

The latter series is convergent since each of its terms is less than the corresponding term of the convergent geometric series.

$$\frac{1}{2} + \frac{1}{4} + \frac{1}{8} + \frac{1}{16} + \cdots + \frac{1}{2^n}$$

Hence, the original series is absolutely convergent.

EXERCISE 10

Test the following series for absolute or conditional convergence.

1. $1 - \dfrac{1}{2} + \dfrac{1}{4} - \dfrac{1}{8} + \cdots$

2. $1 - \dfrac{1}{\sqrt{2}} + \dfrac{1}{\sqrt{3}} - \dfrac{1}{\sqrt{4}} + \cdots$

3. $3 - \dfrac{3}{2} + \dfrac{3}{2^2} - \dfrac{3}{2^3} + \dfrac{3}{2^4} + \cdots$

4. $1 - \dfrac{1}{2!} + \dfrac{1}{3!} - \dfrac{1}{4!} + \cdots$

5. $5 - \dfrac{5}{2\sqrt{2}} + \dfrac{5}{3\sqrt{3}} - \dfrac{5}{4\sqrt{4}} + \cdots$

6. $1 - \dfrac{1}{2^2} + \dfrac{1}{3^2} - \dfrac{1}{4^2} + \cdots$

7. $3 - \dfrac{3^2}{2!} + \dfrac{3^3}{3!} - \dfrac{3^4}{4!} + \cdots$

8. $7 - \dfrac{7}{2} + \dfrac{7}{2^2} - \dfrac{7}{2^3} + \cdots$

9. $1 - \dfrac{1}{2^{1/3}} + \dfrac{1}{3^{1/3}} - \dfrac{1}{4^{1/3}} + \cdots$

10. $1 - \dfrac{1}{3^2} + \dfrac{1}{5^2} - \dfrac{1}{7^2}$

POWER SERIES

DEFINITION: A series of the form $a_0 + a_1 x + a_2 x^2 + \cdots + a_n x^n$ where x is a variable and the coefficients $a_0, a_1, a_2, \ldots$ are constants is called a power series in x.

Similarly, a series such as $a_0 + a_1(x - k) + a_2(x - k)^2 + a_3(x - k)^3 \ldots$ is a power series in $(x - k)$.

Examples:

$$1 + \frac{x}{3} + \frac{x^2}{3^2} + \frac{x^3}{3^3} + \cdots$$

$$1 + \frac{x-1}{2} + \frac{(x-1)^2}{3} + \frac{(x-1)^3}{4} + \cdots$$

If a numerical value is assigned to x, we have a series which may be convergent or divergent. Whether the series is convergent or divergent will usually depend upon the particular value of x assigned. The set of values for x for which a power series converges is called its *interval of convergence*. The interval of convergence is determined by using the ratio test.

Example: Find the interval of convergence of the series

$$1 + \frac{x}{2} + \frac{x^2}{3} + \cdots + \frac{x^{n-1}}{n}$$

▶ $$u_n = \frac{x^{n-1}}{n}, \qquad u_{n+1} = \frac{x^n}{n+1}.$$

$$\lim_{n \to \infty} \left| \frac{u_{n+1}}{u_n} \right| = \lim_{n \to \infty} \left| \frac{x^n}{n+1} \cdot \frac{n}{x^{n-1}} \right| = \lim_{n \to \infty} \left| \frac{nx}{n+1} \right|$$

$$= |x| \lim_{n \to \infty} \left| \frac{n}{n+1} \right| = |x| \cdot 1 = |x|.$$

The series converges for $|x| < 1$ or $-1 < x < 1$.

The ratio test fails when $|x| = 1$, i.e. when $x = +1$ or $x = -1$. For $x = 1$, the series is

$$1 + \frac{1}{2} + \frac{1}{3} + \frac{1}{4} + \cdots + \frac{1}{n}$$

This is the harmonic series which is known to be divergent. For $x = -1$, the series is

$$1 - \frac{1}{2} + \frac{1}{3} - \frac{1}{4} + \frac{1}{5} + \cdots + \frac{(-1)^{n+1}}{n}$$

This is an alternating series which is convergent because the absolute value of each term is less than that of the preceding term and the nth term has the limit zero as n increases without bound.

Thus, the given series is convergent in the interval beginning with -1 and extending to, but not including $+1$. This may be shown as follows

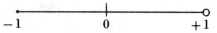

The heavy line shows the interval of convergence. The open circle at $+1$ indicates that $+1$ is not included in the interval of convergence.

Example: Find the interval of convergence of the series

$$1 + x + 2x^2 + 3x^3 + \ldots + (n - 1)x^{n-1} + \ldots$$

▶ $u_n = (n - 1)x^{n-1}, \qquad u_{n+1} = nx^n$

$$\lim_{n \to \infty} \left| \frac{u_{n+1}}{u_n} \right| = \lim_{n \to \infty} \frac{nx^n}{(n - 1)x^{n-1}} = \lim_{n \to \infty} \frac{nx}{n - 1}$$

$$= |x| \lim_{n \to \infty} \frac{n}{n - 1} = |x|.$$

The series converges for $|x| < 1$ and diverges for $|x| > 1$.

For $x = 1$, the series is $1 + 1 + 2 + 3 + \ldots$ and is divergent.

For $x = -1$, the series is $1 - 1 + 2 - 3 + 4 - 5 \ldots$ ard is divergent.

Thus, the given series is convergent in the interval between -1 and $+1$. The values -1 and $+1$ are not in the int rval of convergence. This may be shown as follows.

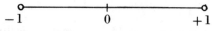

The heavy line shows the interval of convergence. The open circles at -1 and $+1$ indicate that these values are not included in the interval of convergence.

Example: Find the interval of convergence of the series,

$$1 + \frac{x}{\sqrt[3]{2}} + \frac{x^2}{\sqrt[3]{3}} + \ldots + \frac{x^{n-1}}{\sqrt[3]{n}}$$

▶ $u_n = \dfrac{x^{n-1}}{\sqrt[3]{n}}, \qquad u_{n+1} = \dfrac{x^n}{\sqrt[3]{n+1}}$

$$\lim_{n \to \infty} \frac{x^n}{\sqrt[3]{n+1}} \cdot \frac{\sqrt[3]{n}}{x^{n-1}} = \lim_{n \to \infty} \frac{\sqrt[3]{n}\, x}{\sqrt[3]{n+1}}$$

$$= |x| \lim_{n \to \infty} \frac{\sqrt[3]{n}}{\sqrt[3]{n+1}} = |x| \cdot 1 = |x|$$

The series is convergent for $|x| < 1$ and divergent for $|x| > 1$. For $x = 1$, the series is

$$1 + \frac{1}{\sqrt[3]{2}} + \frac{1}{\sqrt[3]{3}} + \frac{1}{\sqrt[3]{4}} + \ldots .$$

This is a p-series with $p < 1$ and is divergent.
For $x = -1$, the series is

$$1 - \frac{1}{\sqrt[3]{2}} + \frac{1}{\sqrt[3]{3}} - \frac{1}{\sqrt[3]{4}} + \ldots .$$

This is an alternating series and is convergent because the absolute value of each term is less than that of the preceding term and the nth term has the limit zero as n increases without bound.

Thus, the given series is convergent in the interval beginning with -1 and extending to, but not including $+1$. This may be shown as follows.

Example: Find the interval of convergence of the series

$$x - \frac{x^3}{3!} + \frac{x^5}{5!} - \frac{x^7}{7!} + \ldots .$$

▶ $u_n = \dfrac{x^{2n-1}}{(2n-1)!}, \qquad u_{n+1} = \dfrac{x^{2n}}{(2n)!}$

$$\lim_{n \to \infty} \left| \frac{x^{2n}}{(2n)!} \cdot \frac{(2n-1)!}{x^{2n-1}} \right| = \lim_{n \to \infty} \left| \frac{x}{2n} \right|$$

$$= |x| \lim_{n \to \infty} \left| \frac{1}{n} \right| = 0.$$

Since $R = 0$ for all values of x, the series is convergent for all values of x.

Example: Find the interval of convergence of the series

$$1 + \frac{x}{3} + \frac{x^2}{3^2} + \frac{x^3}{3^3} + \ldots.$$

▶ $\quad u_n = \frac{x^{n-1}}{3^{n-1}}, \qquad u_{n+1} = \frac{x^n}{3^n}$

$$\lim_{n \to \infty} \frac{x^n}{3^n} \cdot \frac{3^{n-1}}{x^{n-1}} = \lim_{n \to \infty} \frac{x}{3} = \frac{1}{3}|x|$$

The series converges for $\frac{1}{3}|x| < 1$ or $-3 < x < 3$.

For $x = -3$, the series is $1 - 1 + 1 - 1 + \ldots$ which is divergent. For $x = +3$, the series is $1 + 1 + 1 + 1 + 1 + \ldots$ which is divergent. Thus, the series is convergent in the interval from -3 to $+3$, not including the values -3 and $+3$. This may be shown as follows.

Example: Find the interval of convergence of the series

$$1 + \frac{x - 1}{2} + \frac{(x - 1)^2}{3} + \frac{(x - 2)^2}{4} + \ldots.$$

▶ $\quad u_n = \frac{(x - 1)^{n-1}}{n}, \qquad u_{n+1} = \frac{(x - 1)^n}{n + 1}$

$$\lim_{n \to \infty} \left| \frac{(x - 1)^n}{n + 1} \cdot \frac{n}{(x - 1)^{n-1}} \right| = \lim_{n \to \infty} \left| \frac{(x - 1)n}{n + 1} \right|$$

$$= |x - 1| \lim_{n \to \infty} \frac{n}{n + 1} = |x - 1|.$$

The series converges for $|x - 1| < 1$ or $0 < x < 2$. For $x = 0$, the series is

$$1 - \frac{1}{2} + \frac{1}{3} - \frac{1}{4} + \ldots.$$

and is convergent

For $x = 2$, the series is

$$1 + \frac{1}{2} + \frac{1}{3} + \frac{1}{4} + \cdots.$$

and is divergent (the harmonic series).

Thus, the series is convergent in the interval beginning with 0 and extending to, but not including 2 i.e., the interval of convergence is $0 \leqq x < 2$. This may be shown as follows.

$$\begin{array}{ccc} 0 & +1 & +2 \end{array}$$

EXERCISE 11

Find the interval of convergence for each of the following series.

1. $1 + \dfrac{x}{1!} + \dfrac{x^2}{2!} + \dfrac{x^3}{3!} + \cdots$

2. $1 - x + x^2 - x^3 + \cdots$

3. $x - \dfrac{x^2}{2} + \dfrac{x^3}{3} - \dfrac{x^4}{4} + \cdots$

4. $\dfrac{x}{1 \cdot 2} + \dfrac{x^2}{2 \cdot 3} + \dfrac{x^3}{3 \cdot 4} + \dfrac{x^4}{4 \cdot 5} + \cdots$

5. $1 - \dfrac{x}{5} + \dfrac{x^2}{5^2} - \dfrac{x^3}{5^3} + \cdots$

6. $1 - \dfrac{x^2}{2!} + \dfrac{x^4}{4!} - \dfrac{x^6}{6!} + \cdots$

7. $(x - 1) + 2(x - 1)^2 + 3(x - 1)^3 + 4(x - 1)^4 + \cdots$

8. $1 + \dfrac{x}{\sqrt{2}} + \dfrac{x^2}{\sqrt{3}} + \dfrac{x^3}{\sqrt{4}} + \cdots$

9. $1 - \dfrac{2x}{3} + \dfrac{3x^2}{9} - \dfrac{4x^3}{27} + \cdots$

10. $(x - 2) + \frac{1}{4}(x - 2)^2 + \frac{1}{9}(x - 2)^3 + \frac{1}{16}(x - 2)^4 + \cdots$

ANSWERS

EXERCISE 1

1. $\frac{1}{2}, \frac{2}{7}, \frac{1}{5}, \frac{2}{13}, \frac{1}{8}$

2. $1, \frac{3}{4}, \frac{5}{9}, \frac{7}{16}, \frac{9}{25}$

3. $\frac{1}{2}, \frac{4}{7}, \frac{2}{3}, \frac{16}{19}, \frac{8}{7}$

4. $1, \frac{3}{8}, \frac{2}{13}, \frac{1}{16}, \frac{3}{121}$

5. $\frac{1}{6}, \frac{2}{13}, \frac{1}{8}, \frac{8}{69}, \frac{8}{65}$

6. $\frac{1}{2}, \frac{5}{8}, \frac{5}{8}, \frac{17}{32}, \frac{13}{32}$

EXERCISE 2

1. $2n$

2. $\dfrac{1}{n+1}$

3. $\dfrac{n}{n+1}$

4. $\dfrac{1}{n(n+1)}$

5. $\dfrac{n^2}{n+3}$

6. $\dfrac{n^2}{2n-1}$

7. $\dfrac{2^n}{2n+1}$

8. $\dfrac{3^n}{2n}$

EXERCISE 3

1. $\frac{1}{6} + \frac{2}{7} + \frac{3}{8} + \frac{4}{9}$

2. $\frac{1}{2} + \frac{2}{5} + \frac{3}{8} + \frac{4}{11}$

3. $1 + \frac{4}{3} + 2 + \frac{16}{5}$

4. $\frac{1}{4} + \frac{4}{7} + \frac{9}{10} + \frac{16}{5}$

5. $\frac{1}{3} + \frac{1}{8} + \frac{1}{15} + \frac{1}{24}$

6. $\frac{1}{8} + \frac{2}{15} + \frac{1}{8} + \frac{4}{35}$

EXERCISE 4

1. $5n - 3, 5n + 2$

2. $\dfrac{1}{5n}, \dfrac{1}{5(n+1)}$

3. $\dfrac{n+2}{n^2}, \dfrac{n+3}{(n+2)(n+3)}$

4. $\dfrac{n+2}{n^2}, \dfrac{n+3}{(n+1)^2}$

5. $\dfrac{x^{2n+1}}{(n+1)!}, \dfrac{x^{2n+3}}{(n+2)!}$

6. $\dfrac{x^n}{n(n+1)}, \dfrac{x^{n+1}}{(n+1)(n+2)}$

7. $5 + 8 + 11 + 14$

8. $2 + \frac{3}{2} + \frac{2}{3} + \frac{5}{24}$

9. $1 + \frac{2}{3} + \frac{3}{7} + \frac{4}{15}$

EXERCISE 5

1. $\frac{3}{2}$
2. 1
3. 0
4. 0

5. no limit
6. 1
7. 0
8. no limit

EXERCISE 6

1. $\lim\limits_{n \to \infty} \dfrac{n+2}{n+3} = 1$

2. $\lim\limits_{n \to \infty} \dfrac{n^2}{n^2+3} = 1$

3. $\lim\limits_{n \to \infty} \dfrac{n^2+1}{3n}$ does not exist

4. $\lim\limits_{n \to \infty} \dfrac{n^3}{2n+1}$ does not exist

EXERCISE 7

1. convergent
2. divergent
3. convergent
4. convergent
5. divergent
6. convergent
7. convergent
8. divergent

9. divergent
10. divergent
11. divergent
12. convergent
13. convergent
14. divergent
15. divergent

EXERCISE 8

1. convergent
2. divergent
3. divergent
4. convergent
5. $R = 1$, convergent
6. divergent

7. convergent
8. convergent
9. divergent
10. $R = 1$, convergent
11. divergent
12. convergent

EXERCISE 9

1. convergent
2. divergent
3. convergent

4. convergent
5. convergent
6. divergent

EXERCISE 10

1. absolute convergence
2. conditional convergence
3. absolute convergence
4. absolute convergence
5. absolute convergence

6. absolute convergence
7. absolute convergence
8. absolute convergence
9. conditional convergence
10. absolute convergence

EXERCISE 11

1. all values of x
2. $-1 < x < 1$
3. $-1 < x \leqq 1$
4. $-1 \leqq x \leqq 1$
5. $-5 < x < 5$

6. all values of x
7. $0 < x < 2$
8. $-1 \leqq x < 1$
9. $-3 < x < 3$
10. $-1 \leqq x \leqq 3$

Typical Final Examinations

Answer Any Seven Questions

1. Solve for x: (a) $x + \sqrt{x + 6} = 0$

(b) $\dfrac{x + 3}{x - 1} + \dfrac{x + 4}{x + 1} = \dfrac{8x + 5}{x^2 - 1}$

2. Solve for x by "completing the square": $2x^2 + 4x - 3 = 0$.

3. Solve for x and y: $\begin{cases} x^2 + y^2 = 13 \\ xy = 6 \end{cases}$

4. (a) Form a quadratic equation with real coefficients having $2 - i$ as one root.

(b) Find the values of k for which the roots of the following equation are equal:

$$5x^2 - x + 6 = kx$$

5. Find and plot the cube roots of $-4\sqrt{2} + 4\sqrt{2}\, i$.

Express your answers in the form $a + bi$, if possible.

6. Calculate $(1.02)^9$ correct to three decimal places by means of the binomial theorem.

7. Find the roots of $4x^4 - 15x^2 - 5x + 6 = 0$.

8. Express y as the quotient of two determinants and then evaluate the determinant occurring in the denominator:

$$2x + y + 3z = -2$$
$$5x + 3y - z - w = \ \ \ 0$$
$$x - 2y + 4z + 3w = \ \ \ 5$$
$$3x - y + z = \ \ \ 3$$

9. (a) A bowl contains four blue, two red and three white chips. Three are drawn at random. What is the probability that the drawing will contain exactly one of each color?

   (b) From the digits 1, 2, 3, 4, 7, 9, how many odd numbers can be formed, each consisting of three digits? No digit is to be used twice in the same number.

EXAMINATION 1—SOLUTIONS

1. (a) $x + \sqrt{x + 6} = 0$
$$\sqrt{x + 6} = -x$$
Squaring both members, $x + 6 = x^2$
$$x^2 - x - 6 = 0$$
$$(x - 3)(x + 2) = 0$$
$$x = 3, \quad x = -2$$

Check for x = 3
$$3 + \sqrt{3 + 6} = 0$$
$$3 + \sqrt{9} = 0$$
$$3 + 3 = 0$$

Since $x = 3$ does not check it is an extraneous root.

Check for x = -2
$$-2 + \sqrt{-2 + 6} = 0$$
$$-2 + \sqrt{4} = 0$$
$$-2 + 2 = 0$$

Therefore, -2 is a root.

(b) $\dfrac{x + 3}{x - 1} + \dfrac{x + 4}{x + 1} = \dfrac{8x + 5}{x^2 - 1}$

Multiply both members by $(x^2 - 1)$
$$(x + 3)(x + 1) + (x + 4)(x - 1) = 8x + 5$$
$$x^2 + 4x + 3 + x^2 + 3x - 4 = 8x + 5$$
$$2x^2 - x - 6 = 0$$
$$(2x + 3)(x - 2) = 0$$
$$2x + 3 = 0, \quad x - 2 = 0$$
$$x = -\frac{3}{2}, \quad x = 2$$

2. $2x^2 + 4x - 3 = 0$

Divide both members by 2,

$$x^2 + 2x = \frac{3}{2}$$

Complete the square by adding one-half the coefficient of the x-term squared to both members,

$$x^2 + 2x + 1 = \frac{3}{2} + 1 = \frac{5}{2}$$

$$(x + 1)^2 = \frac{5}{2}$$

$$x + 1 = \pm\sqrt{\frac{5}{2}}$$

$$x = -1 \pm\sqrt{\frac{5}{2}} = -1 \pm\sqrt{\frac{5}{2}\cdot\frac{2}{2}}$$

$$x = -1 \pm\frac{1}{2}\sqrt{10}$$

3. $x^2 + y^2 = 13$
$\qquad xy = 6$

$$x^2 + y^2 = 13$$

Multiply both members of the second equation by 2,

$$2xy = 12$$

Add the two equations $x^2 + 2xy + y^2 = 25$

$$(x + y)^2 = 25$$

$$x + y = \pm 5$$

$x + y = +5$	$x + y = -5$
$x = 5 - y$	$x = -5 - y$

Substitute in the second equation

$y(5 - y) = 6$	$y(-5 - y) = 6$
$5y - y^2 = 6$	$-5y - y^2 = 6$
$y^2 - 5y + 6 = 0$	$y^2 + 5y + 6 = 0$
$(y - 3)(y - 2) = 0$	$(y + 3)(y + 2) = 0$
$y = 3, \quad y = 2$	$y = -3, \quad y = -2$
$x = 5 - y, \quad x = 2, \quad x = 3$	$x = -5 - y, \quad x = -2, \quad x = -3$

The roots may be grouped as follows:

x	2	3	-2	-3
y	3	2	-3	-2

4. (a) Since $(2 - i)$ is a root of the equation, then $(2 + i)$ is also a root. The equation of a quadratic equation may be written as

$x^2 +$ (sum of roots with sign changed) $x +$ product of roots $= 0$

$x^2 - (2 + i + 2 - i)x + (2 + i)(2 - i) = 0$

$x^2 - 4x + 5 = 0$

(b) $5x^2 - x + 6 = kx$

$5x^2 - x - k + 6 = 0$

$5x^2 - (1 + k)x + 6 = 0$

If the roots are to be equal then the discriminant $(b^2 - 4ac)$ of the quadratic equation must be equal to zero.

$[-(1 + k)]^2 - 4(5)(6) = 0$

$1 + 2k + k^2 - 120 = 0$

$k^2 + 2k - 119 = 0$

$k = \dfrac{-2 \pm \sqrt{4 - 4(1)(-119)}}{2}; \quad k = \dfrac{-2 \pm \sqrt{480}}{2} = \dfrac{-2 \pm 4\sqrt{30}}{2}$

$k = -1 \pm 2\sqrt{30}$

5. $-4\sqrt{2} + 4\sqrt{2}i$

$r^2 = (-4\sqrt{2})^2 + (4\sqrt{2})^2$

$r^2 = 32 + 32 = 64, \quad r = 8$

$-4\sqrt{2} + 4\sqrt{2}\,i = 8(\cos 135° + i\sin 135°)$

$= 8[\cos (135° + 360°) + i\sin (135° + 360°)]$

or $8(\cos 495° + i\sin 495°)$

$= 8[\cos (135° + 720°) + i\sin (135° + 720°)$

or $8(\cos 855° + i\sin 855°)$

We may find the cube roots of $-4\sqrt{2} + 4\sqrt{2}\,i$ by using De Moivre's Theorem as follows:

$[8(\cos 135° + i\sin 135°)]^{1/3} = 8^{1/3}(\cos 45° + i\sin 45°)$

$= 2(\cos 45° + i\sin 45°)$

$= 2\left(\dfrac{\sqrt{2}}{2} + \dfrac{\sqrt{2}}{2}\,i\right) = \sqrt{2} + \sqrt{2}\,i = 1.4 + 1.4i$

$[8(\cos 495° + i \sin 495°)]^{1/3} = 2(\cos 165° + i \sin 165°)$
$= 2(-\cos 15° + i \sin 15°)$
$= 2[(-.97) + i(.26)] = -1.94 + .52i$
$[8(\cos 855° + i \sin 855°)]^{1/3} = 2(\cos 285° + i \sin 285°)$
$= 2(\cos 75° - i \sin 75°) = 2[(.26) - i(.97)]$
$= .52 - 1.94i.$

Note: The values of $\cos 15°$, $\sin 15°$, $\cos 75°$, $\sin 75°$ were obtained from a table.

The graph shows the roots $2(\cos 45° + i \sin 45°)$,
$2(\cos 165° + i \sin 165°)$, and $2(\cos 285° + i \sin 285°)$.

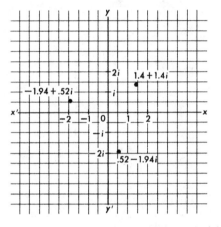

6. $(1.02)^9 = (1 + .02)^9 = (1)^9 + {}_9C_1 (1)^8(.02) + {}_9C_2 (1)^7(.02)^2$
   $+ {}_9C_3 (1)^6(.02)^3 = 1 + 9(.02) + 36(.0004) + 84(.000008)$
   $= 1 + .18 + .0144 + .000672 = 1.195072$
   $= 1.195$ correct to 3 decimal places.

7. $4x^4 - 15x^2 - 5x + 6 = 0$

   All rational roots of the equation are of the form $\dfrac{p}{q}$ where p is a factor of 6 and q is a factor of 4. The possible rational roots are included in the table below.

p	± 1	± 2	± 3	± 6
q	± 1	± 2	± 4	

$$
\begin{array}{rrrrrr}
4 & + \; 0 & - \; 15 & - \; 5 & + \; 6 & \quad\underline{|-1} \\
  & - \; 4 & + \; 4 & + \; 11 & - \; 6 & \\
\hline
4 & - \; 4 & - \; 11 & + \; 6 & \quad\underline{|0}
\end{array}
\qquad
\begin{array}{rrrrr}
4 & - \; 4 & - \; 11 & + \; 6 & \quad\underline{|\tfrac{1}{2}} \\
  & + \; 2 & - \; 1 & - \; 6 & \\
\hline
4 & - \; 2 & - \; 12 & \quad\underline{|0}
\end{array}
$$

Therefore, -1 is a root   Therefore, $\tfrac{1}{2}$ is a root.

The depressed equation is $4x^2 - 2x - 12 = 0$ or $2x^2 - x - 6 = 0$.

$(2x + 3)(x - 2) = 0$

$x = -\tfrac{3}{2}, \qquad x = 2$

The four roots are -1, $\tfrac{1}{2}$, $-\tfrac{3}{2}$, and 2.

8.

$$
y = \frac{
\begin{array}{rrrr}
2 & -2 & 3 & 0 \\
5 & 0 & -1 & -1 \\
1 & 5 & 4 & 3 \\
3 & 3 & 1 & 0
\end{array}
}{
\begin{array}{rrrr}
2 & 1 & 3 & 0 \\
5 & 3 & -1 & -1 \\
1 & -2 & 4 & 3 \\
3 & -1 & 1 & 0
\end{array}
}
$$

$$
\begin{array}{rrrr}
2 & 1 & 3 & 0 \\
5 & 3 & -1 & -1 \\
1 & -2 & 4 & 3 \\
3 & -1 & 1 & 0
\end{array}
$$

Multiply row 2 by 3 and add the result to row 3.

$$
\begin{array}{rrrr}
2 & 1 & 3 & 0 \\
5 & 3 & -1 & -1 \\
16 & 7 & 1 & 0 \\
3 & -1 & 1 & 0
\end{array}
$$

Develop the determinant by minors using column 4.

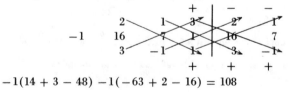

$-1(14 + 3 - 48) \, -1(-63 + 2 - 16) = 108$

9. (a) The number of ways in which one ball of each color may be drawn is
$$_4C_1 \cdot {}_2C_1 \cdot {}_3C_1 = 4 \cdot 2 \cdot 3 = 24$$
The number of ways in which 3 balls may be drawn is
$$_9C_3 = \frac{9 \cdot 8 \cdot 7}{1 \cdot 2 \cdot 3} = 84$$

Therefore, the probability of drawing three balls, exactly one of each color is
$$\frac{24}{84} = \frac{2}{7}.$$

(b) The digits are 1, 2, 3, 4, 7, 9. If the desired numbers are to be odd then the first digit on the right may be chosen in 4 ways (i.e., from the digits 1, 3, 7, and 9). The choice of the other two digits may be made in 5 ways and in 4 ways respectively.

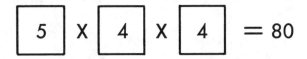

Therefore, 80 odd numbers of 3 digits may be formed.

EXAMINATION 2

Omit One Question

1. Find the four fourth roots of -81 and represent the results graphically.

2. (a) How many four digit numbers can be formed from the seven digits 0, 1, 2, 3, 4, 5, 6 if repetition of digits be not allowed.

(b) Find the fifth term only of $\left(3x - \dfrac{1}{\sqrt{x}} \right)^{13}$

3. (a) Reduce $\dfrac{1-i}{1+i}$ to the standard form $a + ib$.

(b) Given that $\dfrac{-1 + i\sqrt{3}}{2}$ is a root of $x^3 = 1$, what are the other roots?

4. One bag contains 4 white and 6 black balls and a second bag contains 5 white and 9 black balls. If a ball is drawn from each bag, what is the probability that they will be

(a) both white?

(b) the same color?

5. (a) The difference of the roots of $x^2 + 7x + h = 0$ is 3. Find both roots.

(b) Solve for x: $\sqrt{3 - 2x} + \sqrt{1 - x} = 1$.

6. Find all the roots of $2x^4 + 3x^3 + 3x - 2 = 0$.

7. Evaluate

2	5	0	3
3	−2	1	3
1	−1	2	2
4	3	−5	3

EXAMINATION 2—SOLUTIONS

1. $-81 = 81(\cos 180° + i \sin 180°)$

$(-81)^{1/4} = [81(\cos 180° + i \sin 180°)]^{1/4}$

Root 1 $= 3(\cos 45° + i \sin 45°) = \frac{3}{2}\sqrt{2} + \frac{3}{2}i\sqrt{2}$

$-81 = 81[\cos(180° + 360°) + i \sin(180° + 360°)]$

$\qquad = 81(\cos 540° + i \sin 540°)$

$(-81)^{1/4} = [81(\cos 540° + i \sin 540°)]^{1/4}$

Root 2 $= 3(\cos 135° + i \sin 135°) = -\frac{3}{2}\sqrt{2} + \frac{3}{2}i\sqrt{2}$

$-81 = 81[\cos(180° + 720°) + i \sin(180° + 720°)]$

$\qquad = 81(\cos 900° + i \sin 900°)$

$(-81)^{1/4} = [81(\cos 900° + i \sin 900°)]^{1/4}$

Root 3 $= 3(\cos 225° + i \sin 225°) = -\frac{3}{2}\sqrt{2} - \frac{3}{2}i\sqrt{2}$

$-81 = 81[\cos(180° + 1080°) + i \sin(180° + 1080°)]$

$\qquad = 81(\cos 1260° + i \sin 1260°)$

$(-81)^{1/4} = [81(\cos 1260° + i \sin 1260°)]^{1/4}$

Root 4 $= 3(\cos 315° + i \sin 315°) = \frac{3}{2}\sqrt{2} - \frac{3}{2}i\sqrt{2}.$

$\boxed{6} \; X \; \boxed{6} \; X \; \boxed{5} \; X \; \boxed{4} \; = 720 \text{ number}$

2. (a) The first box on the left may be filled by 6 of the given digits (0 may not be used in this box). The second box from the left may be filled by 6 of the given digits (0 may now be used). The third box from the left may be filled by any one of the 5 digits left. Similarly, the last box may be filled by any one of the 4 digits left.

(b) $\left(3x - \dfrac{1}{\sqrt{x}} \right)^{13}$

$r + 1 = 5, \quad$ and $\quad r = 4$

$_nC_r \, a^{n-r} \, b^r = 13C_4 \, (3x)^9 \left(-\dfrac{1}{\sqrt{x}} \right)^4$

$= (715)(19,683x^9) \left(\dfrac{1}{x^2} \right) = 14,073,345x^7$

3. (a) $\dfrac{1-i}{1+i} = \dfrac{1-i}{1+i} \cdot \dfrac{1-i}{1-i} = \dfrac{1 - 2i + i^2}{1 - i^2} = \dfrac{-2i}{2} = -i$

(b) If $\dfrac{-1 + i\sqrt{3}}{2}$ is a root of $x^3 = 1$ then $\dfrac{-1 - i\sqrt{3}}{2}$ is a root of $x^3 = 1$

or $x^3 - 1 = 0$.

The sum of the roots of $x^3 - 1 = 0$ is 0.

Let the third root of the equation be a.

Then $\dfrac{-1 + i\sqrt{3}}{2} + \dfrac{-1 - i\sqrt{3}}{2} + a = 0$

$-1 + a = 0, \quad$ or $\quad a = 1$.

4. (a) The probability of drawing a white ball from the first bag is $\frac{4}{10}$. The probability of drawing a white ball from the second bag is $\frac{5}{14}$. The probability of drawing 2 white balls is $\frac{4}{10} \times \frac{5}{14} = \frac{1}{7}$.

(b) The probability of drawing a black ball from the first bag is $\frac{6}{10}$. The probability of drawing a black ball from the second bag is $\frac{9}{14}$. The probability of drawing 2 black balls is $\frac{6}{10} \times \frac{9}{14} = \frac{27}{70}$.

The probability of drawing 2 white balls or 2 black balls is

$\frac{1}{7} + \frac{27}{70} = \frac{37}{70}$.

5. (a) Let $a = $ the smaller root.

$a + 3 = $ the larger root.

The sum of the roots of the equation is -7, or

$a + a + 3 = -7, \quad a = -5, \quad a + 3 = -2$

The roots are -5, and -2.

(b) $\sqrt{3 - 2x} + \sqrt{1 - x} = 1$

$$\sqrt{1 - x} = 1 - \sqrt{3 - 2x}$$

Squaring both members of the equation,

$$1 - x = 1 - 2\sqrt{3 - 2x} + 3 - 2x$$
$$x - 3 = -2\sqrt{3 - 2x}$$

Squaring both members of the equation,

$$x^2 - 6x + 9 = 4(3 - 2x)$$
$$x^2 - 6x + 9 = 12 - 8x$$
$$x^2 + 2x - 3 = 0$$
$$(x + 3)(x - 1) = 0$$
$$x = -3 \qquad x = 1$$

Check for $x = -3$

$$\sqrt{3 - (2)(-3)} + \sqrt{-3 - 1} = 1$$
$$\sqrt{9} + \sqrt{-4} = 1$$

Since $x = -3$ does not check it is an *extraneous* root.

Check for $x = 1$

$$\sqrt{3 - (2)(1)} + \sqrt{1 - 1} = 1$$
$$\sqrt{1} + 0 = 1$$
$$1 = 1$$

Therefore, the equation has only one root, $x = 1$.

7.
$$\begin{vmatrix} 2 & 5 & 0 & 3 \\ 3 & -2 & 1 & 3 \\ 1 & -1 & 2 & 2 \\ 4 & 3 & -5 & 3 \end{vmatrix}$$

Add column 2 to column 1

$$\begin{vmatrix} 7 & 5 & 0 & 3 \\ 1 & -2 & 1 & 3 \\ 0 & -1 & 2 & 2 \\ 7 & 3 & -5 & 3 \end{vmatrix}$$

Multiply column 2 by 2 and add column 2 to column 3 and then to column 4.

$$\begin{vmatrix} 7 & 5 & 10 & 13 \\ 1 & -2 & -3 & -1 \\ 0 & -1 & 0 & 0 \\ 7 & 3 & 1 & 9 \end{vmatrix}$$

Develop by minors using row 3.

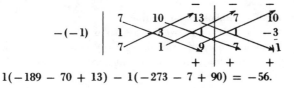

$$1(-189 - 70 + 13) - 1(-273 - 7 + 90) = -56.$$

EXAMINATION 3

Group I—*Omit one question*

1. Solve for x only, using determinants and simplifying your answer.
$$ax + y = a + 1$$
$$x + ay = a^2 + \frac{1}{a}$$

2. Solve for x and check your answers: $\sqrt{2 - x} + \sqrt{4 - 3x} = 2$.

3. Determine the values of x satisfying the inequality
$$b - x < \frac{c}{a} + 2x \qquad (a, b, \text{ and } c \text{ are positive})$$

4. The equation $2x^4 - 4x^3 - 5x^2 - 2x - 3 = 0$ has two rational roots. Find all of the roots.

5. Evaluate
   (a) $_{25}C_3$ and $_6P_3$
   (b) Write down only the 10th term in the expansive of $(2z - b)^{10}$ and simplify.

6. Evaluate the determinant:
$$\begin{vmatrix} 1 & 2 & 3 & 0 \\ 2 & -3 & 0 & 0 \\ 1 & 3 & 1 & 1 \\ 0 & 2 & 3 & -1 \end{vmatrix}$$

7. Determine one cube root of the complex number $-2 + \sqrt{-4}$ in simplified form.

8. Solve simultaneously:
$$x^2 + y^2 - 4y - 5 = 1$$
$$x - 2y = 3$$

Group II—*Omit one question.*

9. (a) How many different sums of money can be made from a penny, a nickel, a dime, a quarter and a half dollar, using as many of the coins at a time as you wish?

   (b) What is the probability that in a row of five seats a certain two people seated at random will be side by side?

10. Find the range of values of k for which the straight line $y = kx - 2$
  (a) will be tangent to the parabola $x^2 = 2y$
  (b) will meet the parabola in no points
  (c) will meet the parabola in two distinct points

11. (a) By considering Descartes' rule of signs and other information obtained without solving, draw conclusions as to the nature of the three roots of $x^3 + 3x - 8 = 0$

  (b) Find the approximate real root of $x^3 + 3x - 8 = 0$ graphically to one decimal place, making use of synthetic substitution.

EXAMINATION 3—SOLUTIONS

1. $ax + y = a + 1$
  $x + ay = a^2 + \dfrac{1}{a}$

$$x = \frac{\begin{vmatrix} a + 1 & 1 \\ a^2 + \dfrac{1}{a} & a \end{vmatrix}}{\begin{vmatrix} a & 1 \\ 1 & a \end{vmatrix}} = \frac{a(a + 1) - 1\left(a^2 + \dfrac{1}{a}\right)}{a^2 - 1}$$

$$= \frac{a^2 + a - a^2 - \dfrac{1}{a}}{a^2 - 1}, = \frac{a - \dfrac{1}{a}}{a^2 - 1} = \frac{\dfrac{a^2 - 1}{a}}{a^2 - 1} = \frac{1}{a}$$

2. $\sqrt{2 - x} + \sqrt{4 - 3x} = 2$
  $\sqrt{4 - 3x} = 2 - \sqrt{2 - x}$

Squaring both members, $4 - 3x = 4 - 4\sqrt{2 - x} + 2 - x$
  $-2x - 2x = -4\sqrt{2 - x}$

Dividing both members by -2, $x + 1 = 2\sqrt{2 - x}$

Squaring both members, $x^2 + 2x + 1 = 8 - 4x$
  $x^2 + 6x - 7 = 0$
  $(x + 7)(x - 1) = 0$
  $x = -7, \quad x = 1$

Check for $x = -7$
$$\sqrt{2 - (-7)} + \sqrt{4 - 3(-7)} = 2$$
$$\sqrt{9} + \sqrt{25} = 2$$
$$3 + 5 = 2$$

  Since $x = -7$ does not check, it is an extraneous root.
Check for $x = 1$
$$\sqrt{2 - 1} + \sqrt{4 - 3} = 2$$
$$\sqrt{1} + \sqrt{1} = 2$$
$$1 + 1 = 2$$

  Therefore, $x = 1$ is the root of the equation.

3. $b - x < \dfrac{c}{a} + 2x$

This inequality may be written as $2x + \dfrac{c}{a} > b - x$

$$2x + x > b - \frac{c}{a}$$

$$3x > \frac{ab - c}{a}$$

$$x > \frac{ab - c}{3a}$$

4. $2x^4 - 4x^3 - 5x^2 - 2x - 3 = 0$

All the rational roots are of the form $\dfrac{p}{q}$ where p is a factor of -3 and q is a factor of 2. The possible rational roots are included in the table below.

p	$+1$	-1	$+3$	-3
q	$+1$	-1	$+2$	-2

$$
\begin{array}{rrrrr|l}
2 & -4 & -5 & -2 & -3 & \underline{-1} \\
  & -2 & +6 & -1 & +3 & \\
\hline
2 & -6 & +1 & -3 & \underline{0} &
\end{array}
$$

Therefore -1 is a root.

$$
\begin{array}{rrrr|l}
2 & -6 & +1 & -3 & \underline{3} \\
  & +6 & +0 & +3 & \\
\hline
2 & +0 & +1 & \underline{0} &
\end{array}
$$

Therefore 3 is a root.

$$2x^2 + 1 = 0$$

$$x^2 = -\frac{1}{2}$$

$$x = \pm i \sqrt{\frac{1}{2}}$$

5. (a) $_{25}C_{23} = {}_{25}C_2 = \dfrac{25 \cdot 24}{1 \cdot 2} = 300$

$_6P_3 = 6 \cdot 5 \cdot 4 = 120$

(b) $(2z - b)^{10}$

$\quad r + 1 = 10, \quad r = 9$

$\quad _nC_r a^{n-r} b^r$

$\quad _{10}C_9(2z)^1(-b)^9 = (10)(2z)(-b^9) = -20zb^9$

6.

$$\begin{vmatrix} 1 & 2 & 3 & 0 \\ 2 & -3 & 0 & 0 \\ 1 & 3 & 1 & 1 \\ 0 & 2 & 3 & -1 \end{vmatrix}$$

Add row 4 to row 3

$$\begin{vmatrix} 1 & 2 & 3 & 0 \\ 2 & -3 & 0 & 0 \\ 1 & 5 & 4 & 0 \\ 0 & 2 & 3 & -1 \end{vmatrix}$$

Develop by minors using row 4. Since -1 is in row 4 and column 4, we have

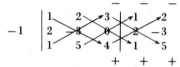

$$-1(-12 + 0 + 30) - 1(+9 + 0 - 16) = -18 + 7 = -11$$

7. $-2 + \sqrt{-4} = -2 + \sqrt{4}\,i = -2 + 2i$

$\qquad r^2 = (+2)^2 + (-2)^2$

$\qquad r^2 = 4 + 4 = 8$

$\qquad r = \sqrt{8}$

$\qquad \theta = 135°$

$-2 + 2i = \sqrt{8}(\cos 135° + i \sin 135°) = 8^{1/2}(\cos 135° + i \sin 135°)$.

Use De Moivre's Theorem to find the cube root

$[8^{1/2}(\cos 135° + i \sin 135°)]^{1/3} = 8^{1/6}(\cos 45° + i \sin 45°)$

$= 8^{1/6}\left(\dfrac{\sqrt{2}}{2} + \dfrac{\sqrt{2}}{2}i\right) = (2^3)^{1/6}(2^{-1/2} + 2^{-1/2}i) = 2^{1/2}(2^{-1/2} + 2^{-1/2}i)$

$= 2^0 + 2^0 i = 1 + i$

8. $x^2 + y^2 - 4y - 5 = 1$

$\qquad x - 2y = 3$

$\qquad\qquad x = 3 + 2y$

$\qquad\qquad\qquad (3 + 2y)^2 + y^2 - 4y - 5 = 1$

$\qquad\qquad\qquad 9 + 12y + 4y^2 + y^2 - 4y - 5 = 1$

$$5y^2 + 8y + 3 = 0$$
$$(5y + 3)(y + 1) = 0$$

$$y = -\frac{3}{5}$$

$$x = 3 + 2y = 3 - \frac{6}{5} = \frac{9}{5} \qquad \begin{array}{l} y = -1 \\ x = 3 + 2y = 1 \end{array}$$

9. (a) Method 1—The coins can be taken 1 at a time, 2 at a time, 3 at a time, etc. These combinations may be represented as $_5C_1 + {}_5C_2 + {}_5C_3 + {}_5C_4 + {}_5C_5$ $= 5 + 10 + 10 + 5 + 1 = 31$ combinations, or different sums of money.

Method 2—The sum of the combinations $= 2^n - 1 = 31$.

(b) The number of ways in which the 5 people may be seated is 5! or 120 ways.

The number of ways in which the 5 people may be seated with a certain 2 people side by side $2 \cdot 4!$ or 48 ways.

The probability is $\dfrac{48}{120} = \dfrac{2}{5}$.

10. (a) $y = kx - 2$
   $x^2 = 2y$

If the straight line is tangent to the parabola, the two equations must have a single common solution. Since $y = kx - 2$ and $x^2 = 2y$, then $x^2 = 2(kx - 2)$ by substitution.

Since there is to be only one solution, the equation $x^2 = 2(kx - 2)$ must have equal roots. The equation $x^2 = 2(kx - 2)$ or $x^2 - 2kx + 4 = 0$ will have equal roots if the discriminant $b^2 - 4ac = 0$; i.e. if $4k^2 - 16 = 0$. Therefore, $k = \pm 2$.

(b) The graphs will not intersect if the discriminant, $b^2 - 4ac$, or $4k^2 - 16$, is negative.

$$4k^2 - 16 < 0$$
$$k^2 - 4 < 0, \quad k^2 < 4$$
$$\text{i.e.} \quad 2 > k > -2$$

(c) The graphs will intersect if the discriminant, $b^2 - 4ac$, or $4k^2 - 16$, is positive

$$4k^2 - 16 > 0$$
$$k^2 - 4 > 0, \quad \text{or} \quad k^2 > 4$$
$$\text{i.e.} \quad 2 < k < -2$$

11. (a) $x^3 + 3x - 8 = 0$

$f(x) = + + -$. Since there is one variation in sign there can be no more than one positive root.

$f(-x) = - - - -$. Since there are no variations in sign there can be no negative roots.

Therefore, there is one positive root and no negative roots.

(b) Let $x^3 + 3x - 8 = y$

```
1 + 0 + 3 - 8   |1
    + 1 + 1 + 4
1 + 1 + 4|- 4
```

x	$y = f(x)$
0	-8
1	-4
2	$+16$

```
1 + 0 + 3 - 8   |2
    + 2 + 4 + 14
1 + 2 + 7|+ 6
```

Since $f(x)$ changes sign for values of x between 1 and 2, the equation $f(x) = 0$ has a real root between 1 and 2.

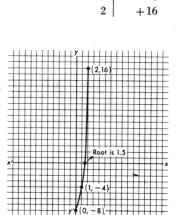

EXAMINATION 4

Answer any SEVEN *questions.*

1. (a) Expand and simplify $(a - 2b)^6$.

   (b) Find the term without x in the expansion of $\left(\dfrac{x^2}{2} + \dfrac{1}{x^2} \right)^{10}$

2. Prove by mathematical induction that
   $$2 \cdot 5 + 3 \cdot 6 + 4 \cdot 7 + \cdots + (n + 1)(n + 4) = \tfrac{1}{3}n(n + 4)(n + 5).$$

3. (a) Find in polar form the cube roots of $4 + 4i\sqrt{3}$.

   (b) Show that $\left(\dfrac{1 - i}{1 + i} \right)^{-2}$ is a real number.

   (c) Multiply $5(\cos 20° + i \sin 20°)$ by $3(\cos 130° + i \sin 130°)$ and write the answer in $a + bi$ form.

4. Find the value of z which is part of the solution of the following system:
   $$\begin{aligned}
   x - 3y + z \qquad &= 0 \\
   2x + y \qquad - t &= 0 \\
   3x - 2y - z - 2t &= 0 \\
   4x - y \qquad + 3t &= 0
   \end{aligned}$$

5. Given the equation $x^3 + x - 4 = 0$.
   - (a) What is the sum of the roots?
   - (b) What is the product of the roots?
   - (c) Write an equation whose roots are each twice those of the given equation.
   - (d) Calculate the real root to the nearest tenth.
6. Given that the roots of $2x^4 - x^3 - 46x^2 - 53x + 38 = 0$ are all real.
   - (a) Before solving determine the number of positive roots and the number of negative roots.
   - (b) Find all the roots of this equation.
7. (a) How many numbers greater than 3000 can be made from the digits 1, 2, 4, 5, 6 if the numbers are to contain no repeated digits?
   - (b) How many chords are determined by seven distinct points on a circle?

8. Prove that if a rational number $\dfrac{b}{c}$, in lowest terms, is a root of
$$a_0 x^n + a_1 x^{n-1} + a_2 x^{n-2} + \cdots + a_n = 0$$
where the coefficients are integers, then b is a factor of a_n and c is a factor of a_0.

EXAMINATION 4—SOLUTIONS

1. (a) $(a - 2b)^6 = (a)^6 - {}_6C_1(a)^5(2b) + {}_6C_2(a)^4(2b)^2$
   $- {}_6C_3(a)^3(2b)^3 + {}_6C_4(a)^2(2b)^4 - {}_6C_5(a)(2b)^5 + {}_6C_6(2b)^6$
   $= a^6 - 12a^5b + 60a^4b^2 - 160a^3b^3 + 240a^2b^4 - 192ab^5 + 64b^6$.

   (b) $\left(\dfrac{x^2}{2} + \dfrac{1}{x^2}\right)^{10}$

The desired term, or $(r + 1)$th term may be represented as ${}_nC_r\, a^{n-r}b^r$.
Examining powers of x,
$$\left(\frac{x^2}{2}\right)^{10-r} \left(\frac{1}{x^2}\right)^r = 1 \text{ or } x^0.$$
$x^{20-2r} \cdot x^{-2r} = x^0, \quad 20 - 4r = 0, \quad r = 5.$

Since $r + 1 = 6$, we must find the sixth term.
$${}_{10}C_5 \left(\frac{x^2}{2}\right)^5 \left(\frac{1}{x^2}\right)^5 = 252 \cdot \frac{x^{10}}{32} \cdot \frac{1}{x^{10}} = \frac{63}{8}$$

2. $2 \cdot 5 + 3 \cdot 6 + 4 \cdot 7 + \cdots + (n + 1)(n + 4) = \frac{1}{3}n(n + 4)(n + 5)$
   PROOF: Step 1—The formula is true for $n = 1$, since
   $$2 \cdot 5 = \tfrac{1}{3}(1)(1 + 4)(1 + 5)$$
   $$2 \cdot 5 = \tfrac{1}{3}(5)(6)$$

Step 2—Assume that the formula is true for $n = k$ i.e.

$2 \cdot 5 + 3 \cdot 6 + 4 \cdot 7 + \cdots + (k + 1)(k + 4) = \frac{1}{3}k(k + 4)(k + 5)$

Add $(k + 2)(k + 5)$ to both sides, to obtain

$2 \cdot 5 + 3 \cdot 6 + 4 \cdot 7 + \cdots + (k + 1)(k + 4) + (k + 2)(k + 5)$

$= \frac{1}{3}k(k + 4)(k + 5) + (k + 2)(k + 5)$

$= (k + 5)[\frac{1}{3}k(k + 4) + (k + 2)]$

$= (k + 5)\left[\dfrac{k^2 + 4k + 3k + 6}{3}\right]$

$= (k + 5)\left[\dfrac{(k + 6)(k + 1)}{3}\right]$

$= \frac{1}{3}(k + 1)(k + 5)(k + 6)$

Thus, we have proved that when the formula is true for $n = k$ then it is true for $n = k + 1$.   Therefore, it is true for all n.

3. (a) From the graph it can be seen that the modulus is 8 and the amplitude is 60°.

$\therefore$ $4 + 4i\sqrt{3} = 8(\cos 60° + i \sin 60°)$

The three cube roots may be represented as follows:

$[8(\cos 60° + i \sin 60°)]^{1/3} = 2(\cos 20° + i \sin 20°)$

$[8(\cos 420° + i \sin 420°)]^{1/3} = 2(\cos 140° + i \sin 140°)$

$[8(\cos 780° + i \sin 780°)]^{1/3} = 2(\cos 260° + i \sin 260°)$

(b) $\left(\dfrac{1 - i}{1 + i}\right)^{-2}$

$\dfrac{1 - i}{1 + i} = \dfrac{1 - i}{1 + i} \cdot \dfrac{1 - i}{1 - i} = \dfrac{1 - 2i + i^2}{1 - i^2} = \dfrac{-2i}{2} = -i$

$(-i)^{-2} = \dfrac{1}{(-i)^2} = \dfrac{1}{i^2} = -1$, a real number.

(c) $5(\cos 20° + i \sin 20°) \times 3(\cos 130° + i \sin 130°) =$

$15(\cos 150° + i \sin 150°) = 15\left(-\dfrac{\sqrt{3}}{2} + \dfrac{i}{2}\right) = -\dfrac{15\sqrt{3}}{2} + \dfrac{15i}{2}$

4. These equations are homogenous linear equations. Obviously, $x = 0$, $y = 0$, $z = 0$, $t = 0$ is a solution (the trivial solution). To determine whether there are other solutions we examine the determinant of the coefficients.

$$\begin{vmatrix} 1 & -3 & 1 & 0 \\ 2 & 1 & 0 & -1 \\ 3 & -2 & -1 & -2 \\ 4 & -1 & 0 & +3 \end{vmatrix}$$

Add the first row to the third row.

$$\begin{vmatrix} 1 & -3 & 1 & 0 \\ 2 & 1 & 0 & -1 \\ 3 & -5 & 0 & -2 \\ 4 & -1 & 0 & 3 \end{vmatrix}$$

Expand in terms of the elements of the third column.

$$\begin{vmatrix} 2 & 1 & -1 \\ 3 & -5 & -2 \\ 4 & -1 & 3 \end{vmatrix} = -70$$

Since the determinants of the coefficients is different from zero, the trivial solution is the only one.

5. (a) $x^3 + x - 4 = 0$

The sum of the roots is zero.

(b) The product of the roots $= 4$.

(c) $x^3 + 4x - 32 = 0$.

(d) $x^3 + x - 4 = 0$.

$f(0) = -4$, $f(1) = -2$, $f(2) = +6$

A real root exists between 1 and 2.

$$\begin{array}{l} 1 + 0 + 1 - 4 \quad \underline{|1} \\ + 1 + 1 + 2 \\ \hline 1 + 1 + 2 \,\underline{|- 2} \\ + 1 + 2 \\ \hline 1 + 2 \,\underline{|+ 4} \\ + 1 \\ \hline 1 \,\underline{|+ 3} \end{array}$$

$$y^3 + 3y^2 + 4y - 2 = 0$$
$$y^3 + 30y^2 + 400y - 2000 = 0$$

$$\begin{array}{l} 1 + 30 + 400 - 2000 \quad \underline{|3} \\ + 3 + 99 + 1497 \\ \hline 1 + 33 + 499 \,\underline{|- 503} \\ + 3 + 108 \\ \hline 1 + 36 \,\underline{|+ 607} \\ + 3 \\ \hline 1 \,\underline{|+ 39} \end{array}$$

$$z^3 + 39z^2 + 607z - 503 = 0$$
$$z^3 + 390z^2 + 60700z - 503000 = 0$$

$$\begin{array}{rrrrr} 1 & + \ 390 & + \ 60700 & - \ 503000 & \underline{\lfloor 5} \\ & + \quad 5 & + \quad 1975 & + \ 313275 & \\ \hline 1 & + \ 395 & + \ 62675 & \underline{\lfloor \quad - } \end{array}$$

∴ The real root $= 1.4$, to the nearest tenth.

6. (a) $2x^4 - x^3 - 46x^2 - 53x + 38 = 0$

According to Descartes' Rule of Signs there are either 2 positive roots or no positive roots, and there are either 2 negative roots or no negative roots.

(b)
$$\begin{array}{rrrrr} 2 & - \ 1 & - \ 46 & - \ 53 & + \ 38 \quad \underline{\lfloor \tfrac{1}{2}} \\ & + \ 1 & + \quad 0 & - \ 23 & - \ 38 \\ \hline 2 & + \ 0 & - \ 46 & - \ 76 & \underline{\lfloor \ 0} \end{array}$$

∴ $\frac{1}{2}$ is a root.

$$\begin{array}{rrrr} 2 & + \ 0 & - \ 46 & - \ 76 \quad \underline{\lfloor -2} \\ & - \ 4 & + \ 8 & + \ 76 \\ \hline 2 & - \ 4 & - \ 38 & \underline{\lfloor \ 0} \end{array}$$

∴ -2 is a root

The depressed equation $2x^2 - 4x - 38 = 0$ may be written as $x^2 - 2x - 19 = 0$.

The equation $x^2 - 2x - 19 = 0$ may be solved by using the quadratic formula

$$x = \frac{2 \pm \sqrt{4 + 76}}{2} = \frac{2 \pm \sqrt{80}}{2} = \frac{2 \pm 4\sqrt{5}}{2} = 1 \pm 2\sqrt{5}$$

The four roots are $\frac{1}{2}$, -2, and $1 \pm 2\sqrt{5}$.

7. (a) The solution may be represented as follows

$$\boxed{3} \ \times \ \boxed{4} \ \times \ \boxed{3} \ \times \ \boxed{2} \ = 72 \text{ numbers}$$

Note that the first box may be filled in only 3 ways i.e. by the numbers 4, 5, or 6.

(b) The number of chords $= {}_7C_2 = 21$.

8. $a_0x^n + a_1x^{n-1} + a_2x^{n-2} + \cdots + a_n = 0$

Let $\frac{b}{c}$ (the fraction $\frac{b}{c}$ is reduced to lowest terms) be a root of the equation.

If we substitute $\dfrac{b}{c}$ for x we obtain

$$\frac{a_0 b^n}{c^n} + \frac{a_1 b^{n-1}}{c^{n-1}} + \frac{a_2 b^{n-2}}{c^{n-2}} + \cdots + a_n = 0$$

Multiply both sides of the equation by c^n to obtain

$$a_0 b^n + a_1 b^{n-1}c + a_2 b^{n-2}c^2 + \cdots + a_n c^n = 0$$

This may be written as $a_1 b^{n-1}c + a_2 b^{n-2}c^2 + \cdots + a_n c^n = -a_0 b^n$. In this equation, c is a factor of the left member. Therefore, c must be a factor of the right member. But c cannot be a factor of b^n because $\dfrac{b}{c}$ is a fraction reduced to lowest terms. Therefore, c must be a factor of a_0.

The equation $a_0 b^n + a_1 b^{n-1}c + a_2 b^{n-2}c^2 + \cdots + a_n c^n = 0$ may be written $a_0 b^n + a_1 b^{n-1}c + a_2 b^{n-2}c^2 + \cdots + a_{n-1}bc^{n-1} = -a_n c^n$. In this equation, b is a factor of the left member. Therefore, b must be a factor of the right member. But b cannot be a factor of c^n because $\dfrac{b}{c}$ is a fraction reduced to lowest terms. Therefore, b must be a factor of a^n.

EXAMINATION 5

1. Solve the system of equations $\quad a^2 - b^2 = 9$
$$5a - 4b = 9$$

2. Expand $(2 - \sqrt{3})^5$ by the binomial theorem and simplify the result to the form $a + b\sqrt{3}$.

3. Find and plot all the cube roots of $-\dfrac{\sqrt{2}}{2} + \dfrac{\sqrt{2}}{2}i$.

4. Noting that $x = 2$ is one root of the cubic equation $x^3 + 2x^2 - 16 = 0$ find the other two roots.

5. By means of the discriminant determine a value of k such that this quadratic expression in x will be a perfect square. Check your answer.
$$(k + 3)x^2 - (k + 8)x + \frac{(k + 18)}{4}$$

6. Solve for x by completing the square, stating results in simplest form.
$$x^2 + kx = 3k + 9$$

7. Solve the equation $\sqrt{40 - 9x} - 2\sqrt{7 - x} = \sqrt{-x}$

8. Evaluate

$$\begin{vmatrix} -1 & 2 & 2 & -1 \\ 2 & -1 & 3 & 2 \\ -1 & 2 & 2 & 2 \\ -2 & 2 & 3 & 2 \end{vmatrix}$$

Omit Two of the Following

9. Find the four roots of this equation by considering it an equation in quadratic form.

$$(x^2 - 3x)^2 - (3x^2 - 9x) = 4$$

10. Approximate graphically the real roots of $2x^4 - 7x^2 + 2 = 0$, showing the curve and a suitable table of values obtained by synthetic division.

11. How many combinations each consisting of four red balls, two white ones and five black ones, can be formed from seven red balls, six white ones and nine black ones?

12. The faces of a cubical die are numbered 1, 2, 3, 4, 5, 6. In a single throw with three dice find the probability of turning up a total of 5.

EXAMINATION 5—SOLUTIONS

1.
$$a^2 - b^2 = 9$$
$$5a - 4b = 9$$
$$a = \frac{9 + 4b}{5}$$

$$\frac{9 + 4b^2}{5} - b^2 = 9$$

$$\frac{81 + 72b + 16b^2}{25} - b^2 = 9$$

$$81 + 72b + 16b^2 - 25b^2 = 225$$
$$-9b^2 + 72b - 144 = 0$$
$$b^2 - 8b + 16 = 0$$
$$(b - 4)(b - 4) = 0$$

$b = 4$	$b = 4$
$a = \dfrac{9 + 16}{5} = 5$	$a = \dfrac{9 + 16}{5} = 5$

There are two sets of roots $(5, 4)$ and $(5, 4)$.

2. $(2 - \sqrt{3})^5 = (2)^5 + {_5}C_1(2)^4(-\sqrt{3}) + {_5}C_2(2)^3(-\sqrt{3})^2$
$+ {_5}C_3(2)^2(-\sqrt{3})^3 + {_5}C_4(2)(-\sqrt{3})^4 + {_5}C_5(-\sqrt{3})^5$
$= 32 - 80\sqrt{3} + 240 - 120\sqrt{3} + 90 - 9\sqrt{3} = 362 - 209\sqrt{3}$

3. $-\dfrac{\sqrt{2}}{2} + \dfrac{\sqrt{2}}{2}i$

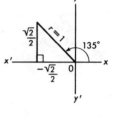

$-\dfrac{\sqrt{2}}{2} + \dfrac{\sqrt{2}}{2}i = 1(\cos 135° + i \sin 135°)$

The cube roots of $-\dfrac{\sqrt{2}}{2} + \dfrac{\sqrt{2}}{2}i$ are

$[1(\cos 135° + i \sin 135°)]^{1/3}$
$= 1(\cos 45° + i \sin 45°) = .71 + .71i$ (A)
$[1(\cos 495° + i \sin 495°)]^{1/3}$
$= 1(\cos 165° + i \sin 165°) = -.97 + .26i$ (B)
$[1(\cos 855° + i \sin 855°)]^{1/3}$
$= 1(\cos 285° + i \sin 285°) = -.97 - .26i$(C)

Note: Values of functions of angles were obtained from a table.

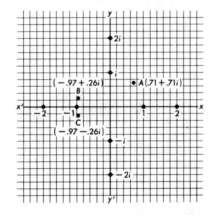

4. $x^3 + 2x^2 - 16 = 0$

$$
\begin{array}{r}
1 + 2 + 0 - 16 \ \underline{\rlap{|}2} \\
+ 2 + 8 + 16 \\
\hline
1 + 4 + 8 \quad \underline{|\ 0}
\end{array}
$$

The depressed equation is $x^2 + 4x + 8 = 0$.

This equation may be solved by the formula,
$$x = \frac{-4 \pm \sqrt{16 - 32}}{2} = \frac{-4 \pm \sqrt{-16}}{2} = \frac{-4 \pm 4i}{2} = -2 \pm 2i$$

5. $(k + 3)x^2 - (k + 8)x + \dfrac{(k + 18)}{4}$

In this case, the discriminant

$$b^2 - 4ac = [-(k + 8)]^2 - 4(k + 3)\,\dfrac{(k + 18)}{4}$$

In order that the expression be a perfect square the discriminant must be equal to zero, i.e. $k^2 + 16k + 64 - k^2 - 21k - 54 = 0$, or $k = 2$.

Check: When $k = 2$, the original expression becomes $5x^2 - 10x + 5$. This is equivalent to

$$5(x^2 - 2x + 1) = 5(x - 1)^2 = [\sqrt{5}(x - 1)]^2$$

6.

$$x^2 + kx = 3k + 9$$

$$x^2 + kx + \frac{k^2}{4} = \frac{k^2}{4} + 3k + 9 = \frac{k^2 + 12k + 36}{4}$$

$$\left(x + \frac{k}{2}\right)^2 = \frac{(k + 6)^2}{4}$$

$$\left(x + \frac{k}{2}\right) = \pm\,\frac{k + 6}{2}$$

$$x = -\frac{k}{2} \pm \frac{k + 6}{2} = 3,\ -k - 3$$

7.

$$\sqrt{40 - 9x} - 2\sqrt{7 - x} = \sqrt{-x}$$

$$\sqrt{40 - 9x} = 2\sqrt{7 - x} + \sqrt{-x}$$

$$40 - 9x = 4(7 - x) + 4\sqrt{-x(7 - x)} - x$$

$$40 - 9x = 28 - 4x - x + 4\sqrt{x^2 - 7x}$$

$$12 - 4x = 4\sqrt{x^2 - 7x}$$

$$3 - x = \sqrt{x^2 - 7x}$$

$$9 - 6x + x^2 = x^2 - 7x$$

$$x = -9$$

Check:

$$\sqrt{40 + 181} - 2\sqrt{7 + 9} = \sqrt{9}$$

$$\sqrt{121} - 2\sqrt{16} = \sqrt{9}$$

$$11 - 8 = 3$$

8.

$$\begin{vmatrix} -1 & 2 & 2 & -1 \\ 2 & -1 & 3 & 2 \\ -1 & 2 & 2 & 2 \\ -2 & 2 & 3 & 2 \end{vmatrix}$$

Multiply the elements of row 1 by 2 and add the results to the corresponding elements of rows 2, 3 and 4, obtaining

$$\begin{vmatrix} -1 & 2 & 2 & -1 \\ 0 & 3 & 7 & 0 \\ -3 & 6 & 6 & 0 \\ -4 & 6 & 7 & 0 \end{vmatrix}$$

Expand in terms of the elements of column 4.

$$-(-1) \begin{vmatrix} 0 & 3 & 7 \\ -3 & 6 & 6 \\ -4 & 6 & 7 \end{vmatrix}$$

Multiply the elements of column 2 by -1 and add the results to the corresponding elements of column 3.

$$+1 \begin{vmatrix} 0 & 3 & 4 \\ -3 & 6 & 0 \\ -4 & 6 & 1 \end{vmatrix} \begin{matrix} 0 & 3 \\ -3 & 6 \\ -4 & 6 \end{matrix} = 1(0 + 0 - 72 + 96 + 0 + 9) = 33$$

9. $(x^2 - 3x)^2 - (3x^2 - 9x) = 4$

$(x^2 - 3x)^2 - 3(x^2 - 3x) - 4 = 0$

Let $x^2 - 3x = A$

$A^2 - 3A - 4 = 0$

$(A - 4)(A + 1) = 0$

$A - 4 = 0$

$x^2 - 3x - 4 = 0$

$(x - 4)(x + 1) = 0$

$x = 4, \quad x = -1$

$A + 1 = 0$

$x^2 - 3x + 1 = 0$

$x = \dfrac{3 \pm \sqrt{9 - 4}}{2}$

$x = \dfrac{3 \pm \sqrt{5}}{2}$

10. $2x^4 - 7x^2 + 2 = 0$

Let $y = 2x^4 - 7x^2 + 2$

$$\begin{array}{rrrrrr} 2 + 0 - & 7 + & 0 + & 2 & \underline{\;\;-3} \\ - 6 + & 18 - & 33 + & 99 \\ \hline 2 - 6 + & 11 - & 33 \,\underline{|+\; 101} \end{array}$$

$$\begin{array}{rrrrrr} 2 + 0 - & 7 + & 0 + & 2 & \underline{\;\;-2} \\ - 4 + & 8 - & 2 + & 4 \\ \hline 2 - 4 + & 1 - & 2\,\underline{|\; +6} \end{array}$$

```
2 + 0 - 7 + 0 + 2    |-1
  - 2 + 2 + 5 - 5
─────────────────────────
2 - 2 - 5 + 5| -3
```

$f(0) = +2$ by inspection

```
2 + 0 - 7 + 0 + 2    |1
  + 2 + 2 - 5 - 5
─────────────────────────
2 + 2 - 5 - 5| - 3
```

```
2 + 0 - 7 + 0 + 2    |2
  + 4 + 8 + 2 + 4
─────────────────────────
2 + 4 + 1 + 2|+ 6
```

Tabulating these results

x	-3	-2	-1	0	1	2
y	101	6	-3	2	-3	6

These are roots between -2 and -1, between -1 and 0, between 0 and 1, and between 1 and 2.

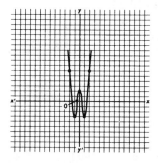

11. $_7C_4 \cdot {}_6C_2 \cdot {}_9C_5 = 66,150$.

12. In a single throw with 3 dice there are $6 \times 6 \times 6 = 216$ combinations. The total 5 may be obtained in the following ways

$1 + 2 + 2, \ 1 + 1 + 3, \ 1 + 3 + 1, \ 2 + 1 + 2, \ 2 + 2 + 1, \ 3 + 1 + 1$.

Since there are 6 ways of obtaining a total of 5, the probability of obtaining a total of $5 = \frac{6}{216} = \frac{1}{36}$.

EXAMINATION 6

Omit one question.

1. Evaluate

$$\begin{vmatrix} 2 & 3 & 1 & 4 \\ 5 & -2 & 0 & 3 \\ 0 & 1 & 2 & -5 \\ 3 & 3 & 2 & 3 \end{vmatrix}$$

2. (a) Complete the theorem, "If r is any constant, and if a polynomial $f(x)$ is divided by $(x - r)$ until a constant remainder is obtained, then. . . ." What is this theorem called? Prove the theorem using the identity $f(x) = (x - r) q(x) + R$.

   (b) Solve for $x = \sqrt{2x + 4} + \sqrt{2x} = 1$.

3. Find all the roots of $2x^3 - x^2 - 6x - 10 = 0$.

4. (a) A box contains 7 different red books and 5 different blue ones. In how many ways can we exhibit six of these books on a row on a shelf, if each exhibit should contain 4 red and 2 blue books?

   (b) Write and simplify the fifth term in the binomial expansion of
   $$(x^{-1/2} - 2x)^{13}$$

5. (a) Simplify:
   $$\frac{3x - \dfrac{2}{y}}{9y^2 - \dfrac{4}{x^2}}$$

   (b) Find the values of k for which the roots of $kx^2 + 4x = 3x$ are real.

6. (a) From a group of 6 men and 8 women a committee of 4 is chosen by lot. Find the probability that the committee will consist of 2 men and 2 women.

   (b) The probability of team A winning a certain game is $\frac{1}{3}$ and of team B winning an independent game is $\frac{1}{2}$. Find the probability that one of them will win and the other lose.

7. Find and plot all the roots of $x^4 + 16 = 0$.

EXAMINATION 6—SOLUTIONS

1.
$$\begin{vmatrix} 2 & 3 & 1 & 4 \\ 5 & -2 & 0 & 3 \\ 0 & 1 & 2 & -5 \\ 3 & 3 & 2 & 3 \end{vmatrix}$$

Multiply the elements of the first row by -2 and add the results in turn to the corresponding elements of the third and fourth rows. The result is

$$= \begin{vmatrix} 2 & 3 & 1 & 4 \\ 5 & -2 & 0 & 3 \\ -4 & -5 & 0 & -13 \\ -1 & -3 & 0 & -5 \end{vmatrix}$$

Expand this determinant in terms of the elements of the third column.

$$= \begin{vmatrix} 5 & -2 & 3 \\ -4 & -5 & -13 \\ -1 & -3 & -5 \end{vmatrix} \begin{matrix} 5 & -2 \\ -4 & -5 \\ -1 & -3 \end{matrix} = -35$$

2. (a) then the remainder is equal to $f(r)$

$$f(x) = (x - r)\,q(x) + R$$

In this identity, let $x = r$

$$f(r) = (r - r)\,q(r) + R$$
$$f(r) = 0 + R, \quad f(r) = R$$

(b) $\qquad\qquad \sqrt{2x + 4} + \sqrt{2x} = 1$
$$\sqrt{2x + 4} = 1 - \sqrt{2x}$$

Squaring both members, $\quad 2x + 4 = 1 - 2\sqrt{2x} + 2x$
$$2\sqrt{2x} = -3$$

Squaring both members, $\quad 4(2x) = 9$
$$8x = 9$$
$$x = \tfrac{9}{8}$$

Check: $\quad \sqrt{2(\tfrac{9}{8}) + 4} + \sqrt{2(\tfrac{9}{8})} = 1$
$$\sqrt{\tfrac{25}{4}} + \sqrt{\tfrac{9}{4}} = 1$$
$$\tfrac{5}{2} + \tfrac{3}{2} = 1$$

The number $\tfrac{9}{8}$ is an extraneous root.

The equation has no root.

3. $2x^3 - x^2 - 6x - 10 = 0$

$$\begin{array}{rrrr|l} 2 \ - \ 1 \ - & 6 \ - & 10 & \underline{\tfrac{5}{2}} \\ \ + \ 5 \ + & 10 \ + & 10 & \\ \hline 2 \ + \ 4 \ + & 4 & \underline{\ 0} & \end{array}$$

$\therefore \quad \tfrac{5}{2}$ is a root.

The depressed equation is $2x^2 + 4x + 4 = 0$

This may be written as $x^2 + 2x + 2 = 0$

Using the quadratic formula

$$x = \frac{-2 \pm \sqrt{4-8}}{2}$$

$$x = \frac{-2 \pm \sqrt{-4}}{2} = \frac{-2 \pm 2i}{2} = -1 \pm i$$

4. (a) The six books may be chosen in $_7C_4 \cdot {}_5C_2$ ways

$$= \frac{7 \cdot 6 \cdot 5 \cdot 4}{1 \cdot 2 \cdot 3 \cdot 4} \cdot \frac{5 \cdot 4}{1 \cdot 2} = 350 \text{ ways}$$

Each of these selections may be arranged in 6! ways. Therefore, the final result is $350 \times 6! = 252,000$.

(b) $(x^{-1/2} - 2x)^{13}$

Since $r + 1 = 5, r = 4$

The fifth term $= {}_{13}C_4 \, (x^{-1/2})^9 (-2x)^4 = 11,440x^{-1/2}$

5. (a)

$$\frac{3x - \dfrac{2}{y}}{9y^2 - \dfrac{4}{x^2}} = \frac{\dfrac{3xy - 2}{y}}{\dfrac{9x^2y^2 - 4}{x^2}} = \frac{3xy - 2}{y} \times \frac{x^2}{9x^2y^2 - 4}$$

$$= \frac{\cancel{3xy - 2}}{y} \times \frac{x^2}{(3xy + 2)(\cancel{3xy - 2})} = \frac{x^2}{y(3xy + 2)}$$

(b) $kx^2 + 4k = 3x$

$kx^2 - 3x + 4k = 0$

The discriminant, $b^2 - 4ac = 9 - 16k^2$.

In order that the roots be real, the discriminant $b^2 - 4ac$ must be equal to or greater than zero.

$$9 - 16k^2 \geq 0, \qquad 9 \geq 16k^2$$

Thus, the roots are real for all values of k equal to or less than $+\frac{3}{4}$ and equal to or greater than $-\frac{3}{4}$. That is, the roots are real for all values of k between $+\frac{3}{4}$ and $-\frac{3}{4}$ inclusive.

6. (a) The number of ways of selecting a committee of 4 from 6 men and 8 women is $_{14}C_4$.

$$_{14}C_4 = 1,001 \text{ ways}$$

The number of ways of forming a committee consisting of 2 men and 2 women is $_6C_2 \cdot {}_8C_2$.

$$_6C_2 \cdot {}_8C_2 = 420$$

(b) The probability that the committee will consist of 2 men and 2 women

$$= \frac{420}{1,001} = \frac{60}{143}$$

7. $x^4 + 16 = 0$

$x = \sqrt[4]{-16}$

When -16 is written in polar form, the nodulus
is 16 and the amplitude is $180°$.

The four roots may be written as follows:

$\sqrt[4]{-16} = [16(\cos 180° + i \sin 180°)]^{1/4} = 2(\cos 45° + i \sin 45°)$

$\sqrt[4]{-16} = [16(\cos 540° + i \sin 540°)]^{1/4} = 2(\cos 135° + i \sin 135°)$

$\sqrt[4]{-16} = [16(\cos 900° + i \sin 900°)]^{1/4} = 2(\cos 225° + i \sin 225°)$

$\sqrt[4]{-16} = [16(\cos 1260° + i \sin 1260°)]^{1/4} = 2(\cos 315° + i \sin 315°)$

Written in the form $a + bi$ the roots are

$$2\left(\frac{\sqrt{2}}{2} + \frac{\sqrt{2}}{2} i \right) = \sqrt{2} + \sqrt{2}i$$

$$2\left(-\frac{\sqrt{2}}{2} + \frac{\sqrt{2}}{2} i \right) = -\sqrt{2} + \sqrt{2}i$$

$$2\left(-\frac{\sqrt{2}}{2} - \frac{\sqrt{2}}{2} i \right) = -\sqrt{2} - \sqrt{2}i$$

$$2\left(\frac{\sqrt{2}}{2} - \frac{\sqrt{2}}{2} i \right) = \sqrt{2} - \sqrt{2}i$$

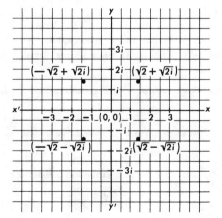

EXAMINATION 7

Answer Seven Questions

1. (a) Find the number of the term and write out the term independent of x in the binomial expansion of
$$\left(x^3 + \frac{1}{x^2} \right)^{40}$$

   (b) Find and simplify the 9th term of
$$\left(x^2 - \frac{1}{2x} \right)^{12}$$

2. (a) Express as a complex number in polar form
$$\frac{(1 + i^9)^5}{(1 - i)^3}$$

   (b) Reduce to $a + bi$ form
$$2(\cos 30° + i \sin 30°) + 2(\cos 150° + i \sin 150°)$$

3. Prove by mathematical induction
$$3 + 5 + 7 + \cdots + (2n + 1) = n(n + 2)$$
   for all positive integers n.

4. (a) Find in $a + bi$ form the cube roots of $27i$.

   (b) Given that the sum of two roots of the equation
$$x^3 - 2x^2 - 5x + k = 0$$
   is 1. Find the other root and the value of k.

5. Solve completely: $2x^4 + 5x^3 - 5x^2 - 11x + 6 = 0$.

6. Given the set of equations
$$\begin{aligned} x + y + z + w &= 2 \\ 2x + 3y - 4z - w &= 0 \\ x - y \qquad\quad + w &= 1 \\ 2x + 3y - z \qquad &= 4 \end{aligned}$$
   Evaluate the determinant of the coefficients of the unknowns and express z as the quotient of two determinants.

7. (a) How many numbers larger than 3400 can be formed from the digits 1, 2, 3, 4, 5 if no digit is repeated in any one number?

   (b) How many committees of 5 can be chosen from 3 girls and 4 boys if each committee must contain at least 2 girls?

8. (a) Solve for x: $\log_{10}(x + 12) - \log_{10}x = 1$

   (b) Solve for x: $4^{-x} = \dfrac{1}{8^{5-x}}$

   (c) Find the value of $\log_a b^2 \cdot \log_b a^2$.

9. Find the real root of the equation

$$x^3 - 3x^2 + 6x - 5 = 0$$

correct to the nearest tenth.

EXAMINATION 7—SOLUTIONS

1. (a) $\left(x^3 + \dfrac{1}{x^2}\right)^{40}$

The $(r + 1)$th term is $nC_r a^{n-r} b^r$. Investigating powers of x,

$$(x^3)^{40-r}(x^{-2})^r = x^0$$

Therefore $120 - 3r - 2r = 0$, $\quad 5r = 120$, $\quad r = 24$.

The term independent of x is the 25th term. The term independent of x is

$$_{40}C_{24}(x^3)^{16}\left(\frac{1}{x^2}\right)^{24} = {}_{40}C_{16}$$

(b) $\left(x^2 - \dfrac{1}{2x}\right)^{12}$

Since $r + 1 = 9$, $r = 8$.
The 9th term is $_{12}C_8(x^2)^4 \quad \left(-\dfrac{1}{2x}\right)^8 = \dfrac{495}{256}$

2. (a) $\dfrac{(1 + i^9)^5}{(1 - i)^3}$

Since $i^9 = i^8 \cdot i = 1 \cdot i = i$, the fraction becomes

$$\frac{(1 + i)^5}{(1 - i)^3}$$

$1 + i = \sqrt{2}(\cos 45° + i \sin 45°)$ in polar form

$1 - i = \sqrt{2}(\cos 315° + i \sin 315°)$ in polar form

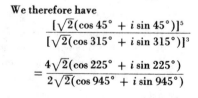

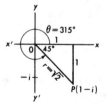

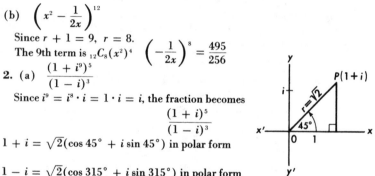

We therefore have

$$\frac{[\sqrt{2}(\cos 45° + i \sin 45°)]^5}{[\sqrt{2}(\cos 315° + i \sin 315°)]^3}$$

$$= \frac{4\sqrt{2}(\cos 225° + i \sin 225°)}{2\sqrt{2}(\cos 945° + i \sin 945°)}$$

$$= \frac{2(\cos 225° + i \sin 225°)}{(\cos 225° + i \sin 225°)} = 2$$

(b) $2(\cos 30° + i\sin 30°) + 2(\cos 150° + i\sin 150°)$

$$= 2\left(\frac{\sqrt{3}}{2} + \frac{i}{2}\right) + 2\left(-\frac{\sqrt{3}}{2} + \frac{i}{2}\right)$$

$$= \sqrt{3} + i - \sqrt{3} + i = 2i$$

3. $3 + 5 + 7 + \cdots + (2n + 1) = n(n + 2)$

This relationship is true for $n = 1$ since $3 = 1(1 + 2)$.

Assume that the formula is true for $n = k$ and prove that it is true for $n = k + 1$.

Add $(2k + 3)$ to both numbers

$$3 + 5 + 7 + \cdots + (2k + 1) + (2k + 3) = k(k + 2) + (2k + 3)$$
$$= k^2 + 2k + 2k + 3$$
$$= k^2 + 4k + 3$$
$$= (k + 1)(k + 3)$$

Thus, we have proved that when the formula is true for $n = k$, then it is true for $n = k + 1$. Therefore, it is true for all n.

4. (a)   When $27i$ is plotted we note that the modulus is 27 and the amplitude 90°. Therefore $27i = 27(\cos 90° + i\sin 90°)$.

The cube roots of $27i$ may be written as follows:

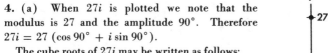

$$[27(\cos 90° + i\sin 90°)]^{1/3} = 3(\cos 30° + i\sin 30°) = 3\left(\frac{\sqrt{3}}{2} + i\cdot\frac{1}{2}\right)$$

$$= \frac{3\sqrt{3}}{2} + \frac{3i}{2}$$

$$[27(\cos 450° + i\sin 450°)]^{1/3} = 3(\cos 150° + i\sin 150°)$$

$$= 3\left(\frac{-\sqrt{3}}{2} + i\cdot\frac{1}{2}\right) = \frac{-3\sqrt{3}}{2} + \frac{3i}{2}$$

$$[27(\cos 810° + i\sin 810°)]^{1/3} = 3(\cos 270° + i\sin 270°)$$

$$= 3(0 - i) = -3i$$

(b) $x^3 - 2x^2 - 5x + k = 0$

The sum of the three roots is $+2$. If the sum of two roots is 1 then the value of the third root is 1.

Since 1 is a root, $x = 1$. If $x = 1$ then $1 - 2 - 5 + k = 0$ and $k = 6$.

5. $2x^4 + 5x^3 - 5x^2 - 11x + 6 = 0$

$$
\begin{array}{r}
2 + 5 - 5 - 11 + 6 \quad \lfloor\tfrac{1}{2} \\
+ 1 + 3 - 1 - 6 \\
\hline
2 + 6 - 2 - 12 \quad \lfloor 0
\end{array}
$$

$\therefore \tfrac{1}{2}$ is a root.

The depressed equation $2x^3 + 6x^2 - 2x - 12 = 0$ may be written as $x^3 + 3x^2 - x - 6 = 0$.

$$
\begin{array}{r}
1 + 3 - 1 - 6 \quad \lfloor -2 \\
- 2 - 2 + 6 \\
\hline
1 + 1 - 3 \quad \lfloor 0
\end{array}
$$

$\therefore -2$ is a root.

The depressed equation $x^2 + x - 3 = 0$ may be solved by using the quadratic formula

$$
x = \frac{-1 \pm \sqrt{1 + 12}}{2} \qquad \text{or} \qquad x = \frac{-1 \pm \sqrt{13}}{2}
$$

Therefore, the roots of the original equation are

$$
\frac{1}{2}, -2, \frac{-1 \pm \sqrt{13}}{2}
$$

6. The determinant of the coefficients of the unknowns is

$$
\begin{vmatrix}
1 & 1 & 1 & -1 \\
2 & 3 & -4 & -1 \\
1 & -1 & 0 & 1 \\
2 & 3 & -1 & 0
\end{vmatrix}
$$

Adding the elements of the fourth column to the corresponding elements of the first, second, and third columns we have

$$
\begin{vmatrix}
0 & 0 & 0 & -1 \\
1 & 2 & -5 & -1 \\
2 & 0 & 1 & 1 \\
2 & 3 & -1 & 0
\end{vmatrix}
$$

Expand this fourth order determinant in terms of the elements of the first row

$$
-(-1) \begin{vmatrix}
1 & 2 & -5 \\
2 & 0 & 1 \\
2 & 3 & -1
\end{vmatrix}
\begin{array}{cc}
1 & 2 \\
2 & 0 \\
2 & 3
\end{array}
0 = 1(0 + 4 - 30) - 1(0 + 3 - 4)
$$

$$
= -26 + 1 = -25
$$

$$z = \frac{\begin{vmatrix} 1 & 1 & 2 & 1 \\ 2 & 3 & 0 & -1 \\ 1 & -1 & 1 & 1 \\ 2 & 3 & 4 & 0 \end{vmatrix}}{\begin{vmatrix} 1 & 1 & 1 & -1 \\ 2 & 3 & -4 & -1 \\ 1 & -1 & 0 & 1 \\ 2 & 3 & -1 & 0 \end{vmatrix}}$$

7. (a) In order to obtain numbers larger than 3400 the first two boxes on the left must be filled by the numbers 3, 4, and 5 arranged in any order. When these boxes are filled we have 3 numbers left for the third box and then 2 numbers for the fourth box. The solution is shown below.

$$\boxed{3} \; \times \; \boxed{2} \; \times \; \boxed{3} \; \times \; \boxed{2} \; = 36 \text{ numbers}$$

(b) The committees required may be composed of 2 girls and 3 boys or 3 girls and 2 boys.

$$_3C_2 \cdot {}_4C_3 = 3 \cdot 4 = 12$$
$$_3C_3 \cdot {}_4C_2 = 1 \cdot 6 = 6$$

Result is 18 committees.

8. (a) $\log_{10}(x + 12) - \log_{10}x = 1$

$$\log_{10} \frac{(x + 12)}{x} = 1$$
$$10^1 = \frac{x + 12}{x}$$
$$10x = x + 12$$
$$x = \frac{4}{3}$$

(b) $4^{-x} = \dfrac{1}{8^{5-x}}$

$$(2^2)^{-x} = \frac{1}{2^{3(5-x)}}$$
$$2^{-2x} = \frac{1}{2^{15-3x}}$$
$$2^{-2x} = 2^{-15+3x}$$
$$-2x = -15 + 3x$$
$$x = 3$$
$$2^{-2x} = 2^{15-3x}$$

(c) $\log_a b^2 \cdot \log_b a^2$

Let $\log_a b^2 = x$ and $\log_b a^2 = y$

Then $a^x = b^2$ and $b^y = a^2$

$x \log a = 2 \log b$ and $y \log b = 2 \log a$

$$x = \frac{2 \log b}{\log a} \text{ and } y = \frac{2 \log a}{\log b}$$

$$xy = \frac{2 \log b}{\log a} \cdot \frac{2 \log a}{\log b} = 4$$

9. $x^3 - 3x^2 + 6x - 5 = 0$

$$f(0) = -5, \quad f(1) = -1, \quad f(2) = +3$$

There is a real root between 1 and 2.

Reduce the roots of the original equation by 1.

$$
\begin{array}{rrrr|l}
1 & -3 & +6 & -5 & \underline{1} \\
  & +1 & -2 & +4 & \\ \hline
1 & -2 & +4 & \multicolumn{1}{|r}{\underline{-1}} & \\
  & +1 & -1 & & \\ \hline
1 & -1 & \multicolumn{1}{|r}{\underline{+3}} & & \\
  & +1 & & & \\ \hline
1 & \multicolumn{1}{|r}{\underline{0}} & & &
\end{array}
$$

$$y^3 + 3y - 1 = 0$$

Multiply the roots of the y-equation by 10.

$$y^3 + 300y - 1000 = 0$$

In the y-equation $f(3)$ is negative and $f(4)$ is positive. Therefore, there is a real root in the y-equation between 3 and 4.

Reduce the roots of the y-equation by 3.

$$
\begin{array}{rrrr|l}
1 & +0 & +300 & -1000 & \underline{3} \\
  & +3 & +9 & +927 & \\ \hline
1 & +3 & +309 & \multicolumn{1}{|r}{\underline{-\quad 73}} & \\
  & +3 & +18 & & \\ \hline
1 & +6 & \multicolumn{1}{|r}{\underline{+327}} & & \\
  & +3 & & & \\ \hline
1 & \multicolumn{1}{|r}{\underline{+9}} & & &
\end{array}
$$

$$z^3 + 9z + 327z - 73 = 0$$

Multiply the roots of the z-equation by 10.

$$z^3 + 90z^2 + 32700z - 73000 = 0$$

Test the z-equation to determine whether there is a root below 5 or above 5.

$$
\begin{array}{r}
1 + 90 + 32700 - 73000 \quad\underline{|5} \\
+\ \ 5 + \quad 475 + 165875 \\
\hline
1 + 95 + 33175\ \underline{|\quad +\quad}
\end{array}
$$

Therefore, there is a real root in the z-equation below 5.

The real root, correct to the nearest tenth, is 1.3.

EXAMINATION 8

Answer four questions from each group.

GROUP I

1. (a) Write

$$\frac{[2\sqrt{2}(\cos 30° + i \sin 30°)]^4}{[2(\cos 25° + i \sin 25°)]^3}$$

as a complex number in the $a + bi$ form.

(b) Find the cube roots of $-8i$ in the $a + bi$ form.

2. (a) Expand by the binomial theorem and simplify

$$\left(\frac{x}{2} - \frac{2}{x}\right)^5$$

(b) Find and simplify the term of the binomial expansion of $\left(x^2 - \dfrac{1}{3x}\right)^9$ which does not contain x.

3. Prove by mathematical deduction that

$$1^2 + 2^2 + 3^2 + \cdots + n^2 = \frac{n(n + 1)(2n + 1)}{6}$$

for all positive integral values of n.

4. Solve $8x^4 + 4x^3 - 18x^2 + 11x - 2 = 0$.

5. Find the positive root of $x^3 - 2x - 7 = 0$ correct to the nearest tenth.

GROUP II

6. (a) From a bag containing 4 white and 5 black balls three are drawn at random. What is the probability that all are black?

(b) If five persons are arranged in a straight line, what is the probability that two particular persons, A and B, are next to each other?

7. For the following set of equations express z as the quotient of two dete - minants and find the value of z.

$$4x + y + 2z + 4w = 14$$
$$x - y - 8z = 7$$
$$3x - 2y - 4z + w = 12$$
$$3y + 4z + w = -3$$

8. (a) If $7.6^x = 2.1$ find the value of x correct to the nearest hundredth.
   (b) If $\log_{10} 2 = a$ and $\log_{10} 3 = b$ find the value of 10^{5a-b}.

9. (a) If k is a real number prove that the equation $(k + 1)x^2 + 3kx + k - 1 = 0$ has real roots for all values of k.
   (b) Without solving find the sum of the squares of the roots of the equation $px^2 + qx + t = 0$.

10. (a) Write as a rational fraction $.545454 \cdots$
    (b) Find the sum of n terms of the arithmetic progression
    $$a + b, \frac{2a + 3b}{2}, a + 2b, \cdots$$

EXAMINATION 8—SOLUTIONS

1. (a) $\dfrac{[2\sqrt{2}(\cos 30° + i \sin 30°)]^4}{[2(\cos 25° + i \sin 25°)]^3}$

$[2\sqrt{2}(\cos 30° + i \sin 30°)]^4 = (2\sqrt{2})^4 (\cos 120° + i \sin 120°)$
$= 64(\cos 120° + i \sin 120°)$

$[2(\cos 25° + i \sin 25°)]^3 = 8(\cos 75° + i \sin 75°)$

$\dfrac{64(\cos 120° + i \sin 120°)}{8(\cos 75° + i \sin 75°)} = 8(\cos 45° + i \sin 45°)$

$= 8\left(\dfrac{\sqrt{2}}{2} + i\dfrac{\sqrt{2}}{2}\right) = 4\sqrt{2} + 4i\sqrt{2}$

(b) $-8i = 8(\cos 270° + i \sin 270°)$
$(-8i)^{1/3} = [8(\cos 270° + i \sin 270°)]^{1/3}$
$= [8(\cos 630° + i \sin 630°)]^{1/3}$
$= [8(\cos 990° + i \sin 990°)]^{1/3}$

The three roots are
$$2(\cos 90° + i \sin 90°)$$
$$2(\cos 210° + i \sin 210°)$$
$$2(\cos 330° + i \sin 330°)$$

These roots written in the form $(a + bi)$ are
$$2i, -\sqrt{3} - i, \text{ and } \sqrt{3} - i$$

2. (a) $\left(\dfrac{x}{2} - \dfrac{2}{x}\right)^5 = \left(\dfrac{x}{2}\right)^5 + {}_5C_1\left(\dfrac{x}{2}\right)^4\left(-\dfrac{2}{x}\right) + {}_5C_2\left(\dfrac{x}{2}\right)^3\left(-\dfrac{2}{x}\right)^2$

$+ {}_5C_3\left(\dfrac{x}{2}\right)^2\left(-\dfrac{2}{x}\right)^3 + {}_5C_4\left(\dfrac{x}{2}\right)\left(-\dfrac{2}{x}\right)^4 + \left(-\dfrac{2}{x}\right)^5$

$= \dfrac{x^5}{32} - \dfrac{5x^3}{8} + 5x - \dfrac{20}{x} + \dfrac{40}{x^3} - \dfrac{32}{x^5}$

(b) $\left(x^2 - \dfrac{1}{3x}\right)^9$

The $(r + 1)$th term $= {}_nC_r a^{n-r}b^r$.

Investigating powers of x and noting that $n = 9$ we have

$$(x^2)^{9-r}\left(\dfrac{1}{x}\right)^r = x^0$$

$$x^{18-2r} \cdot x^{-r} = x^0$$

$$x^{18-3r} = x^0$$

$$18 - 3r = 0, \qquad r = 6$$

Therefore, we must find the 7th term

$${}_9C_6(x^2)^3\left(-\dfrac{1}{3x}\right)^6 = (84)(x^6)\left(+\dfrac{1}{729x^6}\right) = +\dfrac{28}{243}$$

3. $1^2 + 2^2 + 3^2 + \cdots + n^2 = \dfrac{n(n + 1)(2n + 1)}{6}$

When $n = 1$, we have

$$1^2 = \dfrac{1(1 + 1)(2 + 1)}{6} = \dfrac{6}{6} = 1$$

Therefore, the relationship is verified when $n = 1$.

Suppose the relationship is true when $n = k$. Will it be true when $n = k + 1$?

Adding $k + 1$ to both sides we have

$$1^2 + 2^2 + 3^2 + \cdots + k^2 + (k + 1)^2 = \dfrac{k(k + 1)(2k + 1)}{6} + k + 1$$

$$= \dfrac{k(k + 1)(2k + 1) + 6(k + 1)}{6}$$

$$= \dfrac{(k + 1)[k(2k + 1) + 6(k + 1)]}{6}$$

$$= \dfrac{(k + 1)(2k^2 + k + 6k + 6)}{6}$$

$$= \dfrac{(k + 1)(2k^2 + 7k + 6)}{6}$$

$$= \dfrac{(k + 1)(k + 2)(2k + 3)}{6}$$

This proves that when the relationship is true for $n = k$, it is true for $n = k + 1$. Therefore, it is true for all n.

4. $8x^4 + 4x^3 - 18x^2 + 11x - 2 = 0$

The rational roots of this equation are of the form $\dfrac{p}{q}$ where p is a factor of 2 and q is a factor of 8.

$$
\begin{array}{rrrrr|l}
8 & +\,4 & -\,18 & +\,11 & -\,2 & \underline{\tfrac{1}{2}} \\
 & +\,4 & +\,4 & -\,7 & +\,2 & \\
\hline
8 & +\,8 & -\,14 & +\,4 & \underline{\;0} &
\end{array}
$$

$\therefore$ $\tfrac{1}{2}$ is a root.

The depressed equation $8x^3 + 8x^2 - 14x + 4 = 0$ may be written as
$$4x^3 + 4x^2 - 7x + 2 = 0.$$

$$
\begin{array}{rrrr|l}
4 & +\,4 & -\,7 & +\,2 & \underline{\tfrac{1}{2}} \\
 & +\,2 & +\,3 & -\,2 & \\
\hline
4 & +\,6 & -\,4 & \underline{\;0} &
\end{array}
$$

$\therefore$ $\tfrac{1}{2}$ is a double root.

$$4x^2 + 6x - 4 = 0$$
$$2x^2 + 3x - 2 = 0$$

Factoring we have $(2x - 1)(x + 2) = 0$

$$x = \tfrac{1}{2}, \qquad x = -2$$

The roots are $\tfrac{1}{2}, \tfrac{1}{2}, \tfrac{1}{2}$ and -2.

5. $x^3 - 2x - 7 = 0$

$$f(0) = -7, \quad f(1) = -8, \quad f(2) = -3, \quad f(3) = +14$$

A real root exists between 2 and 3.   Reduce the roots of the original equation by 2.

$$
\begin{array}{rrrr|l}
1 & +\,0 & -\,2 & -\,7 & \underline{2} \\
 & +\,2 & +\,4 & +\,4 & \\
\hline
1 & +\,2 & +\,2 & \underline{-3} & \\
 & +\,2 & +\,8 & & \\
\hline
1 & +\,4 & \underline{+10} & & \\
 & +\,2 & & & \\
\hline
1 & \underline{+6} & & &
\end{array}
$$

The equation with roots reduced by 2 is

$$y^3 + 6y^2 + 10y - 3 = 0$$

Multiply the roots of this equation by 10.

$$y^3 + 60y^2 + 1,000y - 3,000 = 0$$

Reduce the roots of this equation by 2.

$$
\begin{array}{r}
1 + 60 + 1000 - 3000 \underline{}\lfloor 2 \\
\underline{+\ \ 2 + \ \ 124 + 2248} \\
1 + 62 + 1124 \lfloor -\ \ 752 \\
\underline{+\ \ 2 + \ \ 128} \\
1 + 64 \lfloor +\ 1252 \\
\underline{+\ \ 2} \\
1 \lfloor +\ 66
\end{array}
$$

$$z^3 + 66z^2 + 1,252z - 752 = 0$$

Multiply the roots of this equation by 10.

$$z^3 + 660z^2 + 125,200z - 752,000 = 0$$

$$
\begin{array}{r}
1 + 660 + 125200 - 752000 \underline{}\lfloor 5 \\
\underline{+\ \ \ 5 + \ \ \ 3325 + 642625} \\
1 + 665 + 128525 \lfloor \underline{\ \ \ \ -\ \ \ \ }
\end{array}
$$

Therefore, the required root is 2.3.

6. (a) The total number of ways of selecting 3 balls is $_9C_3 = 84$.
   The number of ways of selecting 3 black balls is $_5C_3 = 10$.

   The probability of drawing a black ball $= \dfrac{10}{84}$ or $\dfrac{5}{42}$.

(b) The number of ways of arranging 5 men in a line = 5! or 120.

   If we regard the AB and BA combination as 2, the number of arrangements in which two particular persons stand together is $2 \times 4! = 48$.

   The probability that A and B are next to each other is $\dfrac{48}{120} = \dfrac{2}{5}$.

7.

$$
z = \frac{
\begin{vmatrix}
4 & 1 & 14 & 4 \\
1 & -1 & 7 & 0 \\
3 & -2 & 12 & 1 \\
0 & 3 & -3 & 1
\end{vmatrix}
}{
\begin{vmatrix}
4 & 1 & 2 & 4 \\
1 & -1 & -8 & 0 \\
3 & -2 & -4 & 1 \\
0 & 3 & +4 & 1
\end{vmatrix}
}
$$

Evaluating the upper determinant

$$\begin{vmatrix} 4 & 1 & 14 & 4 \\ 1 & -1 & 7 & 0 \\ 3 & -2 & 12 & 1 \\ 0 & 3 & -3 & 1 \end{vmatrix}$$

Add the elements of the first column to the corresponding elements of the second column and then multiply the elements of the first column and add the results to corresponding elements of the third column.

$$\begin{vmatrix} 4 & 5 & -14 & 4 \\ 1 & 0 & 0 & 0 \\ 3 & 1 & -9 & 1 \\ 0 & 3 & -3 & 1 \end{vmatrix}$$

Develop by elements of the second row.

$$-1 \begin{vmatrix} 5 & -14 & 4 \\ 1 & -9 & 1 \\ 3 & -3 & 1 \end{vmatrix}$$

Multiply the elements of the third row by -4 and add the results to the corresponding elements of the first row and then multiply the elements of the third row by -1 and add the results to the corresponding elements of the second row.

$$-1 \begin{vmatrix} -7 & -2 & 0 \\ -2 & -6 & 0 \\ 3 & -3 & 1 \end{vmatrix}$$

Develop by elements of the third column.

$$-1 \begin{vmatrix} -7 & -2 \\ -2 & -6 \end{vmatrix} = -1(42) + 1(4) = -38$$

Evaluating the lower determinant

$$\begin{vmatrix} 4 & 1 & 2 & 4 \\ 1 & -1 & -8 & 0 \\ 3 & -2 & -4 & 1 \\ 0 & 3 & +4 & 1 \end{vmatrix}$$

We may obtain zeros in the fourth column by multiplying the elements of the fourth row by -4 and then by -1 and adding the results to the corresponding elements of the first and third rows respectively.

$$\begin{vmatrix} 4 & -11 & -14 & 0 \\ 1 & -1 & -8 & 0 \\ 3 & -5 & -8 & 0 \\ 0 & 3 & +4 & 1 \end{vmatrix}$$

Develop by elements of the fourth row.

$$+1 \begin{vmatrix} 4 & -11 & -14 \\ 1 & -1 & -8 \\ 3 & -5 & -8 \end{vmatrix}$$

Add the elements of the first column to the corresponding elements of the second column and then multiply the elements of the first column by 8 and add the results to the corresponding elements of the third column.

$$+1 \begin{vmatrix} 4 & -7 & 18 \\ 1 & 0 & 0 \\ 3 & -2 & 16 \end{vmatrix} = -1 \begin{vmatrix} -7 & 18 \\ -2 & 16 \end{vmatrix}$$

$$= -1(-112) - 1(36) = 112 - 36 = 76$$

$$z = \frac{-38}{76} = -\frac{1}{2}$$

8. (a)
$$7.6^x = 2.1$$
$$x \log 7.6 = \log 2.1$$
$$x = \frac{\log 2.1}{\log 7.6} = \frac{0.3222}{0.8808} = .37$$

(b) Since $\log_{10} 2 = a$, $10^a = 2$, $10^{5a} = 2^5$
Since $\log_{10} 3 = b$, $10^b = 3$
$$\frac{10^{5a}}{10^b} = 10^{5a-b} = \frac{2^5}{3} = \frac{32}{3}$$

9. (a) $(k + 1)x^2 + 3kx + k - 1 = 0$

The roots of the quadratic equation will be real if the discriminant is non-negative. Discriminant $= b^2 - 4ac = (3k)^2 - 4(k + 1)(k - 1) = 9k^2 - 4k^2 + 4 = 5k^2 + 4$. If k is a real number $5k^2 + 4$ must be positive. Therefore, the roots of the equation will be real.

(b) $px^2 + qx + t = 0$

Let the roots be r_1 and r_2.

$$r_1 + r_2 = -\frac{q}{p}, \quad r_1 r_2 = \frac{t}{p}$$

$$(r_1 + r_2)^2 = r_1^2 + 2r_1 r_2 + r_2^2 = \frac{q^2}{p^2}$$

$$2r_1 r_2 = \frac{2t}{p}$$

Subtracting, we have

$$r_1^2 + r_2^2 = \frac{q^2}{p^2} - \frac{2t}{p} = \frac{q^2 - 2pt}{p^2}$$

10. (a) $.545454\cdots$

This repeating decimal may be regarded as a geometric progression whose first term (a) is .54 and whose common ratio (r) is .01.

$$S\infty = \frac{a}{1-r} = \frac{.54}{1-.01} = \frac{.54}{.99} = \frac{6}{11}$$

(b) $a + b, \dfrac{2a + 3b}{2},\quad a + 2b, \cdots$

To find the common difference we may subtract the first term from the second, i.e.

$$\frac{2a + 3b}{2} - a + b = \frac{2a + 3b - 2a - 2b}{2} = \frac{b}{2}$$

$$S_n = \frac{n}{2}[2a + (n-1)d]$$

$$S_n = \frac{n}{2}[2(a+b) + (n-1)\frac{b}{2}]$$

$$S_n = \frac{n}{2}[2a + 2b + \frac{nb}{2} - \frac{b}{2}]$$

$$S_n = \frac{n}{2}[2a + \frac{3b}{2} + \frac{nb}{2}]$$

$$S_n = na + \frac{3nb}{4} + \frac{n^2 b}{4}$$

EXAMINATION 9

Answer four questions from each group.

GROUP I

1. Solve for x alone by determinants and simplify your answer.

$$cx + \frac{y}{c} = 0$$

$$3x - cy = 2c^2 + 6c$$

2. Solve for x and check your answers.

$$\sqrt{2x - 2} - \sqrt{x + 3} = 2$$

3. (a) Multiply $3(\cos 10^0 + i \sin 10^0)$ by $2(\cos 20^0 + i \sin 20^0)$ and express your result in the form $a + bi$.

   (b) Find one square root of i and check by squaring your result.

4. Determine whether

$$x + y + z = 1$$
$$3x - 2y + 4z = 3$$
$$x - 2y + 3z = -2$$
$$x \qquad - 2z = 7$$

is a consistent system, without solving for the unknowns, and state the reason for your conclusion.

5. (a) How many different sums involving at least two coins can be made from a penny, a nickel, a dime, a quarter, and a half-dollar?

(b) Numbers with 6 different digits each are to be formed in all possible ways, using 1, 2, 3, 4, 5, 6. What is the probability that the digits 3 and 4 will not occur side by side in one of these numbers, selected at random?

GROUP II

6. (a) Resolve into partial fractions

$$\frac{x + 31}{2x^2 + 7x - 15}$$

(b) Solve for a,

$$a^{-4} + 9 = 10a^{-2}$$

7. Prove the following identity by mathematical induction.

$$\frac{1}{1 \cdot 3} + \frac{1}{3 \cdot 5} + \frac{1}{5 \cdot 7} + \cdots + \frac{1}{(2n - 1)(2n + 1)} = \frac{n}{2n + 1}$$

8. An airplane has just enough gasoline to travel from airport B to point C and return. The distance r from B to P is known as the Radius of Action.

(a) If the speed of a plane on its outward trip is v_1 miles per hour, the speed returning over the same course v_2 miles per hour and the total time of the round trip is t hours, derive a formula for r in terms of v_1, v_2, and t.

(b) Find, correct to the nearest mile, the radius of action of a plane if $v_1 = 250$ miles per hour, $v_2 = 300$ miles per hour, and $t = 3$ hours and 20 minutes.

9. (a) If $\log_e (y + 1) - \log_e (x - 1) = 1$, express y as a function of x.

(b) Solve for x correct to the nearest tenth $5^x = 7$.

10. Solve the equation

$$6x^4 - 25x^3 + 9x^2 + 3x - 1 = 0$$

EXAMINATION 9—SOLUTIONS

1. $cx + \dfrac{y}{c} = 0$

$3x - cy = 2c^2 + 6c$

$$x = \frac{\begin{vmatrix} 0 & \dfrac{1}{c} \\[2mm] 2c^2 + 6c & -c \end{vmatrix}}{\begin{vmatrix} c & \dfrac{1}{c} \\[2mm] 3 & -c \end{vmatrix}} = \frac{-\dfrac{1}{c}(2c^2 + 6c)}{-c^2 - \dfrac{3}{c}}$$

$$= \frac{-(2c^2 + 6c)}{c} \div \frac{-c^3 - 3}{c} = \frac{-2c(c + 3)}{c} \cdot \frac{c}{-(c^3 + 3)}$$

$$= \frac{2c(c + 3)}{c^3 + 3}$$

2. $\sqrt{2x - 2} - \sqrt{x + 3} = 2$

$\sqrt{2x - 2} = 2 + \sqrt{x + 3}$

$2x - 2 = 4 + 4\sqrt{x + 3} + x + 3$

$x - 9 = 4\sqrt{x + 3}$

$x^2 - 18x + 81 = 16(x + 3) = 16x + 48$

$x^2 - 34x + 33 = 0$

$(x - 33)(x - 1) = 0$

$x = 33 \qquad x = 1$

Check for x = 33

$\sqrt{2(33 - 2)} - \sqrt{33 + 3} = 2$

$\sqrt{66 - 2} - \sqrt{36} = 2$

$8 - 6 = 2$

Check for x = 1

$\sqrt{2(1) - 2} - \sqrt{1 + 3} = 2$

$0 - 2 = 2$

Therefore, 1 is an extraneous root.

The equation has one root, $x = 33$.

3. (a) $3(\cos 10° + i \sin 10°) \times 2(\cos 20° + i \sin 20°)$

$$= 6(\cos 30° + i \sin 30°) = 6\left(\frac{\sqrt{3}}{2} + \frac{1}{2}i\right) = 3\sqrt{3} + 3i.$$

(b) $i = 1(\cos 90° + i \sin 90°)$

$$(i)^{1/2} = [1(\cos 90° + i \sin 90°)]^{1/2}$$

$$= \cos 45° + i \sin 45° = \frac{\sqrt{2}}{2} + \frac{\sqrt{2}}{2}i$$

Check: $\left(\dfrac{\sqrt{2}}{2} + \dfrac{\sqrt{2}}{2}i\right)^2 = \left(\dfrac{\sqrt{2}}{2}\right)^2 + 2\left(\dfrac{\sqrt{2}}{2}\right)\left(\dfrac{\sqrt{2}}{2}\right)i + \left(\dfrac{\sqrt{2}}{2}i\right)^2$

$$= \frac{1}{2} + i + \frac{2}{4}i^2 = \frac{1}{2} + i - \frac{1}{2} = i$$

4. A system of $(n + 1)$ linear equations in n variables will have a solution if and only if the determinant of order $(n + 1)$, formed from the coefficients and constant terms, is zero.

$$\begin{vmatrix} 1 & 1 & 1 & 1 \\ 3 & -2 & 4 & 3 \\ 1 & -2 & 3 & -2 \\ 1 & 0 & -2 & 7 \end{vmatrix}$$

This determinant may be evaluated by multiplying the elements of column 1 by -1 and adding these results in turn to the corresponding elements of columns 2, 3, and 4, as follows:

$$\begin{vmatrix} 1 & 0 & 0 & 0 \\ 3 & -5 & 1 & 0 \\ 1 & -3 & 2 & -3 \\ 1 & -1 & -3 & 6 \end{vmatrix} = +1 \begin{vmatrix} -5 & 1 & 0 \\ -3 & 2 & -3 \\ -1 & -3 & 6 \end{vmatrix} \begin{vmatrix} -5 & 1 \\ -3 & 2 \\ -1 & -3 \end{vmatrix}$$

$$= 1(-60 + 3 + 0) - 1(0 - 45 - 18) = -60 + 3 + 45 + 18 = 6$$

Therefore, this set of equations is inconsistent.

5. (a) The number of sums of 2 coins $= {}_5C_2 = 10$

The number of sums of 3 coins $= {}_5C_3 = 10$

The number of sums of 4 coins $= {}_5C_4 = 5$

The number of sums of 5 coins $= {}_5C_5 = 1$

The total number of sums $= 26$

(b) The number of numbers of 6 different digits is 6! = 720.

The number of ways in which the numbers 3 and 4 will occur side by side is
$$2 \times 5! = 240.$$

The number of ways in which the numbers 3 and 4 will not occur side by side = 720 − 240 = 480.

Therefore, the required probability $= \dfrac{480}{720} = \dfrac{2}{3}$.

6. (a) $\dfrac{x + 31}{2x^2 + 7x - 15} = \dfrac{A}{2x - 3} + \dfrac{B}{x + 5}$

$$x + 31 = A(x + 5) + B(2x - 3)$$
$$x + 31 = Ax + 5A + 2Bx - 3B$$
$$x + 31 = (A + 2B)x + 5A - 3B$$

Equating coefficients
$$A + 2B = 1$$
$$5A - 3B = 31$$

Solving this set of equations we obtain $A = 5, B = -2$.
Therefore,

$$\frac{x + 31}{2x^2 + 7x - 15} = \frac{5}{2x - 3} - \frac{2}{x + 5}$$

(b) $a^{-4} + 9 = 10a^{-2}$
$$a^{-4} - 10a^{-2} + 9 = 0$$
$$(a^{-2} - 9)(a^{-2} - 1) = 0$$

$$a^{-2} = 9 \qquad a^{-2} = 1$$
$$\frac{1}{a^2} = 9 \qquad \frac{1}{a^2} = 1$$

$$9a^2 = 1 \qquad a^2 = 1$$
$$a = \pm\tfrac{1}{3} \qquad a = \pm 1$$

7. $\dfrac{1}{1 \cdot 3} + \dfrac{1}{3 \cdot 5} + \dfrac{1}{5 \cdot 7} + \cdots + \dfrac{1}{(2n - 1)(2n + 1)} = \dfrac{n}{2n + 1}$

When $n = 1$, we have
$$\frac{1}{1 \cdot 3} = \frac{1}{(2 - 1)(2 + 1)} = \frac{1}{1 \cdot 3}$$

Therefore, the relationship is verified when $n = 1$.

Suppose the relationship is true when $n = k$. Will it be true when $n = k + 1$?

Adding $k + 1$ to both sides we have

$$\frac{1}{1 \cdot 3} + \frac{1}{3 \cdot 5} + \frac{1}{5 \cdot 7} + \cdots + \frac{1}{(2k - 1)(2k + 1)} + \frac{1}{(2k + 1)(2k + 3)}$$

$$= \frac{k}{2k + 1} + \frac{1}{(2k + 1)(2k + 3)}$$

$$= \frac{k(2k + 3) + 1}{(2k + 1)(2k + 3)}$$

$$= \frac{2k^2 + 3k + 1}{(2k + 1)(2k + 3)} = \frac{(k + 1)(2k + 1)}{(2k + 1)(2k + 3)}$$

$$= \frac{2k + 1}{2k + 3}$$

This proves that when the relationship is true for $n = k$, it is true for $n = k + 1$. Therefore, it is true for all n.

8. (a) The distance on the trip out $= r$

The speed on the trip out $= v_1$

The time on the trip out $= \dfrac{Distance}{Rate} = \dfrac{r}{v_1}$

Similarly, the time on the trip back $= \dfrac{r}{v_2}$

The total time $t = \dfrac{r}{v_1} + \dfrac{r}{v_2}$

Solving for r, $v_1 v_2 t = r v_2 + r v_1 = r(v_2 + v_1)$.

$$r = \frac{v_1 v_2 t}{v_1 + v_2}$$

(b) $$r = \frac{(250)(300)(3\frac{1}{3})}{250 + 300} = 455 \text{ miles}$$

9. (a) $\log_e (y + 1) - \log_e (x - 1) = 1$

$$\log_e \frac{(y + 1)}{(x - 1)} = 1$$

$$e^1 = \frac{y + 1}{x - 1}$$

$$y + 1 = ex - e$$

$$y = ex - e - 1$$

(b) $5^x = 5$

$x \log 5 = \log 7$

$$x = \frac{\log 7}{\log 5} = \frac{0.8451}{0.6990} = 1.2$$

10. $6x^4 - 25x^3 + 9x^2 + 3x - 1 = 0$

$$
\begin{array}{rrrrr}
6 - 25 + & 9 + & 3 - & 1 & \underline{-\tfrac{1}{3}} \\
- 2 + & 9 - & 6 + & 1 & \\
\hline
6 - 27 + & 18 - & 3 & & \underline{0}
\end{array}
$$

Therefore, $-\tfrac{1}{3}$ is a root

$$
\begin{array}{rrrr}
6 - 27 + & 18 - & 3 & \underline{\tfrac{1}{2}} \\
+ 3 - & 12 + & 3 & \\
\hline
6 - 24 + & 6 & & \underline{0}
\end{array}
$$

Therefore, $\tfrac{1}{2}$ is a root

$$6x^2 - 24x + 6 = 0$$
$$x^2 - 4x + 1 = 0$$

Using the quadratic formula, $x = \dfrac{4 \pm \sqrt{16 - 4}}{2}$

$$x = \dfrac{4 \pm \sqrt{12}}{2} = \dfrac{4 \pm 2\sqrt{3}}{2} = 2 \pm \sqrt{3}$$

The roots are $-\tfrac{1}{3}, \tfrac{1}{2}, 2 \pm \sqrt{3}$.

EXAMINATION 10

PART I

Answer all questions in Part I

1. What is the value of k if $x - 2$ is a factor of $x^4 + kx^3 - 3x^2 - 4x + 12$?

2. Express in the form $a + bi$, $\dfrac{2 + i}{3 - 2i}$

3. Determine the value of m for which the roots of the equation
$$x^2 - mx + m + 3 = 0$$
are equal.

4. Find the sum of the first two terms of the expansion
$$\left(1 + \frac{1}{x}\right)^x$$

5. Use De Moivre's Theorem to find the value of $(1 + i)^{10}$.

6. A box with a square base is made of sheet metal. A side of the base is a and the volume of the box is V. Express the height (h) of the box as a function of a and V.

7. A basketball squad consists of 10 men. In how many ways can a team be selected from the squad if only one of the men can play center?

8. The roots of the equation $x^3 - 9x^2 + 2x + k = 0$ are in Arithmetic Progression. Find the value of k.

9. How many gallons of water must be added to n gallons of a solution of salt and water which is $r\%$ salt to reduce it to a solution which is $s\%$ salt?

10. Resolve into partial fractions.

$$\frac{9x - 7}{6x^2 - 7x + 2}$$

PART II

Answer 4 questions

11. Solve for x only by determinants
$$2x + y + z - 4w = -1$$
$$3x - 2y - z + 2w = 3$$
$$5x - 3y + 2z \qquad\;\; = -7$$
$$\qquad\;\; 5y + 3z - 6w = -2$$

12. Solve the inequality
$$2y^2 - 7y + 6 > 0$$

13. In the equation $x^4 + px^3 + qx^2 + 53x - 60 = 0$, find p and q so that two roots will be $2 + i$ and $2 - i$.

14. $2x^2 - 3x + 7 = 0$
    If the roots of the above equation are r and s
    (a) Find the value of $r + s$.
    (b) Find the value of rs.
    (c) Using these results find the value of $r^2 + s^2$.

15. (a) If three arithmetic means are inserted between x and y find the common difference between any two terms.
    (b) Express as a common fraction the repeating decimal $.272727\ldots\ldots$
    (c) If x, y, and z are three consecutive terms of a geometric progression, write x as a function of y and z.

EXAMINATION 10—SOLUTIONS

1. If $(x - 2)$ is a factor of $f(x)$ then $f(2) = 0$
$$f(x) = x^4 + kx^3 - 3x^2 - 4x + 12$$
$$f(2) = 16 + 8k - 12 - 8 + 12$$
$$8k + 8 = 0$$
$$k = -1$$

2. $\dfrac{2 + i}{3 - 2i} \times \dfrac{3 + 2i}{3 + 2i} = \dfrac{6 + 7i + 2i^2}{9 - 4i^2} = \dfrac{6 + 7i - 2}{9 - 4(-1)}$

$= \dfrac{4 + 7i}{9 + 4} = \dfrac{4 + 7i}{13} = \dfrac{4}{13} + \dfrac{7}{13}i$

3. The roots of the equation $x^2 - mx + m + 3 = 0$ are equal if $b^2 - 4ac = 0$.
$$b^2 - 4ac = (-m)^2 - 4(1)(m + 3) = m^2 - 4m - 12$$
$$m^2 - 4m - 12 = 0, \qquad (m - 6)(m + 2) = 0$$
$$m = 6, \qquad m = -2$$

4. $\left(1 + \dfrac{1}{x}\right)^x = (1)^x + xC_1(1)^{x-1}\left(\dfrac{1}{x}\right) = 1 + x(1)\left(\dfrac{1}{x}\right) = 1 + 1 = 2.$

5. $(1 + i)^{10}$
 If $1 + i$ is plotted on a graph, we note that
 $r = \sqrt{2}$, and $\theta = 45°$.

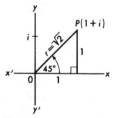

Therefore, $(1 + i) = \sqrt{2}(\cos 45° + i \sin 45°)$
$(1 + i)^{10} = [\sqrt{2}(\cos 45° + i \sin 45°)]^{10} = (\sqrt{2})^{10}(\cos 450° + i \sin 450°)$
$\qquad = 32(\cos 90° + i \sin 90°) = 32(0 + i) = 32i$

6. $V = a^2 h$
 $h = \dfrac{V}{a^2}$

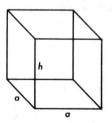

7. Since only one man can play center the choice is a selction of 4 men from 9.

$$_9C_4 = \frac{9 \cdot 8 \cdot 7 \cdot 6}{1 \cdot 2 \cdot 3 \cdot 4} = 126$$

8. $x^3 - 9x^2 + 2x + k = 0$

Since the three roots are in Arithmetic Progression they may be represented as $a - d, a, a + d$.

The sum of the roots $a - d + a + a + d = 9$

$$3a = 9, \qquad a = 3$$

Since 3 is a root, we may substitute 3 for x and solve for k.

$$27 - 9(9) + 2(3) + k = 0$$
$$27 - 81 + 6 + k = 0, \qquad k = 48$$

9. Let $x = $ the number of gallons of water to be added

$$\frac{Quantity\ of\ salt}{Total\ mixture} \quad \frac{.orn}{n + x} = \frac{s}{100}$$

$$s(n + x) = 100(.orn)$$
$$sn + sx = rn, \qquad sx = rn - sn$$
$$x = \frac{rn - sn}{s} \text{ or } \frac{n(r - s)}{s}$$

10. $\dfrac{9x - 7}{6x^2 - 7x + 2} = \dfrac{A}{3x - 2} + \dfrac{B}{2x - 1}$

$$9x - 7 = A(2x - 1) + B(3x - 2)$$
$$9x - 7 = 2Ax - A + 3Bx - 2B = x(2A + 3B) - A - 2B$$

Equaling coefficients $2A + 3B = 9$

$$-A - 2B = -7$$

Solving, we obtain $A = -3, \quad B = 5$

Therefore, $\dfrac{9x - 7}{6x^2 - 7x + 2} = \dfrac{-3}{3x - 2} + \dfrac{5}{2x - 1}$

PART II

11.

$$x = \frac{\begin{vmatrix} -1 & 1 & 1 & -4 \\ 3 & -2 & -1 & 2 \\ -7 & -3 & 2 & 0 \\ -2 & 5 & 3 & -6 \end{vmatrix}}{\begin{vmatrix} 2 & 1 & 1 & -4 \\ 3 & -2 & -1 & 2 \\ 5 & -3 & 2 & 0 \\ 0 & 5 & 3 & -6 \end{vmatrix}}$$

Evaluating the numerator determinant

$$\begin{vmatrix} -1 & 1 & 1 & -4 \\ 3 & -2 & -1 & 2 \\ -7 & -3 & 2 & 0 \\ -2 & 5 & 3 & -6 \end{vmatrix}$$

Add the elements of column 1 to the corresponding elements of columns 2 and 3 in turn. Then multiply the elements of column 1 by -4 and add the results to the corresponding elements of column 4. The resulting determinant is

$$\begin{vmatrix} -1 & 0 & 0 & 0 \\ 3 & 1 & 2 & -10 \\ -7 & -10 & -5 & 28 \\ -2 & 3 & 1 & 2 \end{vmatrix}$$

Expand the above determinant in terms of elements of the first row.

$$-1 \begin{vmatrix} 1 & 2 & -10 \\ -10 & -5 & 28 \\ 3 & 1 & 2 \end{vmatrix}$$

Multiply the elements of the first column by -2 and 10 and add the results to the corresponding elements of columns 2 and 3. The resulting determinant is

$$-1 \begin{vmatrix} 1 & 0 & 0 \\ -10 & 15 & -72 \\ 3 & -5 & 32 \end{vmatrix}$$

Expand the above determinant in terms of elements of the first row. The resulting determinant is

$$-1 \begin{vmatrix} 15 & -72 \\ -5 & 32 \end{vmatrix} = -1(15)(32) - 1(-1)(-5)(-72) = -120$$

Evaluating the denominator determinant

$$\begin{vmatrix} 2 & 1 & 1 & -4 \\ 3 & -2 & -1 & 2 \\ 5 & -3 & 2 & 0 \\ 0 & 5 & 3 & -6 \end{vmatrix}$$

Multiply the elements of column 2 in turn by -2, by -1, and by 4. Then add the results in turn to the corresponding elements of the first column, the third column, and the fourth column. The resulting determinant is

$$\begin{vmatrix} 0 & 1 & 0 & 0 \\ 7 & -2 & 1 & -6 \\ 11 & -3 & 5 & -12 \\ -10 & 5 & -2 & 14 \end{vmatrix}$$

Expand this determinant in terms of the elements of the first row.

$$-1 \begin{vmatrix} 7 & 1 & -6 \\ 11 & 5 & -12 \\ -10 & -2 & 14 \end{vmatrix}$$

Multiply the elements of column 2 in turn by -7 and $+6$ and add the results to the corresponding elements of row 1 and 3. The resulting determinant is

$$-1 \begin{vmatrix} 0 & 1 & 0 \\ -24 & 5 & 18 \\ 4 & -2 & 2 \end{vmatrix}$$

Expand this determinant in terms of the elements of the first row.

$$(-1)(-1) \begin{vmatrix} -24 & 18 \\ 4 & 2 \end{vmatrix} = 1(-24)(2) - 1(4)(18) = -120$$

$$x = \frac{-120}{-120} = +1$$

12. $\qquad 2y^2 - 7y + 6 > 0$

$\qquad (2y - 3)(y - 2) > 0$

The product of these two binomials is positive when the binomials are simultaneously positive or simultaneously negative, i.e. when

$$2y - 3 > 0 \quad \text{and} \quad y - 2 > 0$$

or when

$$2y - 3 < 0 \quad \text{and} \quad y - 2 < 0$$

$$2y - 3 > 0 \quad \text{and} \quad y - 2 > 0$$

when

$$y > \frac{3}{2} \quad \text{and} \quad y > 2$$

i.e. when $y > 2$.

$$2y - 3 < 0 \quad \text{and} \quad y - 2 < 0$$

when

$$y < \frac{3}{2} \quad \text{and} \quad y < 2$$

i.e. when $y < \frac{3}{2}$.

Hence, the inequality $2y^2 - 7y + 6 > 0$ when $y > 2$ or when $y < \frac{3}{2}$.

13. $\quad x^4 + px^3 + qx^2 + 53x - 60 = 0$

Let the other two roots be a and b.

Then, the sum of the roots is

$$a + b + 2 + i + 2 - i = -p \quad \text{or} \quad a + b = -p - 4$$

And, the sum of products of roots taken two at a time is
$$ab + a(2 + i) + a(2 - i) + b(2 + i) + b(2 - i) + (2 + i)(2 - i) = q$$
$$\text{or} \quad ab + 4a + 4b = q - 5$$

And, the sum of products of roots taken three at a time is
$$ab(2 + i) + ab(2 - i) + a(2 + i)(2 - i) + b(2 + i)(2 - i) = -53$$
$$\text{or} \quad 5a + 5b + 4ab = -53$$

And, the product of the roots is
$$ab(2 + i)(2 - i) = -60 \quad \text{or} \quad 5ab = -60$$

The relationships
$$5a + 5b + 4ab = -53 \quad \text{and} \quad 5ab = -60$$
may be used to solve for a and b.
$$ab = -12$$
$$5a + 5b - 48 = -53, \quad 5a + 5b = -5, \quad a + b = -1, \quad a = -1 - b$$
$$b(-1 - b) = -12, \quad b^2 + b - 12 = 0$$
$$(b + 4)(b - 3) = 0, \quad b = -4, \quad b = 3$$
$$a = 3, \quad a = -4$$

Thus, the other two roots are 3, and -4.

Since $a + b = -p - 4$, $3 - 4 = -p - 4$, and $p = -3$.

Also, $ab + 4a + 4b = q - 5$, $-12 + 12 - 16 = q - 5$, and $q = -11$.

14. $2x^2 - 3x + 7 = 0$
$$r + s = \frac{3}{2}, \quad rs = \frac{7}{2}$$
$$r^2 + 2rs + s^2 = \frac{9}{4}$$
$$2rs = 7$$
$$r^2 + s^2 = \frac{9}{4} - 7 = -\frac{19}{4}$$

15. (a) $x, ?, ?, ?, y$
$$l = a + d(n - 1), \quad y = x + d(5 - 1), \quad y = x + 4d$$
$$d = \frac{y - x}{4}$$

(b) $.272727\cdots = .27 + .0027 + .000027 + \cdots$

This repeating decimal may be regarded as a geometric progression with ratio .01.
$$S \infty = \frac{a}{1 - r} = \frac{.27}{1 - .01} = \frac{.27}{.99} = \frac{27}{99} = \frac{3}{11}$$

(c) If x, y, and z are three consecutive terms of a geometric progression then $x : y = y : z$.
$$xz = y^2, \quad x = \frac{y^2}{z}$$

TABLE 1
Common Logarithms of Numbers

N	0	1	2	3	4	5	6	7	8	9
10	0000	0043	0086	0128	0170	0212	0253	0294	0334	0374
11	0414	0453	0492	0531	0569	0607	0645	0682	0719	0755
12	0792	0828	0864	0899	0934	0969	1004	1038	1072	1106
13	1139	1173	1206	1239	1271	1303	1335	1367	1399	1430
14	1461	1492	1523	1553	1584	1614	1644	1673	1703	1732
15	1761	1790	1818	1847	1875	1903	1931	1959	1987	2014
16	2041	2068	2095	2122	2148	2175	2201	2227	2253	2279
17	2304	2330	2355	2380	2405	2430	2455	2480	2504	2529
18	2553	2577	2601	2625	2648	2672	2695	2718	2742	2765
19	2788	2810	2833	2856	2878	2900	2923	2945	2967	2989
20	3010	3032	3054	3075	3096	3118	3139	3160	3181	3201
21	3222	3243	3263	3284	3304	3324	3345	3365	3385	3404
22	3424	3444	3464	3483	3502	3522	3541	3560	3579	3598
23	3617	3636	3655	3674	3692	3711	3729	3747	3766	3784
24	3802	3820	3838	3856	3874	3892	3909	3927	3945	3962
25	3979	3997	4014	4031	4048	4065	4082	4099	4116	4133
26	4150	4166	4183	4200	4216	4232	4249	4265	4281	4298
27	4314	4330	4346	4362	4378	4393	4409	4425	4440	4456
28	4472	4487	4502	4518	4533	4548	4564	4579	4594	4609
29	4624	4639	4654	4669	4683	4698	4713	4728	4742	4757
30	4771	4786	4800	4814	4829	4843	4857	4871	4886	4900
31	4914	4928	4942	4955	4969	4983	4997	5011	5024	5038
32	5051	5065	5079	5092	5105	5119	5132	5145	5159	5172
33	5185	5198	5211	5224	5237	5250	5263	5276	5289	5302
34	5315	5328	5340	5353	5366	5378	5391	5403	5416	5428
35	5441	5453	5465	5478	5490	5502	5514	5527	5539	5551
36	5563	5575	5587	5599	5611	5623	5635	5647	5658	5670
37	5682	5694	5705	5717	5729	5740	5752	5763	5775	5786
38	5798	5809	5821	5832	5843	5855	5866	5877	5888	5899
39	5911	5922	5933	5944	5955	5966	5977	5988	5999	6010
40	6021	6031	6042	6053	6064	6075	6085	6096	6107	6117
41	6128	6138	6149	6160	6170	6180	6191	6201	6212	6222
42	6232	6243	6253	6263	6274	6284	6294	6304	6314	6325
43	6335	6345	6355	6365	6375	6385	6395	6405	6415	6425
44	6435	6444	6454	6464	6474	6484	6493	6503	6513	6522
N	0	1	2	3	4	5	6	7	8	9

TABLE 1—Continued

Common Logarithms of Numbers

N	0	1	2	3	4	5	6	7	8	9
45	6532	6542	6551	6561	6571	6580	6590	6599	6609	6618
46	6628	6637	6646	6656	6665	6675	6684	6693	6702	6712
47	6721	6730	6739	6749	6758	6767	6776	6785	6794	6803
48	6812	6821	6830	6839	6848	6857	6866	6875	6884	6893
49	6902	6911	6920	6928	6937	6946	6955	6964	6972	6981
50	6990	6998	7007	7016	7024	7033	7042	7050	7059	7067
51	7076	7084	7093	7101	7110	7118	7126	7135	7143	7152
52	7160	7168	7177	7185	7193	7202	7210	7218	7226	7235
53	7243	7251	7259	7267	7275	7284	7292	7300	7308	7316
54	7324	7332	7340	7348	7356	7364	7372	7380	7388	7396
55	7404	7412	7419	7427	7435	7443	7451	7459	7466	7474
56	7482	7490	7497	7505	7513	7520	7528	7536	7543	7551
57	7559	7566	7574	7582	7589	7597	7604	7612	7619	7627
58	7634	7642	7649	7657	7664	7672	7679	7686	7694	7701
59	7709	7716	7723	7731	7738	7745	7752	7760	7767	7774
60	7782	7789	7796	7803	7810	7818	7825	7832	7839	7846
61	7853	7860	7868	7875	7882	7889	7896	7903	7810	7917
62	7924	7931	7938	7845	7952	7959	7966	7973	7980	7987
63	7993	8000	8007	8014	8021	8028	8835	8041	8048	8055
64	8062	8069	8075	8082	8089	8096	8102	8109	8116	8122
65	8129	8136	8142	8149	8156	8162	8169	8176	8182	8189
66	8195	8202	8209	8215	8222	8228	8235	8241	8248	8254
67	8261	8267	8274	8280	8287	8293	8299	8306	8312	8319
68	8325	8331	8338	8344	8351	8357	8363	8370	8376	8382
69	8388	8395	8401	8407	8414	8420	8426	8432	8439	8445
70	8451	8457	8463	8470	8476	8482	8488	8494	8500	8506
71	8513	8519	8525	8531	8537	8543	8549	8555	8561	8567
72	8573	8579	8585	8591	8597	8603	8609	8615	8621	8627
73	8633	8639	8645	8651	8657	8663	8669	8675	8681	8686
74	8692	8698	8704	8710	8716	8722	8727	8733	8739	8745
75	8751	8756	8762	8768	8774	8779	8785	8791	8797	8802
76	8808	8814	8820	8825	8831	8837	8842	8848	8854	8859
77	8865	8871	8876	8882	8887	8893	8899	8904	8910	8915
78	8921	8927	8932	8938	8943	8949	8954	8960	8965	8971
79	8976	8982	8987	8993	8998	9004	9009	9015	9020	9025
N	0	1	2	3	4	5	6	7	8	9

TABLE 1—Continued
Common Logarithms of Numbers

N	0	1	2	3	4	5	6	7	8	9
80	9031	9036	9042	9047	9053	9058	9063	9069	9074	9079
81	9085	9090	9096	9101	9106	9112	9117	9122	9128	9133
82	9138	9143	9149	9154	9159	9165	9170	9175	9180	9186
83	9191	9196	9201	9206	9212	9217	9222	9227	9232	9238
84	9243	9248	9253	9258	9263	9269	9274	9279	9284	9289
85	9294	9299	9304	9309	9315	9320	9325	9330	9335	9340
86	9345	9350	9355	9360	9365	9370	9375	9380	9385	9390
87	9395	9400	9405	9410	9415	9420	9425	9430	9435	9440
88	9445	9450	9455	9460	9465	9469	9474	9479	9484	9489
89	9494	9499	9504	9509	9513	9518	9523	9428	9533	9538
90	9452	9547	9552	9557	9562	9566	9571	9576	9581	9586
91	9590	9595	9600	9605	9609	9614	9619	9624	9628	9633
92	9638	9643	9647	9652	9657	9661	9666	9671	9675	9680
93	9685	9689	9694	9699	9703	9708	9713	9717	9722	9727
94	9731	9736	9741	9745	9750	9754	9759	9763	9768	9773
95	9777	9782	9786	9791	9795	9800	9805	9809	9814	9818
96	9823	9827	9832	9836	9841	9845	9850	9854	9859	9863
97	9868	9872	9877	9881	9886	9890	9894	9899	9903	9908
98	9912	9917	9921	9926	9930	9934	9939	9943	9948	9952
99	9956	9961	9965	9969	9974	9978	9983	9987	9991	9996
N	0	1	2	3	4	5	6	7	8	9

VALUES OF THE TRIGONOMETRIC FUNCTIONS

Angle	Sin	Cos	Tan	Angle	Sin	Cos	Tan
1°	.0175	.9998	.0175	46°	.7193	.6947	1.0355
2°	.0349	.9994	.0349	47°	.7314	.6820	1.0724
3°	.0523	.9986	.0524	48°	.7431	.6691	1.1106
4°	.0698	.9976	.0699	49°	.7547	.6561	1.1504
5°	.0872	.9962	.0875	50°	.7660	.6428	1.1918
6°	.1045	.9945	.1051	51°	.7771	.6293	1.2349
7°	.1219	.9925	.1228	52°	.7880	.6157	1.2799
8°	.1392	.9903	.1405	53°	.7986	.6018	1.3270
9°	.1564	.9877	.1584	54°	.8090	.5878	1.3764
10°	.1736	.9848	.1763	55°	.8192	.5736	1.4281
11°	.1908	.9816	.1944	56°	.8290	.5592	1.4826
12°	.2079	.9781	.2126	57°	.8387	.5446	1.5399
13°	.2250	.9744	.2309	58°	.8480	.5299	1.6003
14°	.2419	.9703	.2493	59°	.8572	.5150	1.6643
15°	.2588	.9659	.2679	60°	.8660	.5000	1.7321
16°	.2756	.9613	.2867	61°	.8746	.4848	1.8040
17°	.2924	.9563	.3057	62°	.8829	.4695	1.8807
18°	.3090	.9511	.3249	63°	.8910	.4540	1.9626
19°	.3256	.9455	.3443	64°	.8988	.4384	2.0503
20°	.3420	.9397	.3640	65°	.9063	.4226	2.1445
21°	.3584	.9336	.3839	66°	.9135	.4067	2.2460
22°	.3746	.9272	.4040	67°	.9205	.3907	2.3559
23°	.3907	.9205	.4245	68°	.9272	.3746	2.4751
24°	.4067	.9135	.4452	69°	.9336	.3584	2.6051
25°	.4226	.9063	.4663	70°	.9397	.3420	2.7475
26°	.4384	.8988	.4877	71°	.9455	.3256	2.9042
27°	.4540	.8910	.5095	72°	.9511	.3090	3.0777
28°	.4695	.8829	.5317	73°	.9563	.2924	3.2709
29°	.4848	.8746	.5543	74°	.9613	.2756	3.4874
30°	.5000	.8660	.5774	75°	.9659	.2588	3.7321
31°	.5150	.8572	.6009	76°	.9703	.2419	4.0108
32°	.5299	.8480	.6249	77°	.9744	.2250	4.3315
33°	.5446	.8387	.6494	78°	.9781	.2079	4.7046
34°	.5592	.8290	.6745	79°	.9816	.1908	5.1446
35°	.5736	.8192	.7002	80°	.9848	.1736	5.6713
36°	.5878	.8090	.7265	81°	.9877	.1564	6.3138
37°	.6018	.7986	.7536	82°	.9903	.1392	7.1154
38°	.6157	.7880	.7813	83°	.9925	.1219	8.1443
39°	.6293	.7771	.8098	84°	.9945	.1045	9.5144
40°	.6428	.7660	.8391	85°	.9962	.0872	11.4301
41°	.6561	.7547	.8693	86°	.9976	.0698	14.3007
42°	.6691	.7431	.9004	87°	.9986	.0523	19.0811
43°	.6820	.7314	.9325	88°	.9994	.0349	28.6363
44°	.6947	.7193	.9657	89°	.9998	.0175	57.2900
45°	.7071	.7071	1.0000	90°	1.0000	.0000	

TABLE 2
Amount of 1 at Compound Interest After N Periods: $(1 + i)^n$

n \ i	1%	1¼%	1½%	2%	2½%	3%	4%	5%	6%
1	1.0100	1.0125	1.0150	1.0200	1.0250	1.0300	1.0400	1.0500	1.0600
2	1.0201	1.0252	1.0302	1.0404	1.0506	1.0609	1.0816	1.1025	1.1236
3	1.0303	1.0380	1.0457	1.0612	1.0769	1.0927	1.1249	1.1576	1.1910
4	1.0406	1.0509	1.0614	1.0824	1.1038	1.1255	1.1699	1.2155	1.2625
5	1.0510	1.0641	1.0773	1.1041	1.1314	1.1593	1.2167	1.2763	1.3382
6	1.0615	1.0774	1.0934	1.1262	1.1597	1.1941	1.2653	1.3401	1.4185
7	1.0721	1.0909	1.1098	1.1487	1.1887	1.2299	1.3159	1.4071	1.5036
8	1.0829	1.1045	1.1265	1.1717	1.2184	1.2668	1.3688	1.4775	1.5938
9	1.0937	1.1183	1.1434	1.1951	1.2489	1.3048	1.4233	1.5513	1.6895
10	1.1046	1.1323	1.1605	1.2190	1.2801	1.3439	1.4802	1.6289	1.7908
11	1.1157	1.1464	1.1779	1.2434	1.3121	1.3842	1.5395	1.7103	1.8983
12	1.1268	1.1608	1.1956	1.2682	1.3449	1.4258	1.6010	1.7959	2.0122
13	1.1381	1.1753	1.2136	1.2936	1.3785	1.4685	1.6651	1.8856	2.1329
14	1.1495	1.1900	1.2318	1.3195	1.4130	1.5126	1.7317	1.9799	2.2609
15	1.1610	1.2048	1.2502	1.3459	1.4483	1.5580	1.8009	2.0789	2.3966
16	1.1726	1.2199	1.2690	1.3728	1.4845	1.6047	1.8730	2.1829	2.5404
17	1.1843	1.2351	1.2880	1.4002	1.5216	1.6528	1.9479	2.2920	2.6928
18	1.1961	1.2506	1.3073	1.4282	1.5597	1.7024	2.0258	2.4066	2.8543
19	1.2081	1.2662	1.3270	1.4568	1.5987	1.7535	2.1068	2.5270	3.0256
20	1.2202	1.2820	1.3469	1.4859	1.6386	1.8061	2.1911	2.6533	3.2071
21	1.2324	1.2981	1.3671	1.5157	1.6796	1.8603	2.2788	2.7860	3.3996
22	1.2447	1.3143	1.3876	1.5460	1.7216	1.9161	2.3699	2.9253	3.6035
23	1.2572	1.3307	1.4084	1.5769	1.7646	1.9736	2.4647	3.0715	3.8197
24	1.2697	1.3474	1.4295	1.6084	1.8087	2.0328	2.5633	3.2251	4.0489
25	1.2824	1.3642	1.4509	1.6406	1.8539	2.0938	2.6658	3.3864	4.2919
26	1.2953	1.3812	1.4727	1.6734	1.9003	2.1566	2.7725	3.5557	4.5494
27	1.3082	1.3985	1.4948	1.7069	1.9478	2.2213	2.8834	3.7335	4.8223
28	1.3213	1.4160	1.5172	1.7410	1.9965	2.2879	2.9987	3.9201	5.1117
29	1.3345	1.4337	1.5400	1.7758	2.0464	2.3566	3.1187	4.1161	5.4184
30	1.3478	1.4516	1.5631	1.8114	2.0976	2.4273	3.2434	4.3219	5.7435
31	1.3613	1.4698	1.5865	1.8476	2.1500	2.5001	3.3731	4.5380	6.0881
32	1.3749	1.4881	1.6103	1.8845	2.2038	2.5751	3.5081	4.7649	6.4534
33	1.3887	1.5067	1.6345	1.9222	2.2589	2.6523	3.6484	5.0032	6.8406
34	1.4026	1.5256	1.6590	1.9607	2.3153	2.7319	3.7943	5.2533	7.2510
35	1.4166	1.5446	1.6839	1.9999	2.3732	2.8139	3.9461	5.5160	7.6861
36	1.4308	1.5639	1.7091	2.0399	2.4325	2.8983	4.1309	5.7918	8.1473
37	1.4451	1.5835	1.7348	2.0807	2.4933	2.9852	4.2681	6.0814	8.6361
38	1.4595	1.6033	1.7608	2.1223	2.5557	3.0748	4.4388	6.3855	9.1543
39	1.4741	1.6233	1.7872	2.1647	2.6196	3.1670	4.6164	6.7048	9.7035
40	1.4889	1.6436	1.8140	2.2080	2.6851	3.2620	4.8010	7.0400	10.2857

TABLE 3
Present Value of 1 At Compound Interest After N Periods: $(1 + i)^{-n}$

n	1%	1¼%	1½%	2%	2½%	3%	4%	5%	6%
1	.99010	.98765	.98522	.98039	.97561	.97087	.96154	.95238	.94340
2	.98030	.97546	.97066	.96117	.95181	.94260	.94256	.90703	.89000
3	.97059	.96342	.95632	.94232	.92860	.91514	.88900	.86384	.83962
4	.96098	.95152	.94218	.92385	.90595	.88849	.85480	.82270	.79209
5	.95147	.93978	.92826	.90573	.88385	.86261	.82193	.78353	.74726
6	.94205	.92817	.91454	.88797	.86230	.83748	.79031	.74622	.70496
7	.93272	.91672	.90103	.87056	.84127	.81309	.75992	.71068	.66506
8	.92348	.90540	.88771	.85349	.82075	.78941	.73069	.67684	.62741
9	.91434	.89422	.87459	.83676	.80073	.76642	.70259	.64461	.59190
10	.90529	.88318	.86167	.82035	.78120	.74409	.67756	.61391	.55839
11	.89632	.87228	.84893	.80426	.76214	.72242	.64958	.58468	.52679
12	.88745	.86151	.83639	.78849	.74356	.70138	.64260	.55684	.49697
13	.87866	.85087	.82403	.77303	.72542	.68095	.60057	.53032	.46884
14	.86996	.84037	.81185	.75788	.70773	.66112	.57748	.50507	.44230
15	.86135	.82999	.79985	.74301	.69047	.64186	.55526	.48102	.41727
16	.85282	.81975	.78803	.72845	.67362	.62317	.53391	.45811	.39365
17	.84438	.80963	.77639	.71416	.65720	.60502	.51337	.43630	.37136
18	.83602	.79963	.76491	.70016	.64117	.58739	.49363	.41552	.35034
19	.82774	.78976	.75361	.68643	.62553	.57029	.47464	.39573	.33051
20	.81954	.78001	.74247	.67297	.61027	.55368	.45639	.37689	.31180
21	.81143	.77038	.73150	.65978	.59539	.53755	.43883	.35894	.29416
22	.80340	.76087	.72069	.64684	.58086	.52189	.42196	.34185	.27751
23	.79544	.75147	.71104	.63416	.56670	.50669	.40573	.32557	.26180
24	.78757	.74220	.69954	.62172	.55288	.49193	.39012	.31007	.24698
25	.77977	.73303	.68921	.60953	.53939	.47761	.37512	.29530	.23300
26	.77205	.72398	.67902	.59758	.52623	.46369	.36069	.28124	.21981
27	.76440	.71505	.66899	.58586	.52340	.45019	.34682	.26785	.20737
28	.75684	.70622	.65910	.57437	.50088	.43708	.33348	.25509	.19563
29	.74934	.69750	.64936	.56311	.48866	.42435	.32065	.24295	.18456
30	.74192	.68889	.63976	.55207	.47674	.41199	.30832	.23138	.17411
31	.73458	.68038	.63031	.54125	.46511	.39999	.29646	.22036	.16425
32	.72730	.67198	.62099	.53063	.45377	.38834	.28506	.20987	.15496
33	.72010	.66369	.61182	.52023	.44270	.37703	.27409	.19987	.14619
34	.71297	.65549	.60277	.51003	.43191	.36604	.26355	.19035	.13791
35	.70591	.64740	.59387	.50003	.42137	.35538	.25342	.18129	.13011
36	.69892	.63941	.58509	.49022	.41109	.34503	.24367	.17266	.12274
37	.69200	.63152	.57644	.48061	.40107	.33498	.23430	.16444	.11579
38	.68515	.62372	.56792	.47119	.39128	.32523	.22529	.15661	.10924
39	.67837	.61602	.55953	.46195	.38174	.31575	.21662	.14915	.10306
40	.67165	.60841	.55126	.45289	.37243	.30656	.20829	.14205	.09722

TABLE 4
Amount of An Annuity of 1: $S_n i = \dfrac{(1 + i)_n - 1}{i}$

n \ i	1%	1½%	2%	2½%	3%	4%	5%	6%
1	1.0000	1.0000	1.0000	1.0000	1.0000	1.0000	1.0000	1.0000
2	2.0100	2.0150	2.0200	2.0250	2.0300	2.0400	2.0500	2.0600
3	3.0301	3.0452	3.0604	3.0756	3.0909	3.1216	3.1525	3.1836
4	4.0604	4.0909	4.1216	4.1525	4.1836	4.2465	4.3101	4.3746
5	5.1010	5.1523	5.2040	5.2563	5.3091	5.4163	5.5256	5.6371
6	6.1520	6.2296	6.3081	6.3877	6.4684	6.6330	6.8019	6.9753
7	7.2135	7.3230	7.4343	7.5474	7.6625	7.8983	8.1420	8.3938
8	8.2857	8.4328	8.5830	8.7361	8.8923	9.2142	9.5491	9.8975
9	9.3685	9.5593	9.7546	9.9545	10.1591	10.5828	11.0266	11.4913
10	10.4622	10.7027	10.9497	11.2034	11.4639	12.0061	12.5779	13.1808
11	11.5668	11.8633	12.1687	12.4835	12.8078	13.4864	14.2068	14.9716
12	12.6825	13.0412	13.4121	13.7956	14.1920	15.0258	15.9171	16.8699
13	13.8093	14.2368	14.6803	15.1404	15.6178	16.6268	17.7130	18.8821
14	14.9474	15.4504	15.9739	16.5190	17.0863	18.2919	19.5986	21.0151
15	16.0969	16.6821	17.2934	17.9319	18.5989	20.0236	21.5786	23.2760
16	17.2579	17.9324	18.6393	19.3802	20.1569	21.8245	23.6575	25.6725
17	18.4304	19.2014	20.0121	20.8647	21.7616	23.6975	25.8404	28.2129
18	19.6147	20.4894	21.4123	22.3863	23.4144	25.6454	28.1324	30.9057
19	20.8109	21.7967	22.8406	23.9460	25.1169	27.6712	30.5390	33.7600
20	22.0190	23.1237	24.2974	25.5447	26.8704	29.7781	33.0660	36.7856
21	23.2392	24.4705	25.7833	27.1833	28.6765	31.9692	35.7193	39.9927
22	24.4716	25.8376	27.2990	28.8629	30.5368	34.2480	38.5052	43.3923
23	25.7163	27.2251	28.8450	30.5844	32.4529	36.6179	41.4305	46.9958
24	26.9735	28.6335	30.4219	32.3490	34.4265	39.0826	44.5020	50.8156
25	28.2432	30.0630	32.0303	34.1578	36.4593	41.6459	47.7271	54.8645
26	29.5256	31.5140	33.6709	36.0117	38.5530	44.3117	51.1135	59.1564
27	30.8209	32.9867	35.3443	37.9120	40.7096	47.0842	54.6691	63.7058
28	32.1291	34.4815	37.0512	39.8598	42.9309	49.9676	58.4026	68.5281
29	33.4504	35.9987	38.7922	41.8563	45.2189	52.9663	62.3227	73.6398
30	34.7849	37.5387	40.5681	43.9027	47.5754	56.0849	66.4388	79.0582
31	36.1327	39.1018	42.3794	46.0003	50.0027	59.3283	70.7608	84.8017
32	37.4941	40.6883	44.2270	48.1503	52.5028	62.7015	75.2988	90.8898
33	38.8690	42.2986	46.1116	50.3540	55.0778	66.2095	80.0638	97.3432
34	40.2577	43.9331	48.0338	52.6129	57.7302	69.8579	85.0670	104.1838
35	41.6603	45.5921	49.9945	54.9282	60.4621	73.6522	90.3203	111.4348
36	43.0769	47.2760	51.9944	57.3014	63.2759	77.5983	95.8363	119.1209
37	44.5076	48.9851	54.0343	59.7339	66.1742	81.7022	101.6281	127.2681
38	45.9527	50.7199	56.1149	62.2273	69.1594	85.9703	107.7095	135.9042
39	47.4123	52.4807	58.2372	64.7830	72.2342	90.4091	114.0950	145.0585
40	48.8864	54.2679	60.4020	67.4026	75.4013	95.0255	120.7998	154.7620

TABLE 5

Present Value of An Annuity of 1: $a_{n\rceil}i = \dfrac{1 - (1 + i)^{-n}}{i}$

n \ i	1%	1½%	2%	2½%	3%	4%	5%	6%
1	0.9901	0.9852	0.9804	0.9756	0.9709	0.9615	0.9524	0.9434
2	1.9704	1.9559	1.9416	1.9274	1.9135	1.8861	1.8594	1.8334
3	2.9410	2.9122	2.8839	2.8560	2.8286	2.7751	2.7232	2.6730
4	3.9020	3.8544	3.8077	3.7620	3.7171	3.6299	3.5460	3.4651
5	4.8534	4.7826	4.7135	4.6458	4.5797	4.4518	4.3295	4.2124
6	5.7955	5.6972	5.6014	5.5081	5.4172	5.2421	5.0757	4.9173
7	6.7282	6.5982	6.4270	6.3494	6.2303	6.0021	5.7864	5.5824
8	7.6517	7.4859	7.3255	7.1701	7.0197	6.7327	6.4632	6.2098
9	8.5660	8.3605	8.1622	7.9709	7.7861	7.4353	7.1078	6.8017
10	9.4713	9.2222	8.9826	8.7521	8.5302	8.1109	7.7217	7.3601
11	10.3676	10.0711	9.7868	9.5142	9.2526	8.7605	8.3064	7.8869
12	11.2551	10.9075	10.5753	10.2578	9.9540	9.3851	8.8633	8.3838
13	12.1337	11.7315	11.3484	10.9832	10.6350	9.9856	9.3936	8.8527
14	13.0037	12.5434	12.1062	11.6909	11.2961	10.5631	9.8986	9.2950
15	13.8651	13.3432	12.8493	12.3814	11.9379	11.1184	10.3797	9.7122
16	14.7179	14.1313	13.5777	13.0550	12.5611	11.6523	10.8378	10.1059
17	15.5623	14.9076	14.2919	13.7122	13.1661	12.1657	11.2741	10.4773
18	16.3983	15.6726	14.9920	14.3534	13.7535	12.6593	11.6896	10.8276
19	17.2260	16.4262	15.6785	14.9789	14.3238	13.1339	12.0853	11.1581
20	18.0456	17.1686	16.3514	15.5892	14.8775	13.5903	12.4622	11.4699
21	18.8570	17.9001	17.0112	16.1845	15.4150	14.0292	12.8212	11.7641
22	19.6604	18.6208	17.6580	16.7654	15.9369	14.4511	13.1630	12.0416
23	20.4558	19.3309	18.2922	17.3321	16.4436	14.8568	13.4886	12.3034
24	21.2432	20.0304	18.9139	17.8850	16.9355	15.2470	13.7986	12.5504
25	22.0232	20.7196	19.5235	18.4244	17.4131	15.6221	14.0939	12.7834
26	22.7952	21.3986	20.1210	18.9506	17.8768	15.9828	14.3752	13.0032
27	23.5596	22.0676	20.7069	19.4640	18.3270	16.3296	14.6430	13.2105
28	24.3164	22.7267	21.2813	19.9649	18.7641	16.6631	14.8981	13.4062
29	25.0658	23.3761	21.8444	20.4535	19.1885	16.9837	15.1411	13.5907
30	25.8077	24.0158	22.3965	20.9303	19.6004	17.2920	15.3725	13.7648
31	26.5423	24.6461	22.9377	21.3954	20.0004	17.5885	15.5928	13.9291
32	27.2696	25.2671	23.4683	21.8492	20.3888	17.8736	15.8027	14.0840
33	27.9897	25.8790	23.9886	22.2919	20.7658	18.1476	16.0025	14.2302
34	28.7027	26.4817	24.4986	22.7238	21.1318	18.4112	16.1929	14.3681
35	29.4086	27.0756	24.9986	23.1452	21.4872	18.6646	16.3742	14.4982
36	30.1075	27.6607	25.4888	23.5563	21.8323	18.9083	16.5469	14.6210
37	30.7995	28.2371	25.9695	23.9573	22.1672	19.1426	16.7113	14.7368
38	31.4847	28.8051	26.4406	24.3486	22.4925	19.3679	16.8679	14.8460
39	32.1630	29.3646	26.9026	24.7303	22.8082	19.5845	17.0170	14.9491
40	32.8347	29.9158	27.3555	25.1028	23.1148	19.7928	17.1591	15.0463

Index

689

MAXIMIZE YOUR MATH SKILLS!

BARRON'S EASY WAY SERIES

Specially structured to maximize learning with a minimum of time and effort, these books promote fast skill building through lively cartoons and other fun features. **Each book $8.95**

ALGEBRA THE EASY WAY
Douglas Downing
Written as a fantasy novel, this book includes exercises with solutions, and over 100 illustrations. 352 pp. (2716-7)

CALCULUS THE EASY WAY
Douglas Downing
All principles are taught in an adventure tale. Numerous exercises, diagrams, and cartoons aid comprehension. 228 pp. (2588-1)

GEOMETRY THE EASY WAY
Lawrence Leff
This book focuses on the "why" of geometry: why you should approach a problem a certain way. Each chapter includes review exercises. 288 pp. (2718-3)

TRIGONOMETRY THE EASY WAY
Douglas Downing
This adventure story covers all material studied in high school or first-year college classes. Features practice exercises, explained answers, and illustrations. 288 pp. (2717-5)

LOGIC AND BOOLEAN ALGEBRA
Hilbert and Kathleen Levitz
Charts and diagrams illustrate abstract concepts. Practice exercises with answers. 128 pp., $10.95 (0537-6)

FM: FUNDAMENTALS OF MATHEMATICS
Cecilia Cullen and Eileen Petruzillo, editors
Volume 1 (2501-6) — Formulas; Introduction to Algebra; Metric Measurement; Geometry; Managing Money; Probability and Statistics. Volume 2 (2508-3) — Solving Simple Equations; Tables, Graphs and Coordinate Geometry; Banking; Areas; Indirect Measurement and Scaling; Solid Geometry. Each book 384 pp., $12.95

SURVIVAL MATHEMATICS
Edward Williams
Practical new math concepts for basic computational skills. Numerous practice exercises build basic competency skills. 416 pp., $9.95 (2012-X)

Books may be purchased at your bookstore, or by mail from Barron's. Enclose check or money order for total amount plus sales tax where applicable and 10% for postage (minimum charge $1.50). All books are paperback. Prices subject to change without notice.

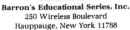

Barron's Educational Series, Inc.
250 Wireless Boulevard
Hauppauge, New York 11788